TRACING TEXTILE PRODUCTION FROM THE VIKING AGE TO THE MIDDLE AGES

TOOLS, TEXTILES, TEXTS AND CONTEXTS

INGVILD ØYE

OXBOW | books
Oxford & Philadelphia

Published in the United Kingdom in 2022 by
OXBOW BOOKS
The Old Music Hall, 106–108 Cowley Road, Oxford, OX4 1JE

and in the United States by
OXBOW BOOKS
1950 Lawrence Road, Havertown, PA 19083

Hardcover Edition: ISBN 978-1-78925-777-9
Digital Edition: ISBN 987-1-78925-778-6 (epub)

A CIP record for this book is available from the British Library

Library of Congress Control Number: 2022930614

Printed in Malta by Melita Press

For a complete list of Oxbow titles, please contact:

UNITED KINGDOM
Oxbow Books
Telephone (01865) 241249
Email: oxbow@oxbowbooks.com
www.oxbowbooks.com

UNITED STATES OF AMERICA
Oxbow Books
Telephone (610) 853-9131, Fax (610) 853-9146
Email: queries@casemateacademic.com
www.casemateacademic.com/oxbow

Oxbow Books is part of the Casemate Group

Front cover: Top, right: Fragment of a pleated tabby-woven fabric of wool, from the grave at Vangsnes in Sygnafylki (Photo: © University Museum of Bergen)
Top, middle: The Veka grave with the buried woman in a coffin (Shetelig 1912)
Top, right: The assemblage of spindle whorls from Ytre Moa (Fylkesarkivet i Vestland)
Middle, top right: Reconstructed garments from Ketilstaðir. Reconstruction by Marianne Guckelsberger and woven by Marled Mader (with kind permission)
Middle, bottom left: Schematic drawing of the warp-weighted loom in front (Source: Hoffmann 1964)
Middle right: Part of a roof painting, Ål stave church. c. 1260–1300
Right, middle: Wool combs from the 10th-century grave at Hyrt, Voss: how they were used, as demonstrated at Osterøy Museum (Photo: the author)
Bottom, right: The long-haired Old Norwegian Sheep, Norw. Spælsau (Photo: Mons Kvamme, with permission)
Bottom, middle: Loom weights of soapstone in a grave at Tokvam, Aurland, Sygnafylki (B 8869) (Photo: the author)
Bottom left: The site at Bjørkum (Photo: Morten Ramstad, with permission)

Contents

Preface

This book has developed over a long time. Besides the obligatory duties connected to my chair, teaching, and supervising Master and PhD students in medieval archaeology, in many respects it also follows on from my earlier research on tools and textile production from the medieval and urban context, and specifically the large-scale excavations at Bryggen in Bergen (1955–1974), which was a theme for my doctoral thesis, and the Kaupang Project (1998–2003), which related to textile tools from the Viking Age town of Kaupang. Both of these projects were based on tools as remains of urban production but from different periods. As ancient farming methods and technology has been another interest and research area of mine, I now wanted to combine these subject areas and dig deeper: what textile production was like in a rural setting of the Viking period and from other social and economic frameworks, such as graves and various workplaces. In rural contexts textile production had to cover all processes related to sheep and arable farming – this study will consider the full production sequence from raw materials of wool and plant to finished products and not only the last processes of spinning, weaving, and sewing as in urban settings. And how was it affected by shifting social and political conditions from the hierarchical rural society of the Viking period into the High Middle Ages? Did it change and in what way? By including tools, textiles, and textual information in their contexts they hopefully would provide opportunities to come closer to the production as an entity and in various settings over time.

This sparked my interest and I set out to pose new questions related to a production of vital importance in this dynamic period of large societal changes – politically, economically, and socially and where textiles played a crucial role: such as the advance of using sails, and all the practical use areas besides their social distinguishing role, to mention some. This was also a period with close westwards connections to the North Atlantic, emigration and colonisation and with strong bonds to western Norway, politically and socially. The local connection to western Norway is therefore extended westwards towards to the North Atlantic and seen from broader geographical but integrated perspectives and in a period represented by both old and new technology. The main area of study relates to central western Norway, covering the same area as once was the core area of the Gulathing Law, ostensibly dating back to the mid-10th century and an area with strong connections to the west too.

The book takes a broad outlook focusing on social, gendered, and technological aspects of textile production in these inter-connected areas within this period of major changes.

Textile production was dependant on farming, both animal and arable farming, indoor and outdoor and on a seasonal and an annual basis and as part of the agrarian complex. This has largely been an overlooked and rather neglected theme related to textile production.

When assessing tools and textiles as an integrated complex I have been fortunate to lean on results from eminent textile specialists and their published works, which so far only has only to a small extent been related to the tools from the same contexts. By approaching tools, textiles, and texts in their varied contexts, we may hopefully also reveal the agents – users and producers – when seen with overarching societal processes over time.

Acknowledgements

There are many to thank for helping me get this far. First and foremost, I would like to thank my department and its leaders from when I started this project, Professors Christhard Hoffmann and Jan Heiret, for support and for grants over the years, not least related to the publishing of this book.

It started as a small scale individual project nearly 10 years ago, collecting and inspecting textile equipment from the collections at the University Museum in Bergen, the National Museums in Iceland, the Faroe Islands, and the Shetlands. There have been many helpers who deserve special thanks: from the North Atlantic area, Guðmunður Oláfsson and Guðrun Gisladóttir in Reykjavik, my good friend and the late Simun Arge in Tórshavn, who is sadly missed, and Brian Smith in Lerwich. From Bergen, I would like to thank curator Sigrid Kaland, in charge of the collections at that time, for giving access to the data and artefacts, and the technician Svein Skare, who served me with all the textile tools I required on trays to my temporary workplace in the museum. Lately, also other colleagues at the museum have been a great help: Hana Lukesová, who is in charge of textiles at the Section of Collection Management and Conservation Department of Cultural History, for giving updated information about some of her analyses. I am also grateful for the positive response to using unpublished data and results from new excavations and research, Morten Ramstad, Head of Department for External Management, and the archaeologists Leif Inge Åstveit and Søren Diinhoff at the same department for giving access to new excavations at Viking Age workplaces related to Viking Age textile production. Adnan Icagic, who is in charge of the photos and photo documentation at the University Museum, has also been more than helpful. I am also glad for having good colleagues, friends, and close family with expertise within different fields related to the themes in this book at hand. I hereby thank them for being good discussion partners and for reading and commenting upon earlier drafts of the manuscript: Else Mundal, Professor of Norse Philology and with broad knowledge and competencies, and Randi Andersen, medieval historian with gender history as a specialisation, and both with keen interest in textile production. My sister and earlier colleague at the Department, Bergljot Solberg, Professor Emeritus in Viking Age archaeology, has also given useful feedback.

My undoubtedly best helper over the years has been graphic designer Per Bækken, who has improved photos, provided illustrations, maps, and overall design in articles since this project started. In this project related to the present book I could not have managed without his support, artistic talents, and graphic competence, for which I am deeply indebted and most grateful. Lastly, but not least I thank Professor Eva Andersson Strand for support and valuable advice. Eva has inspired my own research and was the best reader I could get to comment upon an early draft of the present book.

Ingvild Øye,
July 2021

Introduction

In 1908, the renowned Norwegian archaeologist Haakon Shetelig excavated a large grave cairn at the farm Veka in the community of Voss in western Norway – an extraordinarily richly furnished female grave from the Viking period. While many of the prehistoric burials discovered around the millennium shift of the 20th century were generally poorly documented, the Veka grave, as one of the early professionally excavated graves, provides a unique insight into tools, textiles, technology, and level of textile production in western Norway from the Viking period, AD 800–1050. The huge cairn, about 30 m in diameter, *c.* 3 m high at its upper side and 6 m at the lower surface, revealed the remains of a decomposed female skeleton in a north–south oriented wooden chest with her head to the south. In accordance with Norse pagan rituals, at least in the higher strata, the buried woman was richly furnished with grave goods and dress accessories by the remains of the dress, and with various tools that symbolise female work domains and demonstrate social position.[1] The textile tools gathered in a chest by her feet covered all stages of production – for preparation of fibres, spinning, weaving, and sewing: a pair of wool combs, spindle whorls, loom weights, and a needle case with thin needles. Later analyses of the textile fragments preserved by the brooches have identified them as parts of the funeral dress, a tunic, cloak, and kaftan, woven in different qualities, some in advanced weaving techniques of exceptionally high quality and in various colours.[2]

The case of Veka, as a well-preserved and documented grave, reveals both burial practices of the Viking Age and the potential graves and grave goods offer as sources to textile production in the period and this part of the country. Why was the deceased woman dressed in a high-quality dress, furnished with precious dress accessories and other grave goods, including a whole tool kit for textile production, and commemorated by a huge cairn? Her outfits signal high status above the ordinary, but had she used all these tools herself, mastering all the processes from combing wool, spinning yarn, weaving to sewing? Or did her status rather imply a role as organiser of the production in the hierarchical society of the Viking period?

The Veka burial is one of about 140 grave contexts with textile tools from the Viking period from a central area of western Norway – the two counties, Hordaland and Sogn og Fjordane, united since 2020 into one large county, denoted as Vestland. Even when finds conditions are less optimal than from Veka – with fewer equipment and tools combined with textiles, they offer a multitude of material with a largely unused potential to shed new light on Viking Age textile production at different levels and various modes of production. Finds of textiles and textile tools in graves reflect both social and technological aspects of production in varied environments, including the North Atlantic, reflecting traditions from the ancestral homelands. Additionally, textile tools found *in situ* at settlement sites and work environments from both the Viking Age and the Early and High Middle Ages (AD 1050–1350) – provide a more direct representation of the production in wider and varied contexts and environments, and for broader use areas, and over a longer period of half a millennium.

By the Middle Ages, written records, and to some extent pictorial evidence, supplement the archaeological remains. While written information about textile production reflecting the earliest period appears only casually in various sagas and poems, the sources from the High Middle Ages (*c.* 1150–1350) represent a wider spectre, including both descriptive and normative sources – such as laws, wills, inventories, and gifts as well as regulations, prices, rights, and obligations, subjected to law or as habitual practices of legal force. Although the different source categories alone may seem scanty, they do when combined provide a well of information able to give a broader outlook on production, socially, culturally, and economically – as for technologies, qualities, quantities, extent, as well as gender roles and agency, and changes over time.

Tools and textiles in various archaeological contexts, combined with written evidence covering the period AD 800–1300, constitute the main sources in this study on

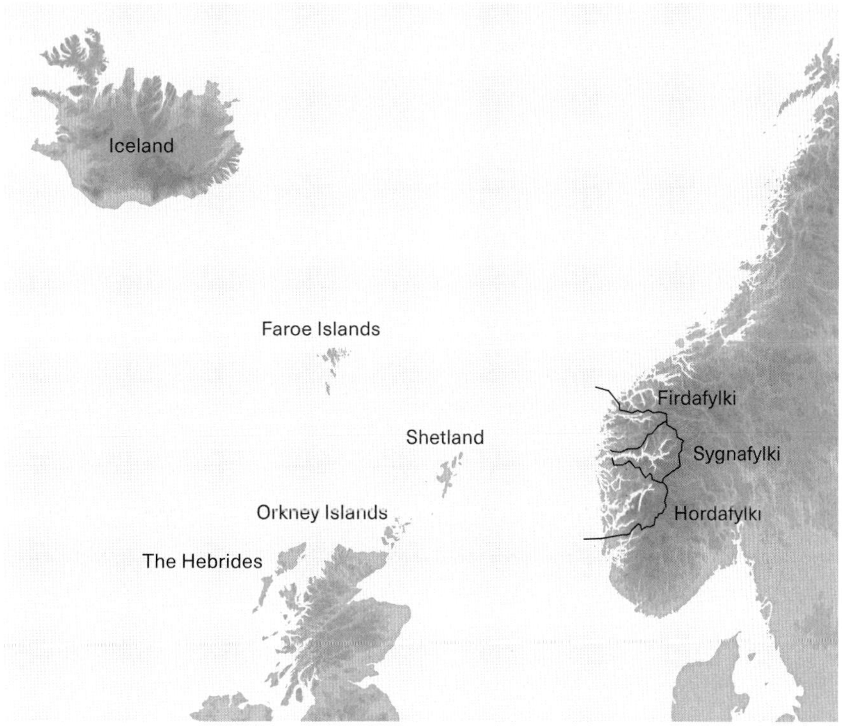

Fig. 0.1: Norway and the North Atlantic region as the macro level, and the focal area western Norway demarcated (Illustration: Per Bækken).

technology and textile production in West Norse environments – here, defined as regions in central western Norway and the North Atlantic, from Orkney in the south, to Iceland in the north (Fig. 0.1). These were areas with historical roots and interaction with Norway, not least western Norway. Together, these areas on the fringes of north-western Europe provide a rich source material related to textile production.

While studies of early textile production have tended to concentrate either on textiles or tools separately and either within a prehistoric or a medieval time frame and by exploring either archaeological or written sources, this study strives to take all available source categories into account, crossing periods and the artificial divide between prehistoric and medieval archaeology. Instead of centring on one element in the process, such as weaving and the finished fabrics, all stages of production have a bearing. When assessed as an integral entity of inter-connected work operations, the complexity and scale of the production becomes clearer and more intelligible. The various stages from raw material to finished fabrics connect to different environments, outdoors and indoors, and to various actors, producers, distributors, and users. Contextually, the traces of textile production are explored as integrated processes and in relation to changing socioeconomic systems of the period and are seen from gender and social perspectives with a focus on local and regional aspects of production.

Textile production in a period of change: from the Viking Age into the Middle Ages

In northern Europe, the transition from the Viking Age to the Middle Ages was a period of major socio-political and socioeconomic transformations. Various records indicate a society marked by population growth, political centralisation processes, and increased social stratification. The first part of the period was marked by migrations, emigration, and from coastal Norway primarily directed towards the the North Atlantic – the Orkneys, Shetland, the Faroe Islands, and to Iceland, as the latest wave of colonisation in the latter part of the 9th century. These trends coincide with the early rise of urbanism in the Viking period in Scandinavia, in towns or prototowns such as Ribe, Birka, Hedeby, and Kaupang, and by expanding international commerce by the Middle Ages, in Norway starting in the 11th century and escalating in the following centuries. However, the North Atlantic region maintained their rural societies. Intensified rural production in the Viking period is often ascribed to agrarian changes in technology and land use, which also characterise the period.[3] As for textile production, it was in a phase of transition between old and new technologies, shifting gender roles, and an expanding economy by the Middle Ages, with new ways of organisation of crafts in urban contexts. In north-western Europe, textile production in the fast growing medieval towns became one of the major urban crafts with textiles as a commodity in demand from larger markets.[4] Altogether, reorganisation of manufacture and marketing became central elements in medieval economy, and where new and more advanced tools, such as the horizontal loom, and new organisations within guilds played a central role.[5] These changes also involved a large-scale interaction between rural and urban economies on demand and supply of raw materials, semi-manufacture, which also relied on an extensive rural work force at greater distances.

Like food and housing, textiles as important and indispensable products in human life constitute one of the most basic human needs. Although the production centred on the household as a nexus of productivity in this period, there was an overlap between domestic and non-domestic production of surplus, often forcibly extracted by the elites, landlords and in the Middle Ages' official authorities.[6] The production could, then, also serve as a medium of exchange to comply with wider communal duties and other obligations depending on both status and demands in society. The production covered multifaceted and multifunctional areas of utilisation – all kinds of fabrics for basic needs within the households for clothing and shelter, such as clothes, blankets, packaging, but also more prestigious clothes within the luxurious segment, as evidenced in the Veka burial. A new and important product by this period was the sail[7] as an essential precondition for raids, trade, and colonisation in the Viking Age, and for the expanding international commerce in the Middle Ages.

The change from a largely domestic household-based textile production to a more commercial urban industry took place at different times and scales in various areas of Europe. The area in focus here followed a different path than areas in Continental Europe and was based on long established methods and technology – hand spinning and weaving on the upright warp-weighted loom[8] and based on customary gender roles and work domains. Still, textile production was not static and uniform, and to some degree, it represented both a production for households' own needs and a specialised production for the elites, as well as a certain amount of obligatory surplus-production. The productive sphere was thus located both inside and outside the domestic sphere on smaller and larger scales. However, the latter part of the period provided easier access to imported goods produced in the larger European urban centres.[9]

The production of textiles was labour intensive with its long chain of work processes in various environments, both human and physical environments indoors and outdoors.[10] Ultimately, the production relied on widely available resources – grazing land and outlying areas for feeding sheep and goats, and breeding and herding on an annual and continuous basis. Cultivation of flax and hemp and harvesting of nettle provided plant fibres needed in the production of both finer and coarser qualities. Agricultural production and animal husbandry were therefore key elements and a precondition for all the other stages in production. Many people with various skills and competence were thus needed when transforming the raw material of animal and/or plant fibres through a meticulous treatment of the fibres and preparing them for spinning and weaving. The amount and quality of the products depended, then, on all the foregoing processes – available environmental resources, varied tools, and an experienced workforce that needed organisation and coordination.

Traditionally, far back into prehistory, women have carried out these work operations, especially spinning and weaving but also preparatory and final processes into finished products of various kinds. The different working stages thus connect to the social representation of technologies, where technical skills and domains of expertise were often not only divided between and within the sexes but also by rank. Both archaeology and literary evidence attest that textile production represents a long recognised female know-how, transmitted within culturally defined gender domains, culturally coded as feminine.[11] Within the hierarchical societal structure of the Viking period and the Middle Ages, the production seems to have attained a ranked structure, where dependents were producing for their landlord/landlady – both for the household and as a surplus industry for wider distribution, and/or more of a luxury production for conspicuous consumption to signify rank. Power and control of textile production is therefore a central issue too, related to access to resources and a workforce of diverse competencies. The production was linked to distribution, consumption, and demand for various textiles at short or longer distances. When seen in larger regional contexts, including the islands in the North Atlantic, networks of both intra-settlements and inter-site exchange and inter-communal and intra-regional cooperation should be looked for.

By the Middle Ages, in the expanding urban centres in north-western Europe, some of these processes, especially the last stages in production, weaving, shearing, and tailoring, became specialised urban professions and male work domains organised within guilds.[12] Here, new technology – the introduction of the horizontal loom – seems to have played a crucial role. It has been regarded as unlikely that the expanding urban weaving centres in England and elsewhere in northern Europe of the 12th century and onwards could have been using anything else than the new treadle loom, and by male weavers.[13] Nevertheless, women constituted an important work force in the production, although often with a subordinate status. However, the change from a domestic household-based production to a more commercial industry took place at different times and scales in various areas of Europe. This shift in technology and gender roles in the Early and High Middle Ages may therefore be exaggerated and more complex than often stated.[14] The issue of gender and gender roles, technology, and organisation also play an important role in this study, being largely interconnected and as integral processes of both genders.

Despite its major utilitarian role, economic and socio-political importance, textile production traditionally has not caught the same research interest as the other spheres of rural and urban production of this period. Still, the varied source material of both archaeological and written evidence makes this a promising field of research. The archaeological evidence from Viking Age burials with gendered grave goods, various types of settlements and working environments, and information from written records, provide new and to a large extent, unused opportunities for exploring textile production in a long-term perspective from the Viking Age into the High Middle Ages. In the present area and period under study, the natural environments and humid climates offered favourable conditions for winter grazing, and thus sheep breeding and high-quality wool production. How these areas in the periphery of the new urban craft centres further south were affected by, or responded to, the new ways of organisation and specialisation in an increasing market economy based on long distance trade by the High Middle Ages, is therefore another issue of special interest. A close study of this broad representation of sources related to textile production from varied environments and contexts in western Norway and the North Atlantic thus offers an opportunity to expand our knowledge and shed new light on this important production and the actors behind it.

The study area: space and time

The primary research area is represented by the two counties in the central part of western Norway, Hordaland and Sogn og Fjordane, recently unified as one large county, named Vestland. In the Middle Ages, this area was divided into three administrative units or counties, in Old Norse (ON) denoted as *fylki: Hordafylki* to the south, with Bergen as the only medieval town in the region; *Sygnafylki* further north, centring around the 200 km long Sognefjord that cuts deep inland; and furthest north, *Firdafylki* (the fjords). The area stretches from the North Sea coast to the inner fjord districts and mountainous areas (cf. Fig. 0.1, see Chapter 5, Fig. 5.1). From the end of the 9th century onwards, this was a core area in the state formation processes in Norway, as it was here the new kings and adherents held important strongholds that could comprise larger territories within their estates.[15] It was also the core area of the Gulathing law province dating back to at least the beginning of the 10th century.[16]

People from western Norway were largely dominant among the emigrants and settlers of the islands in the North Atlantic and of Iceland from the late 9th century on but settlers were also from other parts of the country including northern Norway.[17] They carried with them old traditions and ways of organisation and production as their cultural and mental luggage to earlier partly uninhabited areas, such as the Faroe Islands and Iceland.[18] However, some were also from northern areas of the British Isles, the Hebrides, and areas around the Irish Sea as a secondary wave of settlers, and brought with them slaves and an unfree workforce.[19] Many of them were women, as has been substantiated by genetic evidence.[20] The areas to the south, the Orkneys and Shetland, were already inhabited by the Celtic element but seemingly soon repressed. Culturally, the population then, was not homogenous, but with the sea as a main communication line, people over larger areas were connected – culturally with their ancestral roots, and over time, economically for trade and exchange as well as politically by the end of the period.

The region of western Norway was rich in outlying resources, especially in the inner zone, and able to provide abundant supplies of wool for textile production. This was an expanding industry in the Viking Age and the Middle Ages, not least after the pioneering maritime technology of sailing was adopted and adapted in these northern waters. In Norway as well as in the North Atlantic, production was organised within rural environments and economies with a mainly hierarchical social and political structure.[21] In these settings, the material culture played an important distinguishing role as markers of social status, not least visible through the garments.

All this makes the North Atlantic region an interesting and promising study area seen in a wider comparative perspective – a region where textile production became an important industry and wool, yarn and textiles stand out as trademarks and special properties, also serving as commodity money. In all the islands in the North Atlantic homespun wadmal (ON *vaðmál*) became a standard economic measure and currency for taxation and exchange, however, best documented, and longest lasting in Iceland. Regional and local differences and similarities and changes over time are key issues.

On the west coast of Norway, Bergen emerged as a new town by the end of the 11th century and soon developed into an important international commercial and administrative centre with close relation to the islands in the west.[22] As for textile production, Bergen also became a centre of specialised craft production, with weavers, shearers, and tailors as new professions from the latter part of the 13th century at the latest.[23] The North Atlantic on the other hand, kept an overall rural structure, where Iceland, and to some extent the Faroe Islands and Shetland became primary areas for export of wadmal by the Middle Ages.[24] Hence, western Norway stood in a central position, with textile production in both rural and urban contexts and with close relation to the North Atlantic islands during the entire period of study.

These areas were not homogenous even if they shared important common conditions. Like in western Norway, the North Atlantic region comprised limited cultivable areas but vast areas of grazing land. When the Norse settlers arrived on the islands in the 9th and 10th centuries, they enjoyed a more favourable climate than today, but it became gradually colder by the late Middle Ages.[25] Iceland, as the northernmost area for comparison, situated at about the same latitude as Trondheim/Nidaros in Norway shares a very similar climate (cf. Fig. 0.1). However, the oceanic climate in Iceland went into a relatively cold period around 1300 but was still climatically more like the climate in Norway than the Arctic.[26]

The period is characterised by expansion and an underlying population growth in western Norway – clearing and establishment of new farmsteads in the lowlands and an expanding use of outlying areas for pastures and areas for animal farming. It also involved extraction and exploitation of different resources, such as iron, minerals, and wildlife, converting them into portable commodities for exchange or sale,[27] often based on a subordinate work force,[28] including unfree labour.[29] As noted, textile production may partly have been organised to comply with wider communal duties, stimulated by the state formation processes from the end of the 9th century onwards. By the 10th century, this, among others, was connected to the obligatory and nationwide duty for the naval defence (ON *leiðangr*)[30] providing ships that obviously also needed sails, crew, and food supplies, and earlier probably organised on a smaller scale within and between estates. The new elites attached to the King, and later also to the Church from the 12th century onwards, needed fabrics for many purposes, however, on a smaller scale, such as special furnishings and clothes to signify rank, and as gifts for the maintenance of economic and political relationships, among others. In these settings, the production may have attained another structure of attached craftspeople also producing prestigious textiles on demand.

Within a subsistence economy, textiles were necessary for domestic use and produced within the households. Within the hierarchical societal structure of the Viking Age, textile production may largely have been based on servants or other dependents, including slaves, that required management and organisation. In this setting, the production may also have been more of a luxury-oriented production, with surplus manufacture for wider distribution as well.[31] However, the way work was organised is not always so clear, but it was generally in the domestic frame within a rural household structure for the household's own needs but also production for an elite, often located at central places of *longue durée*. It may also have been influenced by tribute systems with surplus production, and through physical displacement of the workforce at central places and in proto-urban centres, such as Kaupang in Vestfold from around AD 800 onwards[32] and in the Early Middle Ages in towns as new channels for large-scale export and import (see Chapter 9). The identification of places and environments for various types of production is therefore an essential issue in this study, whether at central places or more spread out or dislocated in various types of landscapes. These varied conditions for production also open for comparisons in time and space.

Different contexts

Ancient textile production represented extremely labour intensive and versatile work processes, which required technical skills and expertise, cooperation and larger networks when exceeding the household level. An analysis of the archaeological material as elements of wider socioeconomic contexts may, then, shed new light on textile production, both spatially and in a long-term perspective. The study area represents wide geographical areas – three interconnected socio-political regions within central western Norway and the more distant islands in the North Atlantic but with strong affiliations to western Norway. However, this extensive area is too wide-ranging to be fully covered. Therefore, the study applies multiple scales as an analytical tool, representing interconnected spatial levels – *micro-*, *meso*, and *macro* levels – from local sites to regional units to cover the whole region. The entire study area, western Norway and the North Atlantic, represents the macro level, in which western Norway forms a focal area of research and the three counties represent the *meso* level. This level is delimited to the above-mentioned core area of Viking Age Norway. At this level, individual sites with finds of textiles and textile tools, form *micro* levels, confined to more delimited areas at farm levels, locations, and sites within their respective districts from coast to inland. This tripartite approach is especially interesting related to the socio-political development over time and related to changes within textile production. The comparative North Atlantic region comprises selected contextual studies of tools and textiles from various excavated sites from Iceland and to some extent from the Faroe Islands, Shetland, and the Orkneys that may highlight important similarities or differences.

The finds from these archaeological settings vary and change significantly over time from intentionally deposited burials in the Viking Age to more arbitrary finds of textile tools from different work contexts and lately from more systematic rescue open air excavation caused by new enterprises in protected areas. Stemming from various archaeological sites and structures in both lowland and outlying areas, they reflect diverse loci of production within these areas over time. In western Norway, as to some extent in the North Atlantic, textile tools appear in seasonal settlements in marginal outlying areas, in buildings connected to shielings, and based on extensive exploitation of varied resources. They indicate a surplus production and exchange at longer distances in the wake of urban commercialisation. Urban textile manufacture within the study area is, as noted, only represented in Bergen and for production, basically from the mid-12th century onwards.

In western Norway, grave contexts from the Viking period supply most of the finds until around AD 1000, and mainly from the 10th century in Iceland. Sunken-feature buildings or pit houses form a special category as for textile production and other crafts in the Viking Age Scandinavia,[33] as well as in Iceland[34] and in western Norway but also other types of buildings, such as the traditional longhouses of the period. The finds and their contexts generally provide good chronological frameworks. The traces of the many work operations and associated tools in various environments thus make it possible to come closer to how the production developed, locally and regionally, socially, and economically.

The ritual of supplying the dead with grave goods, and dressed according to their status, signifies social and economic status, and provides another layer of information. Here, tools appear as both functional equipment and symbols, reflecting status, social identity, and gender roles. The way the deceased was dressed and furnished represented an intentional act and for display during the funeral ritual and seems to match their earlier role and position as the heirs wanted and was accepted by the audience. However, these customs were restricted to an upper stratum of the society. Although remains of the dress seldom survive other than in small fragments, they are crucial for evaluation of qualities and modes of production, including social standing and possibly regional variations and changes over time. Still, tools and textiles from grave contexts as symbolic tributes mainly reflect production indirectly.

The find conditions for both tools and textiles thus vary from the Viking Age into the High Middle Ages. More composite tool kits may for example indicate special textile producing environments. Differences in the number and composition of tools at various sites may signify different types of production and household strategies, which again may have had regional economic implications. Not all producers participated in production in the same way and within the same range of activities. The frequencies and quality of tools in each tool category may, for example, indicate that households may have emphasised cloth production. This again, could be indicative of degree of specialisation and/or standardisation. Households – elite and commoner – may have been dependent on each other, and on intraregional exchanges. Here, textile finds can provide a more direct empirical foundation of what kinds of textiles were made and could be produced in time and space. As far as it is possible, tools, textiles, and other sources are therefore assessed contextually and from different angles – technologically, culturally, socially, politically, and economically, including the agents behind them – as producers, consumers, and entrepreneurs/distributors. All these aspects were closely entangled.

The source material

Remains of tools and textiles and their find contexts constitute the main research material – how they relate to other archaeological finds from the same structures, environments, and contexts of which they were part. As concrete remains of the textile technology, both tools and textiles form an important framework for assessment and interpretation of the tools, the character and level of production and organisation in different areas over time. Over the years, a wide range of textile tools from the Viking Age and the Middle Ages has been uncovered from various contexts and landscapes and documented and analysed to a certain extent. However, they have only to a limited extent been investigated as integrated and interconnected elements in wider geographical and socio-economic contexts.

Combining archaeological and written evidence with environmental data, the different source categories have potential for gaining new knowledge about production, as well as socio-technological systems, modes of production, and organisation. The research situation for the period under study takes advantage of the fact that Norway in the late Viking period and the Early Middle Ages started to enter the world of script, and a more abundant body of written evidence is preserved from the following High Middle Ages. The 12th and 13th centuries saw the recording in writing – various genres of saga literature, verse, and important law codes and regulations in Norway and Iceland, shedding light on contemporary as well as past events and conditions, and from the late 13th century, contemporary recorded evidence too. The written evidence supplements the archaeological remains and provides, then, valuable information able to achieve a more comprehensive understanding of the condition for production, based on different and independent source categories.

Particularly important for the development of textile production and its role within farming systems and rural society are provisions of the law codes, as they are not only normative but also describe or reflect factual conditions. Their more programmatic regulations may then be compared with the practices revealed by the increasingly important material record of archaeological findings. Two generations of medieval Norwegian law codes have been preserved: first, the provincial codes, of which the *Gulathing* law related to western Norway, and largely reflect legal situations of the 12th and early 13th centuries. However, it also contains provisions dating further back. The second generation comprises national codes from the 1270s, the so-called Land Law of 1274 (*L*) and the Urban Code (*Bl)* from 1276, originally designed for Bergen. In the following decades, laws were amended and supplemented by royal regulations. Thus, it is possible to establish a rough legal chronology for the period *c*. 1100–1350 with elements of even longer older origins. As for Iceland, the comprehensive law codices *Járnsíða*, *Grágás*, and *Jónsbok,* also from the late 13th century, share traits with Norwegian law from the 1270s but reveal differences that are interesting to look closer at when related to textile production. The later customary norms in Iceland, the so-called *Búalög*, recorded as late as in the 1600s also seem to have older origins, probably from the 13th century.[35] They provide unique information about weaving and working conditions in Iceland. Here, women appear as the main agents in the production, but at the opposite end of the social ladder compared to the woman from Veka.

The archaeological finds of tools and textiles stemming from various contexts, may represent symbolic and prestigious grave goods. The remains of clothing in Viking Age graves, traces of production found in working environments or as discarded remains may additionally reflect former everyday use. Medieval textiles, however, are mainly discovered in urban contexts in Norway but also in various rural contexts in Shetland and especially Iceland, thus representing various working environments. The quality of some of these fabrics of high uniform standard from Viking Age burials, such as fine worsted cloth as was discovered at Veka, and coarser standardised qualities such as wadmal, implies different modes of production too. In the first case, high dexterity and level of organisation, and access to high-quality supplies of wool were required. It has therefore been debated whether these high-quality fabrics had been imported, or whether textile milieus at that level existed in western Norway in the Viking Age (see Chapters 1 and 6).

Altogether, a significant body of textile-making equipment and varied qualities of textiles constitutes the main empirical basis in a wider contextual study of textile production where physical, social, socio-political, and socio-economic environments form parts of their contexts. Here, written sources also provide vital textual information. In this study, identification, and earlier textile analyses of finds from western Norway and the North Atlantic are assessed and compared with textiles and a varied spectre of textile tools. When seen together within a wide time frame and within varied environments, the combined sources make it possible to make a more multifaceted assessment of the development of textile production than is possible in many other parts of Europe.

Aims and research questions

The research questions in this book all relate to the overarching issue of stability or change in textile production as for quality, quantity, and organisation in a long-term perspective of five centuries. By comparing tools, textiles, and contexts in wider time frames and at various spatial levels from the local to the regional, changes and variations within the production become clearer and better explained when seen within larger areas and socio-political trends. While Iceland, the Faroe Islands, and Shetland are known for their standardised bulk production of woollen cloth within a rural domestic setting, there has been little comparison with production from their 'homelands' and with urban production. Some of the key issues to be pursued connect, then, to production time and cultural space: to what extent do tools and technology vary in different environments? Is it possible to reveal similarities or dissimilarities between domestic versus more specialised production, including the actors involved? Do tools and textiles from Viking Age environments differ from corresponding categories from the Middle Ages, both rural and urban? By focusing on the 'small things'– the archaeological remains of tools and textiles – contextually and supplied with information from written sources, the overall aim is to achieve a more holistic comprehension of the processes and levels of textile production on these northern fringes of Europe.

Notes

1 Shetelig 1912, 206–207. See Chapter 5, Fig. 5.19.
2 Bender Jørgensen 1986; Walton 1988; Lukešová 2011.
3 *e.g.* Øye 2005; 2016b.
4 Power 1941; Postan 1972; Munro 2000a; 2000b.
5 *e.g.* Rosser 1997; Soly 2008.
6 *e.g.* Duby 1998 (1968).
7 Andersen *et al.* 1989; Andersen 1995.
8 *e.g.* Hoffmann 1964; Guðjonsson 1994; Johnston 2016a; 2016b.
9 Bruns 1900; Nedkvitne 2014.
10 cf. *e.g.* Andersson Strand 2012.
11 cf. *e.g.* Herlihy 1990; see also Costin 1996; Damsholt 1984.
12 Rosser 1997; Epstein 2008.
13 *e.g.* Walton Rogers 1997
14 Øye 2016a.
15 Helle 1995, 23–25; Iversen 2008.
16 Helle 2001, 25–27.
17 Nielssen 2012.
18 *e.g.* Eldjárn 1956; Byock 1993; Stefánsson 2003; Price *et al.* 2006.
19 *e.g.* Iversen 1997, 95–99.
20 *e.g.* Helgason *et al.* 2001; 2009; Goodacre *et al.* 2005.
21 Lindkvist 2003, 160–167; Stefánsson 2003, 210–14; Sigurðsson 2011.
22 Helle 1982; 2003; Stefánsson 1986; 2003.
23 Helle 1982, 428–429; Øye 1988.
24 Þorláksson 1991; Nedkvitne 2014.
25 Lamb 1995, 165.
26 Júliusson 1997, 176.
27 Øye 2002; 2003; 2009; 2013; Tveiten 2012; Baug 2015; Baug *et al* 2019.
28 Iversen 1995.
29 Iversen 1997; Holm 1986.
30 Lund 1996.
31 Andersson 2008; Øye 2010; 2015a; 2015b; Andersson Strand 2011.
32 Øye 2011; Skre 2011.
33 *a.o.* Madsen 1969; Rasmussen 1969; Strömberg 1971; Andersson 2000.
34 Einarsson 1992; Milek 2012.
35 Júliusson 1997.

1

State of research

Textile production represents a field of research within several disciplines; archaeology, history, art history, anthropology, ethnology, and textile specialisations based on different sources, methods, and perspectives both theoretically and practically. These multidisciplinary conditions also reflect the multiple layers of information that can be derived. Yet until recently, it has been rather uncommon for textile production to be approached transdisciplinary and diachronically by including both prehistoric and medieval sources. Generally, textiles have gained far more attention than the archaeological finds of textile tools. Studying tools and textiles as contextual analyses of the production over larger areas in this part of Europe has also been quite rare.

Within archaeology, textile production largely appears as a specialised field of research based on meticulous analyses of the textile remains that include techniques as for spinning, plaiting, twining, and weaving and its various patterning techniques, dyeing, and sewing, and the various functions of the fabrics. Generally, textile studies are rather specialised, limited in scale and narrow in time and space. However, the pioneering and synthesising works of the Danish archaeologist and textile expert Lise Bender Jørgensen with large-scale overviews of archaeological remains of textiles from Scandinavia and northern Europe and based on wide-ranging and over-arching analyses and collected information, stand out as solid bases for further research.[1] During the last few decades, textile studies have expanded significantly with a rapid development in methods of analyses and increasingly connected to larger international specialist networks. Improved tools and methods, such as digital microscopes, techniques and sophisticated scientific methods for fibre analyses, such as strontium isotope analyses and aDNA, have opened up for wider perspectives and interpretations of, for example, provenance of wool and breeds.[2] Lately, textile research has also been drawn into current discussions in archaeological theory opening up for wider societal perspectives and new approaches, theoretically and methodologically related to tools and contexts.[3] As a result, the value of textiles as sources in overall interpretations of archaeological data has now, to a larger extent, been recognised and drawn into mainstream archaeology.[4]

Nevertheless, the archaeological study of textile production may still appear marginal as it is seldom incorporated as a central theme within archaeology in more comprehensive scholarly overviews. This is largely also the situation for the present study area and period. Textile production has, for example, not engaged the same interest as other spheres of rural production,[5] despite its major utilitarian and socioeconomic role. However, within economic history of the Middle Ages, the study of textile production has been an important field of research since the late 20th century but often restricted to economic perspectives, being an important commodity within the international manufacture and trade.[6] A wide-ranging source material on economic and organisational aspects forms the basis for economic and social studies on international trade, urban crafts, and guilds. Based on written evidence, textile technology has also been a subject of current interest, not least concerning weaving with a shift from the vertical to the horizontal loom, and a transition from a rural domestic production to an urban male industry for the market.

This introductory chapter aims to give an updated overview on research of special relevance for issues taken up in the present study, including a brief outline of sources and both empirical descriptive material and interpretative analytical contributions, serving as a backdrop for analyses and discussions in the succeeding chapters.

Tools in archeological contexts

Textile tools appear in different archaeological contexts over time. In the Viking period in both western Norway and Iceland, they may, as noted in the introduction, occur as intentional deposits in burials, and as left and lost remains in various working environments of deserted settlements and outlying areas appearing as special structures such as sunken-feature buildings. From the Early Middle Ages, various textile tools also appear in the new medieval towns, and within the study area in Bergen from around AD 1100, somewhat later than in the earlier towns of Trondheim and Oslo.[7] In the North Atlantic without any towns in the Middle Ages, textile production was only connected to rural environments.

The role and importance of textile tools, such as spindle whorls and loom weights as archaeological sources, have varied over time – from being largely neglected, not even collected, to being registered and documented in broad categories. However, in later years they have gradually been assessed typologically and functionally with regards to ability to produce threads and cloth of varied qualities. Not least, new insights have been gained through experimental spinning and weaving based on replicas and reconstruction of tools from archaeological contexts. The Swedish archaeologist, Eva Andersson Strand, has made several important contributions to the study of textile tools in Viking Age and Scandinavian medieval contexts, and most important methodologically, their ability to shed light on the textile work in former periods based on replicas. She has studied how differences in the 'tool kits' reflect differences in the textile production and their representation in different types of settlements; in rural, proto-urban, and urban, especially related to Birka, Hedeby, and Löddeköpinge,[8] and related to various types of production and qualities, including sailcloth.[9] The studies are based both on formal analyses of the artefacts and an assessment of their functional qualities based on experimental archaeology related to spinning with the drop-spindle and weaving on the upright loom. The experiments were conducted in collaboration with the Centre for Textile Research (CTR) at the University of Copenhagen. By testing out whorls of different weights and how they affect the quality of the thread and the weave, these experiments have given new insights into how different spindle whorls and loom weights function regarding qualities of the products.[10] The spindle whorls thus have a far greater potential as sources to textile production than earlier recognised.

Such tests have opened for comparing archaeological tools with finds of textiles from the same contexts. The new approaches – assessing tools, textiles, and contexts as integrated elements of principally the same work processes – provide thus added value and a larger potential for interpretation and syntheses. These approaches have inspired my own research[11] and is further pursued in the present study.

Tools in graves from western Norway

In Norway, as in other areas, textile tools in archaeological contexts have long been a rather neglected artefact category and generally paid little or no attention. In the present study, the archeological finds from the Viking period are preserved and stored at the University Museum of Bergen,[12] established in the early 19th century. The older finds of these categories are described rather cursorily in museum catalogues and field reports and have not been studied as material in their own right or assessed in wider contexts. Still, different types of spindle whorls, loom weights, and weaving beaters were earlier represented and depicted in Oluf Rygh's pioneering work *Norske Oldsager* from 1885. This was the first more comprehensive overview of various functional artefact categories and their various representations and is especially renowned for its good illustrations. When later represented, they are often referred to as types according to Rygh's depictions, denoted with a capital R for Rygh, and a number referring to a special depiction in this volume and accessible on the internet. It is still widely used as a classical reference, as is also the case in the present study.

Haakon Shetelig's volume *Vestlandske graver* from 1912, where he, among others, published the above-mentioned Veka burial, is still a useful and first-hand publication when exploring textile tools in Viking Age grave contexts from western Norway, as an overwhelming part of Viking Age graves were discovered and taken care of at an early stage of archaeology (Fig. 1.1; see Chapter 5, Fig. 5.2). Here, some early and important finds related to textile production are described, analysed, and depicted more comprehensively than in the museum catalogues and annual yearbooks.

Jan Petersen's survey of tools from the Late Iron Age (AD 550–1050) in Norway, *Vikingetidens redskaper*, from 1951, describes and quantifies each group of tools according to their main functions, such as spinning and weaving as for textile tools. This comprehensive presentation of tools from the Viking Age in Norway represents a unique and impressive data collection, and it still provides the only more inclusive overview of textile tools on a national level. Here, the different tool categories are described according to shape, measurements, and their broader finds contexts, such as graves, settlements, or stray finds. When found in graves, gender is also referred to, based on find combinations, mostly jewellery related to females and weapons to males. Such finds formed another category, analysed and published by Petersen in 1928. Although the tool assemblages were only updated to the 1940s, his book on tools still comprises the majority of grave finds in this region (see Chapter 5, Fig. 5.2). However, finds related to settlements and other later discovered contexts are clearly underrepresented. Petersen's national overview, although not updated, still serves as an important reference for comparison with the finds from the study area, and of special relevance as for the grave finds. Information about later finds from the study area relies on inventory lists and field reports of the University Museum of Bergen.

Burial customs: social representation, status, and gender

The burial customs from the Viking Age – as reflections of gender roles and social structures within an overwhelmingly

Fig. 1.1: Reconstructed grave mounds at Myklebust in Eid, Firdafylki, excavated and published by Shetelig in the early 20th century. The flattened mound in the foreground contained several graves, of which two contained textile tools (Photo: Anders Sellevold Aaseth, Norwegian Directorate for Cultural Heritage, with permission).

hierarchical society of the period – are aspects that have gained stronger attention in recent research. New perspectives focusing on the explanatory frameworks of landownership and hereditary conventions have provided new insights into gendered and spatial patterns of Viking Age graves. Recent studies by the archaeologist Frode Iversen of the distribution of Viking Age graves in western parts of the county,[13] have revealed a strong correlation between monumental and richly furnished graves and rights to landed property as documented in medieval and Early Modern sources. They indicate that only a small segment of the population was buried in monumental graves furnished with grave goods and that relatively few of these burials were female. As visible monuments in the landscape, often close to the settlement or at farm boundaries, the graves thus seem to have served as symbolic and commemorative markers, where the heirs could signal continued property rights and control over land. This may also explain the relatively low numbers and spatial distribution of graves related to population estimates, regional representations, and a gendered imbalance (see Chapter 5). When only a small segment of the population was buried according to these customs, the graves seem to allude to social identity, and by following conventions of how to display status, roles, rights, and responsibilities. These insights form an important backcloth when assessing the grave finds with tools and textiles contextually in the following chapters.

In a study related to gender representation of graves from various regions in the country from the Late Iron Age and mainly from the Viking period, the archaeologist Bergljot Solberg found that only about one-fifth of the total could be identified as female. This was based on the dress accessories and other female related equipment, since osteological material is rarely preserved.[14] In a national overview of all graves from Norway dated to the Late Iron Age (AD 550–1050), including around 8,000 graves, the archaeologist Frans Arne Stylegar could observe large regional differences, from one-tenth to nearly equal representation. In the present study area of western Norway, however, female burials could comprise about one-third of the graves.[15] These results largely concur with regional patterns of landownership, as documented by Iversen. Female graves furnished with grave goods seem, then, generally to represent a higher stratum in society with certain privileges, probably representing female heirs or widows as guardians of minor children.

Another study by Solberg from the same period and related to gender and status as reflected in the grave goods, differentiates social rank based on the composition of the grave goods, and for women based on the composition of dress accessories and tools.[16] Based on these artefact categories, she defines three social groups, where status group III represents the highest stratum. For this reason, female agency and social status seem to be best represented in female monumental burials, serving as indicators of hereditary rights and roles within a limited social stratum. How such social indicators concur with the assortment of textile tools and the qualities of preserved textiles is therefore a central issue in the following analyses (see Chapter 5). Such indicators – monumentality, range of dress accessories and special tools – are compared with graves in the North Atlantic, which to a large degree reflect the same customs as documented in western Norway, however, smaller in scale and with deviations – often less monumental and relatively fewer in numbers. Iceland, as the largest area in the North Atlantic, also represented the latest settlement and with grave monuments from the last decades of the 9th century and covering the 10th century. However, here the custom of displaying tools did not have the same breakthrough as in Norway – neither for men nor women. Social belonging and status are altogether more firmly expressed in dress equipment, jewellery, and weaponry.[17] Here, the Canadian archaeologist Michèle Hayeur Smith has carried out an interesting study of social representation of female grave goods based on the same criteria as in Solberg's study.[18] With a higher representation of unmarked graves and also with preserved skeletal material due to good preservation conditions, Smith's evidence from Viking Age Iceland thus opens for further comparison. This again offers a broader outlook and on a firmer basis related to gender identification, which is especially useful for areas lacking preserved biological material.

In an analysis of, among others, textile tools in Viking Age burials from the same area as the present study area in western Norway, the archaeologist Anders Rabben, in his unpublished master's thesis from 2002, studied different tool categories, especially textile and kitchen tools, and to some extent agricultural tools in male, female, and double graves, and based then on much of the same source material as in the present study. By focusing on functions and find combinations related to gender, he finds that a relatively large share of textile tools, especially for weaving, connect to male contexts. This may, according to Rabben indicate a male role within textile production, possibly related to sailcloth.[19] As gender is a central issue in the present study too, the information related to finds contexts is therefore closely examined to explore this issue further, since documentation related to many of these early findings often is sketchy and imprecise.

In some recent works, the present author has evaluated and discussed textiles and tools from Viking Age graves and sites in the North Atlantic as case studies.[20] Here, results from the above-mentioned experiments of spinning and weaving and not least inspired by the works by Eva Anderson Strand's studies of tools and textile production in Hedeby and Birka, as well as Löddeköpinge,[21] have been important to get a clearer understanding of special properties of tools in textile production. This approach makes it possible to assess the finished fabrics related to the tools and in varied social and physical environments, which again may have transferable value when assessing similar tools in contexts without textile remains. When tools and textiles are found in closed contexts with artefacts, such as dress accessories and weapons as grave goods, they again may link to social categories and gender.

Textile tools in working environments

Textile tools appear in various rural settlement contexts from both the Viking Age and the Middle Ages, such as the complex of several building remains from the Viking period, documented at the site Ytre Moa in the transitional zone between farmland and mountainous summer grazing areas in the inner parts of the long Sognefjord.[22] In nearby mountains, present-day shielings areas at altitudes from 900–1100 m asl have revealed structures and find contexts for textile production dated from the Viking period into the Early Middle Ages.[23] The largest complex of sunken-feature buildings or so-called pit houses found was agglomerated at the Bjørkum site in Lærdal, also in the transitional area between farming land and high mountains in the inner parts of Sogn.[24] The same type of working environments for textile production, in the form of sunken-feature buildings, have moreover been uncovered at medieval high-status farms within the study area of western Norway[25] (see Chapter 8).

Another but different site is a deserted medieval farm that has been excavated on the island of Sotra in a coastal area just west of Bergen with favourable conditions and vast areas for winter grazing with extensive *in situ* finds of textile tools.[26] These and similar types of sites give concrete and direct information about the production and within longer time frames, and again with parallels in the North Atlantic (see Chapter 9).

Until recently, however, the textile tools in such surroundings have been rather neglected as sources to textile production despite their potential to elucidate production and textile qualities in wider contexts. In an earlier article, the present author has assessed the tools from these sites and environments as case studies and in a comparative perspective, to get a broader outlook of settlements with extensive utilisation of outfields and mountain pastures for sheep and animal husbandry.[27] These studies indicate that a wide range of textile qualities have been produced both at small and marginal farms and summer shielings, and in more specialised milieus. Even limited assemblages offer a potential when assessed contextually as entities and are further assessed in the present study.

Both grave finds and deserted settlements with traces of textile production represent important contexts for assessing both textile production qualitatively, socially, and economically. Altogether, they include a wide range of finds from both newly discovered textile production sites and finds from older but unpublished excavations. A comparison with sources and studies of similar or dissimilar structures in western Norway and the North Atlantic thus provides a wider scope, hopefully to achieve a clearer understanding of the production and the interaction between tools, textiles, and the agents involved.

Urban and proto-urban contexts

In an earlier study, the present author has analysed textile production equipment found at the large medieval Bryggen site in the centre of medieval Bergen, based on more than 1,900 artefacts or parts of artefacts related to textile production. The analysis of implements for fibre preparation, spinning, weaving, sewing, and finishing processes, represents the first analysis of a comprehensive assemblage of textile tools from an urban environment in Scandinavia, covering the period from AD 1150 to 1500.[28] The classification used here has later been applied in other works, including the analysis of textile production equipment from the Viking Age town Kaupang in Vestfold,[29] connected to the large *Kaupang Publication Project*, 1998–2003. The assemblage from the settlement area, as well as the graves surrounding the settlement area, comprises altogether around 1,000 textile related artefacts, or parts of, mainly for spinning and weaving but also needles, needle cases, shears, and smoothing pellets. In the present study, results from these analyses are presented but the scope and perspectives cover both rural and urban production. Here, Eva Andersson Strand's pioneering studies of textile tools from Hedeby and Birka[30] form an important comparative component. Moreover, finds of both tools and textiles from Bergen, Oslo[31], and Trondheim[32] are analysed as integrated entities in a comparative perspective,[33] compared and discussed related to the rural counterparts from western Norway and from the North Atlantic (see Chapters 8 and 9). The overall aim, then, is to gain a wider outlook and clearer comprehension of the production and organisation in varied environments and over time.

Textile production in the North Atlantic

When looking at the textile production and textile production equipment from the North Atlantic, issues related to the natural conditions and opportunities for textile production related to wool and sheep rearing, are particularly relevant. Being based on various source categories studied by historians, archaeologists, and zoo-archaeologists, they provide valuable insights related to textile production[34] and in various environments and of different status.[35] They include both marginal areas, such as shielings in Iceland and the Faroe Islands, and high-status central farms.[36] Additionally, textile production is casually recorded in saga accounts, such as, among others, *Njáls saga* and *Laxdœla saga* and in later records.

Generally, tools are better preserved and more numerous than textiles that require better conditions for preservation. Still, the conditions for preservation are generally more favourable in Iceland than from western Norway and provide altogether a valuable source material.[37] However, no general analyses of textile tools above site level have been published so far from these areas and are only rudimentary even at site level. I have therefore looked closer at tools stored at the National museums in Iceland and the Faroe Islands to get a first-hand impression of the tools – especially spindle whorls and weaving implements from selected sites[38] and to be further elaborated in the present study (see Chapter 9).

Other comparative material

As for archaeological works on special tool categories, the British archaeologist Sue Hamilton has published a study of weaving beaters,[39] which may serve as comparative material. However, more important in this case is the Norwegian art historian and ethnologist Marta Hoffmann's valuable and comprehensive ethnological study of the warp-weighted loom,[40] which is still a standard work for understanding prehistoric and medieval weaving. Archaeologically, it is mainly traceable by the loom weights. However, Hoffmann takes her starting point in preserved looms into the present, in Norway, and related to ethnology and folklore, but also drawing the long lines comparing the historical upright loom with scattered and fragmentary archaeological remains. Recently, weaving traditions related to the warp-weighted loom have been incorporated in a detailed manual giving concrete insight in all the processes in warping and weaving according to old traditions[41] in both western Norway and the North Atlantic.[42]

By and large, information about archaeological textile production equipment – tools and textiles – from both the study area and period under study appear in many site reports and articles, without having a special focus on the artefacts per se. They may still provide important and valuable information and in such cases referred to in the concrete analyses in the following. The same applies to works dealing with textile production and textile tools from other periods or areas when they are related to current issues in the present study.

Tools and textiles in archaeological contexts

The inclusion of textiles and textile tools as important data in archaeological research is a relatively new phenomenon, not least prompted by the appointment of specialist textile researchers at major archaeological museums and departments and use of new and advanced techniques.

Viking Age and medieval textiles from Scandinavia and the North Atlantic have been a topic of numerous scholarly works dating back to the inter-war period and into the present.[43] The pioneer in textile studies of archaeological textiles from Viking Age Scandinavia is undoubtedly Agnes Geijer and her studies of the textiles found in Viking Age burials from Birka.[44] They include some textiles of very fine quality, later denoted as the Birka type – of thin and high-quality fibres, woven in broken diamond twill with a high thread count. Over the years these qualities have recursively been discussed as either imports or possible domestic production on a remarkably high level[45] – an issue that is further pursued in the following.

The production of sailcloth played a crucial socioeconomic role in this period and has therefore been a central issue and topic.[46] Again, Andersson Strand's approach of combining tools and textiles with experiments by using replicas of tools to explore the processes in a concrete manner, has been a rewarding method. The present study thus benefits from these results as for suitable tools for obtaining certain qualities – from raw material, work processes involved and products – and thus of transferable value. Moreover, the production related to various social frameworks: domestic, on a household level, and varying degrees of specialisation at various work stages, either independent or dependent.

Tools and textiles in burial contexts

Textiles found in Viking Age grave contexts from western Norway, from Hordaland in the south to include the county Sunnmøre north of the study area, have been analysed earlier by the archaeologist Inger Marie Holm-Olsen in connection with her unpublished magister dissertation from 1972. Here, she also discusses where the textiles could have been produced, in what kind of environments, and whether they represent domestic production or a more specialised craft – or possibly imported products. The textiles have been analysed as for fibres, spinning (spinning direction), the threads and tightness of the threads in the weave and on what kind of loom. Colour and fibre analyses, however, have been carried out only to some extent.[47] The results from her analysis thus form an important basis for the present study, and where the textiles are related to the textile tools from the same contexts – which represents an approach, which Holm-Olsen in her study also called for as a future task – and an invitation that is hereby accepted (see Chapters 5 and 6).

The Danish archaeologist Lise Bender Jørgensen, as one of the leading specialists within textile archaeology in Scandinavia, has assessed textiles in archaeological contexts from the entire Scandinavian area – where an altogether important corpus of textiles has been preserved. Here, she has added some more finds to Holm-Olsen's corpus that was limited to material belonging to the dress, to also include other textile remains.[48] Her catalogue gives valuable information about technical details for all the textile fragments and includes a list of artefacts and dating material from the same contexts and is an important study and basis for the present study, which also includes finds made after Bender Jørgensen's update from 1986. These analyses, carried out by textile curator Aud Bergli, have lately been re-examined by Hana Lukešová; both textile experts at the University Museum of Bergen.[49] So far, the analyses have only been partly published and thus not completely available for this study. Some of the more prestigious finds with textiles of the highest qualities have been analysed and discussed in detail by other specialists.[50]

The finds of textiles from the Viking Age town of Kaupang in the southern part of the country are all from the surrounding graves. They have been analysed, at least partly, and nearly 150 identified fragments out of nearly 200 have been analysed and published by the archaeologist Anne Stine Ingstad who also analysed the rich material of woollen textiles from the Oseberg burial.[51] Although being found outside the present study area, they are highly important for comparison as they contain textiles in similar weaves and qualities from the same period from western Norway. These textiles are not least relevant related to the issue of local production versus import, as some are of the same quality as the so-called Birka type, or Veka type, as it has recently been renamed[52] – being documented in burials from western Norway, however, not only from Veka. Both the Oseberg burial and some of Kaupang burials contain textile tools too (see Chapter 6). The textiles found in grave contexts mostly belong to the female dress, and thus are represented by a limited spectre of use areas and with a social and gender bias. Some of the burials comprise tools too, which have earlier been studied by the present author along with tools from the working environments.[53]

Analyses of wool and fibres have only to a small extent been carried out within these studies, and the few that have published relate to some of the finest qualities in diamond broken twill, among others from the famous Oseberg grave,[54] and some from burials in western Norway.[55] Generally though, tools and textiles have not been studied together, nor assessed in wider socioeconomic perspectives, which is an overall aim in the present study. By focusing on the textile tools in burial contexts, with and without textiles, makes it possible to fuse the two literary interwoven categories and gain information on the social aspects of production. New scientific analyses of textiles form a promising field of research, however, beyond the scope of the present study. Hopefully though, it can motivate for further research and analyses related to fibres and proveniences.

Tools and textiles in urban contexts

The large-scale excavations at medieval Bryggen in Bergen (1955–1968) uncovered large amounts of textile remains in the wet subsoil in the harbour area. The medieval textile material from these excavations represents a timespan of more than 400 years, from the late 12th century onwards. The analyses of only some of this textile material has so far been published, and the most thoroughly analysed are related to hair products and cordage but also various textiles from the town's earliest period,[56] and only selectively related to the dress.[57] In this respect, the textiles from medieval Trondheim[58] and Oslo[59] are better represented and analysed as for textile qualities. Textile tools in urban contexts have also largely been studied as a separate category, including my own doctoral thesis on textile equipment from the Bryggen excavations in Bergen.[60]

In an article, the archaeological remains of both tools and textiles from all these three towns are compared as for technological similarities and differences as well as textile qualities.[61] The comparisons are based on earlier studies by the archaeologist Karin Gjøl Hagen as for Trondheim and by the ethnologists Anne Kjellberg and Marta Hoffmann as for Oslo.[62] The large assemblages mainly connected to spinning and weaving, comprise more than 2,500 remains of tools and more than 5,200 textile fragments. They open for discussions on stability or change in the urban production from the middle of the 11th century into the Late Middle Ages. In this study, however, covering a longer time span from the Viking Age into the High Middle Ages and in varied environments in western Norway and the North Atlantic, tools and textiles are seen in an even longer time frame, including rural, proto-urban, and urban contexts (see Chapters 8 and 9).

The North Atlantic

Textile remains also relate to different archaeological sites in the North Atlantic, in Iceland as well as in Shetland, and have been analysed and documented by textile specialists as for technical aspects of yarn, spinning and weaving, and made available through online reports[63] and publications. Not least, the archaeologist Michèle Hayeur Smith has analysed textiles from recent excavations in Iceland, however, many dated to the Late Middle Ages – but may still have relevance for the production before *c.* 1300. The results from textile analyses have also been assessed and discussed within broader contexts.[64]

The earliest recorded textiles in the North Atlantic so far, stem from Iceland in one of the pit houses from the site Hofstaðir from the late 900s to the early 11th century.[65] Lately, textiles from a Viking Age grave in Iceland are found to share characteristics with the high-quality textiles from western Norway.[66] However, most of the textile finds date to the High and Late Middle Ages. In some cases, both textiles and tools appear at the same site, such as at the high-status farms The Biggings at Papa Stour in Shetland and Reykholt in south-western Iceland.[67] This has made it possible to compare the textiles with both tools and the later recorded norms and descriptions of different *vaðmál* qualities[68] and textile qualities from other sites in Iceland and from western Norway.

In the present study, tools and textiles are assessed together, as far as it is possible. Here, the written evidence related to the large-scale wadmal production in Iceland, as documented and analysed by the Icelandic medieval historian Helgi Þorláksson, provides important information for the interpretation of both tools and production. In a study of the medieval Icelandic *vaðmál* production and trade, he has made a valuable contribution based on both the normative and narrative sources from the Middle Ages.[69] This information is also essential for the understanding of the archaeological remains, both tools and textiles, and may shed light on production in western Norway too.

Social and gender perspectives

Textile production involves both gender, social and technological aspects, related to producers, organisers, and users. While production, such as spinning and weaving, has traditionally been connected to women, the use areas of textiles relate to both sexes as for their personal properties of clothes, or as common belongings, such as wall hangings, blankets, and more specially sails and others. Whether women were involved in all the processes and all types of textile production has been questioned, and a male participation has been suggested, especially in weaving of sails in the Viking period.[70] However, there are no such masculine indications in the medieval and later sources related to the North Atlantic[71] but is still an option to consider related to gender roles and agency.[72]

The technological change from the upright to the horizontal loom has not least been understood as a change of gender roles in medieval urban contexts represented by professional guilds and related to the introduction of the horizontal loom in urban craft centres (see Chapter 9). However, this issue appears as more nuanced when looking into a broader spectrum of sources, including both descriptive and normative sources, archaeological remains, as well as contemporary pictorial representations, and not only the normative guild regulations.[73] These ambiguities and the possible transgression of gender roles is an interesting issue to be further pursued related to the overall matter of stability and change in technologies and production, as well as organisation.

Terminology and cross-cultural issues

Textile production is largely based on techniques of *longue dureé*, which also cross geographical boundaries. One of the oldest publications on textile production related to the present study area, *Altwestnordische Kleiderkunde*, written by the culturally oriented philologist and linguist Hjalmar Falk,[74] gives a useful overview of Old Norse terminology and sources to textile production in the Middle Ages and of older heritage.

In the last few decades, much effort has been invested in defining and describing terms and terminology for different types of textiles, tools, and techniques across geographical borders and periods of time.[75] These works make it possible to communicate more precisely within an expanding international field of textile studies. A new, thorough, and informative book on the many aspects of textile production, *Encyclopedia of Medieval Dress and Textiles of the British Isles c. 450–1450*,[76] has relevance for Scandinavia too. The new focus on textile production has promoted new perspectives and broadened the scope of textile research, not least initiated by the *North European Symposium for Archaeological Textiles – NESAT* – since 1981 with its many publications touching upon various aspects in this book, thematically, methodologically, and empirically.

The historiography of studies of textile production reveals a multifaceted and complex field of research, comprising several interrelated disciplines and methodologies. The aims of the present study are to take advantage of these opportunities to achieve a more holistic comprehension of textile production in this long period of half a millennium – where regional and local conditions and contexts also could vary considerably in time and space.

Notes

1 Bender Jørgensen 1986.
2 Bender Jørgensen & Grömer 2012.
3 See *e.g.*, Andersson 1998; 1999; Andersson Strand 2012; 2016.
4 Cf. *e.g.*, Good 2001, 210.
5 Cf. *e.g.*, themes presented in the *Ruralia* publications from 1997 to the present.
6 *e.g.*, Power 1941; Carus-Wilson 1969; Postan 1972; Munro 2000a; 2000b.
7 Øye 2015c.
8 Andersson 1999; 2000; 2003; 2008.
9 Andersson Strand 2016.
10 Pfarr 2005; Mårtensson *et al.* 2006a; 2006b; 2006c; Mårtensson *et al.* 2009; Johnston 2016b.
11 Øye 2011; 2015a; 2016b; 2016c.
12 Since the museum was founded in 1825, it has taken different names; initially, Bergens Museum, later Historisk Museum, Bergen Museum, and presently University Museum of Bergen.
13 Iversen 1999; 2008.
14 Solberg 2010.
15 Stylegar 2010, 73.
16 Solberg 1985.
17 Eldjárn 1956; Eldjárn & Friðriksson 2000.
18 Hayeur Smith 2004.
19 Rabben 2002.
20 Øye 2011; 2015a; 2016b; 2016c.
21 Andersson 2000; 2003.
22 Bakka 1971; Lillehammer 1967; 1971; Larsen 1995; Øye 2015b.
23 Bjørgo *et al.* 1992; Bjørgo 2005; Øye 2015b.
24 Ramstad 2010; 2018; Ramstad *et al.* 2011: Cartwright 2015; Øye 2015b.
25 Mortensen 1993; 1997; Diinhoff 2002; 2016; 2017; Åstveit & Melvær 2013.
26 Randers 1981; Øye 2006; 2015b.
27 Øye 2015b.
28 Øye 1988.
29 Øye 2011.
30 Andersson 2003; 2008.
31 Kjellberg 1979; Kjellberg & Hoffmann 1991.
32 Gjøl Hagen [1988] 1994.
33 Øye 2015c.
34 *e.g.* Júliusson 1997; 2010; Edwards *et al.* 1998; Church *et al.* 2005; Vickers *et al.* 2005; Brewington 2010; Harrison 2011.
35 Shetland: Hamilton 1956; Ballin Smith 1999; 2007; Crawford & Ballin Smith 1999; Faroe Islands: Stumann Hansen 1991; 2013; Arge 1997; Mahler 2007; Iceland: Lucas 2009; Milek 2012; Sveinbjarnardóttir 2012.
36 Sveinbjarnardóttir 1991; 2005; Stumann Hansen 1991; Mahler 2007.
37 Sverrisdóttir 2009; Hayeur Smith 2011a; 2011b; 2016; 2018; Walton Rogers 2012; Hayeur Smith *et al.* 2019.
38 Øye 2016b.
39 Harrington 2008.
40 Hoffmann 1964.
41 Andersen & Martinussen 2016.
42 Hoffmann 1964; Johnston 2016a; 2016b; Johnston *et al.* 2016; Juuhl 2016; Larson 2016.
43 *e.g.* Scandinavia: Geijer 1938; 1972; Holm-Olsen 1972; 1975; Hägg 1974; 1985; 1986; 1991; 1994; Kjellberg 1979; Schjølberg 1984; 1988; 1998; Bender Jørgensen 1986; 1992; Ingstad 1999; 2006; Kjellberg & Hoffmann 1991; Gjøl Hagen [1988] 1994; Vedeler 2007; Lukešová 2011.
44 Geijer 1938.
45 Holm-Olsen 1972; Bender Jørgensen 1986; 1992; Hoffmann 1964; Øye 2011, 2015a.
46 Andersen *et al.* 1989; Andersen 1995; Bender Jørgensen 2005; Zagal-Mach Wolfe 2013; Andersson Strand 2016.
47 Holm-Olsen 1972.
48 Bender Jørgensen 1986.
49 Lukešová 2011.
50 *e.g.* Walton 1988.
51 Ingstad 1999; 2006.
52 Bender Jørgensen 1986; Walton 1988.
53 Øye 2011; 2015a; 2016c.
54 Rosenquist 2006.
55 Walton 1988.
56 Schjølberg 1984; 1988; 1992; 1998.
57 Vedeler 2004; 2007.
58 Gjøl Hagen [1988] 1994.
59 Kjellberg 1979; Kjellberg & Hoffmann 1991.
60 Øye 1988.

61 Øye 2015c.
62 Gjøl Hagen 1994; Kjellberg & Hoffmann 1991.
63 *e.g.*, Hayeur Smith 2011a; 2011b.
64 Guðjonsson 1985; 1990; 1994; Hayeur Smith 2013a; 2013b; 2015, 2016, 2018, Hayeur Smith *et al.* 2019.
65 Lucas 2009; Sverrisdóttir 2009; Milek 2012; Øye 2016b.
66 Guckelsberger 2017; Hayeur Smith *et al.* 2019.
67 Walton Rogers 1999; 2012; Sveinbjarnardóttir 2012.
68 Þorláksson 1991; Øye 2016c.
69 Þorláksson 1991.
70 *e.g.* Rabben 2002.
71 Hákonardóttir 2016, 32.
72 Cf. Costin 1996.
73 Øye 2016a.
74 Falk 1919.
75 *NTT* 1974; Anstey & Westen 2005; Coatsworth & Owen-Crocker 2007.
76 Owen-Crocker *et al.* 2012.

2

Perspectives and methods

Interpretative frameworks may often appear as tacit, implicit, and indirect preconceptions, but also based on more explicit and sometimes rather rigid theoretical frameworks. This study, primarily based on empirical analyses related to aspects of textile production – technologies, actors, gender, organisation, and networks at various spatial and temporal levels – also leans on various theoretical and interconnected perspectives on human interactions and behaviour. Some introductory reflections on theoretical positions and methods applied are therefore presented to clarify key conceptions in this study, such as production, technology, contexts, time, and change. An underlying premise is that textile production, like other production, is embedded in political, social, and/or economic systems, formed by both social as well as natural opportunities and limitations in the environments. Production relates, then, to the corresponding socioeconomic systems – distribution, consumption, and demand for the products – and in this case, connected to varied types of textiles. What promotes or delays changes, how material culture and technology spread, and why innovations are taken up in some areas and rejected in others, are issues that need some reflections too. Neither are matters concerning gender and gender patterns, work domains and work identities self-evident. This chapter, therefore, presents and discusses some of the underlying perspectives and methodologies applied in the subsequent analyses.

Perspectives

The complexity of production

Production as a way of transforming raw material, and/or components of such, into usable objects normally represents many stages and ways of organising the work processes. As the American archaeologist Cathy Costin points out, production is not the same as specialisation, which relates to the way of organising it. She distinguishes between a domestic mode for subsistence within the households and specialised work, meaning a regular, repeated provision of some commodity or service in exchange for some other. She further distinguishes between site specialisation, where short-term activity is supported by the entire group to meet its own consumption needs, and specialisation where people gain parts of their livelihood through participation in a special and specialised activity. According to Costin, specialisation is differentiated and often regularised as an institutionalised production system, where producers depend on extra-household exchange as part of their livelihood and then on a regional level. The consumers, on their part, depend on such specialised production for the acquisition of goods they do not produce themselves.[1]

The context of the production may, then, be diagnostic for various modes of production, either for subsistence on a household level or specialisation, either as individual or in nucleated workshops, as part-time or full-time labour, in a nucleated setting for an elite or an institution.[2] This implies that different degrees of specialisation and modes of organisation may change in space over time within shifting socio-political and economic regimes, and in this case from the stratified Viking period into the Middle Ages, and where the production expectedly would vary. As Eva Andersson Strand has suggested, the production may be characterised as: *(i) ordinary household production, (ii) advanced household production, (iii) attached specialist production,* and connected to *(iv) workshops*.[3]

Ordinary household production would be an important element in most households, covering practical and functions to cover the household's own basic needs and based on general skills and available resources. This type of production would characterise the whole period, however, with variations related to status and members of the household.

Advanced household production would include many of the same elements as above, however, generally of higher status and connected to larger farms with a larger work force, relying on work and services from servants within gendered work domains. Expectedly, tools and textiles would also be related to production of wares within the luxurious segment – for conspicuous consumption, gift giving, and within networks.

Attached specialist production would connect to special environments – central places and nodes within estates and probably services and duties within a hierarchical structure, and with access to raw materials at wider distances based on reciprocity and or subordinance. This may be reflected in Viking Age graves with tools, textiles, and other grave goods in the period 800–1000 but also in pit houses and dislocated workstations stretching into the Middle Ages.

Workshops allude to specialists, dominated by many crafts and professions appearing in larger population settings, such as Kaupang, Hedeby, and Birka in the Viking Age and Bergen in the Middle Ages but also central places and/or socially 'neutral' centres but possibly also larger farms and central places in rural environments too.

The organisation and various work stages may also expose various actors involved, and both social and gender representation, since technical skills and domains of expertise are often divided between and within the sexes. Historically, and more or less worldwide, large or all parts of the production chain within textile production seem mainly to have been in the hands of women, forming gender, social, and technological patterns, so-called 'sociotechnical systems'.[4] A disregard of the operational sequences may then lead to a disregard of the role of a large female rural workforce behind the finished product even in urban workshops. Focusing only on the last stage in the process, such as weaving and making the cloth, may therefore result in omission of important roles and, not least, an oversimplification of the production. When being studied as an integrated entity of interconnected work operations, it becomes clear that innovations at one level had wide-ranging consequences for all the other elements in the technological complex.[5] Changes or innovations at one level thus have chain effects affecting all the other work processes. Cultural traditions and stability on the other hand, provide a *longue durée*. These insights are useful analytical tools when assessing seemingly small changes and variations in simple tools, such as the weight of simple spindle whorls when spinning yarn.

The *chaînes opératoire* approach thus helps to identify also the 'trivial' but necessary stages in relation to those that have been regarded as the more 'important' and complicated ones. Tools like wool combs and a wide assortment of spindle whorls were not only basic and obligatory but also important for both quality and quantity of production. Therefore, this study accentuates the details within the material culture of tools and textiles, starting with the well-known tools and practises to the many small variations that together may have constituted new technological environments and opportunities. In this way, the abstract meaning of technology is partly substituted in favour of the concrete and measurable. Yet, this does not imply an underestimation of the overarching levels. It rather represents an integration of social action and technology by emphasising the importance of incorporating routine skills and techniques within larger socioeconomic frameworks. This study therefore combines detailed empirical observations within more broad-ranging comparative perspectives.

As noted, an extensive input of labour and many time-consuming operations were necessary to convert the wool and/or plant fibres into yarn and fabrics, and when taking the whole chain of work operations into account it exposes the importance of all these processes. Although initiated in anthropological studies, this way of documenting technological processes in a detailed step-by-step mode, including social and ritual elements, has been a rewarding method when assessing production in the past too. Breaking the action of production into stages as steps of operations exposes not only the different actors involved but also locations for actions and production. In this way, working stages and various patterns of production connect to the social representation of technologies. The aim is therefore to explore the various production activities and approach the technologies – tools and textiles, as a technological complex and organisation at various levels from the ordinary household to attached households to advanced and specialised production in workshops.

Skills – *know-how* – are also studied through operational sequences, and demonstrate how material, social, and cultural resources link up.[6] The skill level of the actors/producers further correlates with the intensity of specialisation; the more skilled, the fewer mistakes and the more uniform products appears in quality,[7] and in this case, reflected in the textiles. The *chaîne opératoire*-approach is thus a way to approach the organisation of production at either household or specialist levels, or the connection between resources and patterns of production in various environments and at special sites, and indirectly, the primary economic strategies.

Archaeologically, it can be more difficult to follow all these interrelated processes when they have left no or little material evidence. As a perspective – way of thinking – it still serves as a useful analytical tool. It implies that the archaeological artefacts must be assessed contextually not only related to tools and textiles in close find contexts, but include processes connected to wider resource areas, and people in different social and economic environments. Sites, representing households, do not necessarily only indicate domestic subsistence production but possibly also surplus production, for instance to fulfil obligatory duties or for exchange. However, when specialised, production may possibly be for products within exchange networks, which may be illuminated by other artefacts too. Therefore, sites with traces of textile production are assessed contextually, related to wider environments, both natural and social, that may represent various levels of production – common, advanced, or specialised and the latter possibly as part of regional/interregional networks.

Together with the *chaîne opératoire* approach, phenomenological approaches that grasp the actual practice in making and using a tool as a physical experience and inclusion of body techniques are relevant when considering social aspects along with practical dimensions in technological processes. Lately, social agency and engendered approaches to technology have developed this perspective further as an analytical methodology and conceptual framework. The attention focuses on the human agents, those who made and used the tools rather than concentrating only on the artefacts per se.[8] In this way, technology may be understood as embodied know-how and knowledge – as a process as well as objects – where all these elements fuse together.[9] In this context, experimental archaeology based on using tool replicas has been a rewarding approach and path to follow to obtain measurable results in a concrete way.

Social and political contexts

In so-called sociotechnical systems, the connection between the social and technological involved in technological innovations is underlined; how they represent diverse expertise and interest groups, and material and immaterial networks. How production, distribution, and consumption of textiles link to broader social, socio-political, and socioeconomic contexts also makes it better understood. Context in a wide sense involves then the affiliation between the producers and the socio-political and economic component of the demand for their products. Textiles were, as already noted, not only of utilitarian value but also prestigious and costly commodities, representing and displaying wealth and social status, even in the graves.

Over the years, archaeology has used various conceptual models – redistributive systems and control of production, regulating distribution, demanding tributes – a kind of command model with a top-down focus. The production, accordingly, could be carried out by attached specialists, or individual specialists for the market, often of utilitarian products, such as certain textile qualities used by most households, for instance the common quality of wadmal (ON *vaðmál*). Attached specialists, on the other hand, seem generally to have produced more prestigious goods and qualities of limited access, and therefore to a larger extent may have controlled the production, or goods needing special competency, for instance of certain socially distinguishing textiles, such as exposed in the Veka grave.

The geographic concentration of production – how possible specialist production is distributed across the landscapes – assembled, dispersed, or nucleated – may indicate the spatial relationship between producers/specialists and consumers. It is therefore important to identify both the natural and social environments, whether representing attached workers, independent households, or specialists at nodes within production. Generally, there is thought to be a strong correlation between environmental diversity, territoriality, and independent specialisation.[10] When nucleated, it is often because a community can exploit efficiencies associated with local production and resources that are unequally distributed. Production tends to be nucleated in economies with formal markets and marketplaces, since transport costs and risks are moved from the producer to the consumer.[11] This may also be the case in this period, when towns start to emerge along with central places as nodes in smaller and larger estates on a regional level. Scale is therefore another important factor to consider, varying from small individual or family-based production units to production on a larger scale, related to central places and with a dependent workforce and in the new towns, possibly based on a wage labour force or for food and lodgings. Still, both independent and attached producers/specialists could be large or small.

Recent perspectives tend to include both top-down and bottom-up approaches, the latter often with a household perspective, as a basic economic and social unit for production and consumption. In this study, both perspectives are relevant, being largely interdependent and closely connected.

Long-term persistence, change, agency, and regional identities

The causal mechanisms, which promoted change or stability, are often difficult to identify empirically. These changes cannot be explained by one factor alone since they represent complex and integrated processes, caused by both internal and external pressure. Time was marked by slow development of *longue durée* that could be interrupted by faster non-linear shifts promoted by special events and generation – and not necessarily synchronous.[12]

How technology within textile production was spread is yet another aspect to consider. The spatial delimitation of the study area represents regions at various levels and within their locales. However, the conception 'region' is not only a geographically demarcated zone but also subject to changes caused by both changing power systems and political changes. It is also more complex as a historical formation of a distinct identity, with different areas acting as socio-cultural regions at various levels.[13] The focal area under study, western Norway, may have constituted an early political and administrative region, and as noted, corresponding with the core area of the old provincial law area of the Gulathing in western Norway, dating back to the 10th century. This region, divided in three early medieval counties, was thus socially and politically demarcated and a core area within the expanding judicial and political entity of the wider Gulathing law district and part of the over-arching new Kingdom as the result of the ongoing state formation processes from the 9th century onwards, reaching its heyday in the 12th and 13th centuries (Fig. 2.1).

The affiliation with the North Atlantic was both close and distant; in the first phase offering new possibilities in new lands where emigrants took with them their cultural luggage and packages of necessities from their places of origin – and many from western Norway. There were, then,

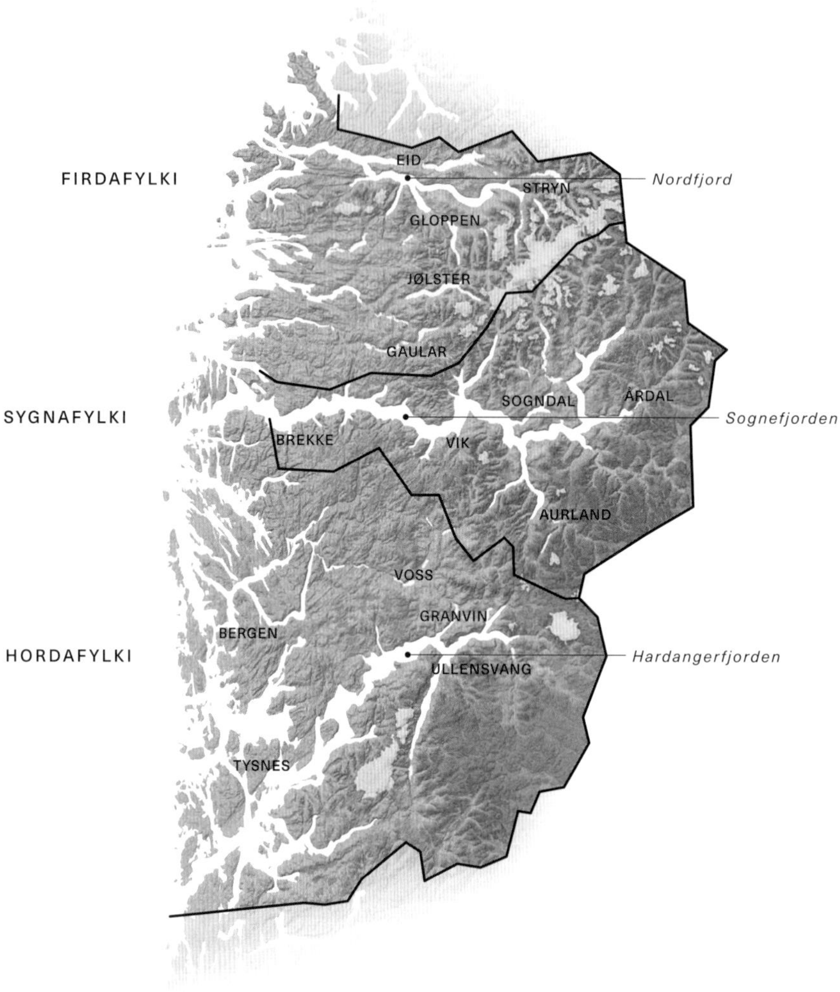

Fig. 2.1: The area of study, Hordafylki, Sygnafylki, and Firdafylki (Illustration: Per Bækken).

familiar, cultural and mental bonds linking the areas from an early stage in the late 9th century onwards. Later in the Middle Ages, under the sovereignty of the King of Norway, the relation was of more political and economic character. However, a not insignificant share of the 'new' inhabitants, largely women, had, as already noted, another ethnic background coming from areas around the Irish Sea, as spouses and servants/thralls, bringing with them other traditions. Hence, a multi-ethnic population is relevant since textile work is regarded as a general female domain and based on an embodied competence achieved since childhood.

Generally, institutions, not least political, are regarded as capable of maintaining and reproducing territoriality and inherent symbolism of *longue durée.* Still, technological, social, and economic development may not follow the political boundaries, since the political landscape in western Norway, both as counties and as a core region, was physically and administratively divided along the degree of longitude, whereas the natural environments offering functional-economic conditions, varied more according to latitude and gradient, from the outer coast to the inner fjord districts and upland and mountainous areas (Fig. 2.2).

The coastal areas relied more on a maritime economy, however, with vast heathlands along the coast, providing grazing land for sheep all year round. The inner fjord districts with a generally drier climate and higher temperatures in the summer, offered better conditions for arable farming and animal husbandry in inland and upland valleys,[14] and with access to other vital resources.

Different areas could produce surpluses based on specific resources and organisation of the production, such as minerals and iron ores, and woodland but also grazing land, which could in turn promote specialisation in varied types of production, and of textiles as one. The interplay between environmental conditions, socio-political structures, regional identities and relations between subsistence, specialisation, and exchange or interchange, is therefore relevant to consider when tracing spatial patterns of production. However,

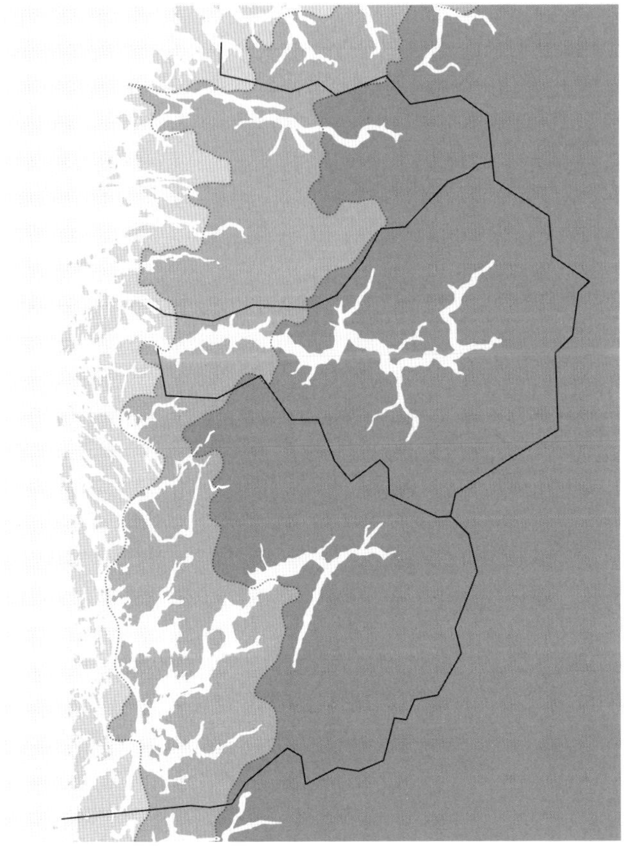

Fig. 2.2: Geographical zones in the study area of western Norway (Illustration: Per Bækken, based on Sømme 1954).

linking textile production to places and regions, does not only concern geography but also political and socioeconomic structures, independent or subordinated within larger estates. This is not least relevant in this period of substantial societal transformations; politically, demographically, and socially, marked by state formation processes, migrations, and mobility. In addition, the period saw religious change and urbanisation with new institutions, such as churches, appearing at central places as local and regional nodes, often at head farms as centres in earlier estates of older heritage, and often marked by grave mounds, both in western Norway and the North Atlantic.

The maritime landscapes in both western Norway and the North Atlantic opened for wider interaction across the Norwegian Sea, and along coastal routes, in-fjord and inland routes and cross-roads, leading to wide resource areas. The various products – including textiles – became commodities within exchange networks and trading centres further off.[15]

Presumably, such changes would create a climate for innovation. Mechanisms of innovation and change do not only include the process of introducing something new but depend on successful acceptance and exploitation of new ideas and methods too, perhaps also new ways of organisation related to larger undertakings,[16] which indeed characterise the period. Therefore, innovative elements in production form another aspect to look closer at related to possible efficiencies in textile production, organisation, and an expanding consumerism of varied textiles.

Textiles as products in demand represented easily portable commodities of widely recognised value and an important medium of exchange. The many stages within production represented a highly labour-intensive manufacture, but able to produce a surplus, which is clearly demonstrated in medieval written records.[17] Surplus production for exchange with intensive exploitation of available resources may therefore also serve as explanatory frameworks for production in special nucleated areas, where producers and organisers of the production were probably connected to exchange networks both in the Viking period and later, on a larger scale in towns.

Gender identities, work domains, and social status

Many studies emphasise women as producers of textiles, and regard gender as a fundamental variable in the processes, especially in small-scale societies. Aspects related to a feminine involvement in technology and textile production are therefore highly relevant. In a review article about gender and technology, the American anthropologist Francesca Bray states that one fundamental way in which gender is expressed in any society is through technology, as skills and domains are generally divided between and within the sexes. Such gender systems appear to be more difficult to change than material technologies – and reveal how technology is also indicated in gender inequalities.[18] This tends to make women 'invisible', since men normally were the heads of households and those who often appear in written records and storytelling. However, such gender attribution can be problematic as stereotypical roles when applied uncritically. Bray argues for instance that new technologies may also promote processes of boundary work and cause renegotiations of what is considered masculine and feminine.[19] The gender issue should therefore be assessed both contextually and critically. Still, the hypothesis in this study is that women's contribution to the general expansion of textile production was crucial and that their role in the new and expanding urban industries is generally underestimated.[20]

Viking Age graves, being contexts of both textiles and textile tools, represent an important source category, not least, for understanding gender roles, social status, and identity. Generally, monumental graves with grave goods represent the upper strata in Viking-Age Norway, marking and symbolising property rights and landholding and therefore fewer graves in this category. Grave mounds and cairns are more numerous in areas based on freehold than in manorial dominated areas based on tenure, where often only the head farm in the estate was marked by monumental graves.[21]

The ritual of honouring the dead with grave goods, jewellery, weapons, tools, and other artefacts signify, then, gender, social identity, social and economic status, and gender roles.[22] This ritual, in the hands of the heirs to mark the shift of generations, thus also reflect their own position in the *erfi* ritual. When appearing as inhumation burials, the graves may also leave traces of textiles, which makes it possible to study them contextually as for gender, status, and tools.

Methodological approaches

The archaeological textile production equipment and textile remains constitute the backbone of this study. However, the multitude of details in the everyday use of techniques is often difficult to explore, as is also how to handle the problem with *e silentio* reasoning, of what is missing.[23] Other relevant source categories are therefore also explored, such as written sources, iconography, and images. A comparison with later ethnological sources is important as the medieval evidence is often sparse and not straightforward when taking the time lapse in consideration. Here, experimental archaeology offers a rewarding method, demonstrating what in fact can be achieved and may then also reveal different local and regional practices over time. A source of error, however, is that people today do not have the same transferred practice and experience as in times when know-how and skills were transmitted from generation to generation as embodied practices since childhood. Still, experimental archaeology adds crucial information that otherwise would be mute. Each source category has its own methodologies, which must be taken into account too.

Various sources and varied methods

The methods used in this study relate primarily to the archaeological remains of textile tools, textiles qualities and their find contexts, both close find contexts and the wider resource areas and socioeconomic environments, as far as they can be traced as structures in the landscape: graves, and various types of buildings serving as workplaces and outlying resource areas of grazing land. Ideally, all remains that can help to reveal the whole production system are relevant: debris, tools and features and product, when trying to illuminate the organisation of production, also in wider contexts. The logistics of distribution may then identify the way in which producers acquired raw material and transferred finished products to the consumers. Searching for spatial patterns and the organisation of production related to special places, environments, resources, and status therefore serves as a basic strategy. They may in best cases include areas and sites for the various stages in production as short-term activities at special locations, or the whole chain of work processes in concentrated areas. Here, historical sources as well as environmental data, such as zooarchaeological and archaeobotanical evidence, may provide important information too. When combining different types of sources, they are related to their specific contexts as for source contribution and validity. As a methodological trail, the various sources are assessed separately at first hand to be confronted with others, since a larger integration of sources may provide a broader framework for interpretation.

In brief, archaeology offers finds of tools and textiles, but generally only as fragments. The finds contexts may reflect working environments, social settings, and gendered space, including graves and other contexts and, not least, for dating. Environmental evidence, especially within zooarchaeology (osteology) and archaeobotany, may reveal information about animal husbandry and plant production of textile fibres and physical conditions for textile production (see Chapter 4). Textile remains leave concrete evidence of qualities and techniques, while archaeological remains of textile tools give information about design, material, and technical details and functional proprieties (see Chapter 6). By testing how replicas of various tools function, experimental archaeology has contributed significantly to the understanding of technology – what is possible and what are limitations.

Written sources in this context often relate to historical accounts; however, related to a wide spectrum of information, both descriptive and normative, such as law codices, diplomas, tax lists, and various literary texts, such as, amongst others, sagas. Altogether, they are best represented in the latter part of the period under study. Generally, there is no textual evidence earlier than from the 12th century from these areas, but various sagas, skaldic poems, and some law texts allude to earlier conditions, at least back to the 10th century and earlier. Place names – toponyms – form another source category of old provenance that can shed light on production and available resources. The sources are therefore assessed source critically as they may be influenced by the time the texts originated. Texts give a social, economic, and chronological context but information about technical details related to textile production is rather rare. Old Norse terminology related to tools and textiles also provides useful information.

Iconographic sources, mainly from the 13th century onwards, sometimes visualise tools and their actors related to spinning and weaving, however there are very few from the present study area, and then mostly restricted to stereotypical Biblical scenes, such as Eve spinning and Adam digging. Still, they show working positions, gestures, and technical details, probably depicted according to the prevailing regional culture.[24] Comparing archaeological and historical evidence with well-preserved and well-described sources may in sum be a rewarding method to achieve a more holistic comprehension, but without disregarding that changes can easily be overshadowed.

Another and larger methodological problem concerns what are missing, and in archaeological contexts depends not least on preservation conditions. To understand technological aspects and change, the absence of innovations must be explained too, as reasoning *e silentio* demands a certain amount of evidence to be able to draw firm conclusions. Often only a few specimens are represented, unevenly spread in time and space, for example, remains of the horizontal loom and its advance. Then it is important to identify where and when they appear and if there is enough evidence related to the presence of other related tools and techniques related to the new and old loom (see Chapter 9).

Most of the archaeological material – both tools and textiles – consists mostly of organic material, which is only found under favourable conditions; in the study area usually in humid cultural layers, bogs or when covered by clay, as well as in chilly climates. Wood and textiles have occasionally been preserved from graves connected to metals and at settlements in humid areas such as in the harbour area at Bryggen in Bergen or in colder and damp climates. For textile production, parts of implements were often made of more durable material such as stone and clay, as used to make spindle whorls that helped the drop-spindle rotate and twist the fibres of wool or vegetal material into yarn, and the loom weights that stretched the warp threads on the vertical loom. From their weights, these finds are still able to illuminate the level and range of textile production and provide vital information concerning chronology, technical level, and to some extent, scale of specialisation.

Contexts at various levels: preservation, representation, and interpretations

Archaeological finds are always bound to time, place, and social contexts but fail if the documentation is poor. It is therefore important to examine the technological and production processes in which the survived archaeological tools formed parts. As far as possible, tools are compared and assessed in relation to finds of textiles from the same contexts, area, and period. When studying tools and textiles at various sites and settlements – rural and urban – the ambition is to assess and contextualise the objects in their social and socioeconomic environments.

The find contexts and conditions for preservation of both textile tools and textiles vary from the Viking Age to the Middle Ages and reflect how they were deposited: intentionally, such as grave goods in graves, or arbitrarily as lost, left, dumped, or abandoned at workplaces, which also reflects social differences in production and at workplaces. The way artefacts were deposited may further indicate whether the finds were arbitrary and reflect work processes and daily routines within production, or intentionally placed as symbols of status and work domains in grave contexts. The physical environments: soil and subsoil, dry or humid underground, sand, clay, acid, or salt affect the preservation. Linen and other cellulosic fibres preserve better in alkaline conditions with a high pH value, whereas animal fibres such as wool preserve better in slightly acidic environments.[25] However, the representations may also reflect various societal conditions and changes in society. Some contexts may show long durability, others may be short-lived, and again reflect stable or shifting socio-political and socioeconomic conditions.

The way graves and other archaeological sites and find contexts have been discovered, excavated, and documented may have special consequences and limitations for interpretation. In western Norway, many prehistoric graves were detected during agricultural works in the infields after a new land reform in the late 1850s split up old common fields.[26] Such early finds and from an early stage of archaeology are therefore often damaged and/or inadequately recorded (see Chapter 5). In many cases, grave finds now stored in the museum may provide an insufficient recorded find context, and some artefacts were not even retrieved at all. Still, a considerable corpus of textile fragments and tools offers important information when taking advantage of using a wide spatial scale, and, as far as possible, assessed contextually.

Although stray finds with uncertain contexts are rather likely remains of a grave, it is difficult to label them as grave finds and confirm their affiliation to gender. Several criteria for categorisation have been used for classification as graves. At their best, they represent professionally excavated sites and finds, making it possible to reconstruct the grave; second best, they include post-excavation of damaged graves, supplied with good or at least sufficient information to decide, or indicate whether the finds are found together, or whether this is unclear. In these cases, they should be regarded as closed finds and not mixed. Further, finds that are found with certainty in a grave context, such as a mound, are also counted as grave finds but, in such cases, it is not always possible to decide whether from one or several burials. However, some finds with little or insufficient information, or chronologically mixed, are regarded as uncertain graves. In this study, these principles are followed and critically applied, then, as for the uncertain categories and the consequent reduced source of value (cf. Appendices to Chapter 5).

The introduction of Christianity ended the old pagan pre-Christian burial traditions with grave goods and therefore also insights into social hierarchies and gender roles in a concrete way. The abolition of slaves by the 11th and 12th centuries must have affected the organisation of work and settlement patterns and physical structures for textile production environments. These shifting socio-political contexts are therefore relevant when analysing graves and sites with such traces.

Traces of textile tools in burial contexts, however, only indirectly signify textile work. The burial customs of the Viking period, both cremations and inhumation graves reflect mostly females of a certain status as main agents of textile production. The tools may therefore signal gender when being found in other contexts too, in both rural and urban environments, however, not necessarily representing the same social layer. Nor should fixed gender roles be taken for granted and therefore an issue to be assessed contextually.

Sites representing working environments have the greatest potential for studying the production directly in various rural environments both in the lowland and at higher altitudes, such as shielings in more distant outlying areas. Textile tools in the present study appear in different production environments in various types of buildings: separate sunken-floored buildings within larger complexes in the early part of the period, and in small house constructions and bigger longhouses. Overall, the production sites and the layout of the features there, debris, and distribution of both finished and unfinished products, as well as the environments and resources outside the regular production contexts, reflect textile work more directly. Some of the sites may possibly signify a kind of community specialisation with dispersed workshops and specialised settings, probably controlled and organised by an elite (see Chapter 8). The urban environments are different and may reflect both domestic production for the urban households and a specialised production (see Chapter 9).

Access to raw materials – wool, hemp, flax – to produce textile fibres, and an estimate of time consumed for the various work processes involved are other vital factors to consider related to the production, locations, and tools. Here, experimental research has given new insights concerning tools and properties related to qualities and time consumed for the various processes. It provides, then, a broader understanding of the requirements and logistics within this labour-intensive production – a production that was also based on farming, both husbandry and arable farming, and documented both archaeologically and historically.

As for wool and sheep farming the oldest historical documentary sources from western Norway are rather late: from the 17th and 18th centuries, however, first recorded in more detail from the late 19th century when the demographic pressure escalated and the pressure on grazing land was at its height. These records may thus indicate a top level of wool and textile production, but not necessarily, considering that textile production probably played a larger role in the Viking period, primarily based on domestic production than in the Middle Ages and later with large-scale imports via Bergen, also including wadmal from the North Atlantic. However, Iceland also provides older and more detailed documentary information about sheep farming dated to the Middle Ages onwards. Here, zooarchaeological material additionally gives detailed information about medieval slaughtering ages, and, indirectly, strategies related to wool production. Altogether, rural sites with tools and textiles are approached contextually according to geographical and topographical conditions, structures, and other contexts, including resource areas and as far as possible in a comparative perspective.

Tools and textiles: qualities, use areas, and contexts

Textiles appearing in these varied archaeological contexts expose a broad range of qualities and techniques reflecting the technological level of textile production, ranging from coarse home-spun fabrics to the finest and softest qualities woven in more advanced qualities, as represented, amongst others, in the Veka grave. The textiles are key sources for exploring and understanding the range of technological level of production and degree of specialisation. In this way, the textiles form integrated and important frameworks when assessing the textile tools, where even small variations in weights of the spindle whorls and loom weights were important to produce the various qualities and where all stages in production had an impact on quality.

When comparing tools with textiles, the study leans on earlier textile analyses, focusing on qualities, fineness of the fibres and yarn (thread counts), spin direction and twist, weaving bindings, and weaving patterns. Different use areas needed both finer and coarser textiles. Clothes and garments were made of qualities according to both practical use and social standing. As noted, the textile analyses from the Viking period are mainly based on Lise Bender Jørgensen's catalogue from 1986 and Inger Marie Holm-Olsen's study (1972), covering the same area in western Norway. Such textile fragments are generally preserved close to jewellery and metal accessories related to the dress, and thus limited use areas, but socially distinctive. Bender Jørgensen's national overview shows considerable regional variations, but the main preserved qualities were plain tabby (with one thread over, the other under), 2/2 twill (two under/two over), in addition to a relatively high share of the more advanced diamond/lozenge twill qualities (Fig. 2.3), and the rest in the category 'other weaving' (see Chapter 6).

However, a few textiles stem from working environments too, as referred to in the previous chapter, and are

Fig. 2.3: Weaving types and techniques indicated by the thread systems in: a) tabby, b) 2/2 twill, c) 2/1 twill, d) broken diamond twill (Graphic design: Per Bækken, based on Bender Jørgensen 1986).

represented by a variety of qualities from different contexts: graves, rural, and urban sites.

Yarn qualities further give information about fibres and spinning techniques and qualities, thickness, and spinning direction, which again add information about skills, specialisation, access to material, and other important factors of textile production and probably culturally transmitted traditions.

Generally, textiles in urban contexts are better preserved than in grave contexts, when found in deeper cultural layers and often in damp and waterlogged areas, such as at Bryggen in Bergen close to the harbour. The textiles found under such conditions are almost all discarded fragments, but generally as larger and better preserved than the small remnants connected to and preserved by the metals of dress accessories in graves, however, limited to funeral contexts. Most of the urban textiles in contrast are found without any identifying functional context, and can often only be categorised according to fibre, spin, and weave, and then related to later information, which may indicate function – sailcloth, for example, was woven in 2/1 and 2/2 twill in tight spun coarse wool,[27] but could also be high-quality textiles, possibly imported.[28] As a general methodological approach in this study, the various textile qualities are, as far as possible, assessed together with tools in the same contexts, although when in funeral contexts without implying that the tools had been used to produce the funeral clothes. As grave goods they rather symbolise work domains but may also indicate use areas and range of qualities.

Textiles and tools also appear in other types of environments, largely connected to workplaces related to varied social strata, and from the latter part of the period of study included both rural and urban settings. Although the textiles from such contexts largely represent discarded rubbish, they may also represent direct traces of production and a wider range of use areas, and ultimately, demonstrate the importance of textile production in wider contexts.

Chain of work processes as a method

As textile production comprises a whole chain of work processes, from raw product to finished cloth meant for different products of different qualities, all these processes are considered, based on archaeological and other evidence. The traces of specific suites of tools are important indicators that multiple steps in the sequence of cloth/fibre production occurred at the site. The frequency of the different tools can then illuminate both presence and level of textile production with differential steps in the process related to varied structures. Hence, both qualitative and quantitative approaches are needed when exploring the spatial distribution of tool categories and the qualities, such as weight and size of spindle whorls and loom weights and again as indicators of qualities, use areas, as well as social contexts.

The tools

Tools reflect work processes, and functional properties related to textile qualities and efficiencies. The various tools are mainly assessed from these perspectives. Typology is generally based on stylistic attributes and physical characteristics and compares the physical attributes and stylistic traits, form, and décor. Spindle whorls, for instance, are categories that are compared according to both quantitative and qualitative characteristics to assess function. Raw material, form, and décor are largely regarded as signifiers of local and cultural qualities. The whorls are compared with thread qualities when found together with textiles to explore possible links between tools and textiles. Not all finds of textile tools are included in the analyses or treated in the same way; only those that can be related to archaeological contexts, and found in combinations, either with various types of tools or several of the same kind. When textiles are found together with implements related to textile production, they are discussed more explicitly and related to their close but also wider contexts. Stray finds of textile tools, or single finds, without any clear or meaningful context are mainly counted and excluded as principal sources. As for the material from Iceland and the Faeroes, the same principles are used, limited to documented and published excavations. However, here, I have also inspected larger quantities of the textile tools to get an impression and to better compare them with tools from the study area in western Norway.

Experimental archaeology

The most frequently found spinning tool from the Viking Age and the Middle Ages is remains of the spindle, *in casu* spindle whorls in archaeological contexts. A spindle consists of a whorl attached to a spindle-rod and a distaff (see Chapter 4, Fig. 4.4 a–c). As the whorl is most often made of more durable material than the wooden rod, only the whorls with some exceptions are represented in the archaeological record from the study area.

Results from spinning tests indicate that the tools, rather than the spinners, influenced the quality of the threads[29] – a result that argues for the importance of detailed descriptions of the spinning tools to illuminate the character of the textile production. This is a method that also characterises this study. A methodological precondition is that they have a potential to shed light on both qualities of the fabric and extent of the production.

Here, experimental research in spinning with the drop-spindle and weaving on the upright loom at the Centre for Textile Research (CTR) at the University of Copenhagen, at the Museum Centre at Ribe and the Museum Village Düppel outside Berlin has given new insight into how different spindle whorls and loom weights function. By testing out whorls of different weights and how they affect the quality of the thread, the experiments have demonstrated that both size and especially the weight of the spindle whorls are important for the quality. Even very small differences of 5 g affect the thickness and quality of the thread, and subsequently also the fabric.[30] When assessing tools and textiles on a macro level as in this study, the weights of the spindle whorls are used as a diagnostic indicator of production.

Dating and chronology

When addressing questions related to stability or change, chronological frameworks are needed. The grave finds from western Norway rely as noted in many cases on insufficient information about the find circumstances. Still, most of them can be dated by the century although some with a closer timeframe, generally based on dress accessories and typological dating. In this study, Oluf Rygh's typology is used (R), basically referred to and used in Jan Petersen's study from 1928 too (JP), also used as a chronological framework in other studies from the study area related to textiles and textile tools.[31]

Here, the youngest datable find is always used as the chronological fix point when found in closed contexts of graves. As grave goods, dress accessories, and other precious artefacts they may even represent more than a life cycle when achieved as heirlooms, and a shorter time span when representing bridal gifts, from their youth and gifts, besides acquisitions as remains from a 'lived life', before ending up in the grave. Still, quite a few grave contexts have left no datable artefacts, such as dress accessories and/or weapons, and sometimes in mixed or incomplete, and altogether unclear contexts that only provide wide and unclear dating within wide timeframes, possibly exceeding the Viking period, possibly the late Iron Age, a period from 200 to 400 years. More recently excavated burials have been supplemented by 14C-dating that would have solved and avoided such wide dating frames. When approaching such mixed and unclear contexts, they are therefore ascribed with limited source value and get less focus when discussing more fine-meshed chronology.

As for rural settlements, the dating methods vary, and in more recently excavated sites using more sophisticated and precise dating methods. As for the urban contexts of Bryggen in Bergen, the chronology is principally based on stratigraphy, and the horizons of datable fire layers based on historically documented fires and combined with datable artefacts and covers a period from the early 12th century to the end of the Middle Ages, *c.* AD 1500. From the period under study, five fire levels form the chronological framework from *c.* 1120 to 1332.[32] As for Kaupang, which forms a comparative proto-urban site from the Viking period, the settlement is dated from AD 800 to around 900, and marked by shorter datable phases,[33] while the burials surrounding the Viking 'town' cover a longer period well into the 10th century.[34]

As for sites in the North Atlantic, radiocarbon dating is a frequently used method combined with typologically dated artefacts. In Iceland, however, volcanic ash layers, tephra, when deposited in historical times can be identified and dated with relative accuracy by reason of written records. Additionally, 14C dating of peat soil immediately above and beneath the tephra layers or by extrapolation, pollen analyses and archaeology provide further information. Within the period under study, three key dated tephra layers stand out, from AD 870, the so-called *landnám* layer, then 1104, and 1362.[35]

This introductory presentation indicates that textiles were important necessities in human life at many levels – functionally, socially, economically, even politically – and as largely interwoven aspects of textile production. How these various features can be traced over time within the study area are subjects in the following chapters – starting with the basic preconditions for textile production – natural and human resources – possibilities and limitations, mainly based on and organised within various farming systems in time and space.

Notes

1 Costin 1991, 1–4.
2 Costin 1991, 8–9.
3 Andersson 2003, 46–62.
4 Costin 1993, 6; Hughes 1987.
5 Bray 2007; Myrdal & Øye 2021.
6 Lemonnier 1992.
7 Costin 1991, 39–40.
8 Dobres 2000.
9 Øye *et al.* 2011.
10 Costin 1991.
11 Costin 1991, 13–14.
12 Braudel 1982; 1996.
13 Øye *et al.* 2011.
14 Sømme 1954.
15 *e.g.* Sindbæk 2007; 2011; 2017.
16 Hjärtner-Holdar 2011.
17 *e.g.* Steinnes [1937] 1982, 6, 68.
18 Cf. Myrdal & Øye 2021.
19 Bray 1997; 2007, 38–42.
20 Øye 2016a.
21 Iversen 1999; 2008.
22 *e.g.*, Solberg 1985.
23 Cf. Myrdal & Øye 2021.
24 *e.g.* Myrdal 2012.
25 Good 2001, 211.
26 Iversen 2008; Øye 2015b.
27 Crumlin-Pedersen 1997, 112.
28 Schjølberg 1992; 1998.
29 Mårtensson *et al.* 2006a, 1.
30 Andersson 1999; 2003, 25–26; Mårtensson *et al.* 2006a; 2006b; 2006c.
31 *e.g.*, Eldjárn 1956; Bender Jørgensen 1986.
32 Øye 1998b.
33 Pilø & Skre 2011.
34 Stylegar 2007.
35 Hafliðasson *et al.* 2000.

3

Farming systems and textile production

Textile production of the period was tightly integrated in farming systems and economic strategies, requiring both natural and human resources, and also linked to socio-political aspects of ownership and management of the resources and production. Access to farming land and extensive grazing land for sheep was a premise for providing enough wool, and arable land for production of textile plants, and for livelihood generally. Favourable climatic conditions for sheep breeding played a crucial role within the study area, although less optimal for cultivation, especially in upland areas.

This chapter gives an overview of various conditions for textile production within the study area and how they varied both physically and socially in this long period of half a millennium. In broad outline, they cover both natural and human resources – means and modes of production, organisation of work, production, and consumption of various types of textiles. As argued in the previous chapter, the production varied within different natural and social environments – within ordinary household production, advanced household production above the ordinary, and attached specialist production, often connected to nodes within larger estates. The sources to elucidate farming systems related to textile production comprise complex multidisciplinary source categories; archaeological, botanical, and zooarchaeological data, as well as written contemporary evidence but also later historical sources, used retrospectively. Altogether, they represent a challenging but also rewarding task, able to illuminate the significance of textile production at many levels, thus providing a backdrop when tracing Viking Age and medieval textile production in the following chapters.

Farms, farming systems, and access to resources

The main unit of production in Norwegian agrarian studies has been the so-called named farm (Norwegian *navnegård*) denoted by its own name regardless of its structure, size, and degree of subdivision into holdings. According to the Norwegian research tradition, the concept of the farm with its own name, however, could cover a whole range of settlement categories. The settlements could vary from small dispersed separate farms or holdings to large subdivided multiple farms, comprising all land used for agricultural purposes: the farmhouses, infields and outfields, and might also include forested and mountain areas.[1] Some appear as nodes and manors within their estates of tenanted farms and were often established with churches at an early stage of Christianity. Certain categories of farm names reflect settlement hierarchies and temporal patterns, as is the case in other regions of Scandinavia and in the colonised areas in the North Atlantic.

Textile production, being closely incorporated in farming systems and economic strategies, is primarily reflected in the animal farming of the period. Nevertheless, the pastoral economy cannot be isolated from the arable sector, as they were joined in a symbiotic relationship. The farming system in both western Norway and the North Atlantic was based on an infield–outfield system where outfields and animal farming played a dominant but still integrated role. Wool from sheep was overwhelmingly the most common raw material for textile production in the whole area. Hair from goats and cows, and even horsehair from the tail were, to a minor extent, used for special purposes, such as strings, cords, and ropes. The share of sheep farming and cultivation of fibre plants within the farming systems of the period, basically flax and hemp, may therefore be indicative of the role of textile production – whether only for household needs or also for surplus, and then, possibly as an attached, advanced, or specialised production above ordinary household levels. These conditions varied in space and over the centuries, caused by population growth, expansion, and clearing of new land and farms, both inland and in the newly colonised areas in the North Atlantic. The expansion relied on other available resources too, not least marine: fishing and whaling, as well as terrestrial resources, wildlife, minerals, and iron ores. The latter, however, were mainly located in the eastern part of Norway and outside the present study area but with access in bordering areas in the inner zone.[2] Socio-political changes in the wake of state formation processes, religious upheaval in the transition to the Middle Ages and succeeding urbanisation also had an impact on production and ways of living. The new secular and ecclesiastical institutions also affected landownership with new types of landowners and elites, as well as new ways of trading and networking, and in turn, the working conditions for their subordinates and tenants for providing a surplus. However, many of these medieval structures also reflect earlier hierarchical structures from the Viking period.

Traditionally, the overall theoretical and methodological framework for Norwegian agrarian historical research has been a high degree of continuity in settlement structure, with a system of self-sufficiency within peasant society. It is also notable that the personal freedom held by medieval peasants has been emphasised – a system without sharp distinctions between tenants and freeholders in legal and economic matters. A largely unspoken premise has been that the family and kin owned their own land and could dispose of it relatively freely with access to commons as a reservoir for cultivation and expansion. Lately, these perceptions have been contested. There are reasons to question whether the early medieval societies in Norway and in the North Atlantic were so egalitarian or belonged to a more hierarchical structure, as also reflected in the saga narratives and burial customs not least. This perspective is also relevant when studying production related to outlying areas, the utilisation and use rights connected to sheep farming and thus wool and textile production, as well as all the other modes and steps in production, also based on a subordinate workforce, including slaves in the early part of the period.[3]

The use of the outfields for animal and sheep farming had social as well as economic and functional dimensions. Estimations on landownership in Norway around AD 1300 indicate that the large majority of farmers were tenants leasing the land: The Crown held 7%, the Church and ecclesiastical institutions about 40%, the secular aristocracy 20% and the remaining *c.* 33% was held by private and freeholding peasants and town burgesses.[4] This implies that tenants with restricted use rights, also in the outfields and wasteland, ran about 70% of the farms in Norway around 1300 – and a system, then, characterised by inequality as for access to land.[5] In Iceland, it has been estimated that as much about 95% of the farmers were tenants around 1300.[6] Nevertheless, they were run within an ordinary household structure for subsistence but also had to produce a surplus for the landowners connected to manors within larger estates, and as documented in written records.

In Norway, recent archaeological studies with the aim of delving beyond the Middle Ages, have substantiated that various kinds of manors with estates of subordinated farms existed already in the Viking Age at the latest. By exploring and combining different types of sources, representing different temporal manifestations, archaeological and medieval written evidence – burial mounds, medieval church sites, and inferences drawn from post-medieval sources on property structure – has made it possible to identify and discriminate such early estates. Grave monuments and other structures embedded in the landscape, such as standing stones and fences, then, are interpreted as physical symbolic manifestations and demarcation of land rights, social stratification, and power. One of the premises is that people without land rights could not normally be given a monumental grave on a property they did not own themselves. Such physical and visible manifestations close to the settlements or borders of the farmland were reserved for landowners and erected by the heirs of the deceased, thus also demonstrating and symbolising their own hereditary rights and power.[7] In his studies of royal estates and farms and property held by royal officials (ON *lendmenn* (pl.)) in western Norway, the archaeologist Frode Iversen found a significant accordance between burial mounds and historically recorded farms owned by freeholders. Correspondingly, there was a general lack of burial mounds on medieval and later tenanted farms, and at that time mostly parts of larger medieval estates, owned by the Church, monastic institutions, and local or regional elites. However, the main farm and nucleus of such estates generally contained large burial mounds – often larger and richer than other mounds at neighbouring settlements. The correspondence between presence/absence of prehistoric grave monuments and historical landownership suggests that medieval landownership has prehistoric roots, going back to at least the Late Iron Age in this part of the country. The predominant absence of visible burial monuments on later tenanted land may either imply that the farms were first cleared after heathen burial customs had come to an end, or that even at an early stage connected to estates were farmed by peasants with subordinate status.[8]

Similar studies in Iceland based on farm and place names, archaeology, locations, and later sources used retrospectively, have correspondingly substantiated an early stratification in settlements of a similar structure as substantiated in western Norway; a pattern of larger farms serving as head farms (*höfuðból*), with outlying farms (*útjarðir*) and tenant farms (*leigustaðir*), often in the vicinity of the head farm. Structurally, they share traits with manors and estates in western Norway, although not necessarily close by. The head farm, then, controlled the others and formed a socioeconomic unit. According to the Icelandic historian Árni Júliusson, this system seems to go back at least to the 12th century and probably to the *landnám* period – the age of the early settlers in the late 9th century. By the High Middle Ages, around AD 1300, farms could expand into large complex estates or so-called *staður*-places, ecclesiastical estates, and earlier chieftain estates, such as the prestigious head farms Reykholt in north-western Iceland and Hofstaðir in Mývatnstveit in the north-western part of the country. Later in the 13th and 14th centuries, according to contemporary sagas, they could expand significantly by strategic acquisitions, such

Fig. 3.1: Places in Iceland, mentioned in the text (Illustration: Per Bækken).

as, among others, the episcopal sees Möðruvellir and Hólar in Eyjafjörður in the West Fjord, by adding farms to their estates (Fig. 3.1).

Judging from information in late medieval sources, there could have been from 50 to 100 such head farms in Iceland at that time.[9] Similar units in Norway can also be recognised in late medieval records, however, of varying size. In Iceland many farms, subdivided in clustered settlements, were also run by freeholders, and like in western Norway, indicated by grave mounds, often as clusters close to the settlement areas.

Human resources and production units

Organisation, division of work, and access to working hands and their work capacity are other aspects to consider when assessing textile production and farming systems in various settings and within social strata: larger landowners, independent farmers and tenants and their respective households. The family farm served as a unit for both production and consumption, basically covering their own needs – representing *ordinary household production*, as referred to in the previous chapter. The sources reflect various types of households and types of production, but principally a division between a landowning elite and subjected tenants renting their land, and therefore obliged to produce obligatory shares as surplus going out of the household's own consumption, opening for an *advanced* or *attached specialist kind of production*. The elite could also promote clearing of new farmland within their territories or in other ways provide a surplus and make products into commodities, and where textiles gained status as 'commodity money' in various transactions.

In Norway as a whole, *c.* 37,800 named farms (farms with their own name) at the beginning of the 14th century, contained between 64,000 and 75,000 holdings.[10] A named farm could be identical to a family holding but could be subdivided into two or more family holdings through inheritance.[11] The notion 'household' (ON *hjón*) included a nuclear family and extended families and employed workers, and in the first centuries of the period under study, also slaves. These structural differences therefore make it difficult to estimate the population in the Viking period as well as in the High Middle Ages with a population peak around AD 1300, before the demographic recession started by the 14th century.

The population and number of farms within the study area is uncertain due to the fragmentary sources. Still, various methods have been combined and used to estimate numbers by what has been regarded as the summit in the High Middle Ages, *c.* 1300. According to the latest estimates, Norway as an entity counted a population of about 512,000 *c.* 1300,[12] or even higher since many deserted medieval farms were still not reclaimed in the census from 1723 when the population counted *c.* 580,000 people. By the middle of the 17th and 18th centuries, only 80% of the medieval farms had been resettled and had still not reached the summit by the High Middle Ages before the demographic recession in the Late Middle Ages. In the census from 1665, the population counted 440,000 people, and as many as 410,000 were living in the countryside. The estimations of people in the study area, of three medieval counties in western Norway, are more difficult to establish. In 1665, the rural population in this area counted around 68,000 people, meaning *c.* 17% of the population on a national level. If this percentage is also a representative share for the High Middle Ages, the region may have included around 87,200 people. This number may again be too low because a hundred years later, in 1768, the population had increased by 28%.[13] These figures can therefore only be tentative as many factors are uncertain, not least the number of holdings and the size of the farms, and accordingly, the households. The tendency of subdivision of farms from the Viking Age into the High Middle Age, generally indicates that the farms and the households were larger in the early period. Marginal settlements in the High Middle Ages also cause problems in such calculations. On a national level, however, there are clear indications that the settlements did not reach the same heights as before the demographic crisis in the 14th century, until well into the 18th century.

In Iceland, the largest area in the North Atlantic, the number of farms has been calculated to around 300 units *c.* 1300, and most densely in the southern part, with about 250 farms. At this time, about 20–30 farms appear as very large, with households of more than 20 people.[14] The Icelandic population around 1300 has been estimated to 60,000–80,000 people, however, without a general agreement on calculations.[15] In the Faroe Islands with their 18 islands, the scattered settlements were clustered in so-called *bygdir*, hamlets or small villages. By the end of the Middle Ages, settlements counted around 85 *bygdir*, each comprising several farms or holdings, so-called *býlingar*. Based on later land registers, the islands may have included nearly 200 holdings or farms and with a total population of about

4,000 people, depending on the estimated size of the households within these units. The archaeological evidence attests that the earliest settlements were located within the infields of such structures, often close to old church sites. At some of these settlement areas, building remains have been dated to the Viking period.[16] The population in Shetland seems to have been higher, counting 10,000–11,000 people when farms and settlements were reclaimed after the demographic crisis in the Late Middle Ages.[17] Of the 70 islands within the Orkney archipelago, only about 15 of them include main inhabited areas. Also, in these islands, habitation and farming largely appear as dispersed settlement patterns and within a hierarchical structure, with high-status farms surrounded by settlements inhabited by a dependent population.[18] However, the number of farms and size of population is uncertain.

Except for Orkneys, mainly outfields and grazing land characterised the cultural landscapes, as only a minor percentage was suited for arable farming. The structuring of the landscapes also shared similar traits, such as fencing the arable infields and with cattle lanes leading the stocks into the wider grazing land and to more distant shieling areas. Another common characteristic is a dispersed settlement pattern, however, with tendencies towards clustering and a settlement pattern of main farms and farms based on land tenure. Within the farming systems based on mixed farming, both animal husbandry and arable farming, a general trend towards specialisation in sheep farming can still be discerned.

The number of people varied with the size of the farms. In both western Norway and Iceland, the smallest units were run by a household of a so-called *einvirki* (ON), representing a single male, married farmer, without other male workforce above the age of 15 in Norway (*G* 131), and in Iceland above 12 years of age.[19] This seems to imply that hired workers were common in both areas, as were female servants, denoted ON *deigja*, f., *seta*, f., or *heimakona*, working for bed, board, and clothing, taking part in daily work, or within the specialised work of weaving, when denoted *vefkona*. So-called *vinnufolk* and *vinnukona* represent paid labourers. In Iceland, more specialised work appears at larger farms, head farms or manors. Larger farms should have at least two workers during the winter according to Icelandic law codex *Jónsbok* from the late 13th century.[20] The farming systems in all these areas under study were based then on gender and age within gendered work domains.

In the Viking period and Early Middle Ages, the rural workforce included unfree labourers too, denoted ON *þræll*, m., or *ambátt*, f. when a female, and both as slaves without personal legal rights and regarded as property. They could count a substantial part of the households and of society, as the Gulathing law even distinguishes between their duties and status, where the two so-called best female slaves were denoted respectively *seta*, f. working indoors and *deigja*, f., with wider responsibilities as a manager of the household.[21] This implies, then, that that there could be more. However, they were all under the auspices of the lady of the house, denoted *húsfrú*. The male counterpart to *deigja* was denoted *bryti*, m., with wider responsibilities than the others. Even tenants had slaves.[22] Altogether, the slaves could count a substantial part of the society, and estimated from one-fifth to one-third of the population in the Early Middle Ages in Norway.[23] In that case, it may have been on approximately the same level as in England, where unfree labour is estimated to have constituted 10–12% of the male population and up to 25% in some areas in the 11th century.[24] These estimates are based on more solid documentary evidence than in Norway. However, the numbers are difficult to quantify and by the 12th century, slavery was gradually abolished as a stepwise process of various stages of freedom.[25] During this process, thralls were denoted as *leysingr*, m., and *leysingja*, f., before reaching full freedom. However, by the High Middle Ages, tenants in both Norway and the North Atlantic were independent and free peasants in a juridical sense, and not attached serfs as on the Continent. They had to pay rent and were under contractual obligation to the landowners but with delimited use rights to resources as stated by law. These varying conditions over time are relevant in the following chapters when assessing textile production in time and space, as reflected in graves and working environments within different settlement structures of farms, manors and inter-connected structures, and networks also crossing borders.

Sheep farming and textile production

Although the topographies of western Norway and the North Atlantic were far from homogenous, the favourable conditions for textile production were a common feature – with their vast areas with suitable grazing land and with more limited cultivable areas. Coastal areas offered special advantages topographically and the milder climates in the lowlands and coastal areas gave excellent conditions for almost all-year-round grazing. In the inland areas, such as in the mid-fjord and inner zones of western Norway with access to outlying grazing land and vast mountain areas, the conditions in the summer season were also well suited for sheep farming. This kind of summer farming is documented in Iceland and the Faroe Islands too.[26]

The so-called Old Norwegian Sheep is the breed with closest connections with the first documented prehistoric sheep (*Ovis aries*). It is a small type of sheep, short-tailed, double-coated with two types of wool: a long, coarser protective outer hair, covering finer, short, soft under-wool, with a fleece varying in colour from almost white to greyish, dark brown and black. The sheep could be both horned and culled. The Icelandic sheep was of the same short-tailed breed group, characterised by a double coat with longer outer hair (*tog*) and downy soft under wool (*Þel*)[27] (Figs 3.2–3.3). This breed tends to moult, and the sheep were therefore plucked instead of sheared.[28] The outer coat on the

Fig. 3.2: Traditional Old Norse Sheep (Photo: Mons Kvamme, with permission).

Fig. 3.3: The long-haired Old Norwegian Sheep, Norw. Spælsau (Photo: Mons Kvamme, with permission).

males has long fibres, from 15 cm up to 30 cm around the neck. Composition of the flocks is therefore relevant when assessing fleece length and weights.

The present Norwegian sheep, so-called *spælsau*, has its origin in the Old Norwegian Sheep that still exists as an all-year-round grazing sheep along the coastal heathlands and is also closely related to Old Icelandic and Faroese sheep. The 'Spælsau' has inhabited Norway and Scandinavia, including Iceland, the Faroe Islands, Shetland, and the Orkneys. This breed is regarded to have been omnipresent until the 18th and 19th centuries when new breeds started to be imported to Norway from England and Spain.[29] It was primarily used for wool, milk, and hides, while the meat was probably less important, at least for older sheep,[30] and the old out-grazing sheep can live long until the age of 12 years.[31]

The Old Norse common notion for sheep, *sauðr*, m., was further distinguished as for age and sex – *lamb*, n. *lamb*, m. (with various notions as for sex), further, *gemla*, f., one-year-old sheep, *gemlingr*, m., one-year-old ram, *ær*, f., ewe, *hrútr*, m., ram, and *geldingr*, m., gelded ram. These categories also varied in price and use as size of the animals, and the age and gender affected the amount of fleece and its weights. Fleece from a ram (castrated ram) weighed more than ewes' and double the weight of fleece from hogs (two-year-old sheep).[32]

The slaughtering age is therefore indicative of use, where older males indicate wool production, while slaughtered at a young age for meat production. Female sheep were crucial for reproduction, wool, and milk, and when slaughtered, meat, hides and skins. The medieval value system, relating commodities, agricultural products, and animal husbandry in Norway and Iceland, share common traits too, where *kúgildi* or *kýrlag*, the value of a healthy matured cow having had her first calf, functioned as a standard norm.[33] In Norway, as well as in Iceland, six sheep were equated with one *kúgildi/ kýrlag*. However, the Icelandic sources related to sheep reveal more differentiated values with regards to age and sex, not least reflecting their role in textile production. They tell, for example, that the number of six could simply be six female sheep ewes, *ær*, but could also be equalled with two so-called *gemlingr*, *i.e.* one-winter-old sheep, and four older fertile sheep that had not been sheared. The same value could also cover other combinations – two *gemlingar* and four older fertile and shaggy sheep, or eight infertile sheep and three-winter-old or older sheep, or eight castrated rams, and so on in even more combinations.[34] Altogether the intricate value system of combinations, underlines the importance of wool within sheep farming and textile production when valuing various wool qualities and the amount. According to the medieval Icelandic law *Grágás*, 20 ewes in total produced one *vætt* (36.2 kg) of wool annually, or 1.8 kg on average per ewe.[35] A range of 1.2–1.5 kg is referred for Shetland sheep,[36] probably like weights for the Old Norwegian types as well. Rams were especially important as they produced more wool, which did not vary in quality due to pregnancy. This may explain a large percentage of rams in both the zooarchaeological analyses in the North Atlantic and written material from Iceland,[37] where these trends can be evidenced. Still, it was the finer wool from one-year-old sheep that achieved the highest price, according to *Grágás*.[38] According to both *Grágás* and *Búalög*, as noted in some later customary laws from Iceland, an average farm needed annually 60 kg wool for clothing, 30 kg for bedding and 10 kg for sails[39] – the latter with a long durability, however, hardly produced annually.

Archaeological skeletal analyses of medieval sheep bones from western Norway have so far been largely limited to finds from urban contexts with good preservation for organic material. Analyses of animal bones in medieval rural and agrarian contexts are less frequent finds. However, a general problem in such analyses is that it can often be difficult to distinguish between sheep and goats based on the bones. Still, the osteological analyses have shown that the sheep were rather small but not much smaller than the present breed of the Old Norwegian sheep, measuring only 60 cm to the shoulder,[40] and the same as English medieval sheep.[41] The age of slaughtering varies and indicates the use areas, whether sheep were kept for consumption of meat and milk or for wool production. Here, low age indicates consumption and high age that they were mainly kept for wool production since older sheep produce more and long-fibred wool.

The animal bone material from medieval urban contexts in Norway shows that sheep were generally slaughtered young, about 24 months or younger, and, based on finds from Bryggen in Bergen, even younger with only 20 to 30% older than one year.[42] When the Urban Code of 1276, as noted originally designated for Bergen, claims that the town levy as an obligatory tax for all kinds of craftsmen, including apprentices and small traders and both male and female, should be paid in sheep and that the carcasses should be cut in two or four parts but not in small pieces,[43] demonstrates then production for consumption. It also seems to imply that townspeople to some extent were engaged in sheep farming, probably for subsistence by using the vast urban commons in the mountainous areas surrounding the town.

The recent emphasis on environmental archaeology and zooarchaeology in the North Atlantic has provided new and vital information about animal husbandry in the Viking period and the Middle Ages and demonstrates that sheep quickly became the dominant domesticate in the North Atlantic. Indirectly, it also sheds light on conditions in the settlers' homelands. Analyses of the faunal material from the site Undir Junkarinsfløtti, at Sandur on Sandoy in the Faroe Islands, have found that caprines (sheep and goats, but mostly sheep) constituted *c.* 60% of the domestic mammal bones in the earliest Viking Age *landnám* phase in the late 9th century and increased to 70–80% in the 11th century.[44] That relatively few were culled while juveniles, only about 2–10%,[45] substantiates the importance of sheep in textile production. The Faroese so-called 'Sheep Letter', *Seyðabrevið,* issued in 1298 as a special regulation for the islands,[46] and a supplement to the Gulathing law, proves that there was extensive sheep farming at that time, since formal actions had to be taken to prevent overgrazing and consequent erosion.

A similar development is evidenced in the zooarchaeological material at several sites in Iceland, such as the above-mentioned high-status site, Hofstaðir, and its probable subordinate farm, Sveigakot, in the Mývatn area in north-eastern Iceland. At Hofstaðir, caprines represented 80% of the domestic animals as early as the 10th century but fell to around 70% in the following century,[47] when the farm declined and was eventually abandoned. At another high-status site, the earlier mentioned Möðruvellir, not far from the medieval trading place Gásir in Eyjafjörður, caprines also dominated among the domestic mammals with 74% of the total, in layers from the 13th to the early 15th centuries.[48]

Altogether, these varied sources of archaeological, zooarchaeological and written evidence from both western Norway and the North Atlantic, clearly document the importance of sheep breeding and wool production as a prerequisite for textile production. Although scattered in time and place, they show that sheep and wool were important common denominators for these areas. How the stocks vary in time and different geographic and socioeconomic contexts, reveal scale, importance, and not least conscious strategies to achieve certain wool qualities for special purposes and for various types of production – as ordinary household production attached more advanced and also specialised production in various contexts, from subordinated farms to manors.

Production and products

The Gulathing law, the provincial law of western Norway representative for the 12th century, however building on older traditions, provides important direct and indirect information related to sheep farming – its importance for wool and textiles, milk, meat, and skins, which also serve as transferable commodities within interconnected value systems of both commodity and monetary value in various economic transactions. Being used for sale, for paying levies and other duties such as tax and tithes, demonstrates that the volume of sheep generally exceeded the needs in the domestic households and that sheep formed a quantitative larger part within animal husbandry – as also verified in osteological analyses in the colonised areas to the west.

In western Norway, goats were of equal value as sheep but seemingly lower estimated, probably because of their more limited use areas – milk and skins but less appreciated for meat and within textile production. The Gulathing law, for example, states that fines should be paid in sheep, not goats.[49] Skins of both sheep and goats were commodities for sale both domestically and for export. Sheep were also subject to taxation, as documented when paying the levy in Voss in the inner zone of Hordafylki,[50] and an area marked by many Viking Age graves with textile tools (see Chapter 5). A special by-law for Bergen issued in 1282 forbade foreigners the right to buy sheep in the surrounding countryside; they should be sold at the urban marketplace.[51] The by-law was evidently an attempt to control and regulate the market as well as prices and perhaps also to supply urban craftspeople with raw material for further processing, such as for shoes and other commodities. The annual produce of wool must be the reason why tithes were also claimed on sheep when mentioned in a special appendix to a tithe regulation in Archbishop Jon's Christian's Codex of 1277.[52] The population in Shetland seems to have been higher, counting 10,000–11,000 people when farms and settlements were reclaimed after the demographic crisis in the Late Middle Ages. These various products were thus used as 'commodity money' for exchange and concrete commodities for export, altogether signifying a surplus production.

Although it is difficult to come to grips with the extent of sheep farming by quantitative measures, the qualitative information that can be derived from various sources leaves the impression that land use related to sheep was so extensive that detailed restrictions on land use were needed. The Gulathing law even has a paragraph[53] stating that during whaling, the whale was not permitted to be placed on a location with grass enough to feed a sheep and lamb during the summer and if doing so would be punished by fines. Another paragraph[54] sets restrictions on animal husbandry, as farmers/tenants were not allowed to keep more animals – not even a lamb more – than could be fed during winter without harvesting another man's property. These sources thus indicate large-scale sheep farming based on an extensive and intensive use of outlying grazing land well into the High Middle Ages.

The advantages of the long outdoor grazing period in coastal areas are clearly reflected in regulations, place names, and various narratives. The above-mentioned tithe regulation in Archbishop Jon's Christian Codex, for example, specifies that tithes on wool production should be paid on all the sheep that are grazing out on the islands '… *af þeim sauðum öllum sem fœðazst i œyum uti…*'.[55] The many place names along the coast of western Norway such as *Sauesund*, 'Sheep sound' and *Saudöy*, sheep islets and sheep islands in coastal areas in Sogn and others, such as *Fårøya*, *Færøya*, *Sauøy*, *Sauholmen*, *Lamøya*, *Verholmen*, among others obviously imply extensive sheep farming in coastal areas (Fig. 3.4).

The danger of wild animals, which represented a major threat for grazing sheep, was additionally reduced at such demarcated places. In areas further inland, where farm animals had to be kept indoors during the winter, it was

Fig 3.4: Grazing flocks of sheep on the island Lygra on the west coast of Hordafylki (Photo: Mons Kvamme, with permission).

crucial to get them out to let them feed themselves as early as possible in the spring and this was regulated by law too. In the spring, sheep should be kept within fences in the outfields (ON *úthagi*) to protect the recently ploughed and sown common infields. Heaths and marshlands, highland valleys, and mountainous areas were nourishing pastures in the summer season too, but the flocks needed herding (Fig. 3.5), as well as seasonal stays at shielings when at distance.

Shielings

According to the Gulathing law (*G* 81), livestock, including sheep, should be sent to outlying shielings after two months of the summer, where the first summer day counted as 14 April, meaning then the middle of June, and to be taken home in the fifth month of summer, from mid-August to mid-September. The law implies that shielings were ordinary and customary structures and although at distances, were tightly integrated parts of the farms (*G* 86). In the sources, through law and litigations, shielings appear as areas of conflicts and indirectly reflect the economic importance of having access and rights to grazing land, sheds, and shelters in the productive summer months. By the 12th century, summer farming connected to shielings had reached a level that made it necessary to specify by law (*G* 19) that summons to thing assemblies should go to the winter houses and not to the shielings.

During the last four to five decades, excavations in high lying areas in western Norway have revealed remains of small building complexes interpreted as prehistoric and medieval shielings. They are situated close to historically used shielings areas at various distances from historical farms. Overall, they indicate that shielings were old and common structures even before the Viking period but expanded significantly in the Viking Age and Early Middle Ages. These excavations have provided new information about their dating and usage, life and work more concretely, also relating directly to textile production (see Chapters 7 and 8).

The two-partite infield–outfield system based on a division of home fields and outlying areas mainly used in the summer as grazing land and production based on animal husbandry, was, as noted, also transferred to the North Atlantic. This is evidenced in Iceland and the Faroe Islands through archaeology, place names, and written records.[56] As in western Norway, the shielings could be situated close to the farms for daily milking or at a distance based on periodical stays of eight weeks or more, in Iceland especially in mountain valleys, green mountains and vast uninhabited lava fields.[57] The distribution of farms and their historically and archaeologically documented shieling areas provide a backcloth when connecting to earlier Viking Age sites and finds of house constructions in upland areas (see Chapter 7).

Fig. 3.5: Section of a frontal from Nes church, Luster in the inner part of Sogn, painted around 1300, now in the University Museum of Bergen. It depicts two shepherds in the fields listening to the gospel for Christmas Day, while herding the flock with a dog – both goats and sheep, all white, with and without horns. Most likely it reflects homely conditions (Photo © University Museum of Bergen).

Table 3.1: Number of sheep recorded in tax lists and public records, 1665 and 1723, within the study area (Source: Aschehoug 1890).

County	*Areas*	*1665*	*Total area: 1665*	*1723*	*Total area: 1723*
Hordafylki	Sunnhordland	16,034	–	17,672	
	Hardanger	6,680	–	9,147	
	Nordhordland & Voss	20,824	–	25,880	
N:		43,538	43,538	52,699	52,699
Sygnafylki	Sogn	18,000	18,000	20,364	20,364
Firdafylki	Sunnfjord	12,373		–	
	Nordfjord	12,769		–	
N:		25,142	25,142	26,968	26,968
N: Total area			**86,680**		**100,031**

Sheep farming and textile production: changes over time and in place?

Judging by these different indicators of scale, there is no reason to assume that animal husbandry and sheep farming played a smaller role in western Norway around 1300 than in the Early Modern period, when the earliest statistical nationwide accounts for taxation appear by the middle of the 17th and early 18th centuries, notably in 1665 and 1723 (Table 3.1). As already noted, the population rates and farms at this time had still not reached the levels predating the demographic crisis and desertion of the Late Middle Ages. Moreover, the number of animals in these tax lists are figured as too low, as they were subject to taxation.

The lists for sheep are also the least exact of husbandry. Furthermore, the numbers are estimated to be higher in the summer season before slaughtering in the autumn and winter-feeding.[58] In this period, wool production seems in fact to have played a smaller role than in the High Middle Ages and earlier and seems to be mainly for subsistence on a household level and when import of textiles had become more available.

Table 3.1 shows the extent of sheep farming as recorded in the three medieval counties in western Norway, and as noted, probably represents minimum numbers. They reveal, however, regional variations both between and within the counties that may be representative. The highest numbers appear in Hordafylki and numerous in both outer and inner districts to the south and north, such as Sunnhordland and Nordhordland and the inner region of Voss. Indirectly, then, these accounts, at least to a certain extent, reflect the extent of wool and textile production within the study area, although hardly representative for the period of study. Nevertheless, that the registered sheep appear to have increased by *c.* 15% in the period of around six decades also indicates that the potential was not fully used, or adequately recorded.[59]

The number of cattle, however, appears as higher than sheep in these records, representing mixed farming mainly for subsistence and not specialisation, contrary to the Viking Age and medieval zooarchaeological data from the North Atlantic. Still, by the late 19th century and before significant changes in farming methods, and when imports of textiles had become more accessible, the number of sheep was even higher. This could imply that sheep farming could also have reached even higher levels in the High Middle Ages if these opportunities were fully utilised, which demonstrates the importance of textile production at an early stage.

The written records related to sheep farming in Iceland are earlier and more detailed than from western Norway, especially as for late medieval records. They corroborate the results from the faunal analyses with more detailed information about the scale and strategies of sheep farming. These sources reveal that more than 80% of the domestic animals within the larger estates in Iceland were sheep, and to a large degree, older sheep. Some of the larger farms had flocks of 500–600 sheep, while smaller and average-sized farms might have had some 60–80 sheep in their stocks.[60] A document from 1340 specifies the stocks belonging to the episcopal see Hólar seem to have comprised 215 ewes, 250 one-year-old sheep, and 65 castrated rams, so-called *geldingar*, altogether 530 sheep, old and young, but then subject to the many tenant farms surrounding the manor.[61] Judging by Árni Juliússon's study, the sheep stocks generally expanded during the Middle Ages, especially related to the larger ecclesiastical but also secular estates. The central farm and church site Reykholt contained some 73 sheep before *c.* 1400 that increased to 105 in the next century. Judging by later written records, the quantitative percentages between sheep and sheep/goats and other livestock range in rather similar percentages as the above-mentioned osteological analyses; at Reykholt from 78% in the first recorded period and rising to 84% in the 15th century, indicating both an absolute and relative increase. When comparing the relation between sheep and cows in written records and excavated osteological assemblages, which has been possible at Möðruvellir, it also matches to a large degree – with 74% in the faunal material and 70% in the written records from the 15th century. According to the written evidence, sheep farming had its largest extent in the western and northern parts of Iceland and in the inner areas, also appearing as the densest shieling areas.[62] This may imply that farming systems in western Norway may have followed different paths from the Viking Age and the Middle Ages, seemingly larger in Norway in its first part of the period, while steadily increasing in Iceland during the Middle Ages.

As medieval quantitative data from Norway generally are more than sparse, it is impossible to pinpoint the scale of sheep farming statistically. However, other written evidence supplies and complements information about sheep farming and of relevance for textile production. Such sources leave enough evidence to substantiate that the production played a significant role in the farm management in western Norway at an early stage and in the North Atlantic through the whole period of study. Sheep farming was multi-purpose, providing milk, meat, and skins, while wool stands out as a long-lasting product that could be harvested annually in the whole life span of the sheep. The importance of wool is demonstrated in conscious priorities of late slaughtering ages – reflecting that wool was a prioritised product in these areas, and not least in the North Atlantic, based both on the normative measures and other information from Iceland. At Reykholt for instance, it could have amounted to *c.* 130 kg to 190 kg wool annually, while at Hólar as much as around 954 kg, representing the largest recorded estate in Iceland.

Textiles as commodities

The extent of processed products as articles for export – skins, wool, yearn, and wadmal – is indicative of the sheep farming as a surplus production. Tax lists of export from Bergen to some English towns from the first decades of the 14th century also give information about such products. Of 252 ships leaving Bergen to east English ports between 1303 and 1311, nearly 5,000 goatskins and more than 2,000 skins of billy goats were in the shiploads and constituted 84% of the total of animal farming products, and 2,000 sheared sheep skins to the same harbours in this period. Wadmal is recorded as cargoes on 36 ships but probably represents Icelandic transhipment via Bergen as wadmal from Iceland, the Faroe Islands, and Shetland were mainly sold via Bergen, being the main commercial centre in the North Atlantic region. Wool is referred to as part of the cargo only on one ship in the same period.[63] This is not surprising since England by then constituted a major area of textile production and sheep breeding in north-western Europe. This also explains why goatskins were a larger export article than of sheep, being of a stronger grain and firmer structure and less stretchy than sheepskin, and in demand for many products when tanned, for example, shoes, gloves, and various kinds of containers. Information about export sheep products to Germany by the end of the century seem, however, to have been larger than to England.[64] These numbers do not refer to textiles. Indirectly, however, they imply larger stocks of both goats and sheep – and as for sheep, they had produced wool for several years – perhaps up to 12 – before ending up as skins for export.

Although the extent of export is elusive, it is obviously based on surplus production in the High and Late Middle Ages. It was largely based on the income of various obligatory duties, taxes, and tithes collected by ecclesiastical institutions, royal representatives, and local elites, altogether, the largest landowners and entrepreneurs engaged in foreign trade, including import of textiles, not least of plant fibres, linen and canvas from the Baltic at least in the Late Middle Ages.[65] The commercial aspects of textile production in the wake of urbanisation in the Middle Ages with import of textiles produced as specialised products in urban settings, also affected the domestic production in the study area. Here, the production of wadmal in the North Atlantic became a competitive product in the market, transhipped via Bergen (see Chapter 9).

Cultivation of textile plants

Flax, the annual plant of *Linum usitatissimum*, was the main vegetal textile fibre in the Viking period and Middle Ages in Norway besides hemp, *Cannabis sativa*, and nettles, *Urtica dioica*, but seemingly of relatively small scale and probably only marginal in the North Atlantic. Flax is known from the Late Bronze Age in Norway and seems to be more common from the Early Iron Age but at this stage probably primarily because of the nourishing seeds.[66] By the Viking Age, flax for textile production seems to have become more common, as also was hemp and nettles. Plant fibres are generally much longer than the animal fibres of only a few centimetres, and flax plants could be more than 1 m and hemp around 3–3.5 m in length and had to go through various processes to become usable as textile fibres (Figs 3.6 and 3.7).

Flax is sown in April/May and the seed capsules develop in July–August. It is ready for harvesting by being pulled green, not fully matured, to get finer fibres. Flax needs intensive weeding, at least three to four times during the growing season, and is therefore far more labour intensive than hemp since weeding is less important for the cultivation of hemp.

Both flax and hemp thrive in self-draining calcareous soils and without demanding high temperatures. Hence, the production of flax can be found on high lying areas and often on sandy terraces.[67] However, it could also be grown in coastal areas, but nitrogen rich soil does not produce good flax fibres. Hemp can also grow on soils with high

Fig. 3.6: Flax fields, with their long stalks up to 1 m, sown in early spring, blossoming in July with blue flowers.

Fig. 3.7: Hemp field with the sturdy and long stalks, about three times longer than those of flax (Photo: Norsk hampeforening, with permission).

pH and muddy soils not suitable for many other crops and without much soil preparation, often in coniferous areas, as has been demonstrated in the southern part of the country from the Viking Age, but also in swampland, marshes, and around lakes.[68]

Remains of fibres of flax and hemp from the Viking Age have been found in a bog in south-east Norway, obviously placed there for soaking and loosening the long stalky fibres.[69] Similarly, hemp has been found in a bog close to the Kvalsund ship, dated to the 8th century, seemingly placed in the bog for soaking. Production of plant fibres was altogether labour intensive through the whole process, from sowing in the spring and lasting during the whole summer season before the many-faceted fibre preparation of retting, clutching, and hackling – all needing coordination and organisation. The production also connects to the more advanced production generally above ordinary household production, representing the more luxurious segment of production.

In the Eddic poem *RigsÞul*a stanzas 21, 21, and 28, products of linen – bridal linen, shirts, and patterned woven tablecloth of shining linen – symbolise the higher social strata. Remnants of linen, including prestigious textiles were found in the well-preserved Oseberg female burial of the highest status, dated to AD 834, together with hemp seeds and dyestuff of woad (Isatis tinctoria).

The use areas of linen and hemp were also different. Flax was used to produce finer cloth, such as undergarments, shirts, veils, beddings, and the like.[70] Hemp, however, being longer and coarser, was used for coarser threads in canvas and as cords, strings, and fabrics, and for sail cloths, among others.[71] However, lately it has been proven that it was also used in finer textiles and tapestries and combined with wool.[72] Nettles, generally shorter and finer than flax and hemp, still have more or less the same use areas as flax,[73] but are generally more difficult to identify in textiles because flax and nettles have the same fibre structure.

In western Norway, archaeological remains of a sunken floor building dated to the Viking period with finds of textile tools and linen fabrics at a high-status farm in Sogn, Stedje in Sogndal in Sygnafylki, are also an early and direct manifestation of flax being taken into use in textile production in a high status context.[74] These types of buildings with traces of textile production are interpreted as buildings of specialised production, possibly by a subordinated workforce, when connected to high-status farms[75] (see Chapter 7).

The Old Norse notions *lín* and *hörr*, and *hampr* denoted both the raw products flax and hemp and the processed products, mostly denoted as *lerept*, originally *línrept*.[76] Farm names dated to the Viking Age at the latest with prefix lín reflect an expanding emphasis on flax cultivation in western Norway, such as Línland in Hordafylki, *Lingjerde* in Sygnafylki, *Línvellir* and **Línsetr*, the latter documented at several places in Firdafylki. *Lín* is also a common prefix in field names, like *línakr*, *-ekra, -land.*[77] Such place names are known from Iceland too – also with suitable climates for such a production. Línakrdalr and Línekrudalr, referring to the early high-status settler Egil Skallagrimsson's domains,[78] are mentioned in the *Book of Settlements*, describing early settlements of Iceland. A system known from early medieval England shows that head farms or manors were surrounded by subordinate farms, based on specialised production, and paid tributes of, among others, flax.[79] It is therefore tempting to interpret the many Norwegian farm names of old age with the prefix *lín*, in such a context, being often connected to smaller and possibly subordinate farms.

While vegetal textile products and production occur rather scattered in sources from the early part of the period, the cultivation seems to have expanded in the Middle Ages. The first mention connects to a special duty on flax referred to in the saga *Ágrip*, recorded in the early 13th century. Allegedly, it was introduced by the Danish king Svein Knutsson during his short period of rule in Norway (1030–1035) as an annual duty. Here, it is denoted as ON *rýgjartó*, deriving of ON *rýgr*, woman or housewife, and *tó*, spinning material.[80] According to the saga, every lady of the house was obliged to pay tribute of a certain amount of flax, hemp, or wool. Although this burden was soon abolished, it shows that this kind of production was far from uncommon. Also, later in the tithe regulation in Archbishop Jon's Christian Codex of 1277, flax and hemp appear among the specified products subject for paying tithes. By the Middle Ages, the expanding urban commerce made it possible to import textiles and import of flax and linen is specifically mentioned by the Norwegian King Sverri in his renowned speech in Bergen in 1186. Here, he praised Englishmen for the import of useful and important foreign commodities,[81] indicating that the domestic production was too small to cover the homely demand. From the 13th century, this import was mainly taken over by the Hanseatic merchants.[82] Linen is mentioned as an import in the Icelandic *Grágás*, costing three times the price of wool.[83]

Based on these varied sources, from osteological and archaeological remains to historical narratives to normative sources from the period of study, they demonstrate that the textile production depended on both sheep farming and vegetal plants in western Norway as well as in the North Atlantic. The later quantitative information about sheep farming also makes it possible to assess the development in a wider timeframe when using it retrospectively and comparatively. The datable archaeological and zooarchaeological evidence, moreover, tend to complement each other and form parallel patterns in both areas. When related to different types and modes of production and organisation – ordinary household production, attached, advanced, and specialised production – the sources also reveal clearer patterns, not least useful when assessing tools, routines, and labour input in all the work processes from fibres to finished textiles of varied qualities and use areas in the following chapter.

Notes

1 Sandnes 1979, 166; Bjørkvik 1981, 625–631.
2 Tveiten 2012.
3 Iversen 1995; 1997.
4 Helle 1995, 54.
5 Øye 2002, 230, 367–368; 2004, 117–119.
6 Júliusson 1997, 26.
7 Skre 1998; 1999; Iversen 1999; 2008.
8 Iversen 1999; 2008.
9 Júliusson 2010, 14.
10 Martinsen 1996; Lunden 2004.
11 Helle 1995, 34; Júliusson 1995, 93; Arge *et al.* 2005, 612.
12 Lunden 1969; 1979; 2004, 159.
13 Aschehoug 1890, 11–12.

14 Júliusson 1997, 84–85.
15 Júliusson 1997, 52.
16 Arge *et al.* 2005, 605.
17 Ballin Smith 2007, 420.
18 Crawford 2013.
19 *KLNM* III, 693–694.
20 *KLNM* XIX, 107–112.
21 *G* 198; Fritzner I, 240; III, 212.
22 Iversen 1997.
23 Sandnes 1983.
24 Duby 1998.
25 Iversen 1997.
26 Sveinbjarnardóttir 1991; Mahler 2007.
27 *e.g.* Guckelsberger 2017.
28 Ryder 1984, 19.
29 Lunden 2004, 184.
30 Hoffmann 1991, 16.
31 P.E. Kaland 1979, 83.
32 Stephenson 1988, 374.
33 Steinnes [1937] 1982, 66–68; Lunden 1978; Júliusson 1997, 48.
34 Júliusson 1997, 48, with references.
35 Þorláksson 1991, 279–80.
36 Owen-Crocker *et al.* 2012, 639.
37 Þorláksson 1991, 297.
38 *Grágás* II, 94, 247; Falk 1919, 1.
39 Þorláksson 1991, 297.
40 Undheim 1985, 71–72.
41 Grant 1992, 176–177.
42 Hufthammer 1987, 63; Øye 1998a, 50.
43 *Bl* III, 8.
44 Church *et al.* 2005, 190.
45 Brewington 2010, 8.
46 *NgL* III, 10.
47 Brewington 2010, 13.
48 Harrison 2011, 15, 18.
49 *G* 223.
50 *DN* XV no. 13; 1329.
51 *NgL* III no. 2, 13.
52 *NgL* II, 355, 474.
53 *G* 149.
54 *G* 81.
55 *NgL* II, 355, 474.
56 Sveinbjarnardóttir 1991; 2005; Mahler 2007.
57 Júliusson 1997, 210.
58 Aschehoug 1890, 50, 53.
59 Aschehoug 1890, 56.
60 Þorláksson 1991, 274–75, 284.
61 Júliusson 2010, 12.
62 Júliusson 1997, 196.
63 Nedkvitne 2014, 58, 59, 174.
64 Bruns 1900, 86–88.
65 Bruns 1900, 60.
66 Myhre 2002, 97, 143; 2004, 56.
67 Høeg 1997, 133–134.
68 Skoglund *et al.* 2013.
69 Høeg 1997.
70 Fritzner II, 534.
71 *KLNM* VI, 108.
72 Skoglund *et al.* 2013.
73 *KLNM* XII, 285; Hoffmann 1991, 17.
74 Mortensen 1997; see also Chapter 8.
75 Skoglund *et al.* 2013.
76 Falk 1919, 3.
77 *NG* XI, 557; *NG* XII, 265, 346, 376, 439, 21, 306.
78 Falk 1919, 2, with references.
79 Faith 1999, 48.
80 *NgL* V, 53.9.
81 Nedkvitne 2014, 214, 30.
82 Nedkvitne 2014, 42, 112.
83 Falk 1919, 3, with references.

4

Tracing the work processes

Textile production involved a lengthy series of time-consuming work operations, both outdoors and indoors, all year round. Tools, knowledge, and technical skills were a prerequisite for transforming raw materials into finished cloth of various qualities. When tracing all the steps from converting raw materials – wool and plants – to finished products, several aspects are relevant: available resources for production – grazing land for the sheep and climate and arable land for plant production, tools, technologies, and people with necessary skills in all the working operations. This chapter aims to explore tools and technology within the whole chain of work processes within textile production to get an idea of the complexity and requirements needed.

The archaeological remains of the tools are concrete and measurable objects. Indirectly, they send wider imperceptible messages through their properties: how they were used in the various stages of production, the workload behind making the tools and what products they were intended for. They may also indicate qualities and competency to achieve various products. Combining the archaeological remains of tools and textiles with results from experimental archaeology, anthropological research, and information from historical and ethnological sources helps to reveal such messages, also related to gender roles and status (Fig. 4.1).

Figure 4.1 illustrates the various work operations, from fibre preparation to finished products, in various environments and different actors behind the operations, mainly women, and for both seasonal and all-year-round work. The archaeological remains relate to various loci and contexts – in graves and working environments – and information collected through written records – some indirect, others more directly concerning textile production. For archaeology, indirect traces relate to all invisible work, needing only hands as necessarily 'tools', and not least related to preparatory operations, such as plucking, cleaning, and sorting wool, however, also reflected in the finished products. Arable farming and cultivation, and the retting and rippling of flax or hemp, are also difficult to trace archaeologically but possible when reflected in palaeo-botanical material and in the textiles. Analysing fibres in preserved textiles makes it possible to deduce such processes. Tools of more durable material, for combing, spinning, weaving, and sewing, have left concrete evidence, often as grave goods, although generally as fragments of durable materials, such as spindle whorls and loom weights of stone, fired clay or lead, and metal tines of the wool combs. Only tools for spinning and weaving, *i.e.* spindle whorls and loom weights, are found in larger quantities. Whorls that once helped the drop-spindle to rotate and twist the fibres into yarn, and loom weights that stretched the warp-threads in the loom, are nevertheless able to illuminate the level and range of textile production by their weights and through experiments. Wool combs, weaving beaters, and different kinds of needles of various metals and organic material for sewing, shears for cutting wool and textiles are also represented in both Viking Age graves and other contexts such as production sites, covering the whole period of study.

Although the preservation conditions vary and are often far from optimal, the assemblages of textile production equipment from the study area in sum, are numerous. When seen in social, spatial, and long-term perspectives, they may indicate variations in household strategies and regional variations in modes and scale over time. Contemporary texts and nomenclature provide important complementary information. Sometimes such information appears as casual and incidental comments in the text or as wider narratives and references, but also as intentional records, expressed as norms, obligations, and restrictions, making it possible to catch important information about production. Archaeological remains of tools and textiles, recorded terminology and information concerning tools and techniques may elucidate work operations, common working loci, various types of products and the people involved. Assessing all available contemporaneous source categories together provides a broader understanding of production. However, younger sources used retrospectively or analogies from other cultures provide useful information too, but still need

NATURAL RESOURCES		HUMAN RESOURCES		HUMAN DEMANDS	
Pastures	Arable land	Labour force	Know-how	Cloth & clothes, special textiles	
Sheep/wool	Plant fibres				
WORK OPERATIONS					
Fibre preparations		**Spinning**	**Weaving**	**Finishing**	
Herding	Sowing	Plying?	Warping	Fulling?	Clipping?
Shearing	–	–	–		
Cleaning	Weeding	–	–	Dyeing?	Sewing
Harvesting			Weaving		Smoothing
Combing	Hackling				
TOOLS: ARCHAEOLOGICAL REMAINS					
Shears	Hoe	Spindles	The loom?	Dyestuff?	Scissors
–	Scythe?	Whorls	Weights		Needles
		Distaff	Beater		
					Smoothers:
					Whalebone plaques
					Glass pellets
Wool combs	Hackle				
TEXTILE REMAINS					
Gender					
Male + female	Female + male	Female	Female	Female	Female

Fig. 4.1: Schematical presentation of the chain of operations in textile production and the archaeological opportunities to follow them in grave contexts. F=Female; M=Male (Illustration: Per Bækken).

a source critical evaluation as for relevance to Viking Age and medieval conditions.

When assessing functional aspects of tools, experimental archaeology, not least, offers a rewarding trajectory to follow, making it possible to gain concrete and measurable results, and to assess the complexity of varied qualities and products, and levels of production. Anthropological studies, moreover, have demonstrated how so-called low technology and simple tools depended on solid competence and skills in all work processes to achieve high quality products. In the following, the basic steps in textile production are presented in both detail and broader outlines, serving as a backdrop for the analyses of tools and textiles in the following chapters.

Preparation of fibres

Before being spun, wool had to be cleaned to remove unwanted particles, sorted in its different qualities, and the fibres preferably stretched as parallel as possible by combing. Shears, combs, or hackles as tools for shearing and preparing wool and plants into textile fibre are generally sparsely represented in the archaeological record. Most frequently, the tools stem from burial contexts and rarely from settlements, where they turn up incidentally as lost, or discarded artefacts, damaged or left objects but probably also due to less optimal preservation conditions for metals. Additionally, written evidence, ethnological information and experimental archaeology supply useful information about their use.

Normally, the sheep shed their fleece naturally in spring – denoted as ON *rýja* – and the wool is harvested by plucking in May or June. The yield of wool could be from 1.5 and up to 1.8 kg from full-grown ewes (see Chapter 3), however, heavily reduced after sorting and combing, although this is not specified. The sheep could also be sheared as referred to in medieval Icelandic sources, specifying lambs, sheep, and herds, ON *klippa reyfi af lambi*; *klippa sauði*; *- hjörð*. The winter wool plucked by hand in spring is finer than the autumn wool grown during the summer, and finer from the abdomen than from the rest of the body.[1] Moulting has the disadvantage in leading to a loss of wool, as some sheep moult prematurely before the flocks are ready for the annual plucking, while others moult later.[2] It was therefore important to pluck at the right time to get a good wool quality and also to avoid a loss. While plucking could be done once a year when the sheep is moulting and the new coat is growing,[3] shearing is possible twice a year, both in spring and autumn (Fig. 4.2).

Today, it takes about 50 minutes to pluck a sheep if the wool is easily removed.[4] It thus represented a high seasonal labour input when having larger flocks, even if the sheep were somewhat smaller in the Viking period and Middle Ages, although not substantially so.[5]

It is often difficult to determine whether the wool is plucked or sheared based on the textiles without carrying out proper textile analyses. The wool in the prestigious textiles from the Oseberg burial, for instance, has been shown to be

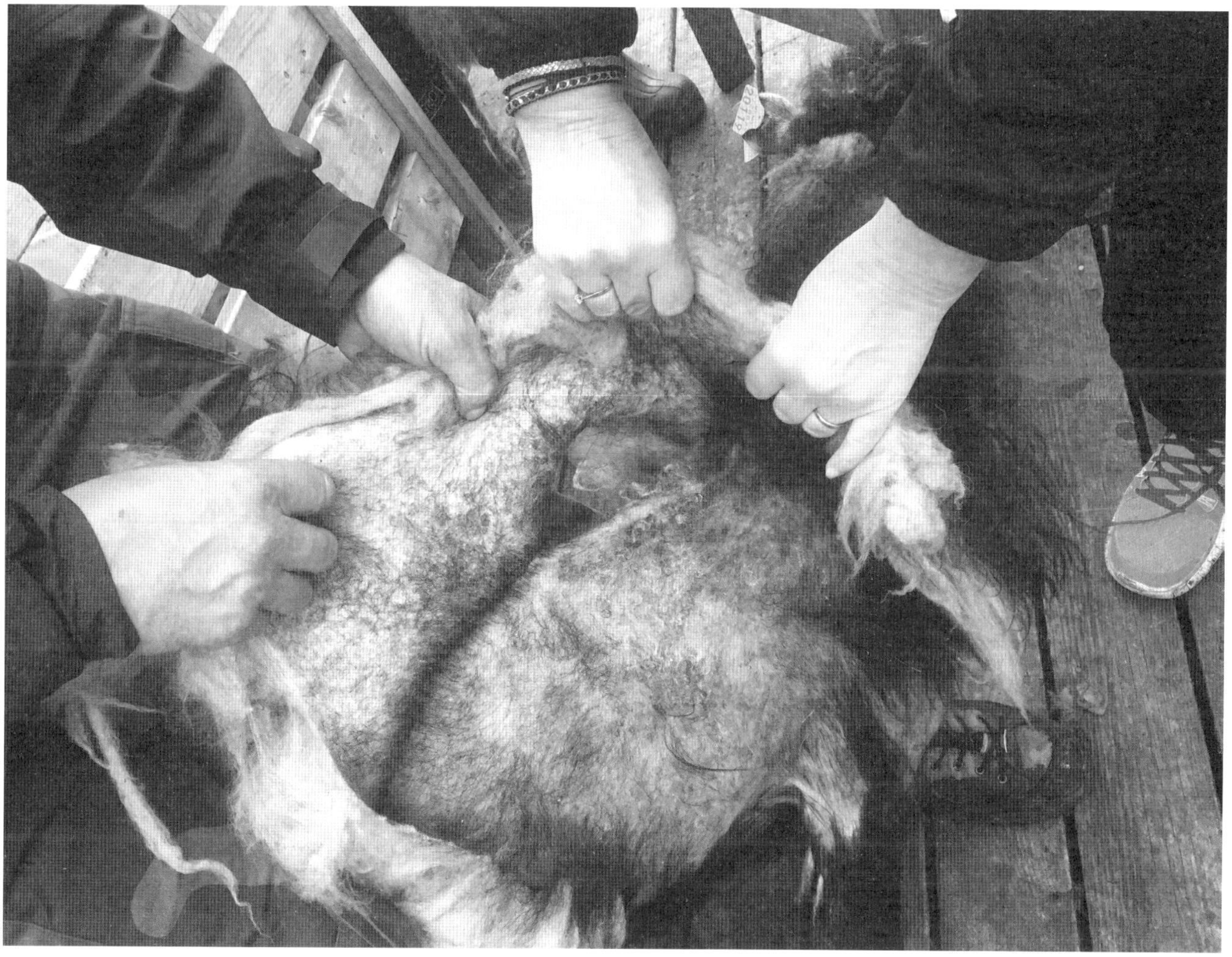

Fig. 4.2: When sheep moult and shed their wool in spring/early summer, it can be plucked and easily drawn off its body (Photo: Mons Kvamme, photo taken in early June).

clipped.[6] Ryder's categorisation of different wool qualities provides a useful analytical framework for wool analyses in this respect, differing between so-called *Hairy-Medium* and *Generalized-Medium* fleeces. Of these, *Hairy-Medium* wool is longer than the finer *Generalized-Medium* fleeces, of which the latter could be plucked. Textile remains from the Iron Age in England show a predominance of prevalence of grey Orkney and Shetland breeds.[7] However, as shown in the previous chapter, the quality of the wool varies not only as for breeds but also for gender and age – lamb, ewes, and rams, and on which part of the body it was grown. Wool from the thighs, for instance, is coarser and longer than the wool from the sides and shoulders.

Wool from each sheep contains three different parts – hair, underwool, and kemp.[8] The underwool fibres are thinner and tend to be shorter than the outer hair and can be spun separate from it and is softer than when mixed with the coarser hair. The latter can also be spun on its own and then into a hard and strong yarn for the warp or be spun with the softer underwool for the weft. The kemp, as the coarsest, has stiffer and more brittle fibres which break easily. Therefore, it was normally mixed. Generally, goat wool is coarser than wool from sheep but can also be very fine.[9] Goat hair, however, lacks crimp and to produce a thread sufficiently strong for weaving, the yarn had to be plied.[10] Strings, cords, and ropes could be made of goat hair, as well as sails.[11]

Sorting was therefore a particularly important skill for the quality of the spun yarn and the final product. The sorting implied more than fineness and the fibres were also sorted according to colour, length, and strength, depending on the use areas of the product, and whether the wool should be used as warp or weft in the weave. The outer wool collected in the summer has long traditions of being called '*varpull*'.[12] The softer underwool was better suited for the weft. The quality of the yarn was thus determined by the raw material and the experience and skills of all the agents involved in these processes both outdoors and

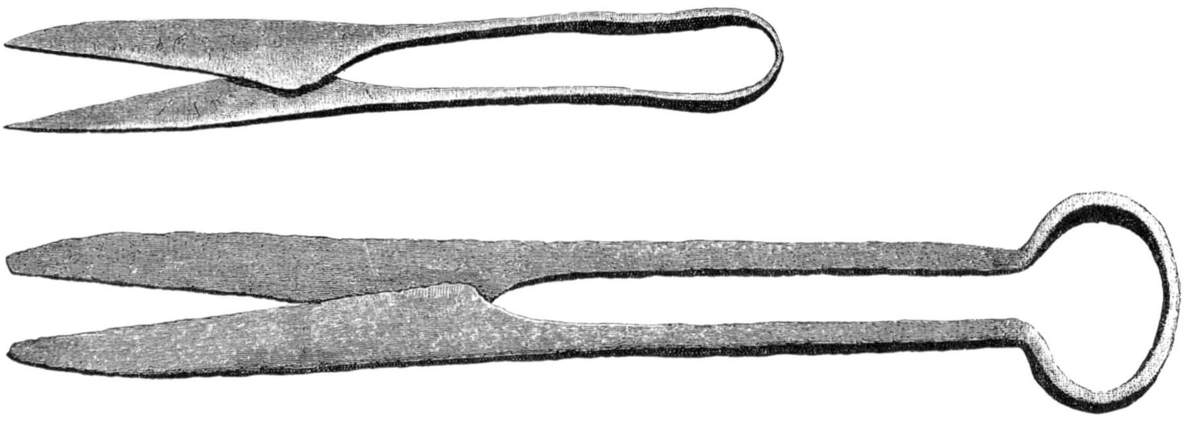

Fig. 4.3: Two types of shears for different use areas; Type R442 (above) and R443 (below) (Source: Rygh 1885).

indoors, not least in the spinning process, where the agents were generally female and called spinsters, eventually to become a marital status, and denoting an unmarried woman from the early 18th century.

The archaeological remains of textiles from the period also testify that wool was sorted according to where it had grown on the sheep and according to its use areas. A conscious sorting could thus produce various qualities. After sorting, the wool was cleaned and hand plucked (ON *rífa ull í sundr*) to remove pollutants and loosen up lumps. In these processes, different kinds of wool could be mixed, for example, different types of wool and hair of sheep and goats.[13] Wool was therefore stored in separate wool boxes (ON *ull laupr*, m.) waiting to be spun and to avoid mixing of qualities, such as for the warp and the weft, or for other defined use areas.

The only fibre preparation likely besides sorting and cleaning, was combing to improve the parallel arrangement of the fibres and separating the two layers of wool. As wool became finer with the change from *Hairy-Medium* to *Generalized-Medium* fleeces, it would have become more difficult to comb. According to Michael Ryder, the yarn produced would therefore have been woollens. Much argument has centred on the length of the staple and whether there was long enough wool suitable for making worsteds – *i.e.* wool prepared by combing. Judging by the finds of wool combs in the Viking Age and the Middle Ages, however, there is no doubt that worsted yarn was produced throughout the period, both within the study area and in other parts of northern Europe.[14] Still, the Viking Age wool combs, with the shorter tines, could be too short to comb the longest fibres.[15] Wool combs (ON *ullkambr*, m.) and combing wool (ON *kemba ullarreyfi*) also appear as common notions in Old Norse.[16]

While the processes of plucking, cleaning, and sorting are difficult to trace archaeologically, shearing and combing have left traces of tools. Shears (ON *skæri*, n.) used for clipping and cutting served multiple functions, where shearing wool of sheep was only one. Shape, size, and find contexts indicate other use areas too. Two main types occur in the archaeological material, R442 and R443, from the study area (Fig. 4.3). The former, up to 38 cm in length and known from the Late Iron Age, have the same shape as known from recent times, and therefore most likely the same function too. They were made in one piece with three functional parts – the blade or cutting edge, the handle or arm, and a loop or bow at the top which formed a spring that opened and closed the two arms or blades. The smaller type (R443) from *c.* 10 cm in length, had a circular, larger, and more distinct loop, and arms with a rounded cross section, and was used for cutting threads and textiles. Some of them have traces of rings too, which indicate that they were attached to a belt. The most common length for both types is 19–25 cm. Scissors with handles pivoted in the middle first appear in the High and Late Middle Ages in Norway as in other north European counties.[17]

The basis for the classification is the shape of the metal rod forming the springing loop at the head of the shears, and the shape of the blade as a secondary element. Types with wide blades are best suited for cutting thick fibres and for wool shearing. The length of the handle gives precise grip for firm control of the gracile blades, which may indicate a need for more accurate work.[18] The shape of the blade, particularly at the junction with the handle/arm has proved to be the most effective criterion for dating the shears, together with the shape of the spring.[19] Altogether, this enlarged loop provides a spring with greater mechanical efficiency than the former type, which is broader at the top with a U-shaped spring, where the latter gets narrower where the slightly curved blades form at the outer part. Attributes related to size and shape are therefore examined related to function when trying to connect them to various use areas: the length of the cutting edge of the blade and the length of the handle and the spring/loop.

Shears are more numerous as grave goods than many other tools, being related to both the initial and finishing

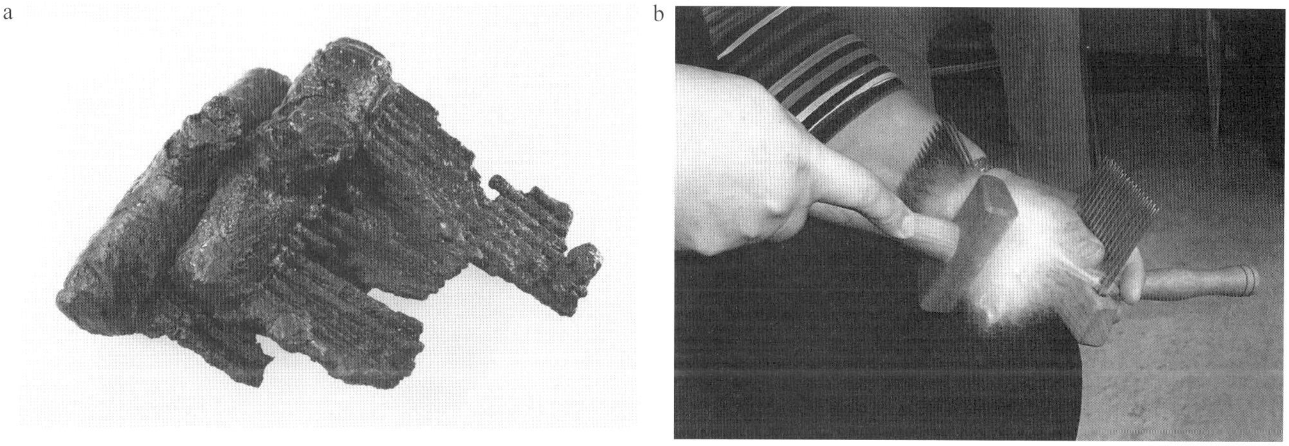

Fig. 4.4: a) Wool combs from the 10th-century grave at Hyrt, Voss (Photo © University Museum of Bergen), b) how they were used, as demonstrated at Osterøy Museum (Photo: the author).

stages of production and thus they represent multifunctional cutting implements. Shears for sheep shearing (R442) occur in both female and male contexts, however, most frequently in the former group.

Wool combs represent a sturdy type of combs for separating and aligning raw fibres to improve the parallel arrangement of the fleeces.[20] When combing, the long fibred wool was mounted parallel on the tines of one of the combs and the free comb was drawn through the wool mass beginning at the tips and working towards the tines. The fibres were automatically transferred to the free comb with each pass, when the combs were exchanged and the process repeated, and finally by drawing the fibres into a narrow strip of wool. As noted, wool combs were essential for preparing worsted fabrics. The later known hand cards for combing wool, appearing in the Middle Ages,[21] are not documented in the present study material.

The comb-like tools appear in two distinctive variants, apparently signifying different functions. However, in his study of Viking Age tools, Petersen relates them both to flax and denotes them as respectively flax combs and flax hackles. The type he refers to as comb, has later been identified as wool combs and to be used in pairs, as two small handheld tools.[22] Each comb has *c.* 20 iron tines fastened to a *c.* 24 cm long wooden piece, strengthened by an iron casing, to which a part of a handle of wood was hafted at one side in an obtuse angle, like a small rake (Fig. 4.4).

Often, and not least in western Norway, these tools, with a few exceptions, have only survived as tines, generally 7–13 cm long with a round section when not too rusty to observe, also when found in pairs.[23] Wool combs from the Viking period, or rather remains of combs as tines, appear generally in female graves. Only three wool combs from the study area have survived in pairs and with remains of the handles. All three are from prestigious 10th-century graves with composite tool kits also leaving remains of the dress – the graves from Veka and Hyrt in Hordafylki and Dale in Fjaler (see Chapters 5 and 6).

The pair of combs at Hyrt, were found with each comb fastened to a *c.* 15 cm long wooden frame with iron covers, connected to an oblique wooden handle. The 11 cm long iron tines, slightly curved inwards, were 20 in number in each when found and when later survived reduced to 15 and 18. The other two better-preserved exemplars, were nearly identical, but with varying numbers of preserved tines, from 19 to 29, and all *c.* 13 cm in length. The comb from a grave at Dale in Fjaler had a *c.* 15 cm long handle made of birch root, slightly curved and was 3.5 cm in diameter.

Recent experiments of various work processes within textile production show that sorting and combing were time-consuming tasks, producing no more than *c.* 14 g of wool ready for being spun per hour for one person according to tests.[24] Experiments related to the preparatory processes of fibres have demonstrated that this labour-intensive process of sorting, cleaning, and combing the wool also leads to a substantial reduction and loss of raw material. As for wool, the reduction amounted to as much as 22 to 50%. Combing 1.4 kg of wool could take about 50 hours.[25]

Flax (*Linum usitassimum*) and hemp (*Cannabis sativa*) were, as noted in the previous chapter, cultivated in special fields, as recorded in written sources and as field names. Both plants are characterised by their long fibres, albeit of different qualities. Flax, renowned for its soft, lustrous, and flexible qualities, was able to produce high quality fabrics. In comparison, hemp, being tensile and stronger, was able to produce strong threads for strings, ropes, and exposed textiles, such as sails, but could be used in finer textiles too.[26]

Flax and hemp, both with bast fibres in their long stalks, needed a whole series of work operations to become converted into textile fibres and usable in textile production, a process that included cultivation, weeding and a strenuous work season of harvesting, pulling the plants up by the roots when they were ripe – and water for retting. The time it takes to pull 1 ha by hand has shown to be as much as 25–30 working days for an experienced worker, if held free of weeds, and otherwise longer.[27] This also demonstrates that both weeding and harvesting have been carried out as teamwork to be able to make the ripe plants into usable textile fibres in time.

Inside the outer layer of the few millimetre-thick stems, the fibres lie in bundles, each with 10–40 single fibres, held together by pectin and lignin. The plants must therefore pass through several biological processes and technical operations to remove the wooden core and loosen the cellulose chains around of fibres – processes that were largely the same for flax and hemp. Since fibres from flax and nettles form a natural s-twist and hemp a stronger z-twist, this biological distinction is now used as an identification marker in analyses of textile remains to distinguish hemp from linen and nettles, which otherwise may be difficult to determine.[28] New methods have also made it possible to identify the even longer long-fibred hops, and now also identified as having been used in textiles,[29] but not yet within the study area.

According to later traditions, the first step after harvesting in late summer/early autumn was to dry the stems, either on poles or lying on the ground. Once dried, it was rippled to separate the seed capsules from the stalks, either by hand or using a ripple comb – a large comb, where the upper part of the stalk was pulled through to remove the capsules and seeds. To loosen the fibres, they had to be soaked in water, starting a fermentation process where the flax should rot slightly. Bundles of stalks were then submerged in the water and weighed down not to float in a pond or slowly running water to dissolve the pectin which fixes the fibres. With the help of microorganisms, bacteria decompose the plant pectin and fungi may destroy and develop the fibres. This process must therefore be monitored closely so that it does not stop too soon or too late. Depending on the weather conditions, retting could take only a few days or up to one or two weeks. An alternative method to water retting was dew retting, where bundles of flax were laid in long rows on the field and turned regularly for the microbes to work regularly on both sides. The latter method took longer: 2–5 weeks, depending on the temperature and moisture. Experiments at Ribe Viking Centre have shown that retting is altogether a very complex process requiring expertise and long-achieved experience to provide a high-quality product.[30] A few finds of retted flax dated to the Viking period have been discovered and documented in Norway too.[31] However, in Denmark sites of retting wells have been recorded in large numbers and scale during the last decade, several from the Viking period and in connection with pit houses for large-scale textile production. The largest found so far, comprised an intricate system of altogether 57 wells or waterlogged pits, and a long canal with two wells with finds of several bundles of flax in the bottom. They were situated at the border of a settlement area counting 69 pit houses or sunken-floor buildings, with traces of textile production and obviously representing an important site for specialised textile production.[32] So far, such finds have not been identified within the present study area, but must have been based on the same principles, when used in lakes and bogs.

After the fibres are loosened, they must be dried again before it is possible to break them and separate the shives from the fibres. Breaking by beating the fibres was done systematically until the wooden parts broke and started to loosen. Experiments show that the stalks had to be well retted before the process of breaking, and that breaking caused a massive loss of fibres. The experiments, however, indicate that small wooden clubs, of the type found in the Oseberg burial and interpreted for breaking flax, were rather unfit for that purpose. Still, they might have served as an additional implement after the breaking.[33]

The next step was scutching. By scraping a large knife-like wooden tool down along the fibres held over a special board, shives from the broken stalks were removed. The very last process before spinning was hackling to remove the last impurities and to straighten the fibres and smooth and separate long and short fibres and make them into bundles ready for spinning. A hackle with several long straight teeth, referred to above as one of the two comb-like tools appearing in Viking Age graves, may have served as an efficient tool. The short fibres, the so-called tow, that were stuck into the hackle could also be spun and used for lines and ropes as well as filling material in buildings and ships.[34]

Relatively few tools used for retting, rippling, breaking, scutching, and hackling are identified in the present study material but in most cases only as probable tools, when only tines are recorded. At Bryggen in Bergen, for instance, some wooden combs with a single row of large teeth are tentatively related to rippling. A hole in the comb part shows that it was connected with a handle, placed in an obtuse angle, like a rake with solid tines, where the stalks could be pulled through.[35] Another implement, interpreted for breaking fibres, stems from the Feddersen Wierde site in the Netherlands, and wooden clubs were used for beating/scutching the fibres after the retting and drying.[36] The aforementioned club, interpreted for beating flax from the Oseberg burial measured *c.* 18 cm long and made of beech wood,[37] however, showed that it was not particularly efficient for beating.

The tool used for hackling, R 433 (Fig. 4.5), earlier mixed up with wool combs (cf. Fig. 4.4a), consists of a rectangular bottom part, *c.* 15 cm long, 4 cm wide, and 2 cm high, comprising of thin iron parts, probably originally riveted to a right-angled quadripartite wooden base that has not survived archaeologically. Two rows of 14–15 straight iron

Fig. 4.5: A hackle from the Viking period (R433) (Source: JP fig. 472).

tines fastened to this frame stick out *c.* 6–7 cm above the frame.[38] According to later traditions, the tines should point away from the hackler who used both hands to guide the flax during the process. The hackle was probably secured to the surface with a peg (Fig. 4.5).

Experiments have further shown that such a hackle requires a certain amount of fineness, which determines the quality of the fibres, and ultimately the threads to be produced.[39] Most of the preparation of flax fibres happened in the autumn after harvesting and altogether a very work-intensive production, which also belonged to the luxurious segment of textiles and with special qualities used in comfortable garments close to the body, as shown in high status graves and preserved textile remains (see Chapters 5 and 6). These textile tools and textiles in themselves also imply a substantial workload and lengthy processes. The finds of hoes, often in female Viking Age graves of high status, may then reflect this laborious work of weeding, which needed management, monitoring, and supervision in long-lasting processes, including repeated weeding during the season (cf. Chapter 3; see Chapter 5).

Spinning

After all these initial and lengthy processes in various outdoor environments in the spring and summer season, the various fibres still lacked the necessary consistency and strength for weaving and had to be spun. From this stage on, however, the same tool categories were used for both plant and animal fibres.

A rod, the spindle (ON *snælda*, f.), provided with a whorl (ON *snúðr*, m. or *snáldr*, m.) of stone, clay, bone, wood, or lead, and of varying weights is used for producing different yarn qualities for varied use areas. In general, the yarn quality depended on the fibre material in question, the whorl weight and way of spinning – the spinning direction and the twist.

Cross-culturally, spinning seems to have been chiefly women's work and work that seldom carried status. However, men could also be involved. A general pattern seems to be that women spun fine threads for clothes and other fabrics, while men spun strings, cordages, and ropes needed at sea and for other kinds of work.[40] Whether this is the case in the period and area under study, is an issue to be explored in Chapters 5, 6, and 9.

Tools and techniques of spinning

When spinning, fibres are twisted into a short thread attached to the spindle – a *c.* 25–30 cm long rod, with a whorl acting as a flywheel. However, experiments at CTR show that the lighter the whorl, the shorter the rod, down to 10 cm for the lightest.[41] The twisting around its own axis makes the thread, and the weight of the whorl affects the gauge of the spinning. The spindle can hang freely to be rotated, at the same time as the spinner draws out the fibres from a distaff, which holds the fibres to be spun. Usually, the combed longhaired wool was placed on a distaff.[42] The distaff (ON *rokkr*, m.) for fastening the combed wool thus makes the arms free while spinning, according to iconographic and ethnological sources (see Fig. 4.6a–c). The distaffs could be long or short, suited for long and shorter fibres. Ethologically, they appear to be *c.* 1.5 m in length and affixed to a heavy wooden base.[43] When found in archaeological contexts, they are much smaller, appearing as *c.* 25–30 cm long wooden sticks carved with notches along the upper part. They were suited for short and finer fibres,[44] and could be held in the belt or under the arm. However, some longer sticks or staffs of iron, earlier interpreted as frying sticks, may have served as distaffs, but is currently an issue of debate (see Chapter 5). The two tools, wool combs, and distaffs, indicate advances in preparation of wool and spinning.[45]

There are two principal types of hand spindles, with either (1) a bottom-weighted 'low whorl', where the whorl is placed on the lower part of the stick, or (2) top-weighted, with a 'high whorl' placed on the upper part of the spindle. The freely rotating spindle is often bottom-weighted and seems to be the most common in both prehistoric and historic north European contexts.[46] Another way is to give the spindle support, often used against the knee, at least in the beginning of the spinning. Spinning according to this method suits top-weighted spindles,[47] as this position affects the way it was set in motion. If the whorl is placed above the middle, it is twirled against the thigh (cf. Fig. 4.6c). A high-whorl spindle is considered more efficient in terms of the length of time the spindle keeps turning.[48]

a b c

Fig. 4.6: a) Spinning with a low-weighted whorl and with the distaff between the knees (part of a roof painting, Ål stave church, c. 1260–1300), b) motif from Setesdal, with the distaff under arm, c) with a top-weighted whorl and sitting with the distaff between the knees (Painting by Lars Osa, c. *1900).*

In Norway, the whorl could be placed both on the upper and lower part, and often in the upper part, as has been documented in the Oseberg burial, found in a chest with various textile tools. This was an intact and complete cylindrical spindle of birch, 29.3 cm and 0.9 cm in diameter at the thickest part near the middle, and the spindle whorl of schist placed near the top of the spindle with a small notch, 5 mm long, in the upper end, to stop the thread from sliding off. The chest contained other spindles too, varying from 28.5 cm to 37.3 cm in length; one of them had a little hook of iron, and the others had traces of similar devices.[49] Old Norse terminology, moreover, reflects top-weighted spindles, when denoting the lower and longest part of the rod as *snælduhali*, f., where the thread wound onto when the spindle reached the ground, contrary to the low whorl. Both methods seem to have been used over time, as demonstrated by a medieval church painting from *c.* 1300 and later illustrations from the late 19th century (Fig. 4.6a–c).

Spindles in ethnological museum collections in Norway have 'high' whorls. In some areas of the western part of the country, this method was still in use as late as in the 1920s.[50] Similar top-weighted spindles from the 19th century are represented in the National Museum in the Faroe Islands (Fig. 4.7).

The thread is made into a ball, or wound up on a reel, or onto a weft bobbin. Although few are preserved, archaeological prehistoric remains of thread bundles show that they were measured and wound into a ball and onto skeins (ON *hespa, þráðariðior*) and smaller units (ON *vinda*) for weft threads.[51]

The fibre length links fleece type and the spinning. The skill of hand spinning lies in drafting – easing the wool needed for the gauge – and twisting in one balanced process. A problem with short-stapled fleece is to achieve sufficient twist to impart strength before the weight of the falling spindle and the spindle whorl create too much draft that breaks the strand of the fibres. Generally, a free-hanging spindle demands a strong and long-fibred material because the gravity acting on the hanging weight of the spindle stretches the strand. The method of the top-weighted spindle with support during the first part of the spinning, does not strain the fibres and is suitable for short-fibred wool and flax, and is preferred in the weft in woollen weaves.[52] When spinning fine and weak threads, the spindle can be supported, moreover, by touching a base, *e.g.* a floor or a bowl.[53] A tight thread needs extra spin to increase the twists per centimetre compared to a loose spun thread. As for flax, the long fibres after the retting and hackling become smooth but also slippery and more difficult to spin than wool. However, it could be spun into extremely fine thread of considerable strength.[54]

The weight of the whorls may thus indicate spinning of different kinds of fleece and qualities. According to Ryder, heavier whorls could spin the longer wools, while *Hairy-Medium* and lighter whorls the shorter wools. However, given only one fleece type, spindle whorls of different weights could spin different thicknesses.[55]

The spinning direction affects the quality too. However, the direction was not dependent on the tool but on the rotational direction the spinner gave the spindle, regardless

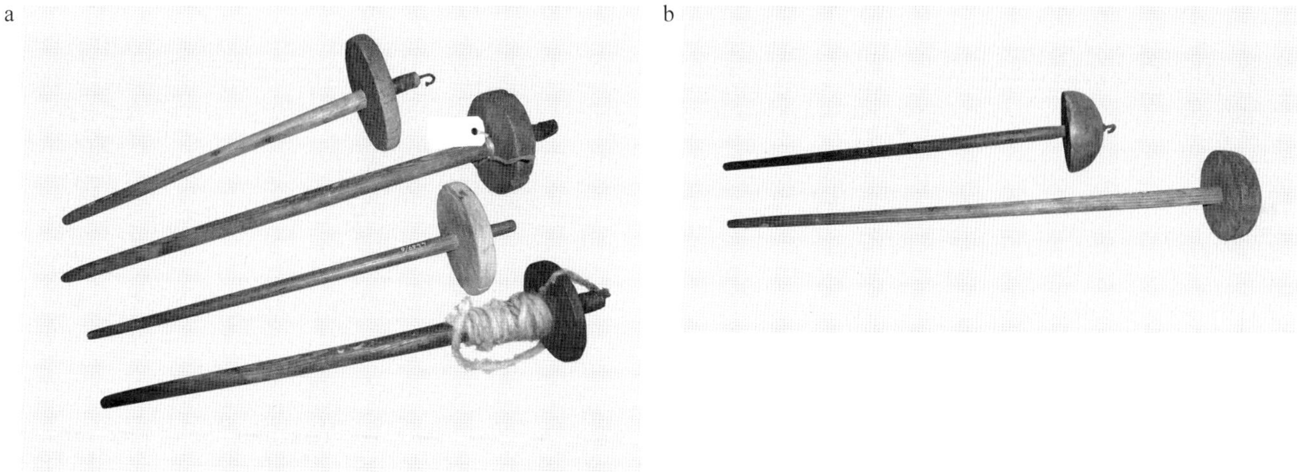

Fig. 4.7: Spindles with whorls from the Faroe Islands: (a) of wood, (b) of stone (Photos: the author).

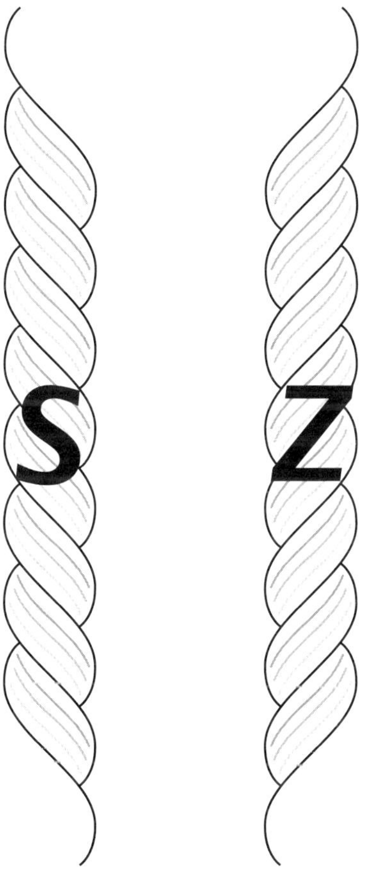

Fig. 4.8: Schematic drawing of z- and s-spinning (right- and left-turned) (Illustration: Per Bækken).

of the type of whorl used. The twist can be to the right, clockwise, so-called z-spinning, or to the left, s-spinning, depending on the quality and strength of the yarn, and according to the intended weave. The natural and easiest way is to rotate the spindle clockwise, giving a z-spun thread with a stronger twist (Fig. 4.8). Hence, the z-spun thread fits well in the warp, as especially strained in the warp-weighted loom. An overwhelming use of z-spinning is documented in the Scandinavian textile material from the Early Iron Age onwards, but from the Middle Ages, s-spinning can also be identified, and seemingly, then, as an intentional preference. Spinning with a high whorl, normally starts by rolling the spindle shaft downwards on the thigh with the hand. The thread will then be s-spun when being right-handed. Conversely, with a 'low-whorl' spindle, starting by flicking the spindle with the hand to the right will result in a z-spun thread.[56]

When plying and two threads are twisted, they can go to the same direction or the opposite, which gives a stronger twist and yarn of greater strength. In general, the yarn is dependent on the material used, the way of spinning, the twist, and the amount of plied yarn.[57] The spinning direction is a deliberate choice independent of the tool. Whether a result of cultural traditions, or deliberately chosen to achieve different qualities is a matter of debate, followed up when assessing the textile remains and their qualities (see Chapter 6).

The spun threads or yarn (ON *garn*, n.) could denote one or twisted two-threaded yarn and warp threads, being often connected with stronger threads and therefore used in the warp, and z-spun of the longhaired strongest summer wool.[58]

Spindle whorls as indicators of textile qualities

In all the area of the present study, spindle whorls dominate among the finds of textile tools and are made of more durable material than the wooden spindle rods, which are rare finds from the study area. In the North Atlantic, spindle whorls are rather sparsely represented in graves with only a few from Iceland,[59] but in contrast, are rather common finds in settlement contexts in the North Atlantic – Iceland, the Faroe Islands, Shetland, and the Orkneys.

Historically, whorls of *c.* 25–35 g were commonly used to spin the wool from the Norwegian short-tailed sheep, and whorls of 50 g for plying the yarn.[60] The varying diameter indirectly indicates varying weights. Experiments at CTR have demonstrated how variations in the diameter and weight of the whorls affect various types of threads. They show that whorls in the common range of 25 g to 35 g in fact could be used to produce a variety of thread gauges, dependent on the dexterity of the spinner, while light weight whorls of *c.* 7 g to 16 g could not readily be used to produce thick threads. The size and shape of the spindle whorls also affected how tightly it could spin. The diameter of the whorl would determine the peripheral speed of its outer edge, which in turn determines the degree of twist imparted. A large spindle whorl does not spin as tight a thread as a small and relatively heavy one, and a conical spindle whorl turns quicker than a disc-shaped one,[61] and so on. The lightest whorls also needed another type of wool – softer and more homogeneous than when spinning with heavier whorls. A spindle whorl of standard type, would have aided the process of achieving a consistent quality for a particular thread type: a limited range of shapes and sizes, and little decorative elaboration would suggest standardised tools for bulk production. To ply two threads into one, it was advantageous to use the power of heavier whorls, 40 g or more, as they then run well after one twist.[62] Altogether, experiments have demonstrated that the greatest significance for the product lies in the weight and its relation to the diameter of the whorl, deciding whether the thread can be thick or thin, and tightly or loosely spun.

The distribution of the different function-related weight groups across the sites has, then, implications for the variation in production activities. Clustering around certain weight groups should thus be a good indication of the type of yarn that was spun, where the lighter was better suited for spinning finer fibres and where a narrower weight range means that they had a more delimited purpose. Detailed studies of these seemingly unimpressive spinning tools may in this way provide information of wider significance, and thus enable the assessment of the level of production from various sites and areas.

Petersen's national overview of spinning implements includes *c.* 430 spindle whorls from 328 burials and sites from the Late Iron Age, AD 500–1000. Altogether 97, or 23% of his total assemblage, stems from the present study area in western Norway[63] and the finds have multiplied since then (see Chapter 5). Although far from updated, his overview still provides a good comparative framework regarding shape and size but unfortunately is only limited to diameter for size and there are no references for weight when trying to compare them with whorls from the study area. Petersen finds that the diameter is normally 30–40 mm, but that smaller and wider also occur, down to 20 mm and up to 52 mm.[64]

According to Petersen's outline, 90% of the spindle whorls from the study area were made of stone, and of these, nearly two-thirds of soapstone, about one-fourth of other minerals, sandstone, schist, and other unidentified minerals, and only a few of fired clay and lead. Distinct regional patterns in the choice of raw materials appear, and depend on access to raw material, but may also reflect differences in preservation and in degree of specialisation or standardisation in proto-urban and urban environments. This is an issue further analysed and discussed in Chapter 8, related to rural and proto-urban contexts and in Chapter 9 including both rural medieval working environments and in Bergen, as the only town in the study area of western Norway and the North Atlantic.

The present analyses of whorls consider both functional and morphological attributes – but overall, weight. The shapes, stereotyped in Figure 4.9, include types represented in the study area too, however, originally based on types from medieval Bergen, and represented mostly in stone, but also metals (lead), fired clay and ceramics, bone, and wood.

The most common types are hemispherical with a flat base, and more or less vaulted on the top side (type A), but some also have a rounded part to the base (type B). Both types comprise altogether about 300 specimens in Petersen's corpus of Viking Age whorls from Norway.[65] Flat types (types E and F) are scarcer, and even more seldom types are formed as a flattened ball (type G), or with conical and biconical shape types (types C and D).

As for functional characteristics, weight, diameter, height, and size of the spindle hole have been measured as means of determining the nature of whorl rotation and varieties of fibre spun. Since weight and diameter are key factors for limiting the type of fibres that could spin, they may indicate the quality of the thread produced. The hole in the whorl also tells the thickness and size of the spindle. When examining tools contextually as cases in the following analyses, histograms and diagrams are largely used to elucidate the nature of the distribution of such features, however,

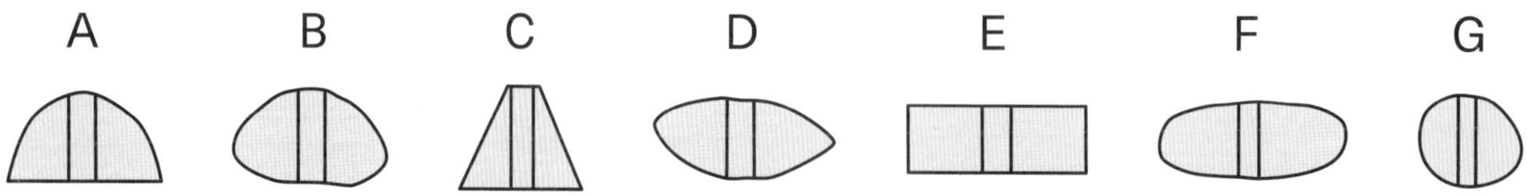

Fig. 4.9: Types of spindle whorls represented in the study material, according to types, defined by Øye 1988.

with a focus on weights able to produce different types of yarns. Of these, the medium weight group is well suited for fine and multipurpose spinning, heavier spindle whorls are suited to make coarser and stronger yarn, while the very light whorls tend to produce the thinnest and finest threads. Various weight patterns thus provide crucial information about production, and not least when found as selected and intentional grave gifts in funeral contexts, and in larger quantities in direct use contexts at settlements and working environments. Since an overarching aim is to assess technological aspects of production as closely integrated in the social, socio-economic, and cultural spheres of production and on a comparative macro level, the focus is directed on spindle whorl weights and their ability to produce textiles in different social and geographic settings.

Tools, threads, and time

Experimental research in recent years has successfully illuminated the issue of how spindle whorls affect the yarn quality and how time consuming hand spinning could be – related to raw materials, both wool and plant fibres, and the spinning equipment. They show that the time taken to convert fleece into usable yarn depended on the quality of the wool and plant fibres, the weight of the spindle whorls, and the dexterity of the spinner. Spinning with very light spindle whorls (4 and 8 g) needed higher concentration than using one of medium weight (18 g) and it was considerably more time consuming to use a light whorl than a heavy one. When using a light whorl, noticeably fewer fibres are used but could produce far longer threads from the same amount of wool than with a heavier whorl. In consequence, the lighter the whorl and the spindle, the less fibres per centimetre and thinner and longer the thread – in contrast, the heavier the spindle whorl, the shorter and thicker the thread. The experiments also showed that it took longer to spin long fibres of linen than of wool.[66]

Of the various whorls in different weight categories in the experiments – a very light (4 g), light (8 g) and medium (18 g) weight – the lightest gave an average output of 35 m of yarn per hour, 40 m for the light, and 50 m of yarn for the medium weight type, which represents a difference of 30% longer threads per hour between the heaviest and lightest spindle whorls. The experiments further showed that spinning and winding could be carried out in between other tasks when using a spindle with a spindle whorl of different material, size, and weight to produce threads of different gauges. However, the lightest could not be combined with multitasking.[67] The experiments further demonstrate that a huge amount of yarn was needed in the production and that only 1 m² of cloth with equal density in the warp and the weft and 10 threads per centimetre required as much as 2 km of yarn.

Table 4.1: Scandinavian whorl weights related to fineness of threads (Source: Andersson 1999, 24).

Weight of the whorl	*Threads per cm of woven cloth*
5 grams	27–37
10 grams	22–30
20 grams	17–22
30 grams	2–15

Various whorls of varied weight have been tested out to measure how fine or coarse the threads they produce could be (Table 4.1), however, different wool qualities may provide other results.[68] These insights are crucial when assessing the tools found in contexts with textiles, mainly in the graves, often of different size and weight signalling production of different qualities and for varied use areas (see Chapters 5 and 6).

The largest enterprise of all in this period was the production of sails – and a huge and wide-ranging production considering all we know about seafaring in both calm waters and open sea in this period. Experiments related to making a sail, based on the few remains left of sails from the period, show that some of the heavier whorls of respectively 38 and 40 g would give coarse yarn, which coincided with the thread count in one of the few remains of medieval sails with 5–6 threads/cm. Using a whorl of 24 g to test out qualities needed for other operations, for instance for sewing the various lengths of the sail together, gave a finer quality with a higher thread count. A whorl of 12 g gave quite another quality again, not suitable in any elements within sail production.[69]

Making a sail undoubtedly represented a large-scale enterprise, how large depended, however, on the size of the boat or ship and consequently the size of the sail. For seafaring in the wave of colonisation of islands in the North Atlantic, with large crews, passengers, cargoes, and even animals, a large ship type like the Gokstad ship must have been necessary.[70] Such ships would have needed sails of 90–100 m² or even larger, and production must have been organised as some kind of collective or even communal enterprise, perhaps not unlike providing the so-called ON *leiðangrsskip*, n., called for as an obligatory nationwide duty ON *leiðangr*, m. by the 10th century. As noted, such duties or obligations may have been organised by local 'chieftains' at an earlier stage, as when crossing the North Sea (cf. Chapter 1). Experiments based on remains of a few Viking Age and medieval sail cloth, show that only spinning and weaving for the largest ships represented more than four years of full-day work for one person when producing a large sail of 90 m². The experiment indicates that spinning alone would have taken more than a year, 385 days, when spinning lengths for threads per cm with a spindle whorl of >30 g.[71] Additionally came all the other preparatory processes, referred to previously.

The remains of boats used as a necropolis on quite a few high-status graves from the study area, both from western

Norway and in the North Atlantic, varied in size but were generally small, around 5–6 m long, and of the type used for five and six pairs of ores and with considerably smaller sails. In these cases, however, the labour input for making sails for both daily use for the household's own needs and when as larger enterprises, clearly extended the household level and in practise an undertaking for more than one household. The experiments thus form a solid and encouraging basis when assessing all these 'small things' as integrated elements in a large-scale production, involving human resources in the many steps of production.

Dyeing

The next stage after spinning could be dyeing, a process, however, above the ordinary and generally restricted to yarn for special garments, largely belonging to the upper social strata. Ordinary fabrics in most cases, were supposedly in their natural colours but could be dyed as a social distinguishing feature. Generally, the yarn was dyed before weaving, which enabled small amounts of the wool or linen yarn to be dyed separately, requiring less effort and capital outlay on equipment than piece-dyeing.[72]

The natural colours of linen were grey (ON *grár linvefr*) and white after bleaching (ON *bleikt lérept*). The natural colours of wool were white (ON *hvítt vaðmal*) or grey (ON *grátt*), in various shades, denoted as *járngrátt* (iron-grey) and *margrátt* (ocean-grey), or it could be brown-red (ON *mórautt*, *móbrunt vaðmál*) or black. The natural colours of the fleeces could still be used to make patterns, especially in diagonal 2/1 twills. However, if cloth was described as blue, *e.g.* ON *bláfeld*, it had been dyed, although *blá* in some contexts could refer to dark /black, such as *e.g. blámaðr*. Green, yellow, and red textiles, also mentioned, are coloured cloth too referred to under the common name ON *litklæði*.[73] However, such cloth could also have been imported.

In some cases, traces of various dyes are identified in the preserved textile through chemical analyses, mostly as vegetal dyestuffs – various types of available moss, lichen, plants, and fruits, and in contemporary sources denoted as ON *litgrös* and *litmosi*. One of the dye plants was woad (*Isatis tinctoria*) – a biennial plant used for dyeing yarn of both wool and linen to be blue. In the second year, the plant blooms with yellow flowers and can be up to 2 m high, and indigotin was extracted from the stalks and leaves. Seeds of woad found in a small casket among the grave goods and textile equipment in the prestigious Oseberg grave,[74] may indicate cultivation at an early stage. Blue dyed textiles also appear in high status graves from the study area, both in western Norway and in Iceland (see Chapter 6). Today, woad seems to have natural habitat in northern Norway (Fig. 4.10).

Madder (ON *maðra*) (*Rubia tinctorum*) and bedstraw (*Galium boreale*) giving colours in reds, orange and yellow, are also known from Norway, and walnut (*Jugland regia*) for producing brown colours, while *Kermes* that gave a red colour too, had to be imported.[75] A plant denoted ON *jafni* (*Lycopodium alpinum*) is mentioned in Icelandic medieval

Fig. 4.10: Wood plants for dyeing textiles blue.

sources and later known for colouring yellow, as birch leaves do too. However, yellow seems altogether to have been less used than blue, green, and red. In Iceland, archaeological finds of lichen, *Pamelia omphalodod*, have been identified in preserved textiles in two shades of brown colours.[76] That these various ingredients of moss and plants are mentioned in medieval sources and appear as common place names show that they have been commonly used for dyeing.[77]

Bleaching with sulphur was used to get natural coloured wool pure white[78] while fabrics of linen traditionally needed no other remedy than about eight days of bleaching the textiles in the sun or snow to become properly white on both sides. Besides, the multifunctional sulphur was used as a pigment component in a bright red colorant known as vermilion, and could be achieved by mixing moss and sulphur, which has been common in Iceland and used to dye cloth too. Being a major medieval export article from Iceland to, among others, Bergen,[79] sulphur may well also have been used for dyeing in wider areas.

Painting on textiles after finishing is a technique denoted as ON *steina*, especially used on tapestries, referred to as ON *steint tjald.* This is a technique known from the study area, such as from a medieval church antependium from Sogndal in Sygnafylki, and then for quite another use area than for dyeing textiles.[80]

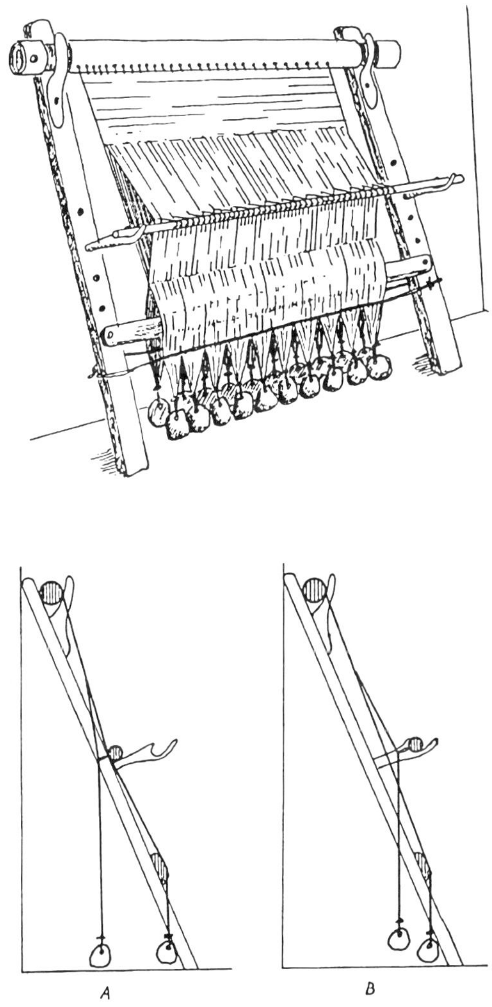

Fig. 4.11: Schematic drawing of the warp-weighted loom in front, and the two cross sections below illustrate the system of changing the natural shed (left) to the counter-shed (right) (Source: Hoffmann 1964).

Weaving

Weaving (ON *vefa*) is the term to describe the process of producing fabric by interlacing warp and weft threads on a loom. The shed is the third requirement – a temporary opening – dividing the warp threads into two planes, between which the weft can pass. The basic requirements to make a weave, then, are the warp, the weft, and the shed, produced not only on the upright warp-weighted loom, but on band looms and as tablet weaving too, primarily for decoration, as bands, belts and the like.

Tools and techniques

The loom that was used in the area and period under study was the upright loom where the warp (ON *varp*, n.) is stretched by loom weights (Fig. 4.11), having given the loom its present name. In the Viking period and the Middle Ages being the main type of loom, it was simply called ON *vefstaðr*, m. or *vefr*, m. – *the* loom. In Iceland, the Faroe Islands, and Shetland, as well as in some rural areas of western Norway, the warp-weighted loom was in use longer than in most parts of northern Europe. In western Norway, it was in fact still in use in some rural areas until the middle of the last century, especially for weaving tapestries.[81] The late use also helps to understand the predecessors.

The warp-weighted loom consists of two slanting uprights (ON *hleinar* pl.), one upper beam (ON *rifr*, m.), and two rows of weights. On the historical known warp-weighted looms from western Norway, the two upright beams were about 2.2 m high and carried a 2 to 3 m long revolving beam with rows of holes (Fig. 4. 11).

The warp – the vertical hanging warp threads and fixed element in the loom and the longitudinal component in a length of a fabric – was kept tight by the two rows of warp-weights (ON *klé*, m. or *klár*, m.) to give tension to the two layers of warp-threads. The shed rod (ON *skapt*), fastened to the two uprights, separates them, and provides a natural shed that is permanently open on the warp-weighted loom. The artificial shed is opened by placing the movable heddle rod in the notches of the holding brackets (ON *yllir*, m.), fastened to the uprights. To divide the threads into different layers, the threads are heddled onto the heddle rods. Tying the warp threads to the heddles must be done carefully to avoid mistakes, and two women therefore often worked together during the warping.

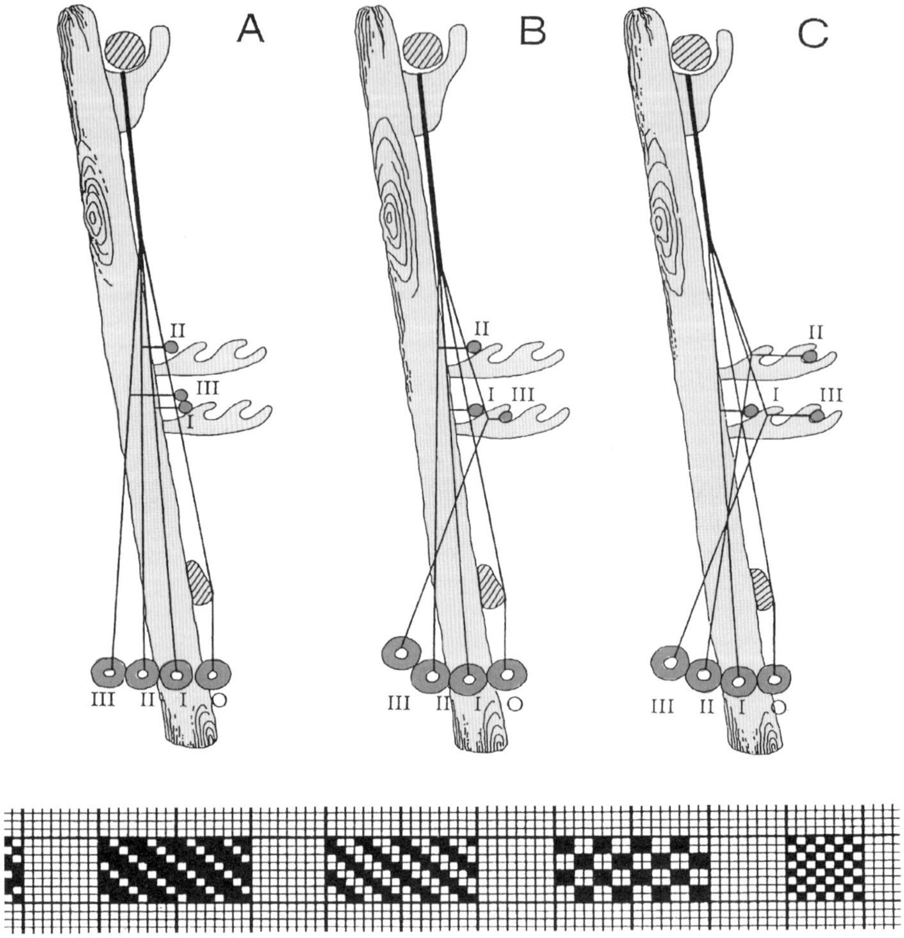

Fig. 4.12: (A) Warp-weighted loom with four rows of loom weights, rest position, (B) Shed 2: heddle rod III placed in the inner notch, (C) Heddle rods III and II placed in the outer notches. Below: weaves that can be woven with the same set up (left to right): tabby, basket weave, 2/2 twills, 3/1 twill, pattern effect (Source: Batzer and Dokkedal 1992, by permission).

The weft threads (ON *veftr*, m.), and the transverse element in the textile are usually parallel to each other and to the upper and lower edges of the textile. These weft threads cross over and under the warp threads, usually at right angles to them.[82] For 2/2 twill qualities there are three heddles, which open the different sheds, plus a fourth natural shed. Each warp thread is tied to one of the three heddles. It is then possible to lift each set of warp threads individually through the shed and pass the weft through it. For 2/2 twill, each weft goes under two warp threads, moving to the right, which creates a diagonal pattern (Fig. 4.12).

This implies that the fabric had to be planned long before warping, as for yarn qualities in the warp and weft and enough as for lengths of the yarn and size of the fabric, and as for patterns.

Finds of double-notched rod-brackets in early medieval urban humid layers from Trondheim, dated to AD 1120–1200, indicate how an extra notch enabled altogether four rows of loom weights, and with the heddle rods in different positions on the inner and outer notches could produce different qualities, from tabby in a rest position, to more advanced weaving techniques able to produce 2/1 qualities on the warp-weighted loom that have earlier been attributed to the horizontal loom.

The separable parts of the warp-weighted loom mean that it could be mounted and set up temporarily in a domestic setting, where weaving was a seasonal or intermittent activity – and a procedure referred to in the provincial law of Frostathing,[83] the law district north of the Gulathing law area.

Warping

The warping depended on the use area of the fabric in demand, as it affected the quality and length, whether it should be loose or tight, coarse, or fine. Hence, the first step before warping was to decide the length, as the length of warp threads would equal the length of the cloth. Furthermore, width and quality, the fineness and density of the threads of the fabric, affected the number of threads of consistent quality when calculating the amount of yarn needed – how much weight that was required per warp thread, and how many threads one would tie to each loom weight, which also depended on the fineness of the yarn. The calculation of weight tension leans on the simple arithmetic principle: the weight of the stone, minus weight required per thread = number of threads per stone.[84] If for example, the yarn needed a tension of 20 g per warp thread, and the loom weight weighed 500 g, the weaver attaches 25 warp threads to this loom weight.[85] Thicker threads and coarser fabrics could take heavier stones, 25–75 g per thread for various fabrics for furnishing and 100–125 g for 2-plyed yarns and rugs, and 125–150 g for 3-plyed yarns and tapestries. The weights also need to be in pairs of equal weight and have a loop of string in each stone of equal weight to give an even tension and be placed so that the selvedges of the fabric would keep an even width. If the pair of loom weights are not roughly of the same weight, the number of threads to each pair must be calculated individually.[86] The warp-threads were tied to the loom weights with fitting weights and size. The complete warping process was thus based on detailed logistics and assessment of qualities for producing weaves of various qualities, and the finer, the more complex and larger amounts of equal quality yarn. Hence, the warping is best reflected archeologically and indirectly assessed in the finished textiles and not least by experimental weaving – as for thread count, pattern, and quality of warp-threads as for fineness and strength, and their twist and spinning direction.

In-situ finds of loom weights in position, and in some cases with remains of strings connecting to bundles of warp threads to the weights, reflect the warping. Historically, a tablet-woven band was tied to the upper beam and was used to arrange the threads in a regular row, forming a starter border. Sometimes starting borders appear on the preserved textiles too. On the historically known looms preserved in museums in Norway, the weights are attached to the warp by means of a looped cord through the hole in the weight and never directly in the hole.[87] The loom weights could thus vary in number and weight according to fineness of yarn and tension per weight, which in practice must have been calculated by hand by an experienced weaver before warping the loom. In fact, instrumental weighing could also have been used, as advanced weighing systems based on units such as *mark* and *pund* existed by the period.[88]

The number of loom weights on historically documented looms varies considerably, from *c.* 13 to 48, mostly between 20 and 40[89] – a number that in many cases concurs with archaeological *in-situ* finds of loom weights. When found *in situ*, they may also vary in weight and accordingly have been subject to calculations related to weight tension on the warp threads. Since different types of yarn needed different tension, it limited how many warp threads could be attached to one loom weight. Hanging side by side, the thickness of the weights is also important as it defines how closely they could hang. In principle, then, loom weights of different weights could be used but had to be calculated as for tension and even distribution among the warp threads[90] and according to the intended qualities.

Generally, the heaviest weights were for the coarsest qualities and the lightest for the finest. Experiments have shown that coarse warp threads for weaving a sail could need as much as 50 g weight per thread. If the warp weight weighed 500 g, 10 threads would then be enough. In contrast, if the thread is finer, it would need only 20 g of tension and 25 threads could be attached to the same loom weight, however, preferably of the same thickness to achieve regular edges.[91]

In experiments and calculations related to production of a large sail, the sail had to be woven in 12 lengths of 1 m, and accordingly the same lengths of warp threads. Each set up to prepare for weaving took two days for two weavers, and in this case, each loom weight of 500 g and with 10 threads per weight, a tension of 50 g.[92] Other experiments have shown that the mounting of the loom with warp threads required 24 hours before the weaving could commence.[93] The time varied, then, in accordance with the textiles to be produced and was more time consuming for more complicated patterns with different sheds[94] but also for weaving of sails.[95] The experiments thus clearly demonstrate the complexity and competency needed in so-called low technology, as Costin also points out as a general observation.

Weaving

After the mounting and warping, the weaving starts by inserting the weft between the warp-threads and after each shift the weft is beaten upwards (ON *slá vef*) by a sword-shaped beater made of wood, bone, or iron to make the fabric tighter. As the loom was placed at an angle, the back warps hang down vertically over the shed rod, creating a natural shed. This prevents the need for heddles on the front threads in plain tabby weave and facilitates some twill weaves, such as plain 2/2 twill and the more complex twills too (cf. Fig. 4.11), such as diamond and lozenge-patterned twill (see Chapter 6). As the weave increased in length, it was wound onto the upper cloth beam (ON *vinda vef*) and the warp further stretched out and elongated until the threads had reached their end (cf. Fig. 4.11).

The experiments, moreover, show that weight and thickness of loom weights were key functional factors, as the weight determines how many threads it could carry to keep it taut and give the needed weight tension without breaking the threads. Generally, more than 30 warp threads per loom weight could cause problems during the setup and the weaving and make it difficult to weave evenly,[96] and keep the warp threads in position. The thickness of the weights and the space they take, then, affect the quality and evenness of the fabric, which should match the width of the fabric and the edges of the selvedge. Another result is that weights of around 300 g and *c.* 3 cm in thickness could produce a greater range of qualities with different

Fig. 4.13: Loom weights of soapstone in a grave at Tokvam, Aurland, Sygnafylki (B 8869) (N:27 of which 25 complete) (Photo: the author).

thread patterns than weights that are significantly heavier or lighter.[97] Generally, light and thin loom weights produce fine cloth, while light and thick weights produce more open but also fine cloth. Heavy and thin weights, again, produce coarser, closed fabrics, while heavy and thick weights give a more open but a coarse fabric.[98] Experiments also show that light weights[99] were generally suitable for weaving fabrics with thin warp-threads, and heavier ones for fabrics with a stronger warp, thus giving the needed weight tension on the threads. Loom weights may therefore indirectly reflect the quality of the textiles. These factors combined, suggest that the loom weights should equal each other as far as possible. However, two loom weights could weigh up for one heavy and the number of threads adjusted to the weight to get an even tension.[100] The number of threads again depended on how thin or thick the yarn was and how loose or tight the fabric should be. Thin threads spun on the lightest spindle whorls (4 g) thus required lighter loom weights than threads spun with whorls of 8 g and 18 g,[101] and the threads would have functioned better in a more open tabby between the warp or in the weft. Altogether, the connection between spinning and weaving is important when assessing the textiles, and loom weights and spindle whorls from the same contexts.

Archaeological remains of the loom

Loom weights are common finds from Viking Age and medieval settlement sites but more limited spatially than the movable spinning tools as the loom was stationary. However, as grave goods they represent the loom only symbolically. Petersen's national survey updated to the late 1940s refers to altogether 449 loom weights from 82 sites/graves, which means that many graves contained more than one weight but usually only a few. Two-thirds of these finds contained 1–3 weights, and five graves seem to have represented all weights in a loom, from 20 to 28,[102] as a loom seems to have required around 15–20 weights, perhaps more.[103] The largest numbers stem from two female graves from the present study area, with 28 and 48 weights, however, of various shapes and weights (Fig. 4.13; see Chapters 5 and 6).

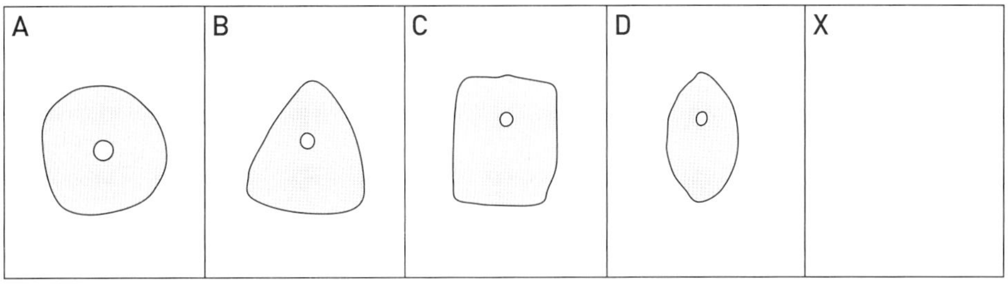

Fig. 4.14: Shape of loom weights – A–D and X of varied shapes – in the three comparative assemblages, Viking Age graves from western Norway, the Viking Age site Kaupang, and the medieval site Bryggen in Bergen (Source: the author).

The weights could be made of different material and shape, where weight seems to be the most important functionally. Some weights, especially, when made of fired clay, could have special stamps and standardised shapes and size, while loom weights of stone varied more but could have incised crosses and may have functioned as indicators of specific weights.[104]

The loom weights from western Norway are mostly of soapstone, but also of other minerals. They can be classified in five main categories as for shape, based on my earlier classification[105] – round and with a central hole (type A); with gently sloping sides (type B); with nearly parallel sides (type C); oval shape (type D). Atypical or unidentifiable weights, denoted as type X, include rough asymmetrical shapes with little or no elaboration except the hole. While type A has a central hole, the holes in types B, C, and D are placed in the upper part of the stone.[106]

It was easier to make standardised weights when made of clay, however, this is not represented in this area of study. The choice of material also varies – natural stone such as soapstone, slate, and other easily workable minerals, dried or fired clay, etc., that altogether may reflect local access or shortage of a given material. The materials do not seem to be transported over long distances. In the North Atlantic, however, weights of soapstone seem to be imported or have followed the colonisers as important tools in their 'integral package of necessities'.

The length of the weights from the Viking period normally varies from 8 to 13 cm, most commonly between 9 and 11 cm while the width varies from 5–8 cm, normally 6–7 cm according to Petersen. He is, however, ambiguous as for thickness, but probably referred to when mentioning 3–4 cm. The largest ones could, however, be up to 15 cm in length and the smallest *c.* 6 cm with the largest width up to 11 cm. In several cases, loom weights have different types of incised marks, often a cross or a circle or a v-figure.[107] However, Petersen leaves no information about weight. My own studies show that they vary between 200 and 900 g, and generally between 200–600 g, but vary in different contexts. Still, the loom weights from the study do not appear as standardised (cf. Fig. 4.13).

In settlement contexts, heaps of loom weights indicate the loom in its working environments – in best cases, the width of the loom indicated by the loom weights, its place and working space, light and working position, and relation to other textile tools. Clusters of weights therefore have a special focus, making it possible to assess qualitative, quantitative as well as spatial dimensions of the loom, functions and working environments. Here, histograms and diagrams also reveal weight patterns and variations within production in time and space. However, often the weights left behind are too few and damaged to discover patterns.

Loom weights from the North Atlantic found at sites in Iceland, the Faroe Islands, and Shetland vary in numbers, from a few up to more than 50 weights. In Iceland, they are often made of unworked stones, water-born basalt pebbles with natural perforations used for the warp threads. However, imported soapstone, probably from Norway, seems to dominate in the 10th century and well into the 11th century along with local material. The loom weights vary in weight at the various sites, from *c.* 140 g to more than a kilo. However, when found in concentrations, they are often of a smaller weight range between *c.* 300 to 600 g (see Chapter 9).

Finds of loom weights are less common in the archaeological record from the Faroe Islands but those found are made of local stone, tuff, and basalt, as well as imported soapstone. The finds from Shetland show a domestically produced technology, which included weights for upright looms made from soapstone of the same types as found in Norway but probably of local steatite, as this is the only area in the North Atlantic with this mineral.

Weaving beaters (ON *skeið*, m.) implements for compressing the weft were also an integrated part of the warp-weighted loom, and best preserved in iron. However, less frequent finds than loom weights, weaving beaters of iron, bone, or wood also appear in the archaeological record but generally as grave goods representing more costly tools. Beaters of whalebone and wood are rather rare finds – probably due to less favourable preservation methods of wood, and inaccessibility for the whalebone beaters, which are mostly found in northern Norway with

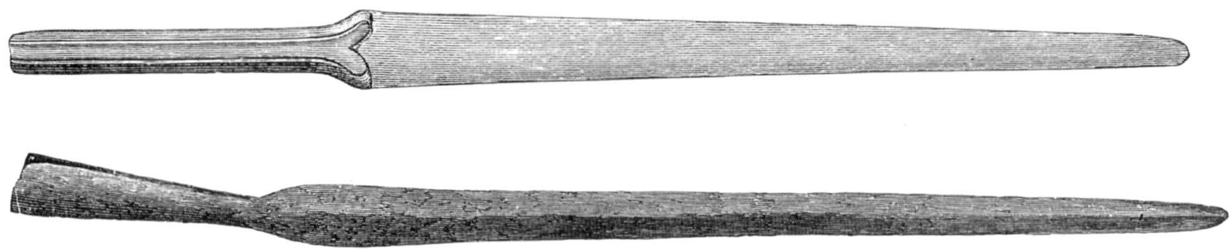

Fig. 4.15: Common shapes of Viking Age weaving beaters, types R439 (above) and R440 (below). Of these, type R440, the most efficient beater, dominated.

more than 70 specimens from graves up to 70° N.[108] The whalebone beaters were probably traded within exchange networks or presented as gifts when appearing further south.

While the weaving beaters of softer organic material were suitable for pressing a thin and fine weft, the stronger and often longer iron beaters were able to weave quicker and press a tighter weft of coarse yarn (Fig. 4.15). The purpose-made hand tool, with the handle tending to counteract the weight of the blade, made this single-handed operation more effective, and is interpreted as a precision tool for making specific densely woven cloth types.[109] Experiments show that iron beaters were 10 times more efficient than those of wood when weaving tight fabrics, such as sails.[110]

The handles must have been covered by now lost wood or horn handles which are sometimes preserved, at least in parts. The operative part was the blade, used for beating up the weft required perhaps as many as 20 strokes per thread.[111]

The overall length, as well as the length of the blade, tip, handle, and the blade section are regarded as the key functional elements. As weaving beaters could be made of different material, iron, bone, and wood, it has been suggested that the iron beaters were also status markers of symbolic significance related to high-status women.[112]

Of more than 300 beaters in Petersen's corpus, altogether 95 or 30% of the total, stem from the present study area – 32 from Hordafylki and 63 from the historical counties of Sygnafylki and Firdafylki.[113] Of the nine long beaters, six are from Sogn. The longest, being 86 cm,[114]and the shortest 49.7 cm.[115] The handle varies in length from 8.5 to 16 cm, with the majority from 9–14 cm. There are, however, some discrepancies. The width of the blade varies from 2.7 to 3.8 cm. Disregarding size, the weaving beaters of R440 are very homogenous, with nearly no variation,[116] and were probably primarily used for making strong fabrics, such as, among others, sails.

A rarer type, with a spit instead of a shaft, comprises 15 specimens according to Petersen, and of these five come from graves from the present study area.[117] They are generally smaller than R440, from 65 to 70 cm. The handle could be of bone, mostly whalebone,[118] or wood.[119] Yet another is rather atypically small, with a blade being only 24 cm in length.[120] Only a few weaving beaters of whalebone (R439; JP fig. 157) relate to graves from the study area. This type is generally longer than R440 of iron, 80–107 cm long, while some are much smaller, 30–64 cm, and with narrow blades of 2.8 and 3.5 cm.[121] Generally, weaving beaters are more common in female contexts than in male burials (see Chapters 5 and 6).

Weaving beaters are rather rare finds in the North Atlantic, and only known from one Viking Age grave from Iceland and made of whalebone.[122]

Weaving bands

Weaving ribbons or bands could be done by using different tools and techniques – by tablet weaving, by using a rigid heddle or by using a band loom. Various types of bands served several purposes, such as decoration of clothing, as belts, straps and not least as starting and/or selvedge bands for textiles woven on the warp-weighted loom.

Tablet weaving is well known and was commonly used in both the Viking period and the Middle Ages, and even long before that – back to the Bronze Age.[123] In tablet weaving, tablets were used for shifting the sheds in the warp and weft. The warp-threads can be stretched between the weaver's belt and some fixed point, such as a peg in the wall. The tablets are all held together, like a pack of cards, parallel with the warp, to be turned together in a single movement, thus altering the relative positions of the warp-thread and forming a series of sheds (Fig. 4.16). The weft is completely covered by the warp threads. A shuttle, a knife-beater, or a small sword-beater of wood or bone is used to beat the weft.

The earlier mentioned Oseberg burial chamber, richly furnished with tools and textiles, contained both elaborately woven bands, a started band, and altogether 52 weaving tablets, some of them as loose tablets, found in a wagon in the forepart of the ship. The Oseberg burial with its exceptionally good preservation conditions and rich assortment of textile tools also contained two largely complete band looms and traces of two more.[124] There are no preserved Viking Age finds of weaving tablets or band looms from the present study area and are only indirectly indicated through preserved textiles (see Chapters 6, 8, and 9). However, the

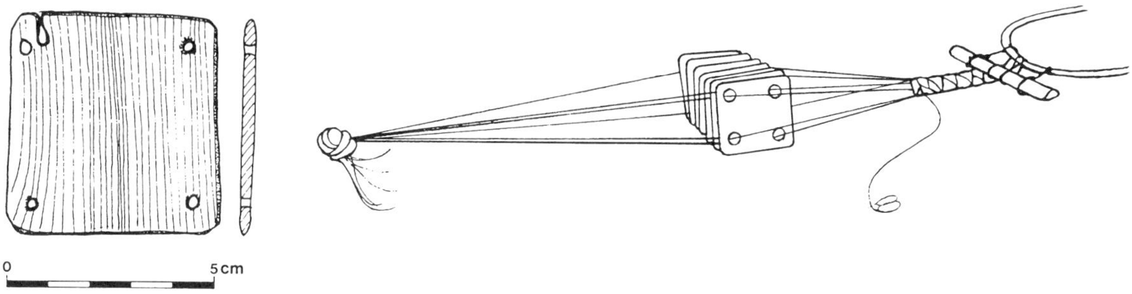

Fig. 4.16: Tablet weaving. To the left: tablet of wood from Bryggen in Bergen (Source: the author).

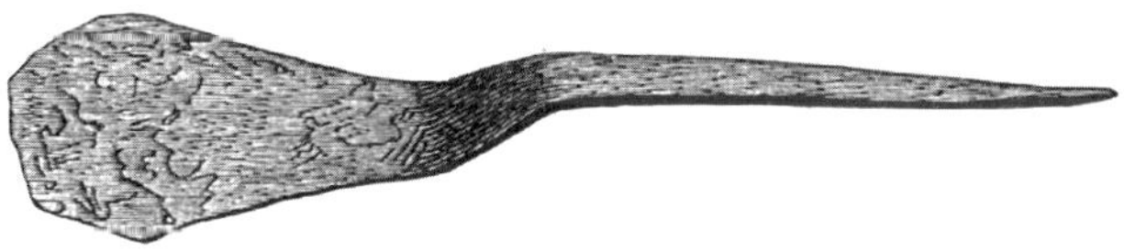

Fig. 4.17: The spatula-shaped tool (R413), possibly used as a small weaving beater in a band loom.

medieval material from Bryggen in Bergen comprised a few tablets for band weaving, along with rigid heddles and band looms for weaving ribbons (see Chapter 9).

The general absence of textile implements made of wood in the graves from western Norway, seems then to be caused by the preservation conditions and that the assortment of tools must have been broader, which is also substantiated by fragments of wooden handles of handheld tools and fragmentary finds of textiles.

A special spatula-shaped tool of iron, R413 (Fig. 4.17) has been much debated as for functions, interpreted as a kind of scraper but by several scholars as a weaving tool and as a kind of beater, possibly for pattern weaving and band weaving,[125] as they generally appear in female graves with textile tools. Its use has been related to small looms and band looms to strike the weft towards the weaver (ON *að ser*). In some publications and in earlier studies, it has been denoted *vevreell*, of ON *hrœll*, deriving from the verb *hrœla*, used for handling the weft in the warp, and often with a bent pin of bone or wood,[126] and thus rather unlike the spatula shaped tools of iron. Petersen refers to 34 such finds, dominantly from female graves, and altogether only 11 from the present study area of western Norway, and now increased to 14, and then combined with other textile tools. They appear in a few Viking Age graves in Iceland too, of which one is in a male grave.[127]

The spatula shaped tools are shorter than the weaving beaters, mostly *c.* 10–15 cm long. Made of iron, they consist of a rounded narrow handle and a lifted obtuse angle to the broader and spatula-shaped short blade up to 4.8 cm wide (Fig. 4.17). A few have remains of a wooden handle, which substantiates their role as small hand tools. Sometimes the blade has a more triangular shape at the end, in contrast to the ordinary weaving beaters and pins tapering towards the end.

However, when confronted with present-day weavers mastering weaving on upright looms and other types of weaving too, this function is doubted, and especially related to the warp weighted loom.[128] Albeit unclear in function, all these tools in the present corpus from western Norway appear in female graves with other textile tools, the tool is included in the further analysis to assess their function in wider contexts, in time as well as socially and spatially.

Rigid heddles or heddle-frames were other tools for making bands. They consist of a frame holding some narrow slats. There is a hole in the middle of each slat and the frame so that the holes and the narrow gaps form a horizontal line across the middle. The warp threads pass alternatively through the holes and the narrow gaps between the slats and are held taut between the weaver's belt and some fixed point. The shed is changed by raising or lowering the heddle-frame, the warp-threads in the holes in the corners of the tablets and drawn past the threads in the gaps. It functions in the same way as a heddle-rod. A shuttle pulls the weft through. When weaving patterns, a knife-beater is used. The only specimens found within the study area stem from medieval Bryggen in Bergen (see Chapter 9).

Weaving and time

Weaving was time consuming. Two people were needed to make the warping efficient, and if the loom was not very narrow, two weavers were needed at the loom as well.[129] To produce a 1 m wide piece of course to medium weight cloth of only about 7 cm length would have taken one 12-hour day, according to experiments. To produce notional two sets of ordinary clothing has been estimated to require 162.5 days for fine wear and 97.5 days for coarser clothing, with much of that time taken up by spinning. Based on these experiments related to hand spinning and weaving on the upright loom, the ratio of spinning to weaving time is estimated to be in the region of 5:1.[130] Later experiments, however, modify these results and show that one could weave approximately 50 cm per day, if the loom functioned well, however, not least depending on the yarn quality, whether coarse or fine, the latter needing many time consuming picks and sheds.[131] Yet other experiments conclude that it takes about three times longer to spin than to weave.[132] As for the experiment of weaving a 2 m wide and 3 m long cloth on the warp weighted loom, the experimenter's one hour's work at the loom produced a 2.7 cm long cloth, of 5.2 cm per square meter. Considering the whole work process, it took 222 hours to spin the warp threads and 135 hours to spin the weft threads, 62 hours to warp and prepare the loom, and 110 hours to weave a 3 m long cloth, altogether 529 hours.[133] Spinning was altogether much more laborious than weaving, taking three to five times longer than the work at the loom – and even longer when the preparation of fibres is included. However, time use varied as for coarseness and fineness of yarn and weaving patterns – and overall, the quality of the fabric.

A few historical sources refer to norms related to qualities, measures, and working hours. One of the most informative and detailed sources in this respect, is the aforementioned Icelandic *Búalög*, a compilation of customary norms as for working conditions, productivity, and standard qualities of wadmal in Iceland in the Middle Ages. It defines standard lengths and widths measured by thread counts and weight and according to various qualities and use areas – whether produced for the household or for export and related to prices and modes of production. It tells that to spin five-mark wool (=1.1 kg) would take around 12–14 hours. Provided with other information given, it would take at least 66 to 77 hours to spin threads to weave a standard piece of *vaðmál* of 20 ells (of 47.4 cm). Such a fabric should, according to the norm, be 9.5 m long and 90 cm wide (=8.55 m^2) and weigh 286 g per ell (=0.423 m^2) or 572 g per square metre.[134] This means that a woven cloth of 20 ells would weigh 6.2 kg and take 4.2 times longer to weave. To spin enough yarn for 20 woven ells would take at least 66 to 77 hours.

This is a far shorter time than indicated by the experiments referred to above, but it is also based on spinning of coarse yarn and weaving with a looser thread count in the warp and weft, and coarser and heavier than the textiles in the experiments, and thereby also faster to produce. However, it may indicate that the medieval textile workers with a long-acquired competence and embodied expertise transmitted since childhood could be more efficient than the present-day spinners. Although not counting as a definite key, such information serves as an important reference when analysing archaeological tools and textiles. These textile tools, long regarded as rather unimpressive artefacts, and not even always taken care of, provide important information that indirectly signify a massive input of labour to provide all kinds of necessities for the household and a surplus, which is important to bear in mind in the following analyses.

New technologies in weaving

During the Middle Ages in northern Europe, the horizontal loom represented an important innovation within weaving technology, working faster and more effectively than the warp-weighted loom. In contrast to the upright loom operated when standing and working upwards against gravity, the new loom was worked in a sitting position. The warp threads are stretched horizontally between two cloth beams, one in front of a seated weaver and another at the far end of the loom. The warp threads are attached to heddles and the shed is changed by means of the foot treadles that are linked to pulleys and shafts that enable selected threads to be raised. The weft thread passes through the shed on a shuttle. Long lengths of warp thread can be wound round the beam, and completed weaving is wound round the front beam, enabling the weaver to weave a piece of cloth much longer than the loom itself.

One of the crucial issues in textile production in Scandinavia as in other parts of Europe in this period relates to the introduction of the horizontal treadle loom. In most parts of northern Europe, the old warp-weighted loom was replaced by the horizontal loom in the Early Middle Ages – earlier in some and later in other regions and mentioned in manuscripts from the 11th century and depicted in manuscripts from the 13th century in England. The archaeological evidence of the whole construction is still meagre, and the first traces of the loom appear as fragments in Scandinavian contexts from the 11th to early 13th century and a heddle rod from the 12th and 13th centuries.[135] As these traces represent an *ante quem* dating, they show that the loom was taken into use, but do not necessarily represent the earliest use. In Scandinavia, traces of possible remains of the horizontal loom are found in early contexts in Hedeby,[136] Lund, and Sigtuna.[137] From the present study area, some finds from Bryggen in Bergen may also stem from one or several horizontal looms, and the earliest from the late 12th century (see Chapter 9).[138]

The new loom thus represented an important competitive innovation that should be considered when assessing the domestic development of a textile production and when assessing textiles as domestic or as possible imported commodities in the Middle Ages.

Finishing processes: sewing and maintenance of textiles

Textile work requires different kinds of needles for sewing, sticking, and binding and pins for pinning things together. Needles of different material for sewing appear in archaeological contexts from the Viking period and later, however, in small numbers. They are rather rare finds probably because of the generally unfavourable conditions for preservation of both metal and organic material. They appear in

Fig. 4.18: Needle-case of copper alloy from a grave at Svingeset, Stryn (B 6483d) (Photo © University Museum of Bergen).

different lengths, sharpness of point, and eye size related to the intended end use and types of materials involved.[139] Material, length, thickness, and size of the needle eye are features regarded as special functional attributes. The quality of the sewing or stitching generally also matches the quality of the fabrics. Sewing is best documented in textiles.

Needles of iron, copper alloy, bone, and antler are all recorded from Viking Age graves, usually female, and in a few cases at occupation sites too. In some cases, generally high-status graves, they appear with tubular needle-cases of copper alloy (Fig. 4.18).

Petersen refers to five finds of small needle cases of metals in grave contexts from the Late Iron Age, and none of bone, varying in length between 4.5 and 6.5 cm, with the smallest and thinnest of copper alloy,[140] some with metal sewing needles inside. The number of needle cases has now expanded and count eight specimens from high-status graves within the study area. The medieval needles from Bryggen also include coarser needles for darning.[141] Most likely, these tools existed in the Viking Age too, but were not found due to preservation conditions.

Bodkins, larger, thicker needles with a large eye and blunt end, could be used for various purposes in working with coarse materials, such as *nålebinding* (looped needle netting or needle binding). Sharp-pointed bone needles were probably used for coarser fabrics such as sails and are best represented in urban settings in humid layers (see Chapter 9). Needles from medieval urban contexts with good conditions for preservation of organic material, such as at Bryggen, are also made of wood, antler, and bones. Coarser needles of wood, bone, or antler with an eye could be used for knotless needling, especially smaller ones of *c.* 6 cm in length. They may have had other functions too such as simple stitching and for netting. Needles were also used as 'shuttles', especially for tapestry weaving.[142] Most likely, these tools existed in the Viking period too.

Finishing processes may also have included fulling of sails and outdoor gear at sea, to increase the thickness and compactness of woven wool by subjecting it to moisture by soaking it in water. This work was most likely carried out in the summer season and these processes can only be observed in preserved textiles.

The finishing process for linen included beating the woven cloth, laundering, beating it again, and then smoothing it. Linen-smoothers – spherical or hemispherical oval or circular pellets of glass as hand-held tools for smoothening linen – were probably heated.[143] The pellets of green or black vary in size between 6.5 and 8.7 cm in diameter, but seldom smaller than 7.5 cm. They represented imported goods, appearing in a few high-status female graves from the present study area, as well as from Kaupang, Hedeby, and Birka (Fig. 4.19).[144]

Some characteristic rectangular-shaped plaques of whalebone from the 9th and 10th centuries, 28.5–39 cm long and up to 27 cm wide, are interpreted as boards for smoothing textiles of linen.[145] Often, the upper ends are decorated, with a pair of animal heads facing each other, and with ring-and-dots and edge lines, but are far from standardised and appear in various shapes (see Fig. 4.20). Lately however, this function has been questioned, arguing that they rather served as boards for other purposes, for instance, chopping meat.[146] However, their contexts as finds in high-status graves, were generally combined with tools for textile production in graves from the northern part of the country and from graves in western Norway. One of the graves from northern Norway from the late 9th century comprised additionally a spindle whorl, weaving beater of whalebone, and another whalebone tool (R450), reminding us of R413 (cf. Fig. 4.17), however, only one of the kinds was from the present area of study and was also in this case found together with a weaving beater of whalebone.[147]

Fig. 4.19: Pellet of black glass (diam. 7.6 cm) and a turned spindle of soapstone from the 10th-century grave at Trå, Granvin (Photo © University Museum of Bergen).

Of the more than the 30 such plaques from Norway, most relate to the northern part of the country with 14 from coastal areas of northern Norway, 12 from the Trøndelag area and seven from the present study area. Both weaving beaters and plaques of whalebone from these northerly waters and with offspring further south and in Ireland and Birka,[148] indicate whaling as a speciality at an early stage[149] but also indicate wide-reaching networks. In later years, the finds of whalebone plaques have multiplied, to include at least 14 finds from the North Atlantic area.[150]

These tools, neither of glass nor whalebone, appear in medieval contexts from the study area. However, rounded, and smooth rolling stones may have served the same purpose (see Chapter 9). Still, these tools relate more to maintenance of special fabrics of linen than production as such (cf. Table 4.1) and are not necessarily related to textiles produced in the area.

A long chain of processes

This overview of the different tools used in the different work processes shows, then, that the tools were made to fulfil several specific requirements – first and foremost by their functional forms and materials to produce textiles of various qualities for different use areas. The tool categories related to the various stages in the textile production, again, represent time-consuming work operations of varied complexity and time involved. Seeing all the steps together as integrated processes, related to time consumption of the various processes and results per hour based on experimental archaeology gives an impression of the workload and varied competency needed for the production related to both animal and plant fibres, seasonal and year-round.

The varied tools for the different stages of production of time-consuming work processes imply a large work force of varied, probably also specialised, competency. The organisational aspects of the production are therefore considered according to different modes of production, referred to in Chapter 2 and in the previous chapter too as a backcloth and a source for understanding the tools and traces they have left and for interpreting these graves in wider societal contexts.

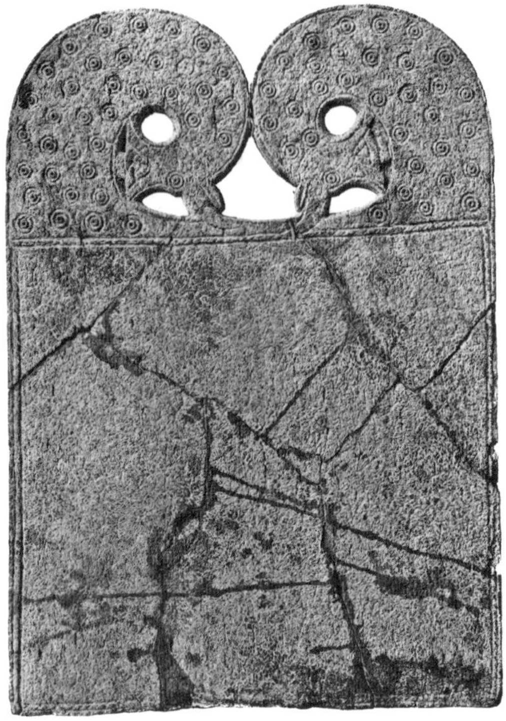

Fig. 4.20: Whalebone plaque, R449, from Meløy, Rødøy (B5393h), 32.5 × 23 cm (Photo © University Museum of Bergen).

Organisation

The textile production in the study area, connected to areas by the North Sea and including areas in the North Atlantic over a period of half a millennium, cannot be treated as one static unit. In this period of larger socio-political transformations, the scale and modes of production were also affected and largely operated within a hierarchical structure. As for textile production, it also varied according to rank and status (cf. Chapter 2), broadly categorised as ordinary household production, advanced household production, and an attached specialist production. Of these, the former as a common mode, basically covered the household's own needs for upkeep of livelihood and based on general skills and available raw materials. An advanced household production, however, would normally be of higher status and generally connected to larger farms, with a larger work force and relying on work and services from servants within a gendered domain. This type of production would need

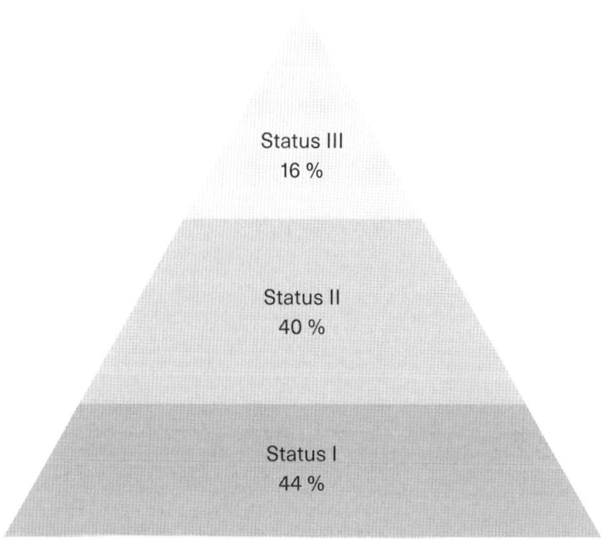

Fig. 4.21: Social status as represented in Viking Age female graves from western Norway (Source: Solberg 1985).

more management and less corporal labour. Expectedly, tools and textiles were also related to production of wares within the luxurious segment – for conspicuous consumption, gift giving, networks and exchange. An attached specialist production based on a subordinate work force of special competency would expectedly also relate to special environments – such as central places as nodes within estates and probably based on services and duties within a hierarchical structure, and with access to raw materials at wider distances based on reciprocity and or subordinance, while *specialist's workshops*, such as in Viking Age centres or early towns, such as Kaupang, and medieval towns, in this case Bergen, formed yet another category. These types of organisation, as earlier promoted by Costin and Andersson Strand, also link closely to the social categories revealed in Viking Age rural graves, in three social status groups, as defined by Solberg, however, as reflected in the whole corpus of Viking Age female graves from western Norway and of diverse status.

The ranked status documented by Solberg as a social pyramid, would expectedly be based on various types of organisation of the production. The majority, those of lower status I and possibly status II, would presumably connect to an organisation of ordinary household production. Female graves of status II and III may rather reflect an *advanced household production* and even with attached specialists in the highest and smallest group of status group III, possibly organised within workshops and at workplaces nearby or at larger distances. (Fig. 4.21).

How the graves furnished with textile tools and with other socially distinguishing grave goods fit into this social hierarchy is therefore a focal issue in the following chapter when assessing the more limited group of some 140 Viking Age graves with textile tools from the three counties. Do they correspond with the general social hierarchy as reflected in the social pyramid illustrated above, or do they to a larger extent connect to the higher strata and represent a more advanced production, possibly with wider networks, based on reciprocity or subordinance?

Notes

1 Falk 1919, 1, with references.
2 Ryder 2005, 126.
3 Barber 1991, 4.
4 Andersson Strand 2012, 30.
5 Øye 1998a, 50 with references; cf. Chapter 3.
6 Hoffmann 1991, 19 with references.
7 Ryder 2005, 125–126.
8 Ryder 1984.
9 Andersson Strand 2012, 30, with references.
10 Crowfoot *et al.* [1992] 2001, 18.
11 Waetzoldt 2007, 113–117.
12 Aasen 1918, 902.
13 Hoffmann 1991, 19.
14 Crowfoot *et al.* [1992] 2001, 157.
15 Eva Andersson Strand, pers. comm. 2020.
16 Fritzner II, 251, 272.
17 Øye 1988, 107–109; Owen-Crocker *et al.* 2012; see also Chapter 9.
18 Harrington 2008, 63.
19 *Medieval Catalogue* 1940, 153–158.
20 Ryder 2005, 126.
21 Owen-Crocker *et al.* 2012, 638.
22 Hoffmann 1964.
23 Petersen 1951, 320.
24 Andersson Strand 2012, 31.
25 Pfarr 2005; Mårtensson *et al.* 2006a.
26 Skoglund *et al.* 2013.
27 Ejstrud *et al.* 2011.
28 Skoglund *et al.* 2013; Lukešová *et al.* 2017.
29 Lukešová *et al.* 2019.
30 Ejstrud *et al.* 2011, 47–49; Troldtoft Andresen & Karg 2011.
31 Øye 2002.
32 Ejstrud *et al.* 2011, 22; Troldtoft Andresen & Karg 2011.
33 Ejstrud *et al.* 2011, 52–53.
34 Ejstrud *et al.* 2011, 11, 55.
35 Øye 1988; see also Chapter 8.
36 Hoffmann 1991, 206.
37 Petersen 1951, 324; Ingstad 2006.
38 Petersen 1951, 319.
39 Ejstrud *et al.* 2011, 56.
40 Barber 2007, 173–174.
41 Eva Andersson Strand, pers. comm. 2020.
42 Crowfoot *et al.* [1992] 2001, 16.
43 Ejstrud *et al.* 2011, 59.
44 Hoffmann, *KLNM* XVI, 499; Øye 1988.
45 Stolcová & Grömer 2010.
46 Wild 1970, 33.

47 Grenander-Nyberg 1990, 76.
48 Owen-Crocker *et al.* 2012, 534.
49 Ingstad 2006, 187–89.
50 Hoffmann 1991, 65.
51 Hoffmann, *KLNM* V, 194, with references.
52 Owen-Crocker *et al.* 2012, 534.
53 Warburg 1976, 78.
54 Owen-Crocker *et al.* 2012, 536.
55 Ryder 2005, 126.
56 Andersson Strand 2012, 33.
57 Warburg 1976, 15.
58 Falk 1919, 14.
59 Eldjárn 1956, 341.
60 Lønning 1976, 20.
61 Warburg 1976, 89.
62 Stolcová & Grömer 2010, 13.
63 Petersen 1951, 304–305.
64 Petersen 1951, 310.
65 Petersen 1951, 302.
66 Ejstrud *et al.* 2011, 62–63; Andersson Strand 2012, 34.
67 Andersson 1999, 9–11.
68 Andersson 2003.
69 Andersson Strand 2016.
70 *e.g.*, Heide 2014.
71 Andersson Strand 2016.
72 Crowfoot *et al.* [1992] 2001, 19.
73 Falk 1919, 40.
74 Ingstad 2006.
75 Walton 1988.
76 Guckelsberger 2017; Hayeur Smith *et al.* 2019.
77 Falk 1919, 42–43.
78 Crowfoot *et al.* [1992] 2001, 19.
79 Mehler 2015.
80 Anker 1969; Øye 1986, 361.
81 Hoffmann 1964; Andersen & Martinussen 2016.
82 Falk 1919; Hoffmann 1964; Owen-Crocker *et al.* 2012, 620; Juuhl 2016.
83 *F* IX, 9 and XI, 49.
84 Juuhl 2016, 148.
85 Mårtensson *et al.* 2009.
86 Juuhl 2016, 148–149.
87 Hoffmann 1964, 37, 42.
88 Steinnes [1936] 1982.
89 Hoffmann 1964, 24–29.
90 Mårtensson *et al.* 2009, 378–80.
91 Andersson Strand 2016, 30.
92 Andersson Strand 2016, 22.
93 Mårtensson *et al.* 2009.
94 Mårtensson *et al.* 2009.
95 Andersson Strand 2012.
96 Mårtensson *et al.* 2009, 389–92.
97 Andersson Strand 2012.
98 Mårtensson *et al.* 2009, 390.
99 Mårtensson *et al.* 2006c II, 11.
100 Hoffmann 1964; Juuhl 2016.
101 Mårtensson *et al.* 2006c, 11.
102 Petersen 1951, 295.
103 Hoffmann 1964; Bender Jørgensen 1991, 75.
104 Militello 2007, 41
105 Øye 1988; 2011.
106 Øye 1988, 58–59.
107 Petersen 1951, 295.
108 Sjøvold 1974, 249.
109 Harrington 2008, 47–48.
110 Andersson 2003, 28.
111 Hoffmann 1964, 40.
112 Owen-Crocker *et al.* 2012, 589.
113 Petersen 1951, 294.
114 B4736 c; Hopperstad, Vik, Sogn.
115 B8390 Fretheimsviken, Flåm, Aurland, Sogn; Petersen 1951, 287.
116 Petersen 1951, 287.
117 B5030g Gutdal, Stryn; B5807h, Myklebust, Eid; B6662, Øvreset, Breim, Gloppen; B3708, Fitje Voss.
118 B5030g; Gutdal, Stryn.
119 B5807h; Myklebust, Eid.
120 B6662; Øvreset, Breim, Gloppen.
121 B5717v Hilde, Innvik, Stryn; B5764g Myklebust, Eid : 64 cm long, and the smallest *c.* 30 cm; cf. Petersen 1951, 288–291.
122 Eldjárn 1956, 96–97.
123 *KLNM* II, 465–467.
124 Ingstad 2006, 186–87.
125 *e.g.*, Petersen 1951, 300; Eldjárn 1956, 342–43.
126 Hoffmann 1964, 135–140.
127 Eldjárn 1956, 342–43.
128 Marta Kløve Juuhl, pers. comm. 2017.
129 Hoffmann 1964, 44.
130 Andersson 1999, 9–11.
131 Eva Andersson Strand, pers. comm. 2020.
132 Pfarr 2005, 59.
133 Pfarr 2005, 63.
134 Þorláksson 1991, 309.
135 Owen-Crocker *et al.* 2012, 345–46; Øye 2016a.
136 Grenander-Nyberg 1984, 145.
137 Geijer & Anderbjörk 1939, 236.
138 Øye 1988, 74–75.
139 Owen-Crocker *et al.* 2012, 381.
140 Petersen 1951, 324-25.
141 Øye 1988, 85–87.
142 Øye 1988, 97 with references.
143 Owen-Crocker *et al.* 2012, 108.
144 Gaut 2011, 230–231.
145 Petersen 1951, 337.
146 Isaksen 2017.
147 B5741, Revikja, Selje, Vågsøy.
148 Sjøvold 1974, 253.
149 Sjøvold 1974, 202.
150 Owen & Dalland 1994; 1999.

5

The tools in the grave: gender, social status, and agency

Sometime during the 10th century, a female from one of the coastal districts in the medieval county of Hordafylki was buried in a large cairn at Toranger in the island community of Austevoll. In 1904, local farmers discovered its remains. In an open space between the stones in its middle, 3 feet or about 90 cm deep, they found traces of cremated human bones, dress accessories, and various textile tools.[1] The Toranger grave is one of some 140 Viking Age graves furnished with textile tools from western Norway. As an entity, these graves, the funeral practices, with tools and other grave goods, provide an important source material with a unique potential related to textile production and social aspects of production – gender, gender roles, and social status. As intentionally deposited artefacts, the tools in the close contexts of graves indirectly elucidate many facets of production – physically, technologically, and practically – in a complex interaction between social and cultural spheres, actors, environments, and available resources.

The term 'grave goods', as used by archaeologists, simply denotes anything found in the grave besides possible human remains. They may include a variety of artefacts – dress accessories, remains of the dress, and intentional depositions of varied items, including sacrificial offerings.[2] According to the burial customs, at least in the upper strata of society, the women were fully dressed and equipped with dress accessories and grave goods signalling social rank and work domains, as in the Toranger grave. Most likely, all these things were selected as emblematic grave gifts by the heirs as a last remembrance and obviously for display as they had been used and were known as integral parts of the deceased's identity and life story. The grave goods of personal apparel and distinguishing items thus visualised and commemorated the deceased's status and roles in life before being buried or destructed on the pyre and covered by a cairn or a mound, or without any superstructure. Cosmology and conceptions of death and an afterlife implied that material goods were important and useful in relation to death and by the transferal of inheritance (ON *erfi*) and was also an economic and lawful act. However, the burial customs were not altogether uniform and varied according to status, gender, and over time and in space.

A main objective in this chapter is to trace patterns in these funeral practices related to textile tools as grave goods – as they are archaeologically documented and represented in western Norway and in the North Atlantic. Such patterns in turn may give a clearer conception of the environmental and socioeconomic conditions behind the textile tools – whether reflecting actors within production at special places and in environments for textile production, either within the luxurious segment, possibly specialised work, or rather within the ordinary household sphere. By analysing the gendered graves and grave goods in time and space, they may reveal relations and interaction at shorter and larger distances, and possibly as part of larger enterprises and wider networks – aspects that are further elaborated in Chapters 6 and 7.

The Toranger grave is only one of nearly 50 Viking Age graves equipped with textile tools in the old Hordafylki. With some 40 such graves in Sygnafylki and more than 50 in Firdafylki further north, they constitute, then, some 140 contexts, which are recognised archaeologically as graves or possible graves within the study area of western Norway (Fig. 5.1; see Appendices 5.1–5.3 to Chapter 5).

In broad outlines, these burial customs were transmitted to the new homelands in the west, and in Iceland as the largest area in the North Atlantic with more than 320 inhumation graves from 170 sites, of which nine were discovered after 2010.[3] They date mainly to the 10th century and the last decades of the 9th century, and largely resemble the Norwegian counterparts, being provided with grave goods in both marked and unmarked graves. However, all were inhumation graves and, unlike graves in western Norway, radically fewer contained textile tools – only five of the graves of the estimated female burials and of these all contained social emblems, brooches and/or other dress accessories and other signs of status.[4] Also, here the females, at least of higher rank, were buried fully dressed, and with similar dress accessories, representing the same social strata.[5] Further south – in the Faroe Islands, Shetland, and the

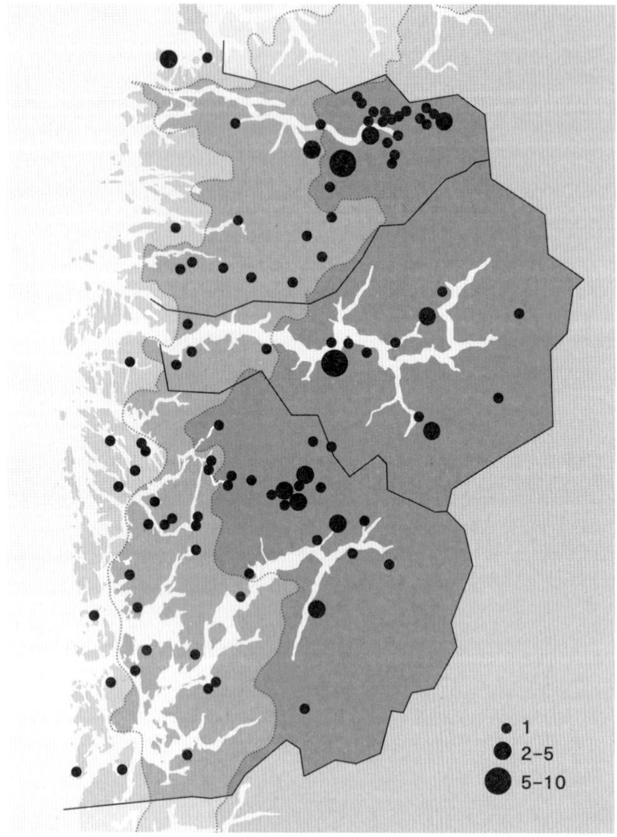

Fig. 5.1: Graves with textile tools in Hordafylki, Sygnafylki, and Firdafylki – presently the counties of Hordaland and Sogn og Fjordane. The area is marked in outer, mid- and inner zones (based on Sømme 1954) and distribution of graves (N:142) (Illustration: Per Bække).

Orkneys – Norse Viking Age graves are few and the burial customs ended earlier, influenced by Christianity at an earlier stage, but reveal similar patterns and burial customs, by also using textile tools as grave goods, as documented among others at Westness on the island Rousay in Orkney.[6]

Viking Age graves and grave goods as archaeological sources thus signify norms and ritualisation of funeral rites. The customs of combining tools and other social markers as grave goods that appear with an exceptional boom in this period in western Norway, offer an important and an altogether underused source category for assessing the social contexts of textile production. Social rank clearly affected praxis of what was possible and acceptable as status markers and grave goods, where dress accessories are typical items in high-status female burials, as weapons are in male graves. This trend of using tools as grave goods coincides with major technological innovations and societal transformations. Population growth, internal land clearing, and external expansion overseas now triggered the need for varied types of tools, and not least for textile production. The burial customs in the higher strata thus seem to reflect this expansion, before they terminated with the conversion to Christianity with other rituals and conceptions of death – and formalised by the early 11th century, some decades later than in the North Atlantic and in Iceland officially approved in AD 999/1000.

When tracing patterns in these burial customs, the graves are analysed at different spatial levels as for contents within individual graves and in a wider comparative perspective: (1) in their nearest surroundings, (2) within various zones from coast to inland, and (3) related to places in the three old *fylki* in western Norway from the beginning of the period *c.* AD 800 until they terminated by the late 10th century. The overall aim in this chapter is to trace and uncover larger patterns of the burial rites that indirectly may elucidate textile production in western Norway and also related to comparable Viking Age graves in the North Atlantic, especially Iceland as the largest area and with the highest number of Norse graves.

The graves as sources

Representation and documentation

The grave at Toranger by the coast of the southern part of Hordafylki, and the earlier mentioned grave at Veka in the inland municipality of Voss, were, as many Viking Age graves, discovered in the early days of archaeology, in the late 19th and early 20th centuries. Changes within farming systems caused by the new *Land Consolidation Act* of 1857 in Norway and by the following restructuring and reallocation of earlier communal farmland, changed both the agrarian and archaeological landscapes. Larger nucleated farms with extensive communal arable fields, and often with a high representation of monumental graves, were more affected by the Act than smaller and more dispersed farms. Not only were old agrarian structures broken, but also grave monuments therein. However, as traditional cultivation methods were still prevalent before heavier agricultural mechanisation set in by the later part of the 20th century, remains of prehistoric graves could still be recognised. The decades after the effectuation of the Act represent therefore a peak of reported graves (Fig. 5.2).

This was followed by *The Cultural Heritage Act* of 1905, by law protecting all archaeological finds in Norway older than the Reformation in 1537, which marked another significant change of what was subject to be reported, collected, and investigated. From that date on, it became obligatory to report and hand in all archaeological finds, in this case to Bergens Museum (now University Museum of Bergen), to be taken care of professionally by the museum staff. Unfortunately, this often happened after the sites had been disrupted and even demolished. The information about finds and find contexts thus varies considerably depending on when, and under which circumstances they were found, reported, and followed up archaeologically. Consequently, many of the early finds can be challenging as sources and archaeological contexts. This was the beginning of professional

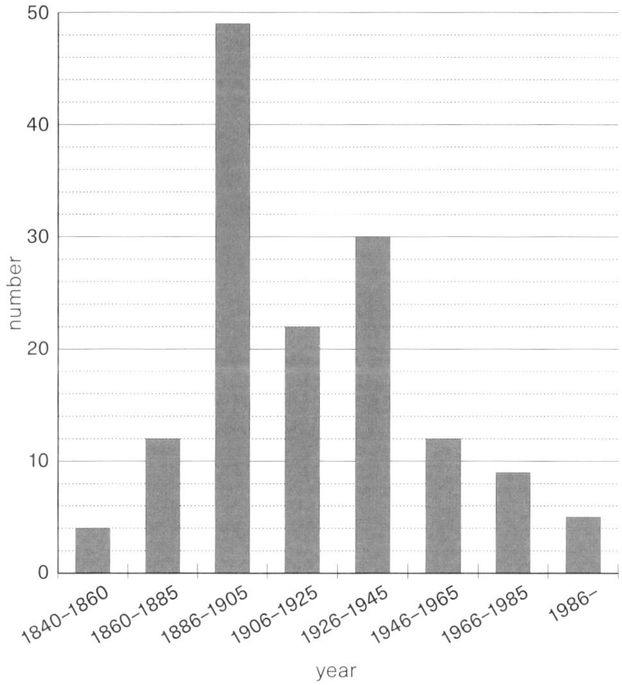

Fig. 5.2: Archaeologically documented graves/possible graves with textile tools from the study area in western Norway, related to periods when being discovered and investigated (N:142).

archaeological documentation and investigations of these monuments, and in western Norway, fortunately in the hands of competent and renowned antiquarians and archaeologists. Nevertheless, many graves are unsatisfactorily documented, and probably incomplete when disturbed and taken care of by amateurs in the first instance. In some cases, this may cause a key problem for identification of status and gender.

Conceptions and representations

The conceptions, cosmology, and reasons behind these varied and coexisting burial rituals of the funeral customs – cremation and inhumation – are less clear. Eddic and skaldic poetry and sagas tell that the dead should be treated according to old customs, '*eptir siðvenju*' or '*eptir fornum sið*' – customs that were well known, and therefore needed no further explanation. They signal that the dead should be treated with respect[7] and in many cases be honoured with monumental graves. Old Norse terminology distinguishes between such monuments: those close to the farmhouses as *heimishaugar*, and grave mounds generally, only denoted as *haugr/haugar*, *dys/dysjar*,[8] or *kumbl*. Judging by the terminology, ON *heygja*, *haugsetja* – to bury in mounds – seems to be more prestigious than building cairns, *lata verpa haug*. Graves unmarked on the surface, simply called *grǫf* or *grǫptr*, must have been the most common type of graves.

The graves with textile tools belong to all these categories, as well as those without preserved visual superstructures. There are several interpretations of cosmology and ideology behind the burial traditions. Cremations could either help or purify the deceased so that the spirit would have an afterlife. It may also have been a way to help and protect the living from the dead and from returning from the grave and haunting them, as hinted in the Eddic poem *Voluspá*, stanza 21. At the same time, they represent ritual and intentional destruction of wealth. A difference between grave goods burnt on the cremation pyre and undamaged objects buried with the cremation remains, may therefore represent different conceptions of death and afterlife. Here, inhumation graves seem to be regarded as a living space for the dead, and a place of ancestral worship.[9] When the Gulathing law forbids offering and worship at pagan burial mounds and offering places …*til haug ok horgi*,[10] this had obviously been a long-lasting tradition.

Placing the dead close to the living was, moreover, a way to underline ancestral rights of ownership of land. The Gulathing law thus connects the funeral rituals to the succession of inheritance (ON *erfi*), celebrated as a ritual feast (ON *erfiǫl*) with juridical implications, and with the heirs as responsible for the burial. The funeral feast thus specifically served to reaffirm and re-establish social relations disrupted by death. Grave goods such as cauldrons, vessels, and food-related utensils may therefore have been used in such ceremonies before they ended up in the graves possibly as food offerings. Even as late as in the Land law of 1274, heritance rights refer to so-called ON *haugsóðal* and denotes the lawful heir as *haugsóðalmaðr*, referring to inheritance rights going back to grave mounds.[11] Landowning thus seems to have been closely connected with monumental graves as physical markers of hereditary ownership of landed estate, in a time when land represented a cornerstone in society, and rights and control over landed property formed basis for legal rights, positions, and power.

Only a small segment of the population, then, received monumental graves furnished with grave goods, and relatively few were female (cf. Chapter 1). According to the Gulathing law, women could inherit and own *óðal* land only as single daughters, without brothers and without close male heirs, as so-called ON *baugrýgr*. If land was transmitted through the female line over three generations, it achieved status as allodial land through the female line. However, if she had both sons and daughters, her eldest son became the heir. As stated by law, when there were male heirs, heritage was only transferred to women in textiles together with other movable goods, while men normally possessed and inherited land.[12] Widows as stewards of minor children, landowning women, and widows as head of the household still held a great deal of autonomy and retained their dowry and bridal gift. Women in that position were, according to the Gulathing Law, also in the position to meet at provincial things and take part in legislative procedures, even to be summoned to local assemblies, and plead their own cases like men.[13] This was the situation in the early Middle Ages, and was probably built on old customs. Hence, heritage rights and the visual manifestations as grave monuments are not only passive reflections of social structures but represented active claims to land and inheritance as a visual demonstration of relations to the dead person by representing status and identity, and for maintaining and re-establishing such relations.

More than a hundred years after the official conversion to Christianity in the 1020s in this part of the country, the Gulathing law still had to forbid people to bury their dead in mounds or cairns (ON *í hauga œða reysa*). If found to be breaking the law, the offender had to remove the grave, bury the body in a churchyard, and additionally pay a heavy fine to the bishop.[14] The need of a paragraph in the statute thus indicates well-established and persistent traditions more than a century after the formal Christianisation. The same law also refers to grave mounds close to the farmhouses as prevailing structures,[15] and indirectly respected, and to a high degree preserved into our recent past. Cremations, referred to mythologically in the younger Edda, there denotes an earlier period, characterised by cremation as a common grave ritual. Archaeology has also confirmed that cremations were more frequent in the early period and became more seldom in Norway by the 10th century[16] and altogether non-existent in Iceland.[17] The Toranger grave thus represents a steadily rarer burial custom this late.

In the early 920s, a member of a diplomatic mission at the Caliphate, Ibn Fadlan, witnessed and described a cremation of a dead Viking tradesman in a ship burial at the Volga shoreline, a conspicuous funeral ritual, lasting for 10 days. It was characterised by heavy drinking and feasting, when lastly one of the female slaves was killed and followed her dead master onto the pyre. Although this story of human sacrifice, so-called suttee, may not represent homely burial customs, as for drama, extravagance, and promiscuity, some central elements may be relevant in a period when masters even had the lawful right to kill their own slaves.[18] Forced suicide or killing as human sacrifice, as known from the sagas too,[19] is therefore an option to consider when assessing graves with high-status weaponry together with simple 'female' artefacts, such as a spindle whorl, or as possible asymmetrical double graves. In the Viking period, households and not only in the upper strata (cf. Chapter 3), depended on a work force of thralls, without legal rights and kept as personal property.[20] Social hierarchies and inequalities are therefore imminent options when assessing tools and other artefacts as grave goods of various status in different contexts, where female slaves as concubines could even follow their master in death as 'grave goods'.[21]

According to the grave ritual reported by Ibn Fadlan, moveable property was divided in three parts following death – one for the heirs, one for the clothes in the grave, and one for beverage at the funeral feast.[22] Apparently, then, there are composite motives behind the burial rituals that are not uniform even when representing high rank. Tools as grave goods may reflect more than the physical work and allude to various roles and responsibilities within production related to both status and gender, either different, reciprocal, or superior roles as organisers. Although it is difficult to ascertain these principles archaeologically, customary laws on heritage and gift giving may explain the wealth in many graves.

All these factors contribute to an under-representation of female graves but also explain their general high status and wealth, only appearing as one or a few female graves among male graves. As graves are often located close to the settlements, at borders and strategic places for both communication and resource areas, and even clustered over generations, however, unevenly, since not all generation shifts were marked in this way. Such manifestations may have been more pronounced in times of conflicts; in needs of demarcation and symbolic demonstrations of rights and control. At central places, such graves may indicate nodes within larger estates of subordinate landholding (cf. Chapter 3).

To achieve a broader understanding of such processes, the graves are assessed both individually and relationally to other graves in their nearest surroundings. Later information – written records related to land owning as *óðal*, or as nuclei within estates with subordinate farms and at central places – is also relevant when assessing graves spatially and relationally, and as far as possible also related to the issue of 'what is missing', as stated in Chapter 2

Spatial distribution: landownership and tenure

Lately, spatial analyses of the distribution of Viking Age graves in western Norway substantiate a correlation between monumental and richly furnished graves and later documented ownership to landed property. Male monumental graves are clearly more numerous in the mid and inner parts of western Norway and, according to medieval and later sources, in areas that coincide with a larger share of freehold. Most likely, then, the graves represent landowners of inherited land and properties. Larger estates surrounded by tenanted land without the same property rights, in contrast, generally connect to coastal areas and coastal estates. Here, the main farm and nucleus in the estate is often marked by monumental graves of high status.[23] Seeing the graves as indicators of various types of landownership – at central places or nodes within estates or other types of farmland, possibly related to *óðal* – serves, then, as a relevant approach.

On a national level, female graves from the Viking period constitute about one-fifth of the total of detected graves from the period. However, in western Norway, the female representation is higher than in the country as a whole and especially related to graves furnished with textile tools. Within the present study area, the proportion of female graves varies from coast to inland, and with a higher representation of such graves in the inner zone but also connected to central places by the coast and in the mid-zone (cf. Fig. 5.1). A contextual analysis of graves at the same sites and other places in the community may thus reveal long-lasting patterns related to landownership, status, and production.

The burial traditions of 'family graveyards' were spread to the North Atlantic too – the earliest in Orkney and Shetland from the early 800s, included female graves that indicate settlements.[24] In Iceland, furthest away and settled later than the southern islands, aDNA analyses of the skeletons in the graves show that many women had Celtic origins – brought to the island either as slaves, concubines, or wives (cf. Chapter 1). As in western Norway, the grave monuments connected to the settlements, often at places by communication lines and at borders, indicating demarcation of land rights. In Iceland, the grave sites could embrace from 1 to 13 graves and of both genders. However, 90% of the sites had less than five graves, with the most numerous at places with early medieval churches, established as private churches at central places and larger farms[25] – a parallel trend to that in western Norway not least from the Late Viking Age.

Graves, gender, status

In Hordafylki, about a quarter of the around 500 documented graves from the period are identified as female: 89, plus 18 probable female graves,107 graves and 352 male graves, and the rest uncertain regarding gender.[26] Of these, 46 graves contained textile tools (see Appendices 5.1, 5.2 and 5.3, Table 5.1). In Sygnafylki, with 230 identified graves from the same period, 151 are identified as male graves and 49 as female, two as double graves and the rest as uncertain as for gender.[27] This implies that around a quarter of the gender-identified graves were female.[28] Of all these, 42 graves contained textile tools, which is altogether a relative higher share than in Hordafylki. In some inland districts in Firdafylki, as much as a third of all known graves from the period were female.[29] Firdafylki also contained the highest number of graves with textile tools, 54 graves in total, and in sum 142 graves with textile tools.

The overall find patterns of gendered objects in closed contexts form the framework when assessing the graves individually and contextually. In inhumation burials, the body could be placed in a wooden coffin but seldom fully documented archaeologically when of organic material, but also in stone tombs or other materialised settings, or without any casing at all. When cremated, ashes and burnt bones were placed in different kinds of containers, or left in fire layers before being covered, often by stones, and lastly by a cairn or mound – or without any long-lasting superstructure.[30] In some cases larger cairns and mounds contained more than one grave – either as double graves or placed in separate areas or as chambers in the mound or cairn. They may have been family graves when appearing of similar status, or when modestly furnished may possibly represent relatives or spouses without the same status.

According to the documentary evidence at the University Museum of Bergen, graves as visible structures in mounds and cairns are best represented, with more than half of the contexts with textile tools. Stone and earth may moreover have covered some of the graves without such information, not least when representing early and disrupted graves structures, many probably flat graves. Many graves, even of high status, had no archaeologically recognised superstructure. This is the case for at least 21, or 15% of the graves with textile tools in the present study area of western Norway, while many are unspecified as regards to structure. Altogether some 15 count as unclear contexts.

Cremations and fire exposed artefacts seem to have been more reported than inhumations when being without human remains and when other information about the structures is scanty. Mounds seem to be more common than cairns, although many graves are without specific information and the burials vary between inhumation and cremation whether found in cairns, mounds, or without any superstructure. Flat graves may therefore be under-represented. Boat graves related to all these categories, whether burnt, unburnt, in a mound or a cairn. Dug into the subsoil, only covered by earth so that the stems may have been visible, is also plausible, when new graves connected to and respected older graves of this kind. Traces of various forms of cremations, registered in some 40 contexts, represent a minimum of 30% of the graves with textile tools.

However, inhumation graves are clearly under-represented, being not specified when skeletal is lacking or other organic material not specified. Mounds and cairns could moreover contain several secondary graves on top or at the sides of the central grave and included more than one corpse or the ashes. Altogether, the burial practices were not uniform and varied, especially related to status. As many as 25 graves with textile tools were boat graves,[31] generally identified by many iron rivets and nails (Table 5.1), and a type of burial of high status for the privileged.[32]

These burial practices were transmitted to Iceland too, however, only as inhumation burials. Here, the

Table 5.1: Graves with textile tools, according to structures and burial customs (N:142).

Area	*Mounds*	*Cairns*	*Flat graves*	*Unspecified structure*	*Total*	*Cremations*	*Inhumations*	*Unspecified*	*Boat*
Hordafylki	21	6	4	15	46	15	7	24	6
Sygnafylki	11	11	6	13	41	7	5	29	9
Firdafylki	18	6	17	14	55	18	16	21	10
Total	50	23	27	42	142	40	28	74	25

better-preserved sex-identified skeletal material show that women without brooches or other gendered accessories could be buried in prestigious mounds too.[33] This may well have been the case in western Norway too, but with generally poorer preservation conditions for biological material. The burial custom of burying people of high rank, both men and women, in prestigious boat graves was transferred to the North Atlantic too, as evidenced in the Orkneys and Iceland.[34]

In western Norway, both the corpse and boats could be burnt, as described by Ibn Fadlan. Sometimes, however, the dead body was cremated and represented in the grave by the remains as ashes and burnt bones but unburnt grave goods. To what extent these burial contexts relate to status and gender, can therefore in some cases be unclear due to insufficient documentation as well as poor preservation conditions.

Gender, gender roles, and status concern both sexes. When human remains are lacking, the identification of sex, as well as gendered status, traditionally relies on dress accessories, not least the brooches fastening the straps of the traditional Viking Age female dress and other accessories, while weapons generally identify male graves, and in both cases people of high status. When the deceased female only has one such brooch and not two, she may possibly have been dressed according to other fashions, civil status, or of lower rank, as promoted by Solberg.[35] Other brooches may also have been lost. Other kinds of grave goods and belongings may indicate gender and status too – such as gender differentiated tool kits or as characteristic single tools, for instance a spindle whorl. As for status, costly dress accessories, brooches of copper alloy, sometimes silver-coated, chains of silver and precious beads of glass and gemstones for necklaces, keys, and various artefacts of foreign origins, count as status indicators. As for males, weapons – their number and composition indicate status too: Status group I, according to Solberg, is represented by only one offensive weapon – a sword, spear, or axe, with the axe as the most prevalent category. Status group II contains two weapons with the combination axe-sword as the most common combination. As a larger percentage of the two-weapon graves are monumental, the graves are more securely identified than for weapons within status group I. Status group III, marked by all three weapon categories, represents the smallest assemblage, often represented in mounds or cairns, in themselves social markers, and securely identified graves.[36]

Although these categorisations cannot count as absolute criteria for social distinctions and gender, they serve as useful analytical tools, besides a contextual assessment of other social attributes and distinctions. Still, not all high-status women wore brooches connected to the burial dress, such as in the renowned Oseberg burial, clearly of high status. The distinction between status, especially as for status categories II and III can be blurred,[37] and depend on the representation of other types of artefacts and tools, but also on the preservation conditions.

In the households as key units for production, the husband was normally the head of the household, financially and judicially. However, this was a role that women could take over as widows or as single daughters without male heirs, or as stewards in the husband's absence, being the second in command in the household, according to the Land law.[38] In the hierarchical society of the period, some households were subordinate to others, and could, then, involve two-way responsibilities and duties. The higher positions, the less physical work and the more seniority, networking, and management. Nonetheless, different types of work were gender divided, also as for what was regarded as proper and honourable for men and women, especially in the upper echelons of society.[39]

Textile production, and especially spinning, appears as an overall female domain, even used symbolically to denote women and not only restricted to unmarried women. Spinning tools therefore served as female symbols, as when the Gulathing law refers to women's rights of heritance and landowning of allodial land, which according to the law happened when land 'three times has come under spindle and whorl' ('*…er jǫrð þrysvar komen undir snuð ok snælldo*').[40] In the Middle Ages, one of the communal duties related to textiles concerned the provision of ships and sails to the national naval fleet, the so-called leiðangr. This duty, probably established in the mid-10th century in Scandinavia,[41] related to the so-called *skipreiða* on a regional level and a duty that also involved textile production, which must have affected both genders within specified districts, to provide a fully rigged ship physically and with manpower, and with the steersman as the overall responsible organiser.[42] As far as for textiles, the obligatory products required specified balls of yarn, ells of woven cloth, ropes and other parts related to the rigging of the sail. They covered, then, various work operations that needed coordination and organisation, but probably based on female producers of what was spun and woven. In an early stage of the period, a similar organisation may have been relevant when furnishing ships for raiding and trading, based on chieftaincy but it ultimately depended on a female work force. In Iceland too, textile production appeared as a female domain of varied status, not least women of lower status for the physical work,[43] and largely invisible through the burial customs.

Altogether, 87 contexts, or 61% of the graves with textile tools from western Norway, count as female graves based on dress accessories or other female adornment and equipment. Graves with dress equipment, weapons, and textile tools, 12 in all, and 9% of all the graves, seem to represent double graves. Whether graves of general high status indicated by the weaponry and with textile tools but without dress accessories, signify single male graves or rather asymmetrical double graves for men and females of unequal status, is ambiguous and a subject for further assessment. They include 16 contexts, or 11% of all. Additionally, 15 contexts

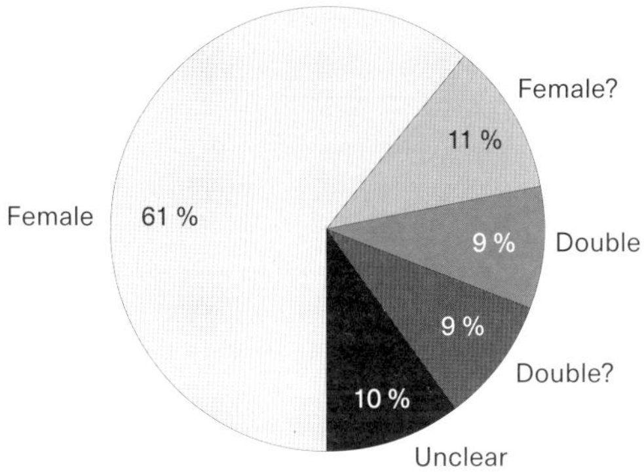

Fig. 5.3: Relative representation with textile tools from female graves from the study area in western Norway – as female graves, possible female graves, double graves, possible double graves, and unclear contexts as for gender (N:142).

with textile tools or 10% of all seem to represent mixed or disturbed burials and appear as unclear as separate graves as closed contexts. Scattered and single tools appearing as stray finds may additionally have represented graves, but are too uncertain to be included, since context is an essential component in this study (Fig. 5.3).

In Iceland, in contrast to graves from western Norway, remains of skeletons have been preserved in the subsoil with more than 180 skeletons both with and without social gendered artefacts and enabling identification of the sex of the buried from both single and double graves. Of these, it was possible to identify 60% for sex, based on both human remains and gendered artefacts. As many as 73 were identified as males, of which 45 are certain and 28 are uncertain, and 35 females – 20 are certain and 15 are more uncertain.[44] However, only a minority of the graves comprised textile tools. Over the period from the *landnám* c. 870 to the change of religions, one third of the graves were female judging by both the skeletal remains and gender-identified artefacts[45] – and about the same relative representation as in western Norway, but then generally based on gendered artefacts.

As death is otherwise equally represented within populations, inheritance and property rights and the symbolic representation of graves best explain this gendered imbalance. Women buried both in monumental and flat graves seem overwhelmingly to reflect the upper strata of Viking Age society in western Norway, and most likely landowning women, as so-called *baugrýgr*, as noted, a woman without brothers or other lawful male heirs. Statistically, this could amount to 20% of a population,[46] but could have been higher in this turbulent period, when both fathers and brothers may have participated in internal conflicts and external raids and women could be acting as stewards in charge of under-aged sons or as widows. For this reason, female agency appears more frequently in such grave contexts and with tools indicating a wide involvement in the various processes of the production and not least as organisers related to textile production within larger households.

The graves and grave goods thus shed light on social aspects – social status, gender roles, with wider implications related to use rights, ownership, power, gender roles and responsibilities. The funeral as a ritual and a rite of passage, and a social event marking the transition from life to death had additionally long-lasting synergy, involving larger financial investments, and was therefore primarily related to the upper echelons of society. Indirectly, the grave goods may indicate wider networks too and not least the many subordinates, as indicated by the tools in the graves, signifying the physical labour in the many stages of production. The mound or cairn as a prominent monument was thus a memento and manifestation of family lines, ownership and use rights and for the generations to come. It marks thus continuity in rights and domains and with a status-enhancing function as a collective material memory. At the same time, the graves and grave goods as indicators of social status and gender roles give insights into the hierarchical structure of the society. The tools as both concrete evidence and as symbols of production, may also reveal different levels of production at special places and in wider environments. Altogether, the graves with textile tools have the potential to shed light on many aspects of a production in both time and space.

On a micro level, women of varied social rank related to various grave types – double graves, separate and single graves, either as primary or secondary graves in different parts of the mound or cairn, or buried without any visible demarcation. The location at farm levels within their districts constitute the meso level, to be compared within larger regional frameworks within the three counties on a macro level. To identify graves as for gender, social status, and former roles within production, needs a careful contextual analysis of the grave goods – where, when, and represented with what kinds of tools and other types of artefacts.

The tools in the graves

The Toranger grave, a clearly visible cairn dated to the 10th century, is one of the few female graves furnished with textile tools this late in the coastal zone. Parts of a tortoise-shaped brooch of copper alloy of a rather common type of the century signal the deceased's gender, social status, and roles, substantiated by the finds of a key and textile tools – shears, a small spindle whorl, a few loom weights and a weaving beater, remains of a smoothing plaque of whalebone, and a sickle. Special tools and dress accessories in Viking Age graves have been interpreted as tokens of status and responsibilities – related not least to food, storage, production and maintenance of clothes and other textiles (cf. Chapter 4). Although the tools as grave goods appear as personal belongings, they may not necessarily have been used by the deceased in the graves and may rather indicate roles, responsibilities, and work domains in the position as female landowners or stewards. Burials and funerary rites thus followed conventions and norms that largely allude to social identity, social and marital status, roles, and responsibilities – and as physical structures in the landscape as the living displayed and marked them[47] as signs of power and heritage with long lasting effect. Appearing mainly in high-status female graves, the tools leave the impression of female work domains as managers of production on a larger scale. Textile production as part of larger societal structures, households of different scale and levels, and within larger or limited networks, opens up for different types of agency – mainly female but possibly also male as the division of labour was not always consistent and may have contained some flexibility between the sexes,[48] at least at lower social levels.

The tools as indicators of production

Tools for all the stages of textile production – shearing, sorting wool, combing, and spinning the fibres, warping, and weaving yarn – were necessary in the production, although when based on moulting sheep, shearing was not obligatory, and sorting needed no tools, only competence and hands – and was only visible in preserved textiles. Combing was necessary for making finer qualities, but not necessarily for the coarser ones, and hackling as for plant fibres. Spinning yarn of various gauges needed spindle whorls of different weights, as also the loom needed loom weights of varying weights and dimensions for stretching the warp, and according to required qualities, fineness, and density of the weave (cf. Chapter 4). Similarly, weaving beaters for beating the weft could be of different materials, preserved in iron and whalebone but wooden beaters would also do for finer textiles, needing softer strokes. Even in small-scale households, based on an ordinary household production, all the basic operations were necessary, and would need more than one or two people at work, depending on the number of people in the household – including slaves – and to cover its own needs for all use areas. The various work processes, outdoors and seasonal – for shearing or moulting and plucking and sorting wool, moreover, sowing, weeding flax and hemp during summer and mowing in autumn – represented only the initial stages of production, and generally time-consuming outdoor activities. Spinning and weaving could be done on an all-year schedule but depended on the foregoing processes and required coordination and administration when based on a large-scale production (cf. Chapter 4). Several people with special competency were crucial within an advanced production – and the finer the qualities, the higher the dexterity.

Tools as grave goods, however, reflecting production indirectly cannot automatically count as a 1:1 expression of production. Nevertheless, as intentionally chosen grave goods, they still have a bearing related to both social status, gender roles, and functional aspects of production – represented by tools for one to all work stages of production – and indicate socially defined work domains, selected for varied processes and qualities. When tools for all processes in the long chain of work stages appear together with other equipment and social high-status emblems, they may signify roles as managers, coordinating and administrating the production. When represented by only one or two loom weights, they may be understood as *pars pro toto,* in the sense as a part [taken] for the whole and thus symbolise weaving, regardless of number. Tools for the initial and final processes, such as spinning and weaving, may likewise indirectly reflect all the other processes. Still, their material and weight reveal functional qualities. Tools for combing, spinning, and weaving offer the best opportunities to assess levels of production and for finer or courser qualities – wool combs for the fineness of the long-haired wool and essential for preparing worsteds; the weight of whorls for spinning yarn of varied qualities, while weaving beaters of different lengths, shapes and materials reflect weaving of finer or stronger textiles, as well as sewing equipment. Tools for treating textiles of plant fibres, hackles, whalebone plaques, and glass pellets for smoothing, reflect production of finer qualities of plant fibres, and mirror generally high-status graves and environments.

The relation to other graves and other environments is a relevant aspect too –for status, locations, access to resources, for communications and control when assessing the graves and the tools therein. From a comparative perspective, the tools may indirectly shed light on agency, roles, and modes of production at various places and in various environments – whether they represent ordinary household production or connect to central places with opportunities for more advanced or even specialised production. Some of these implements are costlier than those made of stone and wood, however, with the latter as the most perishable. Accordingly, tools for making finer textiles within the luxurious segment would be expected as grave goods in high-status graves, such as wool combs or hackles for combing plant fibres, different whorls for spinning and weaving beaters rather than single loom weights for weaving and sewing equipment.

Of the total number of textile tools in the graves, nearly 700 objects in all, the majority (90%) relates to female contexts, when including double graves and probable female graves, while graves with weapons and textile tools but without dress accessories may represent double asymmetrical graves, or male graves, and some unclear as graves. The representation of textile tools as grave goods gives an average of nearly five tools/parts of tools per grave. Tools in female graves, however, have a higher representation with an average of 5.2 tools per grave, and 5.4 in possible female graves and relative highest, 7, in double graves, 3.7 in possible double graves of unequal status or male graves, and

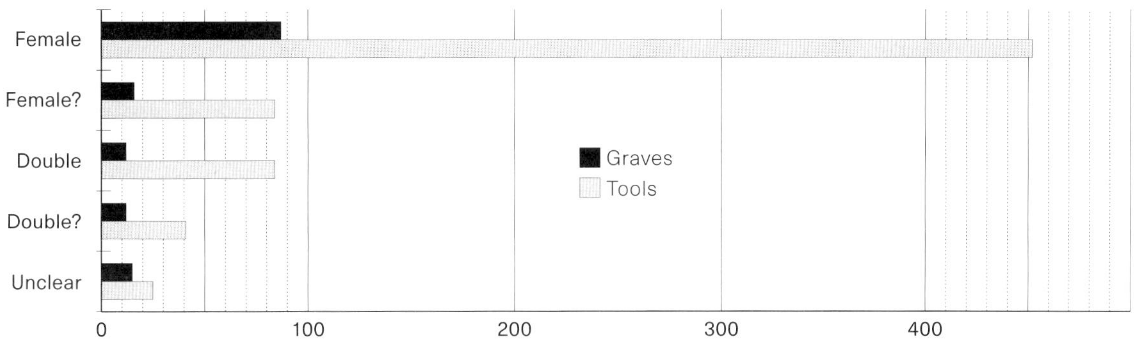

Fig. 5.4: Number of graves with textile tools=black (N:142) and the total number of textile tools therein=grey (N: 686).

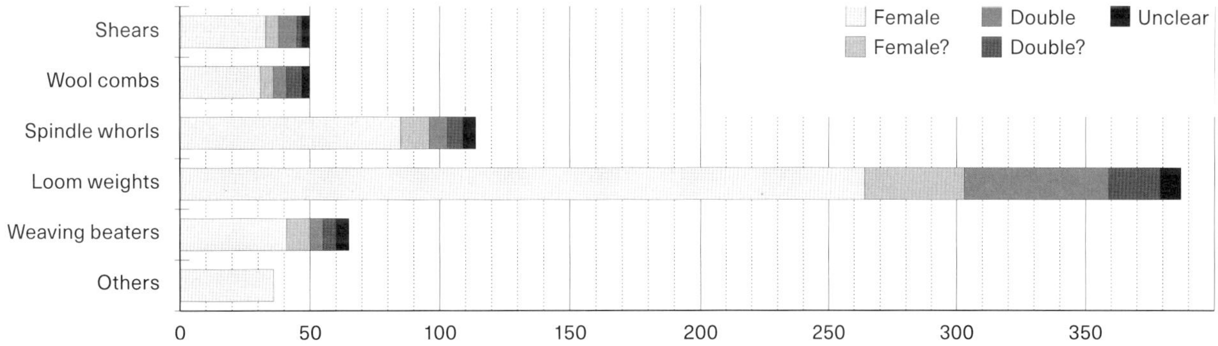

Fig. 5.5: Representation of tool categories related to different grave types and gender. Light grey = female graves, grey = possible female graves, darker grey = double graves, somewhat darker possible double graves, and darkest = unclear contexts (N:702).

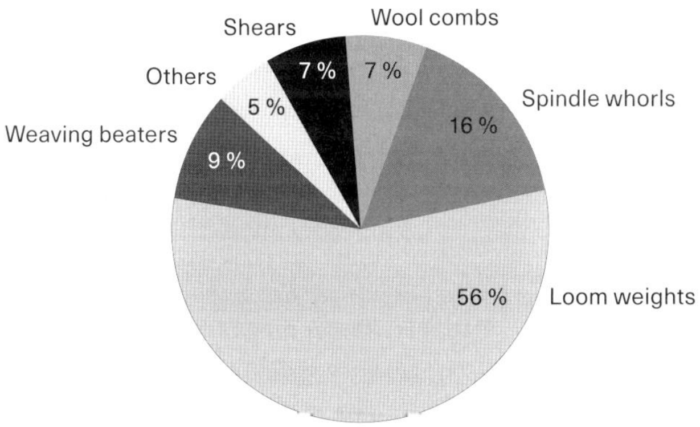

Fig. 5.6: Relative representation of tool categories (N:702).

only 1.7 tools on average related to unclear grave contexts, and an overwhelmingly female representation (Fig. 5.4).

The respective tool categories are represented in varying numbers, with the highest share of loom weights, with 387 items in all, however, only as parts of the loom weights in a loom and 56% of all the tools. They connected mainly to female graves, 29 in all or 33% of the 87 female graves, with from 1 to 48 loom weights, with altogether 246 weights, representing 64% of all the weights. Six, or 37% of the 16 possible female graves included loom weights too, varying in number from 2 to 16 weights and with 40 specimens in all. Four of the 12 double graves with loom weights, varying in number from 6 to 21 and with 56 weights in all. The representation was radically sparser in the more unclear contexts of the 12 possible asymmetrical double graves and the 15 unclear contexts. Of these, only two of the 11 graves contained loom weights, three in all, and four of the unclear contexts with eight loom weights (Figs 5.5 and 5.6).

Spindle whorls make up the second largest group counting 114 objects, representing 16% of the textile tools and weaving beaters, 65 in all and 9%. Shears and wool combs comprise equally 7% of the tools (Fig. 5.6). The group 'others' represents small assemblages of needles, needle cases and tools for smoothing, and constitute no more than 5% of all textile tools.

Of these, spindle whorls have the strongest archaeological affiliation to women, represented in 56 graves, or 64% of the 87 female graves with textile tools. They vary in numbers from one to four whorls, with 81 whorls in all. Seven of the 16 possible female graves, represented by one to three whorls, contained 11 whorls in total. Five of the 12 double graves, with either one or two spindle whorls, comprised eight whorls in all, while six of the possible asymmetrical double graves contained only one whorl each, like in five of the unclear contexts (Fig. 5.7). In Iceland, the four graves with spindle whorls were all female.

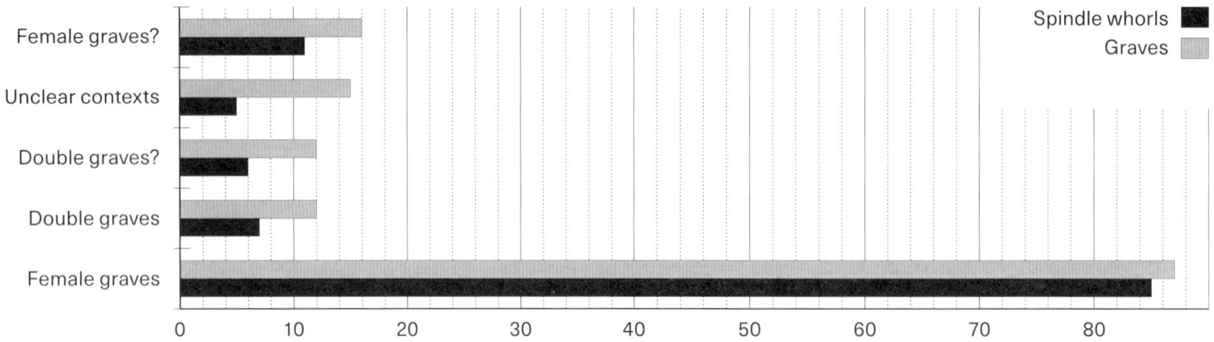

Fig. 5.7: Spindle whorls (black) related to graves according to grave types and gender (grey) (N:114 whorls).

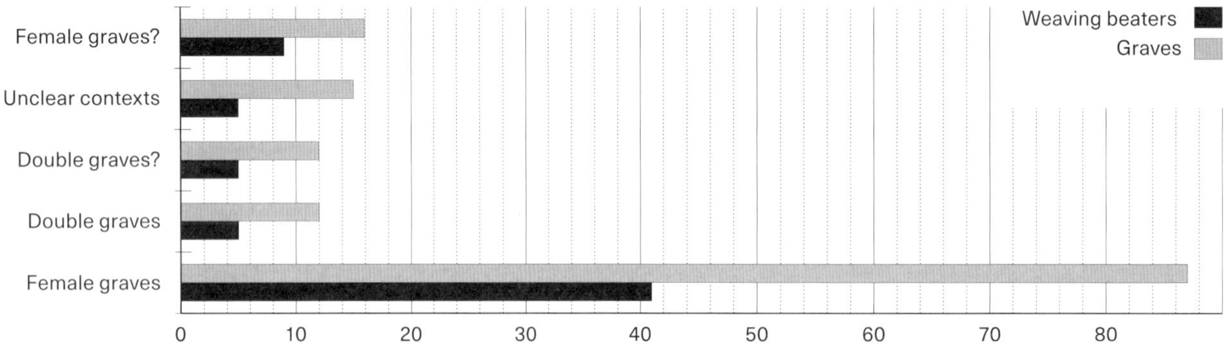

Fig. 5.8: Weaving beaters (black) related to graves according to grave types and gender (grey) (N:65).

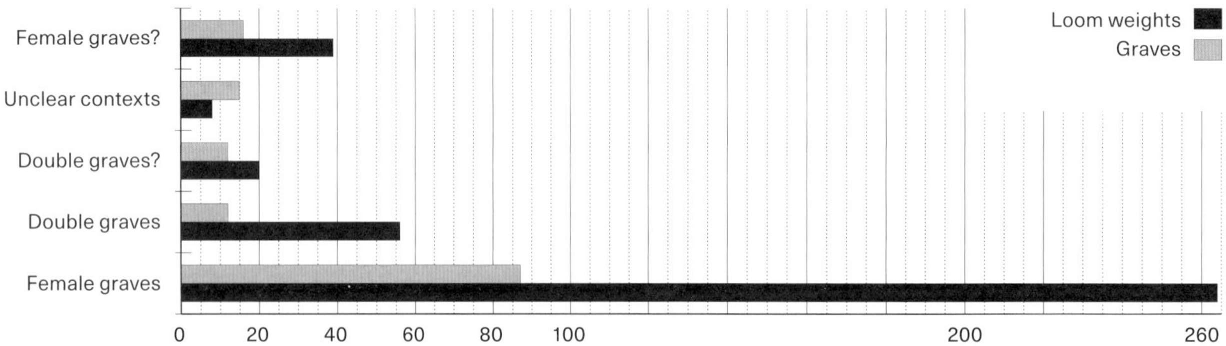

Fig. 5.9: Loom weights (black, N: 387), related to graves according to grave types (grey).

Weaving beaters as grave goods, represented by 65 specimens, show a similar pattern, also with the highest representation in female graves with 41 beaters and with nine in possible female graves. Five beaters relate to double graves, five to possible asymmetrical double graves or male graves, and yet another five to unclear contexts (Fig. 5.8). Of all these, four beaters were made of whalebone, all the others of iron. To what extent the representation of weaving beaters of iron, as far more costly tools than loom weights, primarily connects to status or signal various production of textiles, such as sails, needs a closer inspection, as for size, material, and shape. In Iceland, renowned for its extensive textile productuction of *vaðmál*, only two of the graves included weaving implements, and one of them, a prestigous female grave, with a weaving beater of whalebone,[49] and hardly efficient for weaving *vaðmál* qualities, such as, among others, for sail cloth (Fig. 5.9).

Quantitatively, loom weights form the largest group, with 387 weights in all. However, as they mostly appear in clusters, and generally with more than one weight and up to 48 weights in one of the graves – and dominantly related to female contexts. About two-thirds of all the loom weights, 264 specimens, related to 39 of the 87 female graves. Of these, 31 graves contained more than one weight and nearly half of them, 16, with more than five loom weights. Additionally, six of the 16 probable female graves contained

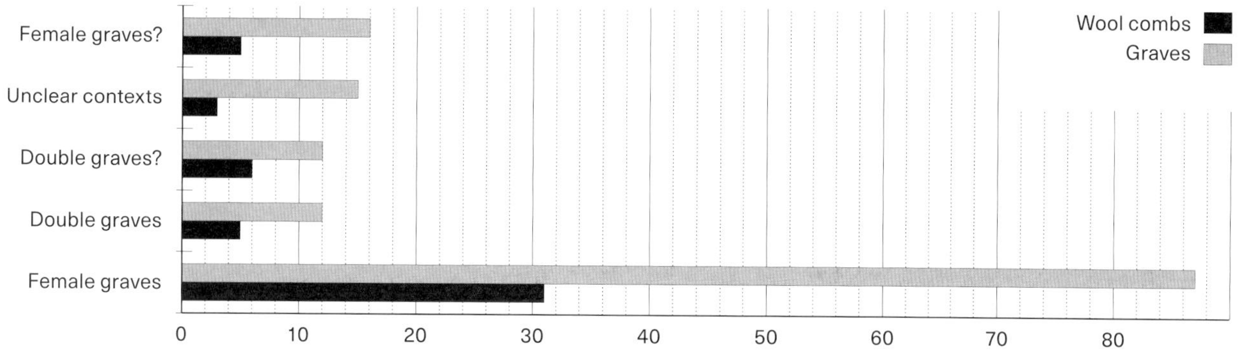

Fig. 5.10: Representation of wool combs (black, N:50) in graves according to gender and grave types.

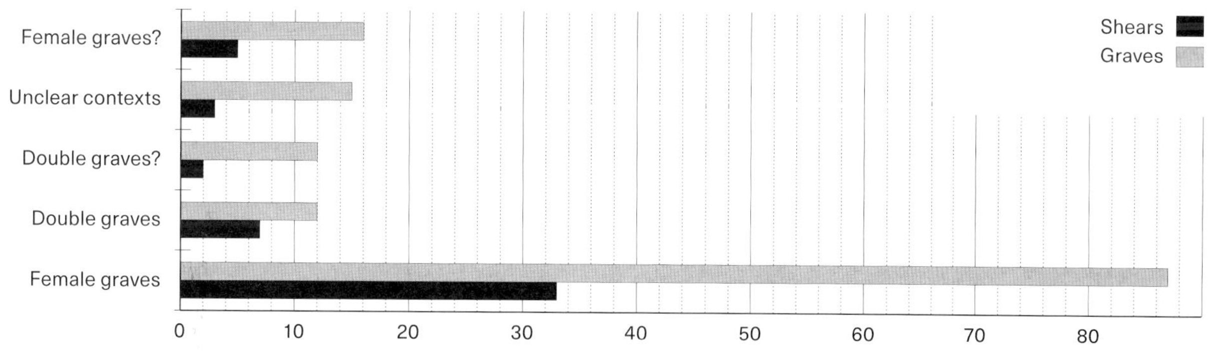

Fig. 5.11: Representation of shears (black, N:50) in graves according to gender and grave types.

loom weights, with 39 weights in all, moreover, four of the 12 double graves contained 56 weights, while three of the 12 possible asymmetrical double graves or male graves contained 20 in all, and eight weights stemmed from four of the unclear contexts (Fig. 5.9).

The representation of wool combs and shears as grave goods form largely parallel patterns – both in numbers and related to grave types, with remains of equally 50 shears and wool combs (Figs 5.10 and 5.11). Both shears and wool combs are represented in 31 of the 87 female graves, and with five specimens each from the 12 double graves and represented in the 11 possible asymmetrical double graves or male graves, and four of 16 in probable or possible female graves (Figs 5.10 and 5.11).

In contrast to shears for sheep shearing (R442), the type of shears for cutting yarn and textiles (R443), 18 in all, only connect to female contexts of high status, as many as 13 in female graves of status III; three in high-status double graves, one in a probable asymmetrical double grave with a multiple tool kit and one in an unclear context. In Iceland, the seven graves with shears, generally for sheep shearing, were all female and one from a double grave.

The small group of 'others', includes, as noted, needle cases and needles, tools connected to sewing, and pellets of glass or whalebone plaques for smoothing, and nearly all are related to female high-status graves. The grave at Rousay in Orkney and a grave in Iceland, both high-status female graves contained whalebone plaques too,[50] as in the Toranger grave, besides six other graves from the present study area and all of high status. One of them was a prestigious boat burial at Folkedal in the inner Hardangerfjord and a high-status female grave at Seim in the same district and the coastal grave at Fosse in Hordafylki too. The spatula-shaped tool (R413), appearing in 11 female graves with textile tools and one possible female grave, plus two double graves, and tentatively connected to band weaving, form a special group of tools to be assessed too.

Overall, this quantitative assessment of tool categories demonstrates a systematic overrepresentation of textile tools in female contexts, and is therefore hardly accidental. Spindle whorls and loom weights – spinning and weaving – show the strongest female affiliation, along with implements relating to seasonal and outdoor work of sheep shearing and combing. These patterns may be even clearer when seen contextually in their wider environments and correlated to status and as part of wider tool assemblages in the graves, either for all or nearly all main processes in textile production, or only a few.

Female graves: tools, work domains, and status

Female graves were not homogenous either for status or for composition of textile tools – use areas and in numbers.

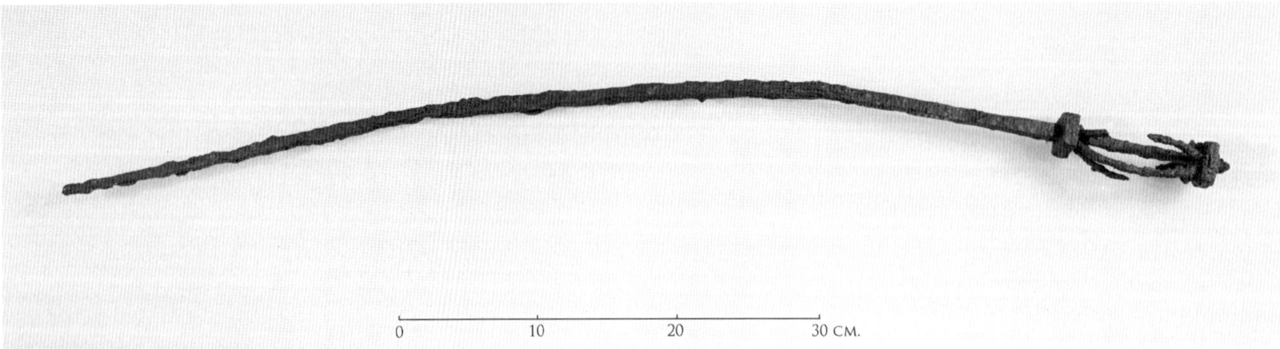

Fig. 5.12: Long iron stick with a handle – a tool or a sign of status? (B6657, Trå, Granvin) (Photo: © University Museum of Bergen, with permission).

Table 5.2: Female graves with textile tools and related to number of tool categories and status groups (N:87).

Tool categories	*No. of graves*	*Status III*	*Status II*	*Status I*
6	3	2	–	1
5	13	11	2	
4	7	6	1	–
3	13	10	2	1
2	31	15	9	7
1	20	7	9	4
	87	51 (59%)	23 (26%)	13 (15%)

Comparing social symbols, connected to dress with the remains of tools and other grave goods may indicate work domains and roles in production. This may give a clearer conception of ways and modes of production – either as ordinary domestic production or as managers at a large-scale and advanced level, needing many hands and based on a larger work force, or on a specialised level, and based on special competence.

One of the most varied tool kits comes from a richly furnished grave in a large mound at Trå in Granvin, in the inner zone of Hordafylki.[51] Here, in a secondary grave in the outskirts of an old grave mound, a high-status woman, buried in a wooden chest, had tools for all stages in textile production – shears, wool combs, two spindle whorls, a weaving beater and several loom weights, a needle case, and a glass pellet for smoothing (see Chapters 6 and 9). The grave goods included, moreover, tools for food processing and storage: a vessel of soapstone, a frying pan, a ladle and a kettle of iron, pieces of another ladle, and a bronze cauldron of Insular origin (see Table 5.2), and also a long spit or staff of iron, earlier interpreted as a frying or roasting spit or a skewer (Fig. 5.12).[52] However, lately other interpretations have been launched – as signs of status,[53] as a ritual symbol of the so-called *vǫlva*, a female soothsayer and of witchcraft,[54] or simply a practical distaff[55] used for holding and stretching the wool when spinning with a drop spindle (cf. Fig. 4.5). Their contexts as grave goods are therefore relevant, both as for status and as tools (Fig. 5.12).

Bronze mountings of a chest and remains of horse gear – bridle bits, bridle ring-fittings, and strap mountings – were other prestigious grave goods in the grave. This snapshot from the imposing burial finds illustrates the importance of assessing the tools contextually.

Within the study area as much as 51 of the female graves with textile tools, or 59% of the 87 graves in this category belong to status group III, 23 graves or 26% to status group II (or in-between status FII–III), and 13 or 15% to status group I when including the in-between group FI–II (Table 5.2; cf. Appendix 5.1).

Table 5.2 shows that graves with textile tools from western Norway are highly overrepresented compared with the overall representation of social status female graves. According to Solberg's comprehensive study of status in Viking Age graves, only 16% of the female graves connects to the highest status group III, 40% to status group II, and the rest, 44% to the lower status group I. However, the representation in female graves with textile tools forms quite another social structure – in fact, an inverted pyramid, with 59% of these graves of the highest status III, 26% of status II and only 15% of the graves of the lower status group I and clearly not a normal social representation in a stratified society (Fig. 5.13).

This upturned pyramid illustrates thus a significant correlation between high status and textile tools, which highlights the importance of textile production as such. The number of tools and tool categories is clearly highest in female high-status burials both quantitatively and relatively (cf. Table 5.2), where as much as 59% of all the female graves with textile tools represents graves of status III. Indirectly, then, the tools in these graves signify superior and wider roles than as producers on a household level, and rather roles as managers, directing and coordinating the production on a larger scale, and carried out by a group of people, from slaves to other subordinates and largely invisible in

Fig. 5.13: The overall representation of female graves groups from the study area of western Norway according to status (left) and representation of graves with textile tools from the area of study according to status (right) (N:87).

Table 5.3: Number of textile tools in female (F) graves according to status (N:473).

Status	*Shears. R442 (R443)*	*Wool combs*	*Spindle whorls*	*Loom weights*	*Weaving beaters*	*R413*	*Needle-case/ needles*	*Smoothing tools*	*N: Tools*	*N: Graves*
FIII	27 (13)	22	56	171	27	7	6	6	322	51
FII	5	6	15	35	8	1	–	1	71	23
FI	1	2	10	62	3	1	1	–	80	13
N	33	30	81	268	38	9	7	7	473	87

the archaeological record. To some degree, the tools in the graves included farming tools too, especially hoes and sickles, and altogether tools for the labour-intensive processes of weeding and mowing textile plants of flax and hemp. Considering the stratified-structured Viking Age society, the upturned pyramid of high-status graves 'hides' and implies a large and archaeologically invisible workforce without representation in preserved graves.

Comparing social markers connected to dress with the remains of tools and other grave goods, and their general higher status, they also indicate work domains and roles in production as producers and/or organiser, and then possibly on an advanced or even specialised level. This aspect becomes even clearer when considering the extensive amount of work involved in the many work processes and stages of production, as shown in Chapter 4.

The representation of textile tools as parts of tool kits or as single tools – combined with dress accessories, imports and other status-indicating artefacts in female contexts, and weaponry in male contexts – reflects both social and gendered hierarchies. Most female graves with textile tools for more than two work stages belong to status III. Altogether 28, or 54% of the 51 female graves of that status group contain composite tool kits, with from three to six various tool categories related to textile production. Still, graves with only one or two tool categories appear as the most common – 51 female graves, of which 24 and 16 connect to status groups III and II. Generally, graves of lower status have fewer textile tools.

Tools for weaving as the last major stage of production, represented by loom weights and weaving beaters, dominate in status III-graves, and with weaving beaters in 34 graves of status III or II. The tools for the seasonal preparatory processes of shearing and combing are more frequent than tools for spinning, with wool combs as most common in graves of status group III and indicate the significance of preparing wool to get fine long-fibred wool to achieve high quality textiles, spun into yarn and for weaving worsteds. Spinning as obligatory before weaving, however, is relatively best represented in status groups I and II (Fig. 5.14).

The smaller groups of tools as grave goods and related to the last stages of production – shears for cutting yarn and textiles (R443), needles and needle cases related to sewing, the spatula-shaped tool possibly used for band weaving

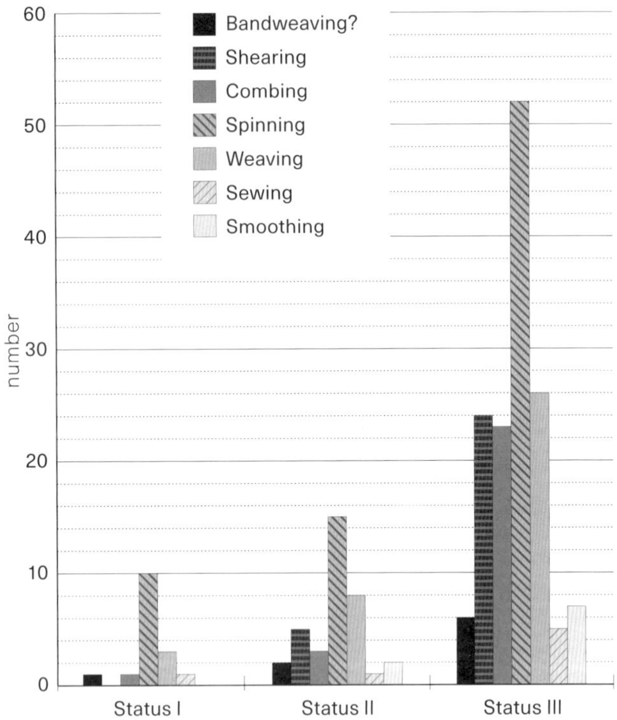

Fig. 5.14: Work processes indicated by textile tools in female graves according to status (N:446 tools).

(R413), and pellets and plaques for smoothing the woven and finished textiles of linen – largely connect to high-status burials and often at central places and in textile producing environments (Fig. 5.14: cf. Appendix 5.1).

All these processes from shearing, moulting, or plucking, and followed by combing, spinning into various yarn qualities and lengths in the warp and the weft to produce the wanted fabrics when weaving and sewing the fabrics – clearly needed organisation and coordination. The various combinations of equipment chosen as grave goods may then directly and indirectly, reflect various degrees of involvement – whether as administrators of the production as such, or related to selected work areas, such as the basically indoor and inter-connected work of combing, spinning, and weaving. Graves represented with tools for spinning and weaving, constitute the most common combination, documented in 12 of the female graves with two tool categories. Eight graves, with one tool are equally represented for weaving or spinning and relate to status III only in two cases. Overall, composite tool kits are most common in the female graves of high status and with weaving equipment of loom weights and/or weaving beaters and spindle whorls as the most common implements.

Two of the high-status graves with composite tool kits are the earlier mentioned Trå and Veka graves, with tools for all stages, from shearing to weaving and sewing. In the Veka grave (Fig. 5.15) all the tools were gathered in a chest by the feet – wool combs, two spindle whorls, several loom weights, the spatula-shaped tool (R413), and a needle case of copper alloy with sewing needles. The tools included moreover agricultural tools; two hoes and a sickle. Like in the Trå grave, it contained a long spit or staff/distaff, by her side, 77 cm long and with a ring and handle (Fig. 5.15).[56]

Overall, then, textile tools as grave goods primarily link to high-status women and especially when represented by tool kits for the whole chain of work operations, they indicate a superior role within the production. Still, more than half of the 87 female graves contained fewer tools and most frequently for spinning and weaving. As this combination often appears in burials of status II or I, it may be tempting to interpret them as signs of active producers of yarn and fabrics for the household, and/or possibly as more specialised producers connected to the household. Still, it is problematic to interpret the grave goods as 1:1 manifestation of the deceased's roles and actions in former life.

In comparison, the few graves with textile tools from the North Atlantic were also connected to high-status women judging by the dress accessories, but only seldom included the same broad range of tools as the many graves from western Norway. However, the earlier mentioned female grave from Rousay in Orkney from the early 9th century – the young high-status woman of around 20 years old and buried with her little child, was furnished with a tool kit as grave goods, with tool for shearing, combing, weaving, and smoothing.[57] One of the high-status graves at Daðastaðir in Iceland had an inclusive tool kit too: shears, wool combs, two spindle whorls of steatite, and a sickle.[58] Another high-status female grave at Kornsá in north-western Iceland contained tools for shearing and a weaving beater of whalebone, besides a cooking vessel of iron and a special find of a scale of copper alloy.[59] Yet another grave of high status, at Ketilstaðir in eastern Iceland, only had tools for sheep shearing and spinning.[60] Shears are moreover found in five grave mounds in Iceland, of which at least four, and probably all, were female graves.[61] Of these, four were for sheep shearing (R442), while one of the high-status female graves had shears for cutting textiles (R443) – and a grave at Álaugarey, that also included the special long iron spit/staff/distaff, 78 cm in length,[62] like the staffs/distaffs in the graves at Trå and Veka too. Although tools as grave goods did not have the same break through as in western Norway, they show parallel trends for gender and status.

Tools as grave goods reflect in this way gendered roles, social status, and social identity but also indicate various modes of production. The graves and grave goods should therefore be assessed as an entity, practically, functionally, and as for agency, also reflecting the social aspects of production.

Keys as symbols of the landlady?

Traditionally, keys with their ability to lock and unlock, have served as a strong symbol, even juridical, of power and responsibilities, protecting property, stock, and storage of various goods, such as food, textiles, tools, jewellery,

Fig. 5.15: The Veka grave with the buried woman in a coffin (Shetelig 1912), and with a reconstruction to the right. The textile tools are in a small chest by the feet and with a long iron stick/staff in front of the deceased and a sickle by the hips and two keys below the brooches (Reconstruction by Neil Price and Þórhallur Þórarinsson, with permission).

and the like. The power of the key in daily life seems to have belonged to the landlady/housekeeper (ON *húsfrú*), and as a precondition for a lawful marriage, according to early medieval laws in Norway. The Eddic poem Rigsþula 23, tells that Karl, symbolising the farmer, *húsbóndi*, as a prototype, brought his new wife to the farm with hanging keys.[63] As signs of the role of overall managers of households, keys in female graves may symbolise responsibilities within the households and add further evidence to the issue of roles and gender roles in production. However, as keys sometimes occur in male graves too, they cannot count as absolute gender criteria.

The Toranger grave, introduced in this chapter, was one of 21 female graves with textile tools and keys, and nearly all within high-status III and II and with the Toranger grave as one in the in-between group II–III. The number of keys could vary from one to four, but mostly only one or two. One of these graves was the grave Folkedal with three keys without the common brooches, but richly furnished with textile tools and other prestigious grave goods. Another female grave in this category at Øvreset in Breim, was also well-equipped with tools. Except three female graves, one of status II, and three of status I, they all represent the highest status III, and eight with tools from one to five tool categories for textile production and with altogether 46 keys (Table 5.4). Keys may then count as a distinguishing sign of superior roles in households and as grave goods generally related to the upper strata. Keys appear moreover in six double graves of similar high status (Table 5.4; cf. Appendices 5.1–5.4).

The keys differ in shape, size, material, and function – used for locking smaller or larger containers or controlling access to buildings. Hooked iron keys of various sizes and locking devices represent the most common key type of the period for locking caskets or chests (R459). With one to three key-bits, they indicate various degrees of security of the locks. Small turning keys of bronze, with a relatively large and often decorated head as a handle and a short stem and various key-bits, are more seldom (R454, 455) (Fig. 5.16).

Of the six double graves with tools and keys, all but two were of equal high-status III or II for the female and male. Of these graves, three contained only one tool category, for either shearing, combing, or weaving, while two had tools for three processes, respectively combing, spinning, and weaving: the other for shearing, sewing, and the special spatula-shaped tool. Only one of these double graves with keys had tools for all basic processes – shearing, combing, spinning, and weaving. Two probable asymmetrical double graves had tools for two processes: respectively combing and weaving and spinning and sewing. The tools in these graves substantiate, then, a female presence but in the possible asymmetrical double graves with a general male superiority.

In many cases, graves with textile tools comprised both keys and chests, but more often a chest without a key, and sometimes only keys. Still, many chests have traces of locks, and indirectly, then, keys, such as the Trå grave with a chest with a lock for a hooked key. Altogether, there is a strong connection between keys, locks, lockable chests, and high-status female graves with textile tools. Keys and chests were grave goods in Iceland too, appearing in seven graves, however, both male and female.[64]

Textile tools, combined with food-related tools too – for cooking, frying, serving, and storing – signify status too. The grave at Trå contained as many as seven types of tools for cooking, frying, and serving, and as noted, a long iron stick or staff with a handle, whether related to frying or spinning or for no practical reason, it undoubtedly symbolised high-status. This grave is only one of 41 graves with textile tools combined with both textile and food-related equipment and the majority, 32 of status II and III and only seven of status I. Still, one and two tools of this kind dominated in all status groups (Table 5.5).

Textile equipment and agricultural tools is another frequent combination, especially hoes and sickles in arable farming, and as grave goods often combined with shears, wool combs, or hackles – representing the initial stages of textile production outdoors. Weeding related to the production of flax was especially demanding, as also was mowing the ripe cereals as another labour-intensive season. These processes have traditionally been female work domains, needing many hands, both within an ordinary household production and when on a larger scale, they to a larger degree also needed coordination and organisation. Agricultural tools, hoes, and especially sickles connected

Table 5.4: Graves with keys and various categories of textile tools: female graves (N:21); other graves with keys (N:7); six double graves, and a possible asymmetrical double grave.

	Female graves with keys	*Status III*	*Status II*	*Status I*	*Other graves with keys*	*N: Graves*	*N: Keys in total*
5	4	2	1	–	1	6	8
4	5	5		–	1	6	12
3	2	1	2	–	1	3	5
2	7	3		1	2	9	17
1	3	2		–	2	5	4
N:	21	17	3	1	7	29	46

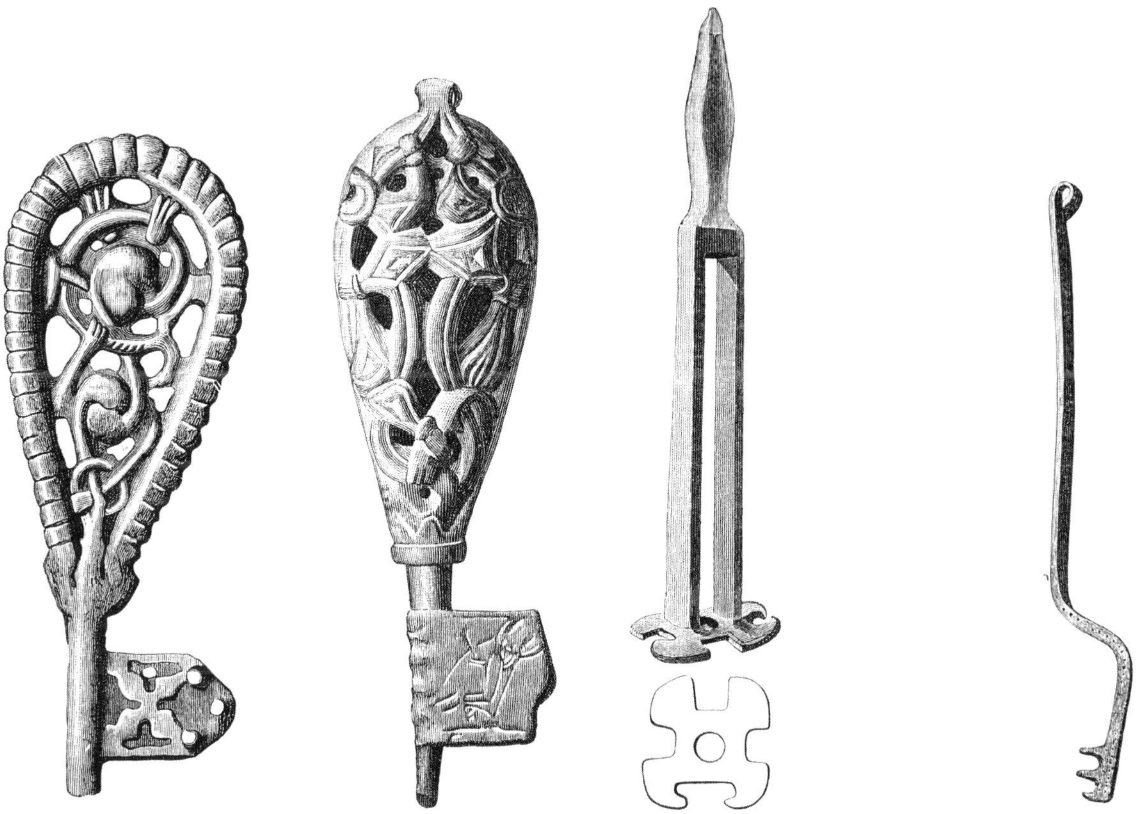

Fig. 5.16: Hooked key (R459), push key (R456), turning keys of bronze (R455, R454) (Rygh 1885).

Table 5.5: Female graves with textile tools and various food related artefacts for cooking, frying, serving, or storage (N:41).

Tools status	*1 category*	*2 categories*	*3 categories*	*4 categories*	*5 categories*	*6 categories*	*7 categories*	*N:*
III	11	5	4	2	2	1	1	26
II	5	3						8
I	4	3						7
N	20	11	4	2	2	1	1	41

frequently in the high-status female graves with textile tools and are represented in about half of these graves but only in a few of them combined with hoes in the same grave (Table 5.6). When dominating in high status graves they altogether indicate a superior role within production (Table 5.6).

Horse gear, not least bridles, for riding and other transport, appears as a status indicator for both men and women, and in as many as 24 of the female grave contexts with textile tools, of which 18 are of status III and three of status II. The combination of textile tools and status-indicating artefacts, such as keys and horse gear thus seem to underline the social aspects of production with a strong visible and symbolic effect, seemingly as managers, in charge of goods at stock and possibly on horseback when for instance inspecting outdoor production.

Table 5.6: Female graves with both textile tools and agricultural implements.

Status	*N: Graves*	*Sickles*	*Hoes*	*N: Tools*
III	51	25	7	32
II	23	2	1	3
I	13	1	2	3
N	87	28	10	38

A recent study of horse gear in the Viking Age graves from the two northern areas Sygnafylki and Firdafylki, focusing on their role as both status markers and for utility purposes, revealed that altogether there were 127 graves with such grave goods and of general high status. Of these, 21 were female graves or 17% within this category,[65] and

Table 5.7: Graves with textile tools, combined with other types of tools (N:163) in female graves of varied status.

Status	*N: Graves*	*Keys*	*Chests*	*Kitchen utensils*	*Strike-a-light*	*Sickles*	*Horse gear*	*Stick/staff*
III	51	31	10	27	8	25	21	9
II	23	2	2	8	1	2	5	–
I	13	1	2	7	1	1	–	–
N	87	34	14	42	10	28	26	9

of interest in this context that all of them comprised textile tools either in separate female graves or double graves. When 26 graves, or 29%, of the 87 female graves with textile tools and of general high status include horse gear, it also underlines their special status (Table 5.7). This is further corroborated when including probable female graves with textile tools but without the traditional dress accessories and double graves of both equal high standing and seemingly asymmetrical status in the following sections.

Seeing all these status indicators combined and related to status, the conception of the key as a general sign of the role of housewife may be overrated as they appear only in a smaller segment of the female graves, with 46 keys from 29 graves and nearly of general high status, II and III. As both keys and chests generally connected to high-status women and more often to textile production than to other kinds of production, they may possibly distinguish a certain stratum of managers of textile production.

The combination of grave goods and contexts show that most of the graves with such tools belong to women from the upper strata of Viking Age society, and as noted as much as 59% of the female burials link to the highest stratum of status III, 26% to a mid-position of status II, and 15% of the lower status I (cf. Fig. 5.13). In some cases, these women have been buried in secondary graves, and may represent a new generation of landowning women, as single daughters, as *baugrýgr*, or widows with status-enhancing roles, such as stewards for minor sons. In several cases, the same farm site comprised more than one grave with textile tools of different status and with grave goods and tools in varying combinations. In such cases they may possibly represent different roles or a new generation of women in this position. When found together with contemporary male weapon graves close-by, they also signal places of power and control.

The broad spectrum of tools as grave goods, which included food-related equipment, 61% of these graves connected to the graves of highest status III, while graves with agricultural tools, and especially sickles for mowing, nearly all related to that social stratum (Table 5.7). Such graves may indicate roles as managers of a broader spectrum of work operations and domains. As shown in Chapter 4, the textile tools connect to both seasonal outdoor activities, shearing of wool, weeding in the summer, and mowing the arable fields in the autumn, and the year-round indoor work of spinning, weaving, sewing, and maintaining clothes and other textiles for broad use areas, including sails. Considering the work amount these labour-intensive work operations involved, they represented no one-woman job and would have needed a considerable workforce at larger farms and/or within larger estates on both a seasonal and an annular basis. Since only a minority of women achieved monumental burials or other prestigious burials with costly grave goods, the female and social aspects of textile production become even clearer, as they indirectly hide a large archaeologically 'invisible' work force at the bottom of the social ladder (cf. Fig. 5.13), without archaeological traces and representation in the inverted pyramid reflecting graves of varied social status. These anonymous categories referred to in written sources, not least various categories of slaves, serve as an indispensable workforce. The social emblems, tools, and general affluence of grave goods in these graves, however, indirectly reveal women's ability to produce such wealth, and represented from each end of the social ladder although archeologically, only indirectly.

Graves with textile tools without social emblems: female graves of lower status?

The prestigious boat burial in a large mound at Folkedal in Granvin, presented earlier in this chapter – without the traditional female dress accessories but furnished by other status-indicating grave goods and a substantial tool kit – four spindle whorls, a weaving beater, and a smoothing plaque of whalebone and additionally two large, hooked keys, food processing equipment, agricultural tools, and horse gear – represents an interesting case. This grave is an example of graves without the gendered social emblems of brooches that can still represent a female high-status grave judging by other grave goods – of which the famous Oseberg burial is the most prominent case. A female grave, in this case at Hopperstad in Vik, represents another case: a boat burial without the traditional Norse brooches but adorned by precious beads of various material and gemstones for a necklace, and distinguished by keys and special artefacts, many of foreign origin. Brooches were then not the only female signs of status.

When as many as 16 graves with textile tools to the category of graves without the traditional gender defining dress accessories but furnished with other female related adornment, they most likely represent female graves too. Still, being of variable prominence, they count as probable graves and in some cases as possible female graves (see Appendices 5.2 and 5.3). Three of the probable female

Table 5.8: Representation of textile tools in the category probable/possible female graves (N:16).

No. of graves	Shears	Wool combs	Spindle whorls	Loom weights	Weaving beaters	R413	1–2 tool categories	3–6 tool categories	N: Tools
16	4	4	11	33+	9	1	11	5	62

graves were located close to each other at Setre/Øvreset in Breim in Firdafylki. One of them, a grave in a cairn contained tools for all the basic work stages, besides a key, and horse gear but, as noted, no dress accessories – however, it was undoubtedly a female grave of some status. The other 15 in this category were more heterogeneous and with various textile tools to cover the various work stages, mostly for one or two of the work processes but some with agricultural and cooking implements too. Tools for weaving dominate – loom weights and as many as nine weaving beaters, connecting to 10 of these 16 contexts. Spinning tools relate to half of them and some with more than one whorl and of different weights to produce different yarn qualities. Tools for combing and shearing relate to six of these graves, weaving beaters in nine, and one of whalebone from a coastal grave in Firdafylki, facing the North Sea. All in all, the tools are clearly sparser in this group (Table 5.8).

When adding these 16 probable/possible female graves to the 87 gender-identified graves by the dress accessories and/or other status indicating artefacts, they amount to 103 female graves. In Iceland too, females buried in monumental mounds did not necessarily contain dress accessories but were securely identified as female by the preserved skeletal material.[66] Tools as grave goods were generally sparser.

One of the female graves in a cairn at Tokvam in Aurland offers strong proof that social emblems do not necessarily correspond with the representation of tools. While the deceased in the central grave with prestigious dress accessories, contained a single spindle whorl as grave gift, the woman in the secondary grave burial was furnished with costly beads for a necklace, supplied with tools for all processes, from shearing to weaving. This again is a reminder that there is not a 1:1 relation between social status and tools. Still, when seen relationally and in larger numbers, the tools and other grave goods reveal patterns that indicate both gender and status. One should also bear in mind that perishable materials are under-represented, not only the skeletal remains. To avoid a reasoning *e silentio* needs therefore a certain amount of information and trends to draw firmer conclusions. In this case, the lack of social gendered artefacts does not exclude a female presence and when assessed contextually may be further corroborated by other indicators.

Double graves: partners or subordinates, or possibly single female graves?

In the stratified society of the Viking period, where monumental graves allude to social status and hereditary rights, graves were marked in different ways (cf. Table 5.1). Burials in mounds or cairns could be primary and secondary graves of different gender or double graves. Based on the grave inventory, such as weapons, tools, and other artefacts, they may reflect either symmetrical or asymmetrical relations. An important contention, then, concerns equality and simultaneity when several burials/individuals relate to the same mound or cairn and whether gendered artefacts, in the case of weapons, textile tools, and social emblems, indicate symmetrical or asymmetrical relations. Or is it possible that weapons could have belonged to women too? As a widow and as head of the household with the right to meet at local thing assemblies, she had, like men, the right to carry weapons.[67] Still, when being without female emblems or other gender-related grave goods this can hardly be proven.

One of the graves with textile tools, connected to a large grave mound at Ytre Arna a central place in the mid-zone of Hordafylki – a cremation burial identified as a double grave of equal high-status judging by the grave goods, and strangely enough with a double set of double-edged swords and two axes, besides prestigious dress accessories (cf. Fig. 5.1). It represents one of five cremation graves; the other seven are all inhumations (cf. Appendix 5.3). The grave goods included three keys for different locks, mountings to a chest, horse gear and three sickles, which often appear in female grave contexts too. The textile tools, although limited, were of a kind that most often appear in female high-status graves – shears of the type R443, related to the finishing processes, a needle case of copper alloy and the special spatula-shaped tool, R413.

As it is rather seldom that couples die simultaneously, one may wonder how this happened, and the find information is often too general to decide the circumstances. Here, burnt bones and ashes and unburnt artefacts lay gathered without any order as for gendered tools and artefacts. Judging by the weapons, dress accessories, and other grave goods, the couple, presumably cremated in the same pyre and later buried in the same cairn, seems to represent equal high standing. As for cremations, where the bodies were destructed on the same pyre, it calls to mind Ibn Fadlan's description of human sacrifice of a female suttee victim. This was hardly the case at Ytre Arna or for three of the four cremations appearing as double graves of status II or III, while two – both boat burials in mounds at Ytre Hauge in Gloppen and Flo in Oppstryn – could possibly indicate asymmetrical relations. In the other graves, dress accessories and jewellery in at least eight of these 12 graves interpreted as double graves, indicate high status that largely match the male in the same grave (Table 5.9; cf. Appendix 5.4).

Table 5.9: Number of tool categories in double graves according to status, related to both genders.

No. of tool categories	*No. of graves*	*Status III*	*Status II*	*Status I*	*N: Weapons categories/status*
5	1	1	–	–	3
4	1	1	–	–	5
3	1	1	–		2–3
2	6	5	–	1	1–3
1	3	–	3		2
N:	12	8	3	1	

For all these grave contexts, status indicating dress accessories generally identify the graves as female, and often equipped with tools that largely concur with the grave goods represented in female graves or could be gender neutral, such as horse gear, for example. Horse bits appear in six female graves, five of high-status III and one of status II. Considering the later known inheritance customs, it is possible that widows or spouses without hereditary allodial rights, based on *óðal*, may have been buried with their partners by reopening the mound, while high-status women in casu *baugrygr*, or as widows as tutors of minor children may possibly have been commemorated with symbolic grave goods, possibly even equalling male status by weapons as grave goods. Still, the graves and grave inventories seem rather to represent a couple when commemorated with weapons and textile tools, among others. This is clearly documented in one of the large Myklebust mounds, and as presented in Chapter 1 (Fig. 1.1).

The tools for textile production as grave goods vary, mostly covering tools for one to two processes, shearing, combing and/or weaving and in one case four processes, from shearing to weaving. The double grave as inhumation burials, generally in mounds and of high status for both genders, contained textile equipment for two to four processes, and frequently with other kinds of tools too. Still, in many cases, the female-related grave goods do not match the male's status and may represent a woman of inferior rank, possibly an *ambátt*, a female slave/mistress, as referred to by Ibn Fadlan and in Norse sagas too.[68]

Compared with the burial customs in the North Atlantic and Viking Age graves in Iceland, due to better preservation conditions the buried could be gender-identified by their human remains. The burial practices reveal both similarities and differences not least as for double graves, which seem to be less common than in western Norway. However, like the Myklebust grave (see Fig. 5.19), mounds were reopened to include new graves. It could be for a spouse or of the same biological sex, probably as close relations of varied status and age. Other examples from Iceland show that a mound with a male grave had been reopened to bury a middle-aged high-status woman.[69] It could also be graves without social emblems, however, identified as for age and biological sex by the skeletal remains.[70] Larger mounds could contain from two to several male graves of different age and prestigious grave goods too, with complete weapon gear as well as practical tools.[71] They are of the same kind then as evidenced in western Norway, and with several mounds at the place often close to settlements. One grave, although incomplete, revealed traces of a combined female and male grave of high status: the female with dress accessories, remains of an oval brooch, 26 beads of varied colours and shape for a necklace, and the male with two spearheads, an axe, and parts of a shield boss,[72] and again of the same kind as symmetrical double graves from western Norway. The female graves from Iceland, either identified by their dress accessories and/or skeletal material, appear most frequently as single female graves of various status, and all without weapons, like in western Norway too.

Textile tools in male contexts?

Textile tools in weapon contexts and without other traces of female indicators are ambiguous, and open to various interpretations. They may represent asymmetrical relations, a high-status male with a subordinate female, as referred to above – or could they possibly indicate a masculine involvement in textile production, as proposed by Rabben? If the latter be the case, they could either allude directly to men as producers, or indirectly as administrators of domestic or communal enterprises, for instance related to production of sails and as part of various private or common enterprises connected to seafaring of the period along the coast or crossing the North Sea.

Altogether, 12 contexts from western Norway fall into this category of probable/possible asymmetrical double graves, while 15 contexts with both textile tools and weapons appear as unclear grave contexts (cf. Appendices 5.4 and 5.5). The textile tools are generally fewer both in number of tools and tool categories compared with the other and better identified double graves but may still substantiate a female presence. These graves are generally represented by one or two tool categories and only two contexts had tools for more, such as at Hilde at Innvik by the inner Nordfjord. Five graves included only one textile tool: a single spindle whorl in three cases, and wool combs and a weaving beater in two.

Additionally, five of the 15 unclear contexts contained weaving beaters, and in four cases supplied with other textile tools: one with three tool categories – wool combs, a spindle whorl and two loom weights. Another contained two categories, shears, and wool combs, while the other two contained only wool combs or a loom weight. In the other seven contexts, the textile tools included one to two tool categories – such as remains of wool combs, and in two cases represented by a single spindle whorl. Another two had spindle whorls and additionally a smoothing pellet and a needle case, otherwise only linked to female contexts and work domains and of general high status. This combination of a spindle whorl and a smoothing pellet of glass, and a spindle whorl and a needle case and comb of bone, and combined with keys in another two cases, represent artefacts that dominantly connect to female grave contexts. However, only one blurred context contained four tool categories: shears, remains of wool combs, a spindle whorl, and 17 loom weights and probably a disrupted grave, possibly a boat grave. The weaponry in nearly all these contexts in contrast generally match with male status I or II, and only two of status III judging by the weaponry, which in nine cases were swords – the highest status symbol of all weapons. The weapon gear in these contexts is substantial: six represented by full armament, three with two offensive weapons, sword/axe, and arrow/spear, axe/arrows respectively, and in case of status II, according to Solberg. Only one context contained one weapon category, an axe.

Still, single tools, such as a spindle whorl, as documented in two such cases, may possibly represent a commemorative gift from a spouse and not necessarily a buried woman. Still, it seems rather unlikely that the heirs and organisers of the funeral would furnish the same person as distinguished by status-indicating weapons with such gender-biased artefacts that could represent an insult.[73] In a hierarchical society where rank played a significant structuring role, asymmetrical relations between the buried as for gender, status and as producers are more feasible options. Altogether, these graves rather seem to represent asymmetrical double graves than men involved in textile production. Reaching this conclusion, another 12 double graves, however asymmetrical, may be added to the other 12 possible double graves (cf. Table 5.9; Appendices 5.4 and 5.5).

Food-related tools appear in three of these graves and sickles or hoes in three. Horse gear that appears in both male and female contexts is represented in five of them. Still, the weapons undoubtedly mark an overall male supremacy. At the same time, the assortment of keys and other female-related tools do not indicate inferiority. Keys, caskets and both food-related implements and farming tools, besides the more special textile tools listed above, are of the same kind as found in high-status female graves, and to some degree documented in graves in the North Atlantic too.

Altogether, these 15 contexts with textile tools and weapons appear as unclear both as possible asymmetrical double graves and as male graves. Many of them are early finds, found during cultivation and with sparse documentation (see Appendix 5.5). While the males in these contexts overall represent status I and II and in three cases male status III by their weapons, the textile tools are limited to one or two textile tool categories, and only a few with other types of tools. As they often lack precise contextual information, their status as graves and their connection to the tools is rather uncertain, it must be admitted. Seven of them cover only one weapon category – in two cases, an arrowhead, another two with a spearhead, one with a shield boss and two with a sword. In at least two cases, the artefacts seem to stem from separate graves in a mound or cairn, as some objects are fire-exposed, others not. In five cases, they appear as mixed or incomplete grave finds and may represent more than one grave, of which at least two as cremations. Another six cases represent unclear status related to other graves in mounds or cairns, or appear as a damaged/incomplete grave, or with imprecise locations. Generally, they contain few textile tools and in 11 cases are only represented by one tool category – four with a single spindle whorl, three with a weaving beater, and another two with a loom weight each, one with a needle case and one with wool combs. Two contexts contained two textile tool categories, two with three, and only one, as noted, with four tool categories (cf. Appendix 5.5).

More than half of these unclear contexts contained small artefacts – single spindle whorls, single loom weights, and a needle case – all easily lost objects and may have been lost and rather represent stray finds than graves. In cases of larger and costlier objects, such as weaving beaters, however, they seem to indicate intentional depositions, most likely graves. There are five such contexts, of which three have only a single tool; the other two contained wool combs and/or shears. The weapons, in contrast, comprised from one to three offensive weapons. As the contextual affiliation to the weapons is difficult to ascertain, the tools count as mixed contexts that may stem from two or several graves.

Lately also women, identified biologically by their human remains, have been found with weaponry in Viking Age graves,[74] which opens for a wider agency. Still, it is difficult to ascertain when found in unclear settings. Some graves appear as mixed finds, of which three contain fire exposed artefacts, and as referred to above, one in a mound with several cremation burials, and with a needle case of copper alloy that otherwise connects to female graves with status-indicating jewellery. Due to all these uncertainties, they count as unclear grave contexts but not necessarily male. Still, some of these mixed and unclear contexts may represent male graves and related to weaving, when five, or one-third of them, contained weaving beaters. In contrast to the more securely identified female graves, they appear as single tools, while the remaining two either have shears or wool combs.

Compared with the 87 female graves, nearly half, 39 of the graves, contained, as noted, weaving beaters and mostly part of multiple tool kits with textile tools from two to four tool categories, but in 11 cases only one type of tool. Weaving beaters as grave goods in double graves were generally parts of tool kits too, with tools for one to four processes and representing the same assortment as in single female graves. The uncertain double graves, in contrast, seldom contained more than one or two tool categories, in one case three, and mostly for combing and in one case for combing and spinning. Weaving beaters were still represented in both double graves and possible double graves in five cases in each and mainly differ as for the representation of tools (cf. Figs 5.3 and 5.4).

Altogether, the highest representation of weaving beaters connects to female contexts, as indicated in female graves and symmetrical double graves. This is also the case judging by the graves from Iceland. The find conditions, the composition of tools in weapon contexts, interpreted as male graves, may rather imply that the tools linked to separate female graves or double graves, that appeared as mixed and unclear contexts when discovered, possibly due to earlier destruction and before the archaeologists arrived at the sites. Of these contexts, only one included more varied tool categories, while the others comprised one or two related to weaving, and in two cases, combing and weaving, and for sewing and smoothing. Status indicators, such as horse gear, were found in six of these unclear contexts; four of them indicate status III and two status II based on weaponry, and in at least two cases appear as cremations and were supplied with food-related tools and agricultural equipment, indicating earlier graves.

All in all, the assessment of textile tools in assumed male contexts does not support the theory of a change in gender roles related to weaving. Weaving implements too tie mostly to female graves, both high-status female graves and those of lower or unclear status. Neither do other sources, written or archaeological from Iceland with an extensive textile production, leave information about weaving as a male domain on a practical level.[75] Still, the graves connected to weaponry and textile tools are not uniform. They appear as double graves of similar social standing, possible asymmetrical double graves, and as separate graves of varied status, however, sometimes damaged, and appear as mixed or incomplete finds when discovered.

This review and considerations concerning status, gender roles, and tools serve as a backdrop for the further analysis when assessing the tools as grave goods in their various environments in time and space. Overall, the graves, burial practices, and norms related to gender and social status, show that it was members of the highest social strata who set the clearest imprints in the landscape.

Regional patterns: changes over time?

How people acted and interacted when burying their dead – where, when, and why they supplied the graves with certain kinds of tools – are central issues in the following section. The aim is to detect and trace spatial patterns and patterns over time – how the graves connected to special or diverse environments at places in various geographical zones from coast to inland and from south to north. By examining the tools and artefacts in the graves, they may indirectly reveal various types of agents, as for gender, status, and work domains in various environments and social settings from the beginning of the Viking period until the burial practices terminated by the late 10th century.

A diachronic view and comparative approach serve as a methodological trail when assessing the graves relationally, spatially, and over time. As their locations often connect to earlier settlements and graves, earlier structures may have had an impact when erecting new graves by re-establishing and confirming old traditions – or breaking them. New enterprises, new technology for exploitation of resources, could create new opportunities, traditions, and alliances at different levels, either as reciprocal partners or within a hierarchical framework, or at household level. Entrepreneurs in one part of a community, often at central places, could operate in close contacts with equal partners, at similar places and at further distance – including upland/inland areas with access to resources able to produce a surplus. The graves, grave goods, and location may, then, count as indicators of whether representing reciprocal or hierarchical relations.

The earliest known administrative units – (all ON) *fylki, skipreiða, herað,* and *sókn* – represent different spatial scales and socially shaped divisions of old heritage. Of these, *fylki*, corresponding to a county, was the largest.[76] On a meso-level, *skipreiða* represented the local units within the national naval defence by the 10th century – obliged to build and furnish a ship, provide crew, weaponry, and food and to be operative when called upon. The districts of a *herað* were generally smaller, some could constitute one fourth (ON *fjorðungr*) of a *skipreiða*, but were larger than *sókn*, a parish as an ecclesiastical division. Such divisions served as fine-meshed structures for people seeking church within a defined area and obliged to certain duties such as tithes, first established during the 12th century. However, in many cases, they connected to older cult centres, later private churches and at central places, often with Viking Age graves and earlier graves close by. Within the *skipreiða* system, according to the Gulathing law and established as a military structure on royal initiative in the Late Viking period, Hordafylki should provide 24 ships, Sygnafylki

16 and Firdafylki 20. These geographical levels of *sókn, herað, skipreiða*, and not least *fylki* may altogether have older origins,[77] and especially in western Norway with its broken topography, where the physical environments formed natural boundaries but also links of communication by water and land routes that bound people together. This also explains their long durability. Supposedly, they built on older structures and were organised by local chieftains. As in Iceland and the Faroe Islands, the earliest churches, in the 11th and early 12th century, were often built as private churches in the first instance and in western Norway within smaller or larger estates of older heritage. As these units largely coincide with topographically formed landscapes in western Norway, they serve to some degree as analytical frameworks when assessing the graves spatially. Here, the *fylki* represents the meso-level when assessing the graves spatially and the geographical zones from coast to inland as central elements for assessment – and how they related or interrelated.

The trend of burying the dead with their tools and social emblems – largely female graves with textile tools – that escalated in the Early Viking period, built on earlier traditions too, from the Early Iron Age, especially the Migration period (AD 400–550). However, these customs faded out and were temporarily broken in the transition to the Merovingian period, AD 550–*c.* 800. The decades before AD 800 thus marked a striking change in the way of burying the dead and their locations, and there was a boom in the early 9th century (Fig. 5.17).

Of these 142 contexts a large share dates to the 9th century and especially from the first part of the century, at least 39 and 37 date to the Late Viking period, with some

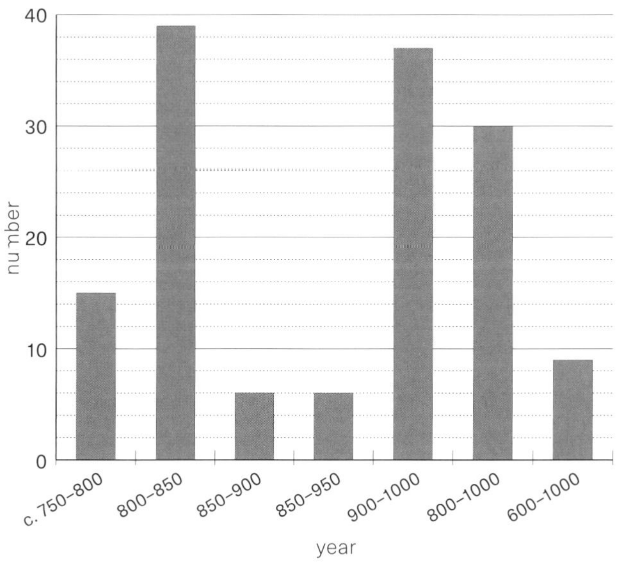

Fig. 5.17: Temporal representation of graves with textile tools as far as they can be dated more accurately (N:142).

in-between groups. Those without datable artefacts can only be dated broadly within these time frames. Fifteen of the graves predate AD 800 of which three are rather uncertain.

Some of the very early female graves from the 6th and 7th centuries into the 8th reveal both long traditions and new trends related to gender, status, textile tools and grave goods, and environments. Some of them even connected to the same central places as later in the period, demonstrating status and continuity into the 9th and 10th centuries (see Appendix 5.0). Still, the later Viking Age graves of this type are far more numerous, with a broader range of grave types, grave goods, and not least tools, reflecting technological advances and expansion at many levels. But the end of the Early Viking period, however, experienced new progress, demographically, alongside socio-political transformations and technological innovations. They were not least related to innovations within iron production and forging with wide-reaching effects: clearing of new land, expansion within farming and intensification of arable farming, exploitation of inland and upland resources for wood working, forging, and boatbuilding. They were also based on improved and strengthened tools – axes, hoes, spades, and ard and plough for cultivation, as well as other specialised tools. These changes involved not least textile production, based on wool and vegetal fibres, which again depended on both wide grazing land and expansion of cultivated arable land – and because of the many use areas at land and sea, practical as well as ceremonial, and prestigious garments for the elites.

The later Viking Age graves are, with a broader range of grave types, grave goods, and not least tools, reflecting technological advances and expansion at many levels. The technological differences concern both weaving and the preparatory stages. The notably shorter weaving beaters with a short blade, some pointed, and a shaft connected to a tang at the same plane, differed from the later longer weaving beaters with an obtuse angle from the blade to the shaft, which made it easier to beat upwards and adapted to weaving broad, tight, and strong weaves. The spindle whorls retrieved in the graves, judging by their size, material, and weight, were adapted for spinning varied qualities, from very fine, medium, to coarser yarn. They obviously served the needs for products of varied functions, both within the luxurious segment of textiles and the coarser practical use areas. These changes, not least related to weaving, represent important advances for efficiency and for producing strong and tight textiles, needed for diverse qualities, from fine garments for the elite to coarse and tight weaves needed for, among others, sails – a new product in demand and a laborious prerequisite for the sea-oriented expansion of the Viking period. How these grave categories and tools spread over time, in physical and social environments, may, then, indirectly shed light on various modes of production and possible interaction between people at these places, and thus provide a broader outlook on the social aspects of production.

Established patterns as reflected in the burial customs socially and spatially, reveal a concentration to central places for communication and for access to resources that could be long-lasting. Still, in this turbulent period of exploring, raiding, trading, and emigrating that characterised the early 9th century, conflicts within the elites and struggles between petty kings, chieftains, and their allied could disrupt old structures and strengthen new bonds. The ongoing and accelerating centralisation processes and boom in colonisation westwards in the late 9th century and escalating in the 10th century serve as a backcloth when assessing the graves contextually.

At which places, in what kind of environments, and how the graves vary for status and tool assemblages over time, are main issues in the following section, with the aims to elucidate the actors within the varied modes of production – practically, socially, and economically.

The beginning of the Viking period and the earliest graves with textile tools

Records of 'Viking' raids, in Anglo-Saxon, Irish, and Frankian annals from the 790s onwards provide the first written evidence of the westbound expansions from these areas, recorded by the literary clergy, who saw their churches and monasteries plundered and ravaged by seafaring warriors. This was the case when men from Hordafylki arriving in three ships attacked the monastery at Lindisfarne on the east coast of England in the year 793. This wave of attacks spread into the Irish Sea with particular force in the first part of the 9th century to then decrease by the latter part and change character when settling down and establishing camps and diasporas abroad, such as in Dublin, thus forming strategical alliances, strengthened by gift-giving and exchange of commodities.[78] Such raids, trade, and alliances have left traces in the graves in the present corpus, especially from this expansive period of the 9th century. The broader archaeological evidence and other sources provide a more nuanced view of the expansion, related to settlements, workmanship, trade, and alliances across borders.[79] Iron production, quarrying, woodworking, and forging was all necessary for boatbuilding and a prerequisite for the seaward expansion. So was textile production, for producing sails and all kinds of gear for rigging, wear, covers, and for other practical purposes, besides garments of daily wear and tear and prestigious textiles for demonstrating status.

When as many as 12 to 15 graves with textiles may stem from the very beginning of the period, around AD 800, and some decades earlier, they reveal a remarkable concurrence in time with the westward expansion, followed with a boom of such graves in the following decades, many with Insular artefacts, most likely reflecting such raids, which signal links and common enterprises too (see Appendixes 5.1 and 5.5; Fig. 5.18).

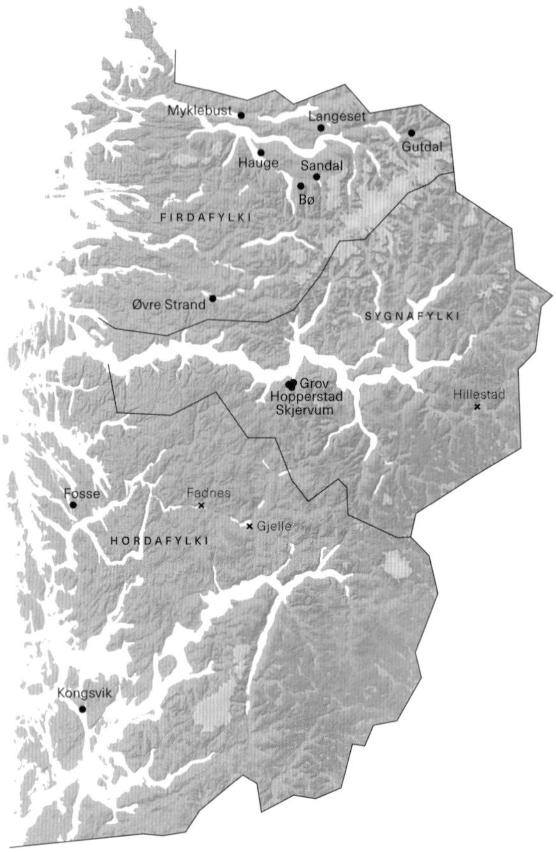

Fig. 5.18: Graves with textile tools from +/- AD 800 (with uncertain dating marked) from Hordafylki, Sygnafylki, and Firdafylki (Illustration: Per Bækken).*

In Hordafylki to the south, two of these early graves connected to strategic places by the coast at junctions for communication and access by the inner and protected sailing route. One was a double grave furthest south-west in Hordafylki, by a protected harbour at Øvre Kongsvik in the archipelago of Austevoll. The other was a high-status female grave further north in smooth waters by the inner sailing route at Fosse further north. They differed both for tools and prominence. While the tools in the double grave were limited to a pair of shears and some loom weights, the other contained a broad range of textile tools, besides exquisite dress accessories,[80] four keys of copper alloy, and other tools, there were also artefacts of Insular origins. The two others, and more unclear graves as for dating located to the inland Voss district, Fadnes one at a junction between fjord and land routes to the west and the other at Gjelle in the main Voss district, represented by weapons and a few textile tools, possibly remains of double graves.

In Sygnafylki, three similar early graves all related to the community Vik, mid-fjord by the long Sognefjord on the southern side, and as one of the richer agricultural communities in this county and all in the central area of

Vik. Here, two graves were erected at the low terrace by the wide bay, one at Hopperstad and the other a double grave at Grov, at the other side of the river. The third grave in Vik at Skjervum, not far from Hopperstad, connected to other graves at the site, a high-status female buried in a mound with two other female graves and a weapon grave. Of these, the Hopperstad grave was the most prestigious as a burial in a *c.* 6 m long boat, buried with dress accessories and tools for spinning and weaving, a sickle, and three keys. The two other graves in Vik contained only a single weaving beater each (cf. Fig. 5.18).

In Firdafylki with six or seven early graves of this kind, two connected to the mid-zone, two were centrally located by the larger fjord systems, Gloppenfjord and Nordfjord in the northern part of the county. The grave furthest south connected to Øvre Strand by one of the larger lakes. One was a possible but uncertain asymmetrical double grave at Hauge/Ytre Hauge by the inner Gloppenfjord, while the other was the earlier mentioned high-status but atypical double grave in the large Myklebust mound by the head of a side arm of the long Nordfjord. Both these places represented then gateways by the fjord systems, connecting to the sea and inland areas. The possible but uncertain double grave at Ytre Hauge was a prominent burial in a more than 8 m long boat in a large mound, however, only the male was distinguished socially and with both weapons and tools for woodworking and forging. A single spindle whorl, three beads, mountings to a casket, and a hoe were the only artefacts that could indicate a female presence, and in case of unequal status and a social mismatch, possibly a slave, a so-called *ambátt*. The other graves located in the districts Breim and Stryn with respectively two in Breim, at Bø and Sandal. The most prestigious and the earliest was the female grave at Sandal, in an upland area 300 m.a.s.l. It appeared as one of two separate high-status graves of a woman and a man in in the same mound – the woman with tools for all processes from shearing to weaving, besides various food-related equipment and a sickle, a key, bridles, and a strike-a-light. Also, here the male grave at the site contained tools for woodworking, forging, and/or agricultural work. The double grave at Bø in the main agricultural district of Breim was a few kilometres north-east of the large lake Breimsvatnet, dated somewhat later around the turn of the 9th century. Here, the female was buried with social emblems and textile tools for four work stages and the male with full weapon gear.

Three of the four earliest graves with textile tools in Stryn connected to Gutdal in the very northern part in the district Oppstryn, furthest north-east in Firdafylki. The fourth grave was located furthest west in the Stryn district, at Langeset in an upland area, *c.* 200 m.a.s.l. in the border zone between Stryn and Hornindal to the west. These inland graves represent the very earliest graves of this kind in the district and with three graves in a cluster at Gutdal, a rather flat side valley leading eastwards, however, surrounded by steep mountains. This is an area with land routes to inland areas in the county Oppland in eastern Norway, renowned for early iron extraction.[81] The female graves at Gutdal, all cremation burials, were of nearly equal status judging by the grave goods. Two were secondary graves in mounds, and the third a flat grave. The textile tools varied, for respectively two, three and four work stages, and supplied with keys and/or chests. Appearing as secondary graves, the male graves in the mounds, and at least two similar early as the female burials, were equipped with tools too. Here, access to iron obviously offered new opportunities, stimulating the secondary production of forging and for production of various types of tools, among others, for textile production. The other inland and upland grave in Stryn at Langeset to the west and by another central land route connected to waterways, the large lake Hornindalsvatnet and with further land routes to the outer parts of Nordfjord and to in-fjord/inland Stryn.

At least five of the seven graves in the inner zone and inland areas of Firdafylki indicate textile production environments close to inland resources and combined with other types of production. That many of them related to central and strategic places for communication too, by the inner, sheltered sailing route, at strategic and connecting land and waterways – and those two of the three graves appeared symbolically and physically as boat graves by the main fjords, Sognefjord, Gloppenfjord, and Nordfjord – is another link. These graves were connected to other graves without textile tools that were close to inland and upland and to wide pastures and varied resources, and also to graves with tools for differentiated work areas, which were indicators of production places and resources above ordinary household production. They were seemingly interconnected, judging by the grave inventories and locations.

Of all these early graves in the northern part of Firdafylki and in the mid zone, Myklebust stands out as a node for communication and a link between Sandal in Breim to the south and inland Stryn furthest north, as well as to the west and east. Here, in a large mound a woman was buried outside the coffin of a man, after reopening the grave, probably her late husband, buried with his weapons and tools – a single edged sword, a shield boss, arrow heads, a hammer, and a sickle.[82] The woman was furnished with a single brooch as adornment and a few textile tools, for hackling and weaving, a knife, and a casket. Dress accessories and tools indicate both status and work domains, supplied with tools for combing and weaving, a knife, and a casket as other grave goods (Fig. 5.19).[83] The mound included moreover an earlier male grave with tools for agricultural work, woodworking, and forging – a scraping iron, drill, hammer, saw, and forging tongs.

Another double grave of equal high standing, at Bø in inland Breim, however, as noted, was dated to possibly a

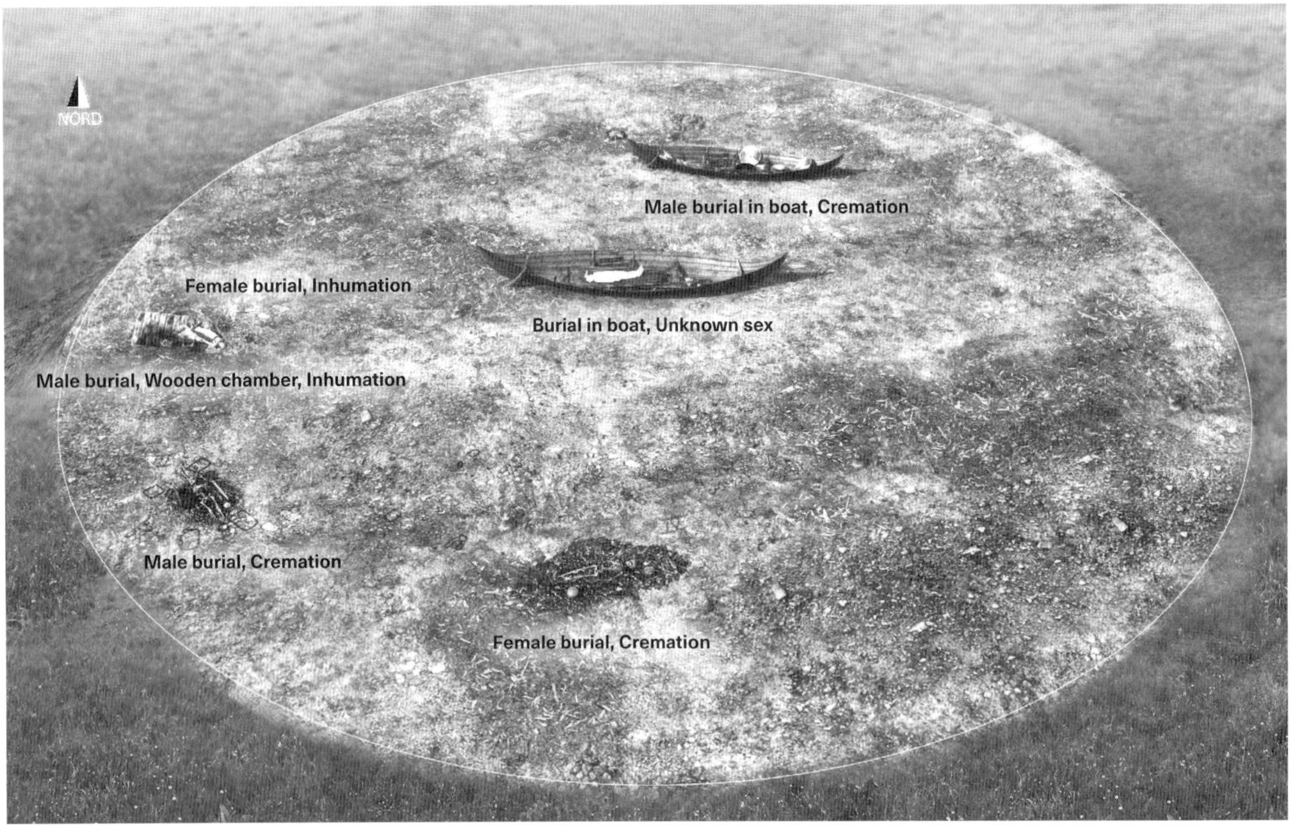

Fig. 5.19: Reconstruction of the graves in the large Myklebust mound, c. *32 m across, with the early 'double grave' to the west. Model of the mound with all its graves is based on Shetelig 1905 (Production: Arkikon, with permission).*

few decades later than the others. Additionally, two weapon graves of high status at Bø contained tools for forging and farming.[84] When textile tools appear in double graves or together with male graves in the same mounds or as separate graves with both weapons and other tools, they seem then to reflect gendered and reciprocal roles of production. When representing only the last stage of textile production, weaving, the tool indirectly presupposes all the other stages and a broader range of tools used by various actors. The tools may thus signal an overall management, also substantiated by their high status. Many of the early graves with only a few textile tools still distinguish themselves socially, while others with composite tool kits may have less defined status based on dress accessories (Table 5.10).

Still, tools as grave goods do not necessarily represent a 1:1 relation as for production. However, when graves of high-status women contained more inclusive tool kits and connected to special places and over time, they may have represented production on an advanced or specialised level. To test this hypothesis, their connection to places and environments with favourable conditions for production is further explored. Do they link to other graves at places nearby, or appear as nodes in their communities and at wider distances? Do they distinguish themselves socially, indicating capacity and means to organise, sustain, and support the necessary work force?

Table 5.10: Tool categories in the Early Viking Age graves from +/– AD 800 from the whole study area (N:14).

Number of work processes	*1*	*2*	*3*	*4*	*5*	*N: Graves*
Status III	1	1			2	4
Status II	1	1	1		1	4
Double grave FII–III /MIII		1				1
Double? MII–III+F?	5					5
N:	7	3	1		3	14

When following the *chaîne operatoire* approach, tracing the tools and tool combinations from the initial steps to the final stages of production, and what kind of tools were chosen to commemorate the deceased, it may reflect the deceased's role and responsibilities in various milieus (Table 5.11).

Five of these early graves of status II and III, also comprised the most wide-ranging tool assemblages, and all with textile equipment for more than two work processes. One of them related to a coastal area in Hordafylki and four to inland and upland Firdafylki (Table 5.12).

Table 5.11: Tools in the early graves from +/- AD 800, related to status (N:15).

Tools/status	*Shears*	*Wool combs*	*Spindle whorls*	*Loom weights*	*Weaving beaters*	*N: Graves*
Status III	1	3	5	10	3	4
Status II	2	4	2	6	1	4
Double	2	1	–	6	2	2
Double?/Unclear	2	1	3	5	2	5
N	7	9	10	27	8	15

Table 5.12: Graves from +/- AD 800 with traces of tools textile tools (N:15). P = preparation, S = spinning (No), W = weaving, P = shearing, P** = combing, P*** = shearing and combing, beater, W* = loom weight(s), W** = weaving beater, W*** = weaving beaters and loom weights, F = finishing (sewing), M = maintenance, M* = smoother of glass, M** = whalebone plaque, H = Hordafylki, S= Sygnafylki, F= Firdafylki.*

No.	*Place/County*	*District*	*Fjord*	*Coast*	*Upland/ Inland*	*N: Tool categories*	*N: Work processes*	*Keys*	*Foreign goods*	*Gender/ status*
B5807	Kongsvik H	Austevoll		x		2	2:P*+W			MIII/FII
B12695	Fosse H	Meland		x	–	5	4: P***+ S (3) + W***, M**	4	x	FIII
B7814	Fadnes* H	Voss				1	1:p**			MII/F?
B9061	Hopperstad S	Vik	x			2	2: S+W	3		FIII
B6500a-w	Skjervum S	Vik	x			1	1: W		x	FIII
B8143	Grov S	Vik	x			1	1: W			FII/MII
B13628	*Øvre Strand F	Gaular			x	1	P**	–	–	MII/F?
B5150	Ytre Hauge F	Gloppen	x			1	1:S			MIII/F?
B8501	Bø F	Breim	–	–	x	5	4: P*** +S+W***	–		FIII/MIII
B 9014	Sandal F	Breim			x	4	4: P*** +S (1) +W*	1	x	FIII
B5807	Myklebust, F	Eid	x			2	P**			FII–III
B5144	Gutdal F	Stryn			x	5	3 P***+ S (1)	1		FII–III
B6846	Gutdal F	Stryn			x	5	4: P*** +S (1) +W	–		FII
B5909	Gutdal F	Stryn			x	1	1: W			FII
B6035	Langeset F	Stryn			x	2	2: p**+W	–	–	FII

Five of the 15 early graves had tools for more than two processes, and four for four work stages. Four contained status symbols of keys too – the prestigious graves at Fosse in Hordafylki with four keys, and the Hopperstad grave in Vik with three, and two with one, the grave inland/upland graves at Gutdal and Sandal, with one each. That they contained foreign artefacts too, is another link to trace.

Table 5.13: Spindle whorls in weight groups in graves related to status, +/- AD 800 (N:15, of which 13 measurable).

Weight groups in g	*10–15*	*16–20*	*21–25*	*26–30*	*31–35*	*36–40*	*41–45*	*46–50*	*Not available*	*Total N*	*N: Graves*
Status III	5	1	–	3	–	–	–		–	9	4
Status II	–	–	1	1	–	–	–		–	2	2
Double?	–	1	–	–	–	–	–	1	2	4	2
N	5	2	1	4	–	–	–	1	2	15	8

All five comprised tools for the preparatory stages of production – shears of the type for sheep shearing and/or wool combs for preparing the longhaired fibres. All except the coastal grave at Fosse had shears and all with the type for sheep shearing.

- Fosse, Meland, status FIII: wool combs
- Bø, Breim, status FIII/MIII: shears (R442), 27 cm, wool combs
- Sandal, Breim, status FIII: shears (R442), more than 20 cm, wool combs
- Gutdal, Stryn (B5144), status FII–III: shears (R442), 24 cm, wool combs
- Gutdal, Stryn (B6846), status FII: shears (R442), 28 cm long, wool combs

Four other early graves, however, with tools for one or two processes, contained equipment for the initial processes too: two with shears and three with wool combs. One of the graves with shears was the high-status double grave at Ytre Kongsvik furthest south-west in Hordafylki, the other the uncertain double grave at a central junction to Voss, and both combined with tools for weaving, respectively loom weights and a weaving beater. One of the two other graves with wool combs and combined with a weaving beater, was the grave in the large Myklebust mound (cf. Fig. 5.19). The other and less prestigious upland grave at the fringes of Stryn to the north-west, contained remains of wool combs and a single loom weight.

Spindle whorls of varied weights for spinning yarn of different textile qualities, counted 15 whorls in all from eight of the graves, of which six were of the highest status groups FII and FIII, and two were from high-status double graves but also two in a possible asymmetrical double grave. They range in weight from 11 g to 30 g, but two were unavailable for measuring (Table 5.13). Nearly all the whorls, 12, were parts of multiple tool kits from six graves, and clearly an overrepresentation.

The largest assemblages connect to the high-status female grave at Fosse with three whorls within the weight groups 11–15 g and 26–30 g for spinning both very fine threads and rather ordinary yarn for many use areas. The spindle whorls from four of the graves from Firdafylki, the lightest of 11 g, relates to the high-status female grave at Sandal, while the double grave at Bø with two whorls were of rather different weights, 19 g and 46 g, and the whorls from two of the graves from Gutdal, respectively 24 g and 27 g, represented flexible use areas.

- Fosse, Meland, status FIII: 3 spindle whorls, 2: 11–15 g, 1: *26–30 g (*estimated weight)
- Bø, Breim, double grave, status FIII/MIII: 2 spindle whorls, 19 g and 46 g
- Sandal, Breim, status FIII: 1 spindle whorl, 11 g
- Gutdal, Stryn (B5144), status FII–III: 1 spindle whorl, 27 g
- Gutdal, Stryn (B6846), status FII: 1 spindle whorl, 24 g

These early graves with composite tool kits can further be compared with graves represented by single spindle whorls or combined with one other tool category – three graves in all, of which two are identified as possible asymmetrical double graves – one from the boat grave by the Gloppenfjord, however, the spindle whorl is no longer available. The other from the rather unclear grave in the inner zone of Sygnafylki and within the common weight group, 25–30 g. The third was the high-status boat burial at Hopperstad in Vik, with a spindle whorl of 18 g, besides remains of spindles and a weaving beater (Table 5.13 and Fig. 5.20).

Of all these graves, the spindle whorls in the graves at Fosse and Sandal comprised whorls for producing the finest but also medium fine and coarser as well as rather ordinary qualities. The two whorls from the grave at Bø, the heaviest of 46 g, could have been used for spinning coarse yarn or also twining yarn.

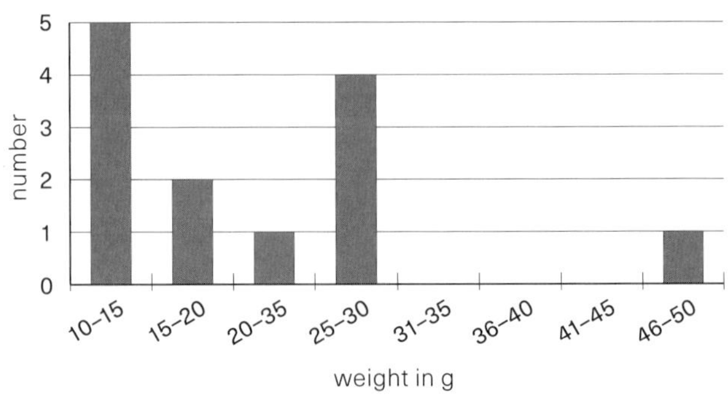

Fig. 5.20: Measurable spindle whorls in these early graves from +/- AD 800 whole area and according to weight (N:13).

Four of these graves with tool kits included tools for weaving: the grave at Fosse, Bø, and Gutdal, while two of them also had loom weights and all had multiple tool kits.

- Fosse, Meland, status FIII: 4 loom weights, fragments of a weaving beater (R440)
- Bø, Breim, status FIII/MIII: 1 weaving beater (R440)
- Sandal, Breim, status FIII: 6 loom weights
- Gutdal, Stryn (B6846), status FII: 1 loom weight, 1 weaving beater (R440), *c.* 57 cm long

Three of the other graves contained weaving equipment too, combined with another tool category and two as single tools. Two were, as noted, the central graves at Myklebust at Eid and at Hopperstad in Vik, combined with respectively wool combs and a spindle whorl, however, weaving beaters of different types, at Myklebust, with a beater with a straight short blade with a spit and with a handle of wood, while the Hopperstad grave had the traditional Viking Age type, but rather short, 58 cm long, about the same size as the beater in the Gutdal grave. The two other early graves in Vik, at Skjervum and Grov, contained weaving beaters of the same type too, although somewhat longer. Altogether, the weaving beaters in these early graves are relatively short. Two of the early but unclear grave contexts from Voss included a weaving beater, and the other some loom weights, possibly also shears and spindle whorls.

The more rare tools for smoothing – whalebone plaques, originating from northerly waters and imported glass pellets – represent maintenance rather than production, but still indicate production of textiles made of plant fibres. That the grave at Fosse, strategically located as for maritime communication in Hordafylki, had a whalebone plaque too, implying contacts further north.

Keys and foreign artefacts, as other indicators of status, are represented in four of the five early graves with multiple tool kits. Of these, the Fosse grave had three keys, and the upland/inland graves at Sandal and Gutdal in Firdafylki with one each. Two of these graves – at Fosse and Sandal – comprised foreign artefacts too (see Chapter 7), which like the whalebone plaque indicate wider networks.

Altogether, these early graves distinguish themselves for status and range of tools. They form an interesting spatial pattern in rather different landscapes, connecting to both strategic places for communication and to resources at larger distances – from the southern part of Hordafylki, protected by islands along the inner sailing route. Further north, the fjord systems lead into the long Sognefjord to Vik, and in Firda with Dalsfjorden in the southern part, Gloppenfjord and the long Nordfjord to the north, give access to land routes and wide resources of grazing land and access to iron not far off.

Since tools as grave goods reflect production symbolically rather than concrete production, a qualitative assessment of the grave goods and with fewer tools for selected stages should also be considered and related to both status and location. We see, then, that five of the early graves contained tools for both the initial and last stages of production – from shearing/combing/spinning to weaving, and indirectly reflect textile production as such. The more rare tools as grave goods – for sewing and maintenance – constitute a special and more limited category related to high-status graves, often connected to central places. Two of the graves with tools for the initial and final stages, respectively combing and weaving and spinning and weaving, connected to Myklebust at Eid in Firdafylki and Hopperstad in Vik in Sygnafylki – both central places located mid-fjord along the long fjords and fjord systems, connecting land routes to places by the coast to inland and upland districts.

The expansion of graves with textile tools in the 9th century

By the 9th century, the number of graves with textile tools expanded significantly, with some 38 new graves with textile tools in the first part of the 9th century, with at least 20 in Firdafylki, eight in Sygnafylki and 10 in Hordafylki. Additionally, some graves with the wider dating frame of the Viking period may stem from the early period too, judging by types of tools and locations. This massive extension related largely to the same environments as the earlier graves but also within wider areas and other districts (cf. Fig. 5.17; see Appendix 5.6).

In Firdafylki, with the largest representation of graves with textile equipment, all seem to be related to the first part the 9th century. Like the earliest graves, they varied socially and for number of tools and tool categories. The majority, 17 of graves, when including the four graves with broader dating frames, marked high status, status II and III. As a total, disregarding status, the graves, and grave goods varied. In graves with tools for one or two work operations, weaving equipment dominated, followed by tools for spinning, combing, shearing, and with sewing equipment in two graves only. In high-status graves and double graves of equal high status between the genders, tools for spinning and weaving were the most common categories and with a higher share of weaving beaters.

Ten of the graves connected to the district Stryn, three to Eid and four to Gloppen but seven when including the three of the same character but with broader dating frames. Five, possibly eight, related to the districts in the southern part of the county. As for the earlier graves, they varied socially and as for tools, for one to five work operations. The majority, 17 of all 25 graves, represented high status. As a total disregarding status, the graves varied considerably, both in number and for work areas and grave goods (Table 5.14).

Now, only four of all the graves with textile tools from Firdafylki contained tools for more than two work processes,

Table 5.14: Tools in 9th-century graves in Firdafylki related to status (= tools of whalebone) (N:25).*

Status	*Shears*	*Wool combs*	*Hackles*	*Spindle whorls*	*Loom weights*	*Weaving beater*	*R413*	*Needle(s)-case*	*N: Graves*
F: Status III	3	4	1	1	2	4		2	11
F: Status II				4	18+	1			6
F: Status?	2	2		4	30	3 + 1*	1		5
Double grave	3	1			9	1	1		2
Double M+F?	1	2							1
N	9	9	1	9	59+	9+ 1*	2	2	25

Table 5.15: Firdafylki, 9th-century graves with textile tools for more than two work operations. P = preparation, S = spinning, W = weaving, P = shearing, P** = combing, P*** = shearing and combing, W* = loom weight(s), W** = weaving beater, W*** = weaving beaters and loom weights, W**** = R413, F= finishing (sewing), M = maintenance, M* = smoother/pellet of glass, M** = whalebone plaque.*

No.	*Name*	*District*	*Central/ strategic places*	*Upland/ Inland*	*N: Tool categories*	*N: Work operations*	*Keys*	*Foreign goods*	*Status*
B5030	Gutdal	Stryn	–	x	4	4: P** + S+W*	1	x	FIII
B6662	Setre/ Øvreset	Breim	–	x	6	4: P**+S+W****	1	–	?
B7810 II	Setre/ Øvreset	Breim	–	x	5	4: p** +S + W***	–	–	?
B11413	Sandal	Jølster	x	x	5	4:P+S+W***+M			FIII

compared with the early graves. Once again, one of them located in Gutdal in inland Stryn, one at Sandal in the inland district Jølster from the latter part of the century, while the two of unclear status, the earlier mentioned graves at Øvreset/Setre inland Breim in two of the three neighbouring mounds and buried without socially distinguishing dress accessories and dated broadly to the Viking period, but most likely to the 9th century (Table 5.15).[85]

The new grave at Gutdal – a chamber grave in a mound – represented higher status than the earlier graves at the site and with an all-inclusive tool kit for four work stages (cf. Table 5.15). It contained two types of shears, for the initial sheep shearing and additionally the smaller type for the finishing process, wool combs, a spindle whorl, and a weaving beater with a short blade, 43 cm of the same short type as in the earlier Myklebust grave. The mound contained moreover a high-status weapon grave with a full tool kit too – a hammer, file, drill, grindstone, hone, and shears,[86] which corroborates Gutdal's status as a central production place.

Additionally, four other graves with textile tools related to these inner parts of Stryn. Of these, one connected to a neighbouring place to Gutdal at Folva – and again a high-status female grave in a mound and in this case furnished with textile tools for shearing and spinning. Also, this mound contained a weapon grave, in fact matching in time with the earlier Gutdal graves. Additionally, four other graves with textile tools related to these inner parts of Stryn.

Two other graves, one at each side of the large and 26 km long east–west-oriented lake Oppstrynvatnet, 29 m.a.s.l., a female grave at the headland Fosnes/Nes and a probable double grave at Flo at the northern side of the lake, had tools for, respectively, weaving and combing. Of these, the double grave was a boat grave, judging by the more than 140 rivets. Besides the wool combs, it contained farming tools and tools for wood working – a hoe, a lea, and sickle, besides a special knife, a chisel, and a planer, probably for boat building. The female grave at Fosnes on the southern side of the large lake and a central place for communication was situated by the end of the long Stryn valley and by the lake. This grave was surrounded by high-status weapons graves, of which one was probably a boat grave, and was supplied with tools for boat building too, four special knives (R408, R410), a hammer, a planer, a drilling tool, a chisel and 45 iron rivets, and remains of an iron cauldron. The finding of a payment ring of gold at the site further underlines status and possibly surplus production and/or trading. This is further substantiated by the finding of a depot of 53 iron bars (R438)[87] at a site further in-valley of Oppstryn – probably traded from the neighbouring iron-production county Oppland to the east. Altogether, the graves in Oppstryn indicate expanding production environments of varied kinds from the very beginning of the Viking Age onwards, and seemingly above household levels.

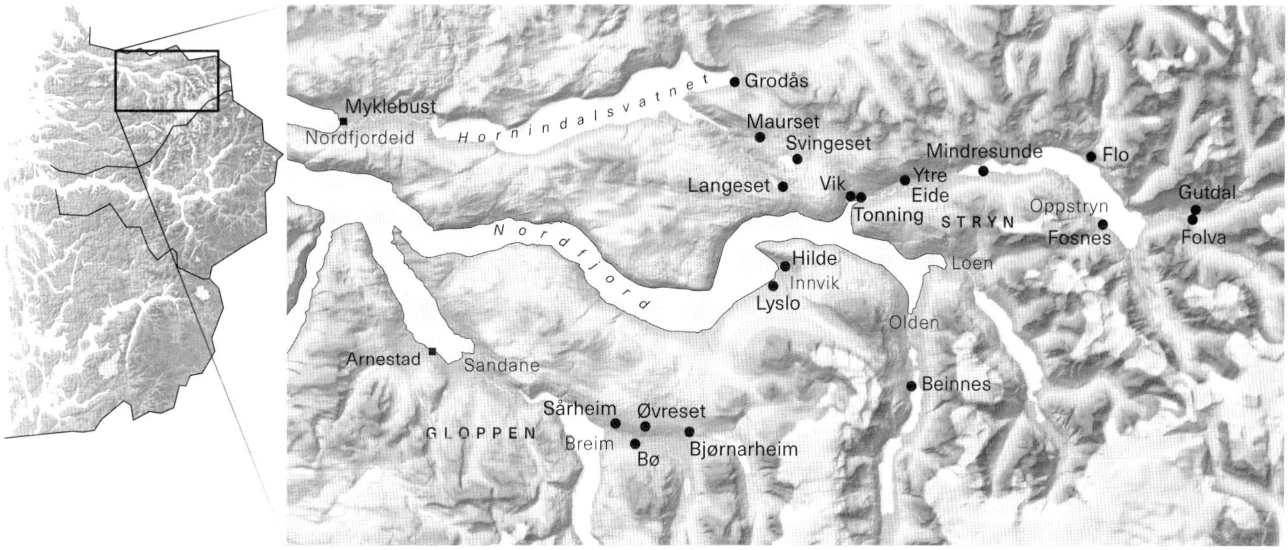

Fig. 5.21: Graves from the 9th century with textile tools in the Stryn area in the northern and inner part of Firdafylki (Illustration: Per Bækken).

Similar double representations of high-status female and male graves relate to the main Stryn valley too, such as at a strategic place for communication at a narrow strait at Mindresunde by the largely sailable river from the lake, and able to control access to the inner valley and Oppstryn. Here, a high-status female grave co-located with three male graves. While the female grave had tools for spinning and weaving plus an artefact of Insular origins (see Chapter 7), one of the male graves comprised a bundle of arrowheads and a small payment ring of 1.5 g gold, which may corroborate its position as a place of control. Two other female graves in the same neighbourhood, at Ytre Eide an isthmus by the river, had tools for, respectively, spinning/weaving and shearing/spinning and once again with a weapon grave close by, with tools for woodworking – a drill, awl, and a planer – and similar grave goods then as in the weapon graves by the lake (Fig. 5.21).

Close by the head of the fjord and entrance to the valley, at Vik, a double grave formed as a long ship-shaped stone setting and marked a double grave in a boat – the male in full weapon gear and with tools for woodworking and forging, like in graves in-valley. The female also of high status in contrast, held only a pair of shears and the spatula-shaped implement, R413, supplied with equipment for cooking and farming, sickles, a key, and casket. Another grave close to the fjord, and a single female grave at Tonning contained textile tools for spinning and weaving.

This concentration of early 9th-century graves with their various tools, thus indicates production milieus related to textile production, woodworking, and forging, based on varied resources – wood and wool and iron – and with access to iron not far off. Similar gendered representations of graves are found in lowland areas and by the fjord and at the fringes of Stryn, at Beinnes, and Lyslo with two female graves and with tools for, respectively, combing/weaving and spinning/weaving. However, the male weapon graves in these environments in contrast all lacked tools. A single dispersed but high-status female grave in an upland area, *c.* 300 m.a.s.l., at Svingeset by the land route from Stryn westwards to Hornindal, contained, as noted, some special tools for sewing, besides spinning.

Of all these early 9th-century graves in the northern and inner part of Nordfjord, Gutdal stands out with its many graves as for status, gender representation, and tools – as early and special workplaces with access to varied resources. The graves and spectrum of grave goods seem to indicate interaction between people in upland/inland areas, mid-valley and by the fjord, not least marked by a high-status double grave at Vik by the fjord in Stryn as a harbour and node for wider sea-bound communication.

In Breim further south (cf. Fig. 5.21), with the two somewhat early graves – the double grave at Bø and the upland grave at Sandal, some decades earlier – four female graves with textile tools stem from the early part of the 9th century and all with textile tools for one or two processes but also surrounded with male weapon graves with various tools. One was a high-status female grave at Sårheim in the main valley but in contrast to the graves in Stryn, it contained only a single weaving beater for textile equipment, besides a sickle and three keys. The grave was surrounded by three weapon graves, richly equipped with tools for forging, woodworking, agricultural work, cooking, and shearing, and one of them probably a boat grave judging by the many rivets. Another female grave in the same district further up-valley, at Bjørnarheim, and a possible boat grave too but of lower status judging by the dress accessories, contained tools for spinning and weaving. Yet another grave, Eide, located close to the Gloppenfjord and at a place of early power, had tools for combing and weaving. This grave was co-located with at least four weapon graves, of which three were boat burials.

Two of the three graves with textile tools at Øvreset/Setre in the central agricultural district of Breim are, as noted, more ambiguous for status and dating as they lack status- and gender-indicating dress accessories. One of them, buried in a cairn, had tools for four processes from shearing to weaving, with two types of weaving beaters (cf. Table 5.15), besides a key and horse bits and some beads, which indicate a woman of some status. One of the two other graves without social emblems represented by four tool categories for four processes, while the third of these interconnected graves contained only a few loom weights. That two of them contained tool kits for three and four work stages for textile production, and with a high-status male burial with tools in another mound at the site, with a lea, a hoe, and a hammer, indicate a working environment like many other places in this district.

Like in Stryn, the graves in Gloppen form a special spatial pattern – represented by both female and male graves, mainly related to inland areas with access to vast grazing land and woodland, and with tools for utilisation of the same kind of resources but for different use areas. Here, the graves at Øvreset/Setre with composite tool kits may indicate a special place for textile production. Like the earlier graves from the turn of the 9th century, one of the new graves connected to the fjord and once again with few tools but still of high status. Three of the seven graves with textile tools in the area, then, were buried in boats, and another two graves with textile tools close by, and four boat graves in all. Altogether, Gloppen represents the district with the most boat graves in the northern part of Firdafylki, with 15 boat graves in all and 10 from the 9th century,[88] and nearly a quarter of them were graves with textile tools.

At the central place Myklebust mid-fjord by the long Nordfjord, the large mound at the isthmus Eid was re-opened and probably extended to give place for a second female grave – this time a cremation burial of a high-status woman, furnished with tools for shearing and hackling, besides costly food-related equipment among the grave goods (cf. Fig. 5.19). Two other high-status graves were in the northern part of Firdafylki, in the mid-zone of the inland and upland area between Eid and Stryn close to the border to Hornindal, at Maurset and Svingeset with tools for one or two processes, like the somewhat later grave at the central place Grodås in Hornindal by the large lake Hornindalsvatnet, and both were equipped with tools for sewing.

Of all these graves in the northern parts of Firdafylki – by the fjord in Stryn, at Eid, and in Gloppen – now many connected to central places for communication and status, and at places marked by contemporary weapon graves. The female graves contained fewer tools than the earlier graves, often with multiple tool kits, indicating special environments for production. Another grave at Naustdal in the southern part of the county, a strategic and central place by the Førdesfjord, was a high-status female grave, however, with a single spindle whorl as grave gift. Yet another grave central place, Holmedal further west, possibly a double grave, contained only a few rather light loom weights as grave goods. These sites differed from the graves in the northern part of the county, also without corresponding male graves.

Only four of these 9th-century graves from the northern part of Firdafylki now had tools for more than two work processes, one at Gutdal in Stryn and two in Gloppen, and one from the southern part, Sandal in Jølster and all from inland areas and textile production environments, and all with shears for sheep shearing (R442).

- Gutdal, Stryn, status FIII: 2 shears (R442, R443), 20 and 24 cm, wool combs
- Øvreset/Setre, Breim, status ?: shears (R443), wool combs
- Setre II, Breim, status ?: shears (R442), wool combs
- Sandal, Jølster, status III: wool combs

As for wool combs and hackles, only wool combs related to all the graves with multiple tools. Like shears, wool combs seldom appear as single tools in the graves, except in the possible double grave buried in a boat at Flo in Oppstryn. Two graves from the fringes of Stryn had two tool categories for textile production, combined with tools for weaving, one with loom weights and the one of higher status with a weaving beater. In Gloppen, wool combs related to a high-status grave Eide closer to the Gloppenfjord together with a weaving beater. In the Myklebust grave, however, the combs were rather for hackling plant fibres.

Now nine graves contained spindle whorls, 18 in all, scattered from coast to inland in Firdafylki when including the graves with tool kits. They weigh from 15 to 40 g and are thus generally heavier than in the earliest graves from the late 8th century (Table 5.16; Fig. 5.22).

The four graves with composite tool kits all contained spindle whorls – with from one to four whorls. Two graves had only one whorl, one with two, within the common weight group 20– 25 g, and the grave with four whorls ranged in weight from 18 to 40 g. They were thus able to produce a variety of qualities, however, not the very finest.

- Gutdal, Stryn, status FIII: 1, 20 g
- Setre, Breim, status ?: 2, 20 g and 31 g
- Setre II, Breim, status ?: 1, 20–25 g
- Sandal, Jølster, status III: 4, 18 g, 21 g, 38 g, 40 g

Table 5.16: Spindle whorls in weight groups in graves in Firdafylki related to status, 9th century (N:11 graves; spindle whorls: 18).

Weight groups of 5 g	*15–19*	*20–25*	*26–30*	*31–35*	*36–40*	*40+*	*N*	*N: Graves*
F: Status III	3	1	1		1	1	7	4
F: Status II	1		2	1	1		5	3
F: Status?	2	2	1	1			6	2
N	6	3	4	2	2	1	18	9

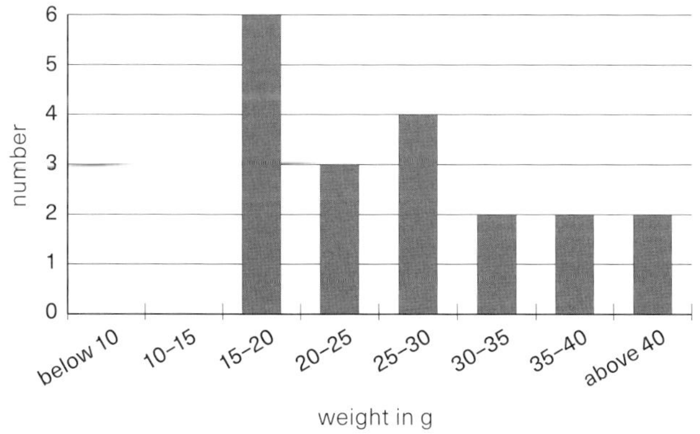

Fig. 5.22: Spindle whorls in weight groups in graves in Firdafylki from the 9th century (N:9 graves; spindle whorls: 18).

The other whorls appearing as grave gifts range from 15 to 37 g – the lightest related to a grave at Folva neighbouring to Gutdal and the heaviest in a grave at Mindresunde at the narrow sound mid-valley in Stryn. The other whorls belong to the weight groups 20–25 to 35–40 g, and were thus able to spin medium yarn qualities, such as the whorls from the two graves by the inner Nordfjord at Hilde, Tonning, and Lyslo close to the fjord and Langenes at the other end, facing the North Sea (Fig. 5.22).

As for weaving, all the four graves with composite tool kits comprised beaters as well as smaller and larger assemblages of loom weights – from a single loom weight to 20, although not all retrieved in the museum. Of the two graves at Øvreset/Setre, without dress accessories, one comprised two weaving beaters; the other one of the traditional type with an obtuse angle, while for the one with two, one was of the same type as found in the early Myklebust grave – straight, short, and with a small spit.

- Gutdal Stryn: 1 loom weight + 1 weaving beater, R440
- Øvreset/Setre, Breim (B7810 II): 7 loom weights preserved of originally 20, from 230 g to 516 g in three weight groups, 200–250 g, 400–450 g and 500–550 g + 2 weaving beaters (R440; R150), 1 R413
- Setre II, Breim (B6662): 2 loom weights + 1 weaving beater
- Sandal, Jølster (B11413): 10 loom weights + 1 weaving beater (R440), 82 cm

The six other graves in Stryn had tools for two work stages. The whorls were combined with tools for weaving, and in one case, at Svingeset, combined with tools for sewing. Compared with the whorls in the earlier graves in the county, they represent weights for producing varied qualities, but not the very finest.

Both spindle whorls and loom weights indicate production of medium fine to coarser qualities. Three of these four graves contained tools for the finishing processes too – shears for cutting yarn, and for smoothing and maintenance of textiles made from plant fibres, linen, nettles, or hemp, represented in the grave from Jølster:

- Gutdal Stryn (B5030): shears (R443)
- Øvreset/Setre, Breim (B7810 II): shears (R443)
- Sandal, Jølster (B11413): 1 smoothing plaque of whalebone

Places of production?

While the graves in the northern areas of Firdafylki had textile tools for varied work processes according to status and environments, the tools in the graves to the south were fewer and, except the Sandal grave in Jølster, generally limited to two work operations, often spinning/weaving, and with spindle whorls able to produce rather ordinary yarn. Nine other graves from Firdafylki included weaving beaters, besides the four graves with multiple tools and all were of general high status. Two high-status female graves in the inland/upland Hornindal and Stryn, at Maurset and Svingeset at its border, stand out from the others with tools for sewing – one with two small sewing needles of copper alloy, besides a small spindle whorl of lead and the other with a needle case of bronze with iron needles and matched in a nearby grave at Grodås with a needle case of copper alloy.

So, is it possible to encircle places of production and textile production environments based on the grave goods in these graves from the 9th century? Compared with the very early graves in Firdafylki from the late 8th century, the number of graves with composite tool kits, was the same in the earliest part from the very late 8th century than later in the 9th century, both with four such graves, however, of varied status, with the two early high-status graves in Breim, the female grave at Sandal and the somewhat later double grave at Bø, while the two graves at Gutdal in Stryn may indicate lower status.

Table 5.17: Measurable spindle whorls according to weight (28) in graves from the late 8th century and the 9th-century graves in Firdafylki (N:20).

Weight period	*10–14 g*	*15–19 g*	*20–24 g*	*25–29 g*	*30–34 g*	*35–39 g*	*40+*	*g?*	*Measurable*	*Total N*	*N: Graves*
750–800	5	2	1	4	–	–	1	2	13	15	8
800–900	–	4	3	3	2	2	1	1	15	16	12
N	5	6	4	7	2	2	2	3	28	31	20

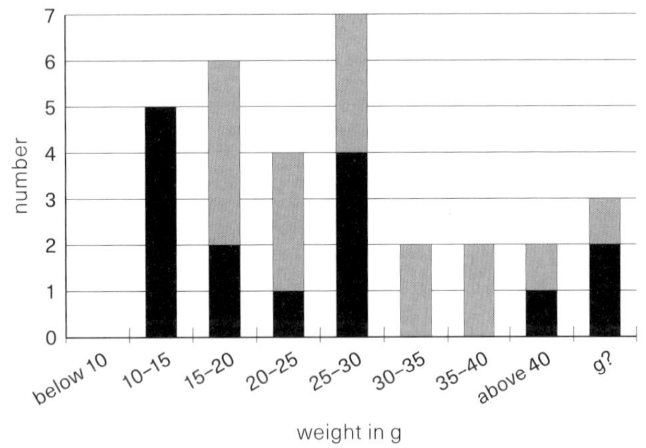

Fig. 5.23: Spindle whorls according to weight in graves from the late 700s (black) and the 9th century (grey) in Firdafylki (N: measurable 28).

The graves were more concentrated in the northern parts of the county and districts in the inner zone of Oppstryn and Breim with access to wide upland resources. In most cases, they appeared in areas and at places together with contemporary high-status weapon graves and with tools connected to upland resources: wool and wood, based on both animal and arable farming, as well as forging tools of iron. Stryn and Breim/Gloppen have therefore been characterised as forging districts, based on the tools in the graves.[89]

Keys as status indicators also related to many of the same graves with multiple tool kits and to other graves the same districts. Seven related to inland areas – one to Gutdal, three more to the same district Oppstryn, Folva, and Svingeset and others to inland Breim, at Sandal and Sårheim. All but two were female graves of status III – the others, a double grave at Flo in Oppstryn, Mindresunde mid-valley, and the prestigious boat-shaped double grave at Vik, an inlet and by the head of the long fjord Nordfjord and entrance to the Stryn valley. This implies that nearly half of all the graves with both keys and textile tools from the century (cf. Table 5.4) related to inland areas in Stryn, mainly from the first half of the 800s – an altogether imbalanced but also significant representation. The tools in the graves belonging to the upper echelons of society and of both sexes, indicate a diverse production that would have needed special competence and a large work force. Indirectly, the graves shed light on organisation and coordination of a large group of probably dependent workers that were, however, invisible archaeologically.

Another characteristic is that many of the graves connected to production environments and to central and strategic places – at junctions for communication and at pivot points for access and control. In Stryn, the double grave at Vik by the head of the fjord shares traits with the inland graves in Oppstryn, and notably Gutdal, and with male representation too with tools for woodworking and forging. A similar dual pattern appears in Gloppen and Eid, with graves and monuments that demonstrate power and control on a higher level. The tendency that women were buried in double graves, and as secondary graves in mounds, or at sites with male weapon graves, further underlines the aspects of power, control, and demarcation of resources and reciprocal work domains. The use of tools as grave goods thus indirectly corroborates the importance of the inland/upland resources and textile production, and as it appears through the grave goods, at an advanced or possibly even specialised level, needing organisation and coordination, reflected in high-status graves marked with composite tool kits and socially distinguishing artefacts.

When using spindle whorls in the graves as indicators of textile qualities, we see that whorls in the graves from the turn of the century represent a wider weight spectrum – between 11 and *c.* 46 g – than the graves from the 9th century, ranging in weight groups from 15 to 40 g. Of the spindle whorls in the graves at Gutdal, the whorls in two of the earliest graves were heavier, 24 g and 27 g, than the later and more prominent grave with a whorl of 19 g, and another grave in this area, Folva, with a whorl of 15 g (Table 5.17; Fig. 5.23).

The graves with textile tools in Firdafylki from the transition between the Merovingian period and the Early Viking period, reveal an interesting pattern, reflecting both internal and external socio-political conditions, and sea-oriented activities. It was marked by intensification and expansion of production at many levels, involving people at many levels, both men and women. The tools and other grave goods reflect the production and the agents behind producing a variety of textile qualities for the luxurious sphere and coarser qualities for practical use areas and represented in higher numbers in the first part of the 9th century. By the late 9th century, the graves seem to

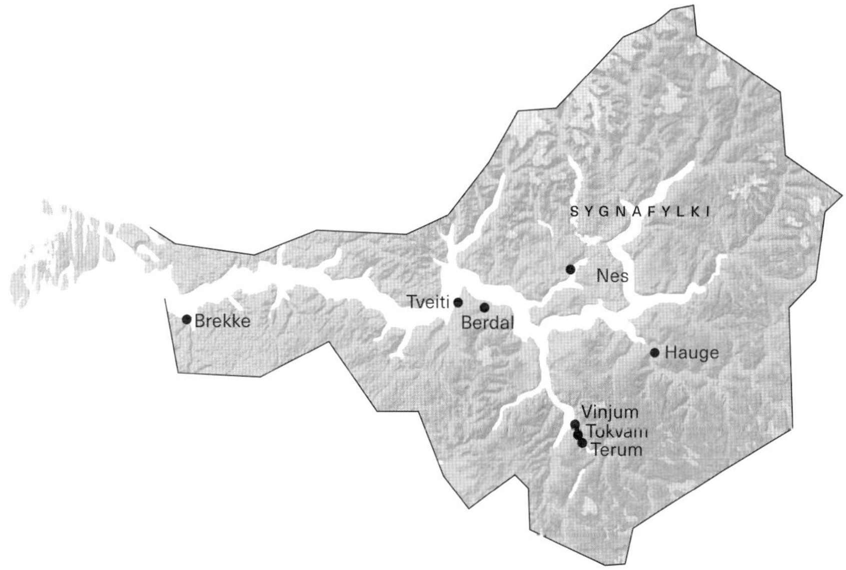

Fig. 5.24: Textile tools from 9th-century graves in Sygnafylki (Illustration: Per Bækken).

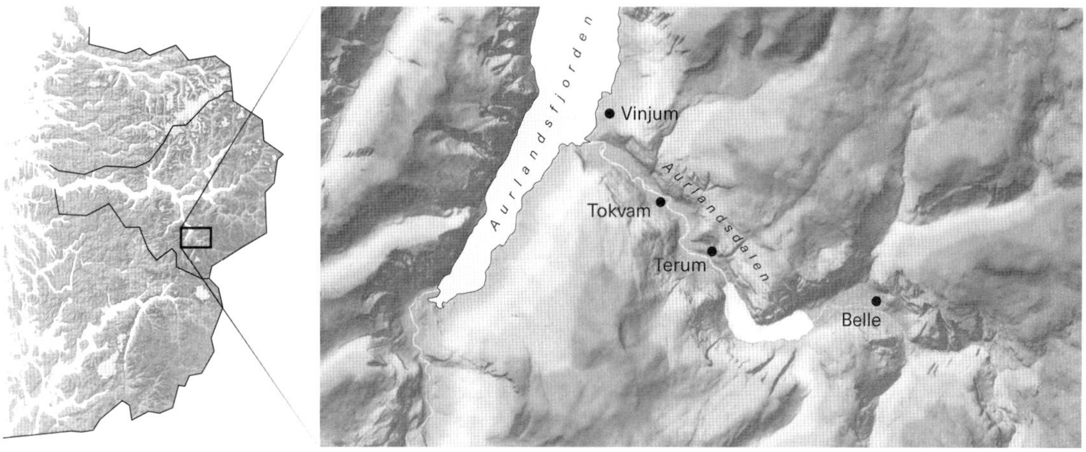

Fig. 5.25: Graves with textile tools in Aurland from the 9th century (Illustration: Per Bækken).

concentrate in special areas, around and at central places and without the same spread as in the earlier centuries. Of the four later graves from the second half of the 9th century in Firdafylki, two indicate high status – the one furthest north-east and the earliest at Gutdal, and the other furthest south-east at Sandal in Jølster from the latter part of the century.

In Sygnafylki, the number of such graves expanded from three or four by the late 700s to 14 new such graves by the end of the 9th century, and within wider spatial environments than earlier (Fig. 5.24). The highest concentration of graves was in the community of Aurland with six such graves from the early 9th century, of which one with a later dating frame, two in Sogndal, and five other more unclear contexts in the districts Aurland and Sogndal. Ten other contexts with a wider dating frame of the Viking period may stem from this century too, and in such case, 23 contexts.

A grave by the entrance of the Sognefjord at Brekke with an earlier grave from the turn of the century is another context with a wide dating. Three other graves related to 'new' places, and another three with broader dating frames, and key and a third with an iron bar besides the weapons, like the graves in Stryn and Breim. Of these, the grave at Brekke with a strategic place for communication, control, and access to the in-fjord districts, was a place with several Viking Age graves, however, mainly weapon graves. One of the cairns with two graves, a male grave and a female, the latter with a single spindle whorl for tools. Some of the other graves at the site contained tools too: one for agricultural work and forging and another with a tong, hammer, file, and whetstone.

Now, Aurland, further in the Sognefjord by the inner fjord arm, Aurlandsfjorden, was the district in Sygnafylki with the largest representation of graves with textile tools from the 9th century – altogether six female graves and the seventh from its latter part at the central place Vinjum at the entrance of the valley (Fig. 5.25).

Nearly all the graves connected to the agricultural landscapes in the long rather flat west–east-oriented Aurland valley, before the valley rises steeply up to the large Hardanger mountain plateau with access to vast upland resources. The two graves furthest in-valley, at Belle, on the eastern side of a long and large lake Vassbygdvatnet, were both boat burials, dated broadly to the Viking period, of which the supposedly earliest at the site had remains of a damaged male grave too. Two grave mounds at Terum on the western side of the lake comprised female graves too – one without socially distinguishing emblems, the other with dress accessories, indicating the highest status III and both with tools for the initial stages of textile production. However, the one without dress accessories also contained cooking equipment, agricultural tools, and three keys, which does not indicate low status.

A large cairn at Tokvam, some 3 km further down-valley, also comprised two graves and both with textile equipment but still rather different. While the central grave of a high-status woman buried with brooches and a cruciform Irish brooch had only a single spindle whorl as grave gift, the secondary grave in the outskirts of the cairn contained similar grave goods and tools as appearing in graves of high status, however, differed by being without the traditional social emblems of brooches. However, buried in a monumental grave, supplied with tools for all stages of production and furnished with 38 fine beads for a necklace, it does not indicate inferior status, although it was somewhat ambiguous as for status.

In Aurland, there was only one new grave of this kind, from the late part of the 9th century and later than the other graves with textile tools in the valley.[90] It was located strategically on a terrace at the place by the entrance of the valley, at Vinjum and at a central place that became a medieval church site, referred to as *kirkian á Ryg*,[91] the terrace.

The Vinjum grave stands out as one of the most prominent of the Viking Age graves in Sygnafylki, with a broad range of tools for textile production, besides food-related equipment, agricultural tools, and other prestigious grave goods. It contained one of the most composite tool kits of all graves in the district and the only grave in Aurland with tools for all stages of textile production, shears for both shearing wool and cutting yarn, and wool combs and hackles, two spindle whorls of differing weights, 48 loom weights, of which only 18 remain, and the spatula-shaped tool R413 besides other distinguishing grave goods. In many respects the grave fit into the pattern of other prominent graves from the 10th century, both for its central location and prominence and multiple tool kit. The buried woman in the grave at Vinjum was moreover distinguished by a long stick/staff/distaff, like in the graves at Veka and Trå in Hordafylki (Fig. 5.26 and cf. Fig. 5.15).

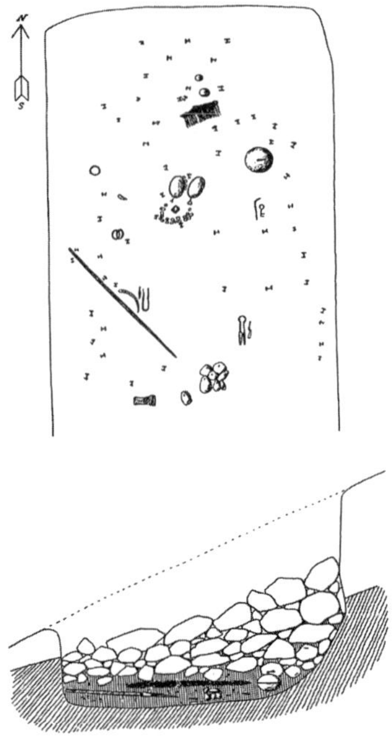

Fig. 5.26: Plan and section of the Vinjum grave. The brooches and a necklace of beads can be discerned in the upper end of the grave, with the stick/staff on her right side (Source: Bøe 1926, pl. 1).

In Sogndal, two high-status female graves stem from, respectively, the first and late part of the 9th century too and both were located close to each other at a headland by the entrance to the main Sogndal valley at Nes, and both were equipped with tools for three work stages. A possible third grave was there, also with textile tools, but was more uncertain as a closed grave context, and was possibly a double grave. The graves at Nes comprised also a male weapon grave with tools for boatbuilding and other equipment, however, not preserved as a closed context.[92]

Altogether, the community of Sogndal represents a relative high share of unclear grave contexts, with three others, dated broadly to the Viking period. They all connected to strategic places – one by the entrance of the Sogndalsfjord, at Lomelde another by the headland and entrance to the main Sogndal valley, at the head of the fjord, and the district Kaupanger, connected to the main Sognefjord by a large and protected harbour. The textile tools therein were few: one with tools for weaving; the others for shearing and combing.

In Vik, mid-fjord, the graves with textile tools represent unclear contexts and date broadly to the Viking period but possibly from the 9th century. One of them related to an earlier excavated mound at Hopperstad but only as vague remains, the other connected to a mound at Hove at the same terrace as Hopperstad, and the third grave at Tryti, as a secondary grave in a mound further up-valley (see Fig. 5.24 and Fig. 5.35). The textile tools in these rather unclear

Table 5.18: Sygnafylki, 9th century status III with traces of tools for more than two work operations. P = preparation, S = spinning, W = weaving, P = shearing, P** = combing, P*** = shearing and combing, W* = loom weight(s), W** = weaving beater, W*** = weaving beaters and loom weights, F= finishing (sewing), M = maintenance, M* = pellet of glass, M** = whalebone plaque.*

No.	Name	District	Central place, communication and control	Fjord	Upland/ Inland	Tool categories	Work operations	Keys	Foreign artefacts	Status
B9064	Nes	Sogndal	x	x		3	P*+S+ W** B4841 (3)			FIII
B9065	Nes	Sogndal	x	x		5	P***+W** (3)		x	FIII
B3972–86	Nes	Sogndal	x	x		3	p**+S+W			M+F?
B8669	Tokvam II	Aurland			x	5	P***+S+W+M(4)		x	F ?/III
B7731	Vinjum	Aurland	x	x		6	p**+p***+S+W*		x	FIII

Table 5.19: Tools in 9th-century graves in Sygnafylki related to status (N:11).

Status	Shears R442+R443	Wool combs	Spindle whorls	Loom weights	Weaving beaters	R413	Shears R443	N: Graves
Status III	3+1	3	2	15	2	1	1	5
Status II	–	–	3	7 (21)	2	–	–	4 + 1 MII/FII
Status I	1	1	2	28	2	2	–	2
N	4 + 1	4	7	50 +(21)	6	3	1	11

Table 5.20: Tools in graves in Sygnafylki from unclear contexts, dating and status (N:10).

Unclear contexts	Shears	Wool combs	Spindle whorls	Loom weights	Weaving beaters	N: Tools
N: 10	3	3	1	4	6	17

contexts varied from one to three processes – respectively with a weaving beater, wool combs, and a spatula-shaped tool, R413, and the third context at Tryti with shears, wool combs, and a weaving beater, now missing.

The three other 9th-century graves with textile equipment in Sygnafylki appeared as single graves of this kind in their districts, of which two related to the mid-fjord, close to Vik with tools for, respectively, one and two processes – and one at a dispersed area at Tveiti by the fjord outside the main habitation in Vik. The other was a high status female boat grave in a large cairn in an upland area at Berdal *c.* 200 m.a.s.l. The third was a double grave in a mound in the inner part of the county, at Hauge located mid-valley in the long Lærdal valley, at a central and strategic place for communication. Two uncertain contexts, at Hillestad in Lærdal, but possibly one of the oldest, predating *c.* AD 800 and an unclear grave at Moa in Årdal in the inner parts of the valley had tools for one to three processes (see Fig. 5.35).

Compared with Firdafylki, the Early Viking Age graves in Sygnafylki were fewer both in numbers and as for tools. Four or five of the 9th-century graves contained tool kits for more than two work processes, of which four connected to central and strategic places in their districts, Sogndal and Aurland (Table 5.18).[93]

Three of these places, Nes in Sogndal, Tokvam, and Vinjum in Aurland, each with two graves with textile tools, and at Nes possibly three, of which one dates to the late part of the 9th century, like the Vinjum grave. The two female high-status graves at Nes, from the first and second part of the 9th century respectively, had tools for three and four processes and were represented by three and five tool categories, while the unclear grave at Nes had tools for three work processes and all with tools for the preparatory stages.

- Nes (B9065), status FIII: shears (R442) + wool combs
- Nes (B9064), status FIII: shears (R442)
- Nes (B3971-86), status ?: wool combs
- Tokvam (II), status ?: shears (R442) + wool combs
- Vinjum, status III: 2 shears (R442 and R443) + wool combs

All but the uncertain grave at Nes had shears of the traditional type for sheep shearing, while the Vinjum grave had shears for both wool sharing and cutting yarn and textiles. Three contained wool combs, including the uncertain context at

Table 5.21: Spindle whorls in 9th century (N:13) in graves from Sygnafylki related to status (N:8).

Weight groups in g	*15–19*	*20–24*	*25–30*	*30–35*	*35–40*	*40+*	*N: Whorls*	*Not available*	*N: Graves*
Status III	2	2	2	1	–	–	7	1	3
Status II	–	1	1	–	–	–	2	–	2
Status I	–	1	–	1	–	–	2	–	1
Status?	–	–	1	–	–	1	2	–	2
N	2	4	4	2	–	1	13	1	8

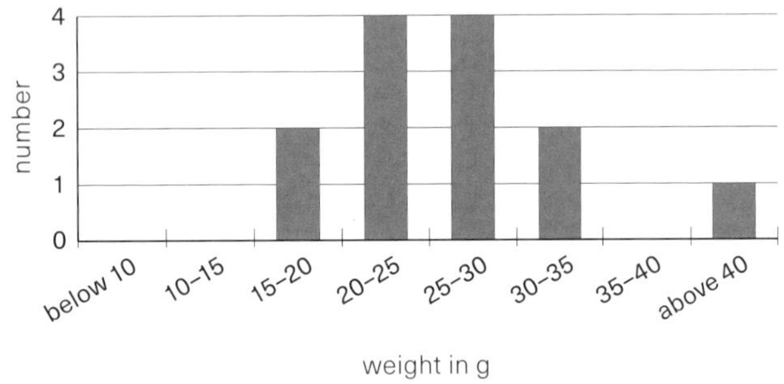

Fig. 5.27: Spindle whorls in 9th-century graves in Sygnafylki, spread within weight groups of 5 g (N:13, measurable 12).

Nes. Compared with the graves with textile tools from the whole county, including 10 from rather unclear contexts, the four graves with composite tools stand out as tools for the preparatory stages, shearing and combing (Tables 5.19 and 5.20).

All but one of the graves with composite tool kits had tools for spinning, and three of them with more than one whorl and of varying weights and weight groups, weighing from 15–20, 25–30, and 30–35 g able to produce all kinds of yarn qualities, from relatively fine to rather ordinary qualities. When including the uncertain double grave at Nes represented by a whorl of 46 g, they were able to produce coarse and strong yarn too, including twined yarn.

- Nes (B9064), status III: 3, 22 g, 29 g, 15–20 g
- Nes (B9065), status III: –
- Nes? (B3971/86?), status M/F?: 2, 46 g + 1 very light
- Tokvam (B 8869 I+II), status III, status I?: 3, I: 25 g, II: 25–30 g, 30–35 g
- Vinjum (B7731), status III: 2, 19 g, 28 g

The eight other graves with whorls but with fewer tools, all had one whorl each within the same overall weight pattern, from *c.* 15 to 35 g and only one above 40 g and with altogether 13 measurable whorls (Table 5.21; Fig. 5.27).

All the graves with tool kits contained tools for weaving – two with weaving beaters and three with both weaving beaters and loom weights – from 15 to 28 weights, although not all were retrieved.

- Nes (B9064): 1 weaving beater
- Nes (B9065): 1 weaving beater, 15 loom weights, 1 R413
- Nes? (B3971/86?), status M/F?: 1 weaving beater
- Tokvam II: 1 weaving beater, 28 loom weights, 1 R413
- Vinjum: 48 loom weights, 1 R413

The loom weights from the Tokvam grave weigh from 140 to 450 g but mostly within the range of 300–400 g with a mean of 298 g (see Chapter 6).

The spindle whorls in the graves from Sygnafylki as indicators of yarn qualities, concentrate on weights between 20 and 35 g and for producing medium fine to coarse yarn and were generally coarser than indicated by the whorls from Firdafylki. The graves with textile tools in Sygnafylki also differ from those in Firdafylki as for type of locations, largely to lowland areas by the fjord and concentrated to two districts, Aurland and Sogndal, with fewer in-valley and upland areas with a larger share of unclear contexts. The double representation of male and female graves with tools for complementary work as in Firdafylki, however, is less pronounced in the burial customs. A common characteristic, however, is the concentration of mainly female graves of high status and with tool kits for varied work processes. In the three best represented districts, Vik, Aurland, and Sogndal, they relate to central places for communication and status, such as Brekke, Nes, and Vinjum, and with tools and special and exquisite grave goods indicating a production above ordinary household level.

In Hordafylki, 16 of the graves with textile tools date to the 9th century (see Table 5.23a), of these at least nine to its early part and six to its latter part, possibly stretching into the 10th century. Still, many are unclear as closed contexts and as for a closer dating. Five thus date broadly to the Viking period and at least another six to the Late Iron Age, although most likely to the Viking period (Tables 5.23b and c). This implies that possibly 25–28 contexts stem from the Viking period – and about half, 16, from the 9th century. Four graves connected to coastal areas, eight to the mid-zone and the majority with broader time frames to the inner zone with a concentration to Voss and with two in Granvin, at, respectively, Folkedal and Seim. Based on the

Table 5.22: Textile tools in female graves from Hordafylki, from the 9th century and with extended time frames of tools dated to the Viking period and Late Iron Age (N:28).

Status/tools	Shears	Wool combs	Spindle whorls	Loom weights	Weaving beaters	Smoothing plaque*	Needle case	N: Graves/ contexts
Status FIII	5	2	10	10	5+1*	2	–	7
Status F?	–	1	6	8	3	–	1	10
Double	1	1	1	–	1*	–	1	3
M+F?	–	2	3	9	4	–	–	8
N:	6	6	20	27	12 + 2*	2	2	28

social emblems, six female graves from the early 9th century count as status III, three of status II, three of status I and three without dress accessories, one of them, however, of high-status judging by the grave goods and probably dating to the first part of this century, as well as the other grave from this district. Only one of the assumed double graves, a grave from Lunde in central Voss, had social markers for both genders – dress accessories and weapons. Graves with weapons and textile tools but without other female status markers count as possible double graves of unequal status, where in six cases, the weapons indicate high status – status III, in four cases of status II and two of status I. This insecure identification as grave contexts and with a close dating is more prevalent in the inner zone than for those by the coast and the mid-zone.

Four coastal graves related to strategic and central places along the inner sailing route, Hodneland and Votno, like the two early graves from the turn of the century or earlier. Two of them connected to places known as nodes in estates of the early Kingdom and as central places for communications, at Fitjar and Lygra. Another four female graves of high status related to the mid-zone, two of them in the district Fana, both buried in mounds but in rather different physical and social landscapes. The more prestigious with a central location with access to the fjord and a large lake giving access to inner and upland areas. This was a place that later became a medieval church centre at Fana and one of the 12 medieval royal chapels in the country. The other was a dispersed place in an inland/upland area, at Gimmeland, however, not so far off when using the lake for transport. The third grave at Vinnes had a strategic location in the fjord systems further south. The fourth, also to the south at Netteland, is more uncertain as a closed context. These graves by the coast and in the mid-zone differ from the graves from the inner zone, not least from the agricultural area Voss, seemingly disturbed by more early and more intensive farming and unclear as for identification of gender, status, and dating – and as a result also with fewer tools and other artefacts.

Two of the graves relate as noted to the inner district Granvin by the Hardangerfjord, the earlier mentioned boat grave at Folkedal and Seim and both cremations and high-status graves, one as a boat burial in a long mound, the other in a mound. While the boat grave had no dress accessories, except a few beads, it still included the largest assemblage of grave goods of them all, among others, three keys, a frying pan, an iron cauldron, two horse-bits, and with tools for all work stages and a special whale bone plaque, like the one found in the early coastal grave at Fosse as also in the neighbouring Seim grave.

Two of the contexts from Voss count as double graves, located in the central agricultural area and at each end of the large lake Vangsvatnet. One was a double grave in a mound by the western end of the lake at Lunde, and an inhumation burial with grave goods indicating equal high standing, and as for tools furnished with shears, a single spindle whorl and horse gear. The other to the east was one of several graves at Bryn, and was in a mound with weapons but without female social emblems, however, with gendered tools – wool combs, a weaving beater of whalebone, two keys – much like in the Folkedal grave. Another high-status female grave from the Voss area, at upland Stalheim, nearly 400 m.a.s.l, comprised a weaving beater, a frying pan, and a hoe. Five other contexts were from the late part of the 9th century, possibly into the 10th century, of which three related to Voss, from both central areas and the border zones. The two others were from the inner zone of Hardanger and one from the mid-zone in the southern part of the county. Generally, they appear as uncertain as closed contexts – and earlier discussed as possible or probable double graves (see Appendix 5.6). The tools are generally few, for one or two work operations and in four cases, only with one tool category – either for spinning, weaving, or, as in one case, sewing. Five others dated broadly to the Viking period, fall into the same spatial pattern – three from Voss in both central areas and at its fringes, one in inner Hardanger and another in the mid-zone, all with tools for one or two processes. The six others with even broader time frames, all relate to the inner and mid-zones and the same districts as the other graves, and possibly from the 9th century too. Being without clear female identification and weapons, except an axe, the textile tools may indicate female graves. When they appear as graves in mounds and in two cases as boat burials, such as at Bolstad and Sørheim in, respectively,

*Table 5.23a: Tools in the 9th-century graves related to status from Hordafylki (N:16 graves) * = made of whalebone.*

Status/tools	*Shears*	*Wool combs*	*Spindle whorls*	*Loom weights*	*Weaving beaters*	*Smoothing plaque*	*Needle case*	*N: Tools*	*N: Graves/ contexts*
Status FIII	4	1	10	10	5+1*	2	–	33	7
F Status?	1	–	2	8	2	–	1	14	3
Double M+F	1	1	1		1*	–	–	4	1
M+F?		1	2	1	2	–	–	6	5
N	6	3	15	19	9 + 2*	2	1	57	16

Table 5.23b: Tools in the graves in Hordafylki, dated broadly to the Viking period (N:6).

Status/tools	*Shears*	*Wool combs*	*Spindle whorls*	*Loom weights*	*Weaving beaters*	*Needle case*	*N: Graves*
F: status?	1	3	4	–	1	–	3
Double F+M	–	–	1	–	–	1	1
M+F?				1	1	–	2
N	1	3	5	1	2	1	6

Table 5.23c: Textile tools in the female grave in Hordafylki, dated to the Late Iron Age (N:6).

Status/tools	*Shears*	*Wool combs*	*Spindle whorls*	*Loom weights*	*Weaving beater*	*Needle case*	*N: Graves*
Female	–	–	4	–	1	–	4
Double? M+F?	–	–	1	7	1	–	2
N	–	–	5	7	2	–	6

Table 5.24: Spindle whorls in graves from Hordafylki, 9th century, according to weight and related to status.

Weight groups in g	*10–15*	*16–20*	*21–25*	*26–30*	*31–35*	*36–40*	*41+*	*Unavailable*	*N: Weights*	*N: Graves*
Status III	5	1	–	2	1	–	1		10	5
Status II	1		–	–			–	1	1	2
Double grave	–	–	–	–	1	–	–		1	1
N	6	1	–	2	2	–	1	1	12	7

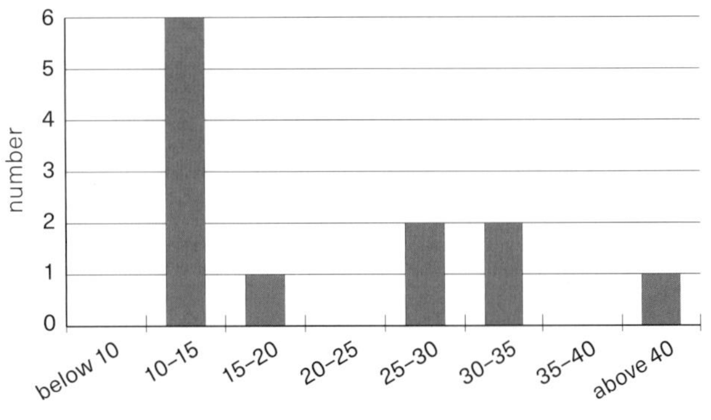

Fig. 5.28: Spindle of measurable whorls from Hordafylki 800–900 according to weight (N:12).

Voss/Evanger and Etne they do not indicate low status. The textile tools, however, are limited, for one or two processes in all these cases.

Differences in status and environments

When assessing these 28 contexts as an entity, the female graves of high status or of uncertain status fall into the same spatial pattern as for representation of tools and grave goods but with a smaller repertoire in graves with wider dating frames (Table 5.22).

Two of the graves by the coast had shears for sheep shearing and loom weights or weaving beaters, one of them even the high-status grave at Fitjar. The most common combination – spinning and weaving – is represented in

Table 5.25: 9th-century graves from Hordafylki with tools for more than two processes.

Seim, Granvin, status FIII	*Folkedal, Granvin, status FIII*	*Gjerdskvål, Voss, status FIII*
B8635	B 7114+ B7149 +7867	B5914
Shears	Shears	shears
wool combs	wool combs	Wool combs
1 spindle whorl 14g	4 spindle whorls 2: 15 g, 2; 26 g	2 spindle whorls 15, 26 g
–	1 weaving beater	–
1 whalebone plaque	1 whalebone plaque	–
2 keys	3 keys	–

nearly all 15 graves from the 9th century and in three of the 12 graves with broader dating frames (Tables 5.23 a, b, c).

Altogether, seven of the graves contained spindle whorls, with eight available whorls and in varied weight groups, ranging from 11 g to more than 40 g indicating then a production of different yarn qualities, from the finest to the course for wide use areas. Although the tools show no clear correlation to status and special environments, the lighter whorls, 11–15 g, tend to connect to graves of higher status (Table 5.24; Fig. 5.28).

The two high-status graves in impressive mounds in Fana in the mid-zone were both cremation burials. Of these, the high-status grave from the central place comprised only two spindle whorls as tools, both light, 14 and 16 g, for producing yarn within the luxurious segment, which is further substantiated by prestigious grave goods of foreign origins, a bronze cauldron, and remains of melted glass. The other grave further inland, in contrast, contained shears for sheep shearing and a single whorl in the common weight group, *c.* 30 g, for producing yarn for coarser qualities and with a hoe for tilling. The high-status female grave at Seim in Granvin, as one of two graves with tools for more than two processes, contained shears for sheep shearing, wool combs, a light whorl of 14 g, and a whalebone plaque for smoothing. It contained additionally two keys, remains of a casket, food-related equipment, strike-a-light, and bridles, all signalling high status. This whorl was about the same weight as in the double grave at Lunde in Voss with a whorl of 33 g and in the same combination of shears. In contrast, some female coastal graves that seem to represent elite milieus were modestly equipped with tools, limited to tools for shearing and weaving, representing the first and last steps in production, such as in the grave from Fitjar. One of them, and yet another high-status grave at Vatne/Votno, had two weaving beaters of respectively iron and whalebone, obviously for different use areas and for weaving finer and coarser qualities. Another high-status grave located at the aforementioned central place, Lygra, with tools for spinning and weaving, included a light spindle whorl of 12 g, eight loom weights and remains of a weaving beater. In contrast, another grave nearby and again of high status, contained two spindle whorls of 31 g and 42 g and eight loom weights, that signal production of coarser yarn and fabrics, possibly sails.

The graves with broader time frames fall into a pattern of few tools too, and with a concentration of weaving equipment, and five graves with spindle whorls but seldom with other grave gifts (cf. Table 5.23c). In two possible graves from Voss with weapons equating status II, all the tools were for weaving: one with a weaving beater and a sword, the other with a single loom weight. Another grave of this kind in the inner zone of Hardanger with weapons, revealed a needle case, a spindle whorl of unknown weight, and a key, which indicate a female presence and possibly an asymmetrical double grave. Three other contexts that may have been female or double graves represent a broader spectre of tools, especially for spinning but not weaving (cf. Table 5.24).

Tool kits are more rare as grave goods than in the counties to the north; only three graves of all the 28 contexts had tools for more than two work processes and all of them were from the inner zone, the districts of Granvin and Voss at, respectively, Folkedal and Seim in Granvin in the inner zone of Hardanger, and as richly furnished as some of the graves in the counties further north. They shared many common traits, with tools for shearing, combing, spinning, and both with a whalebone plaque for smoothing – the Folkedal grave also had four spindle whorls of differing weights as well as a weaving beater, besides food-related equipment, a strike-a-light, two bridles, and three keys, thus emphasising status and broader roles, probably extending an ordinary household production. The third grave with a tool kit for three processes, related to a high-status grave in a mound at Gjerdskvål in Voss with tools for shearing, combing, and spinning (Table 5.25).

Except the graves from the coastal and the mid-zones, all the graves were concentrated to the inner zone and from about the same areas, and generally with tools for one or two processes and mostly for spinning and weaving. One of the graves, a possible double grave at Bryn centrally located in Voss, had wool combs and a weaving beater of whalebone,

Table 5.26: Measurable spindle whorls in weight groups of 5 g (N:20) from 20 graves/contexts from Hordafylki, from the 9th century and of wider unspecified dating (N:20).

Weight groups in g	*10–15*	*16–20*	*21–25*	*26–30*	*31–35*	*36–40*	*41+*	*Unavailable*	*N: Graves*
c. 800–900	6	1	–	2	2	–	1	1	12
800–1000	1	1	–	1	–	1	–	1	3
Late Iron Age	1	1	2	–	–	–	–	1	5
N	8	3	2	3	2	1	1	3	20

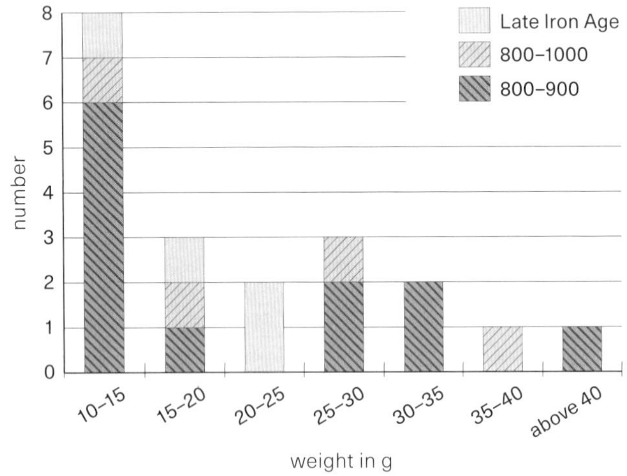

Fig. 5.29: Measurable spindle whorls in Viking Age and Late Iron Age graves from Hordafylki (N:20).

and may indicate production of finer textiles, needing softer strokes, and in contrast to the female grave the fringes at Stalheim with a long weaving beater as a single tool.

The measurable spindle whorls in graves with broader dating frames fall into the same weight pattern as in the better-documented graves from the century. Of 17 measurable whorls, about half, nine whorls, relate to the lighter weight groups from 10 to 20 g. When represented in pairs, they belong to different weight groups used for spinning different qualities (Table 5.26).

More than half of the measurable spindle whorls, 20 in all, weigh less than 20 g, and were able to produce very fine yarn and few were above 30 g, for spinning coarser qualities. When comparing the most common categories – spindle whorls and 20 measurable in all – they comprise a relatively larger share of light whorls than in Sygnafylki (Fig. 5.29).

The composition of the tools seems to reflect graves of varied status and generally fewer in high-status graves compared with graves in the counties to the north. This difference may indicate other modes and range of production, and when not connected to the same high status rather indicates production on a household level, especially in the inland areas, with the many graves of this kind in a district such as Voss and not connected to central places or as nodes in estates, like in Firdafylki and Sygnafylki. The coastal graves show another pattern, resembling the high-status female graves from the turn of the 9th century and like graves in the counties further north as for status and tool assemblages.

The relative high share of female-related tools in both double graves and possible double graves of unequal status (cf. Tables 5.23a–b), however, such as spindle whorls, smoothing plaques, weaving beaters of whalebone, and a needle case, besides weaving equipment does not support weaving as a new male working domain, as suggested by Rabben. When mainly found in unclear contexts they cannot count as evidence – but in such cases, possibly symbolising a role as organisers rather than producers, being otherwise represented by few tools.

Concentration and centralisation?

The burial customs of using textile tools as grave goods continued in the Late Viking Age in both the same areas as earlier but also at new places, including central places in the lowlands but fewer in the outskirts. While these burial customs lost ground in the northern counties by the 10th century, they expanded in Hordafylki. This shift to new places and environments and as for status and tools, may indicate changes within production too, socially, and technologically. Did they now rather concentrate in central places as centres within larger estates and new or of older heritage?

In Firdafylki, the custom of using tools as grave goods decreased in Stryn and Gloppen by the turn of the 10th century and nearly ceased in this northern part. From Firdafylki as a whole, they now count 14 contexts, compared with 27 from the 9th century, and with a shift from inland/upland areas to agricultural landscapes in the southern part of the county, in the districts of Gaular, Fjaler, Førde, and Jølster and many at central places as junctions between land routes and waterways (Fig. 5.30).

The tool assemblages reveal both similarities and differences to the earlier graves. Although fewer, they are of generally higher status and have a broader range of tools: 14 of status III, nine with tools for three, four and five work processes, and respectively five and three graves with tools for one or two stages of production (Table 5.27).

In Stryn, now only two graves with textile tools related to Oppstryn, but they still represent high status. One once again was located in Gutdal, and furnished with costly,

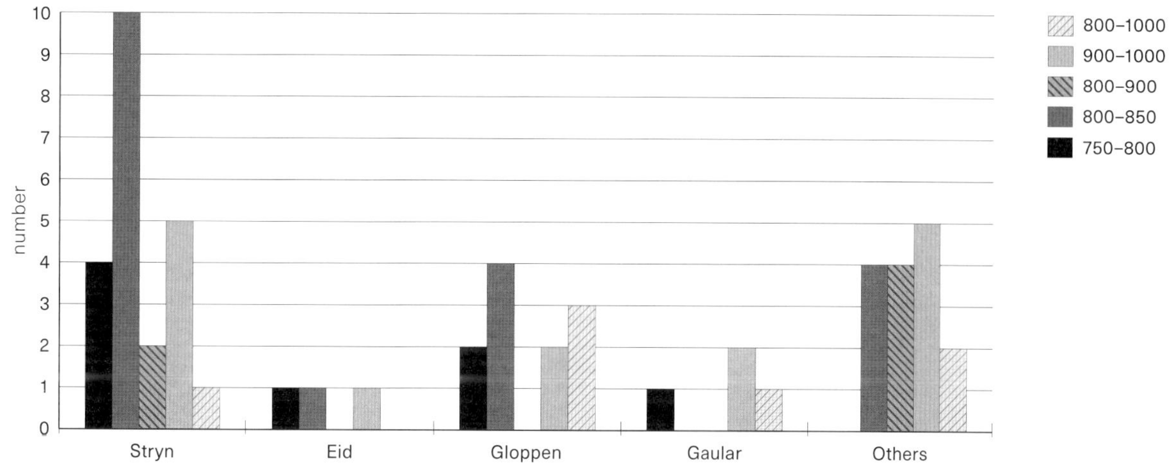

Fig. 5.30: Temporal distribution of graves with textile tools in Firdafylki over time (N:55).

Table 5.27: Number of work processes in graves from the 10th century in Firdafylki according to status (N:18 graves).

Work processes	*1*	*2*	*3*	*4*	*5*	*N: Graves*
Status F III	2	5	1	5	1	11
Status FII		1	–	–	–	4
Double grave (F+M)			1	1		2
Status ?	1					1
N:	3	6	2	6	1	18

prestigious, and rare dress accessories – a tongue-shaped brooch of copper alloy (R664), a pendant of copper alloy, like a snake (JP167), two similar rounded pendants of copper alloy, and beads of glass, carnelian, crystal, but furnished only with two small spindle whorls of fired clay, besides food-related equipment, remains of a grip for a vessel, and various unidentified iron remains, a hooked key, and a long iron spit or staff with a handle with a ring, discussed as being for various functions, among others a distaff, but obviously also a sign of status.

The other and less distinguished grave at Brekke in Oppstryn had tools for combing and weaving. Two graves mid-valley in Stryn, yet again connected an earlier grave of this kind although some decades older, and at a place that became a medieval church centre, *kirkiu á Æiði*,[94] now denoted Kirkeide. Another high-status female grave at Loen, was located at the head of the long Nordfjord at a site marked with a so-called 'holy white stone' of older heritage, predating the grave, and surrounded by several contemporary weapon graves. A stone cross at the site also predated the medieval church,[95] *kirkian j Loda*.[96] Except the Gutdal grave, all these graves had tools for two processes, for shearing, spinning, and weaving in different combinations but the graves at Gutdal and Loen were also supplied with food-related equipment as grave goods.

The two most prominent graves in the wide Stryn district for status and tools are a high-status female grave and a double grave – located at the central place Hilde further out in the fjord by the wide bay Innvik. This was also a site for a medieval church, *Hildin kirkiu* also denoted as *kirkian j Innvik*,[97] and probably a node in an earlier estate, as indicated by fees and land rent registered in the cadastre and was a common trend in fact for all the early churches. Except the two graves in Oppstryn, all these 10th-century graves in the wide Stryn area thus related to central and strategic places for communication and power. In Loen and Innvik, both burials in mounds excel for status and tools. The two graves both in mounds at the central place Hilde, the high-status female was cremated and buried in a boat, while the double grave close by represented seemingly unequal status, being without socially distinguishing female emblems. However, both graves had textile tools for three work processes, besides other tools. The female grave at Hilde contained the special spit/staff/distaff, 73 cm long, like the one in the high-status grave at Gutdal. It also comprised a weaving beater of whalebone, of the same type as found in a detached grave mound by the coast at Langenes by the entrance of the fjord and facing the North Sea.

At Eid, the new female grave in another mound at Myklebust than the one with two of the earlier graves, presented earlier. This later grave was again a high-status woman, placed in the mound after it had been erected and as the last of the graves in the mound, 16 m × 19 m N–S/E–W, and *c*. 3 m high (see Fig. 5.31). The deceased was furnished with some fine brooches, and with an exquisite collection of beads, a large silver bead decorated by a thin silver thread, 30 beads, some of chrystal, carnelian, but mostly of glass, red, blue, yellow, and white, of different size and all kind of tools for a prominent lady, before she was cremated in a boat with all her treasures. Among these, she had similar iron stick/staff with a handle and ring too, like in the high-status female graves at Gutdal and Hilde. The staff was placed at her right side, and with textile and food-related tools scattered around. The tools comprised, among others items, a sickle for harvesting, besides food-related equipment, such as an iron cauldron, a key, and horse gear, altogether signalling high status and broad work domains. The tools for three processes also shared traits with one of the graves at Hilde, with, among others, a weaving beater of whalebone.[98] The graves at Myklebust thus marked continuity over generations – from around AD 800 into the 10th century.

The two graves in Gloppen, at Arnestad and Mardal, located in the good agricultural landscapes by the inner

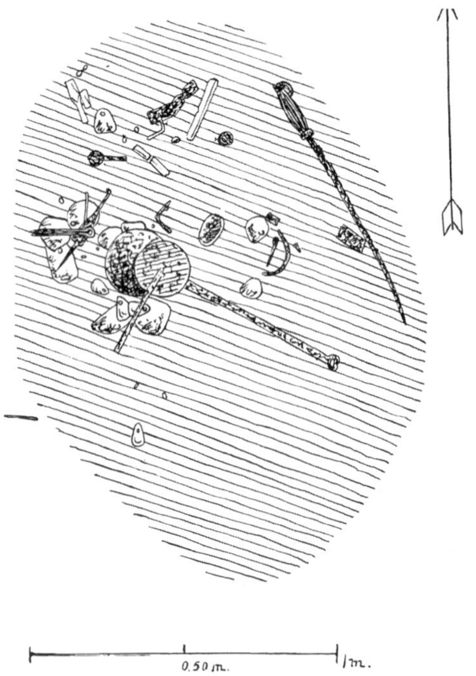

Fig. 5.31: The grave at Myklebust with preserved grave goods from the 10th century B6754 (Source: Shetelig 1912, fig. 438).

Gloppenfjord in the mid-zone are the only graves with textile tools in this area this late. Both were richly furnished with grave goods – tools for four textile processes, as well as food-related equipment and farming tools. At Arnestad, the prestigious female grave in a long mound had textile tools for all processes, except sheep shearing and was additionally furnished with a long stick/staff/distaff of iron, 83 cm long, like in the three prominent graves at Gutdal, Hilde, and Myklebust in Firdafylki and was from the same period.

Like the earlier graves in the district, the male graves at many of the sites contained tools for forging. They reveal, however, a spatial and temporal shift in the burial customs compared with the graves from the beginning of the period lasting well into the 9th century when they largely related to inland and upland Breim. The six other graves with textile tools in Firdafylki now connected to the southern part of the county in the mid-zone, such as Dale in Fjaler and Fyling and Korsvoll in Gaular, of which Dale and Fyling had composite tool kits. Three graves located to the community of Førde: one at Gjelsvik by the fjord, the other two at Gjerland, to an inland/upland area at 300 m.a.s.l., co-locating with other prehistoric graves.[99] However, all these graves had tools only for one or two processes, shearing, spinning, and/or weaving, and with spindle whorls within the common weight groups, between 23 and 30 g. Seven of the graves, however, had composite tool kits – five female graves and two double graves, all of high status – from both 'old' and 'new' areas, and by now overall related to lowland and central areas by the fjord, in Eid, Innvik/Stryn, Gloppen, Fjaler, while the double grave in Gaular was located in an inland/upland area by the large lake Viksdalsvatnet (Table 5.28).

The grave at a central place, Dale furthest south-west and a place that became a medieval church centre, *Dalskirkiu*,[100] and another place with several grave mounds, again indicating

Table 5.28: Firdafylki, 10th-century graves with traces of tools for more than two work operations. P = preparation, S = spinning, W = weaving, P = shearing, P** = combing, P*** = shearing and combing, W* = loom weight(s), W** = weaving beater, W*** = weaving beaters and loom weights, F = finishing (cutting, *sewing**), M = maintenance, M* = smoother of glass, M** = whalebone plaque.*

No.	*Name*	*District*	*Central place*	*Fjord*	*Inland*	*Tool categories*	*Work operations*	*Keys*	*Foreign artefacts*	*Status*
B5717+B5755	Hilde	Innvik/Stryn	x	x		4	(3) S +W***+F*	1	x	FIII
B5588	Hilde	Innvik/Stryn	x	x		4	(3) P**+S+W***			MIII/FI?
B5764	Myklebust	Eid	x	x		4	(3) P**+W***+F*			FIII
B7683	Arnestad	Gloppen	–	x		4 (5)	(4) P**+ S+ W**+F		x	FIII
B4605	Mardal	Gloppen	–	x		3	(4) P**+ S+ W*		x?	FIII
B5910	Dale	Fjaler	x	x		6	(5) P**+ S+ W*+F**, M*	1	x	FIII
B11470	Fyling	Gaular	–		x	3	(3) P***+S+W	1	x	MIII/FIII

Table 5.29: Distribution of tools in graves, from 10th-century Firdafylki, according to types of work operations (N:17).

Operations	*Shearing*	*Combing*	*Spinning*	*Weaving*	*Smoothing*	*Sewing*	*Cutting*	*Graves*
1	–	–	1	2	–	–		3
2	4	2	3	3	–	–		6
3	–	1	2	2	–	–	2	2
4	1	4	3	4	–	1	1	5
5	–	1	1	1	1	–	1	1
N:	5	8	10	12	1	1	4	17

Table 5.30: Measurable spindle whorls in weight groups of 5 g (N:15) in graves related to status in 10th-century graves from Firdafylki (N:11).

Weight groups in g	*10–14*	*15–19*	*20–24*	*25–29*	*30–34*	*35–39*	*40–45*	*Not available*	*N: Graves*	*N: Whorls*
Status III	1		2		1			2	7	6
Status II	1		1	1					1	3
Double F+M	1	1	2			2	1		2	7
Double? M+F?			1						1	1
N	3	1	6	1	1	2	1	2	11	17

a node of *longue durée*. It contained tools for five processes, while the other inland grave in this part of the county – the double grave at Fyling – had tools for four processes. Altogether, many of these high-status graves connected to 'new' and central places as for this type of graves, but also demonstrating continuity, such as at Myklebust. The graves may thus signify a centralisation of power – a trend discerned in the latter part of the 9th century such as at Vinjum and this escalated by the 10th century. An issue to pursue, then, is whether the new places indicate specialisation of production too?

When comparing the representation of the tools in these seven graves with all 17 graves from the century, we see that the tools for combing, spinning, and weaving are the most common in all the contexts (Table 5.29).

While the number of graves decreased radically in the inland/upland areas by the late 9th and 10th centuries, the number of graves in the lowland increased. Judging by the tools, they varied not only in number but also for status and location, now mainly connected to central places and nodal points in their respective communities, Eid, Gloppen, Fjaler, and Gaular. This new pattern substantiates their higher status and as possible nodes in textile production too.

The seven 10th-century graves with multiple tools from Firdafylki – six female and one double grave of high status – had tools for nearly all textile processes, besides other status indicating grave goods. They distinguish themselves, not only for tools and status but also for locations – with three graves at central places – Hilde in Innvik, Myklebust at Eid, and Dale in the southern part – the others in the agricultural area such as Arnestad and Mardal close to the Gloppenfjord.

All of these graves contained tools for fibre preparation – four for wool combing, however, the Myklebust grave rather for hackling plant fibres, and the grave at Fyling also shears for sheep shearing:

- Hilde, status FIII: wool combs
- Hilde, status MIII/FI? wool combs
- Myklebust, status FIII: hackles
- Arnestad, status FIII: wool combs
- Mardal, status FIII: wool combs
- Dale, status FIII: wool combs
- Fyling, status MIII/FIII: shears (R442), wool combs

All except the Myklebust grave, contained spindle whorls, and all but one with two whorls or more – 13 whorls in all, but only seven were retrieved. Most were of steatite, one of the lightest of fired clay and another of lead, now unavailable, but clearly light and heavy respectively.

- Hilde, status FII: 2 whorls, 10–15 g (fired clay), 25–30 g
- Hilde, status MIII/F1?: 1 whorl 20–25 g
- Arnestad, status FIII: 2 whorls, 11 g, 20–25 g
- Mardal, status FIII: 2 whorls, unavailable g? (cannot be estimated or weighed)
- Dale, status FIII: 2 whorls, 33 g + g?
- Fyling, status MIII/FIII: 2 whorls, 20–25 g, g? (lead)

Compared with the 21 whorls in the total assemblage (cf. Table 5.18), they fall within the same weight categories, ranging in weight from 11 to 33 g, and lightest in the female graves at Arnestad and Hilde. However, the six graves with two whorls appeared in differing weights, but not all measurable, but still underline production of varied qualities (Table 5.30; Fig. 5.32).

Table 5.31: Distribution of tools in the graves from the 10th century in Firdafylki according to status (N:17 graves).

Status	*Shears R442*	*Hackle*	*Wool combs*	*Spindle whorls*	*Loom weights*	*Weaving beater*	*R413*	*Shears R443*	*R450*	*Smoothing pellet*	*N: Graves*
Status FIII	1	1	6	12	15	2; 1*	1	5	1	1	14
Status FII	–	–	–	–	1	–	–		–	–	1
Status?	–	–	–	1	–	1*	–		1	–	1
Double			2	1	2	2	–		–	–	1
N	1	1	8	14	18	4; 2* (6)	1	5	2	1	17

Table 5.32: Spindle whorls for weight from Viking Age in Firdafylki, g? = not measurable (N:33).

Weight period	*Below 10 g*	*10–15 g*	*15–20 g*	*20–25 g*	*25–30 g*	*30–35 g*	*35–40 g*	*40+*	*g?*	*N*	*N: Graves*
800–900		–	6	3	3	2	2	2	1	19	11
900–1000		2	1	4	1	4	2	1	3	18	18
N		2	7	7	4	6	4	3	4	37	29

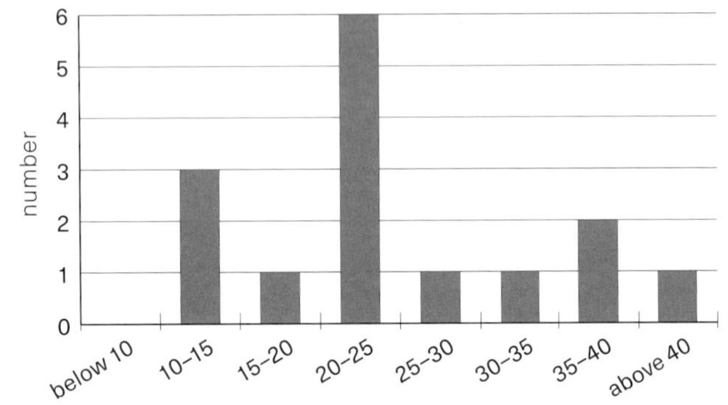

Fig. 5.32: Measurable spindle whorls in weight groups of 5 g (N:15) in graves related to status in 10th-century graves from Firdafylki (N:11).

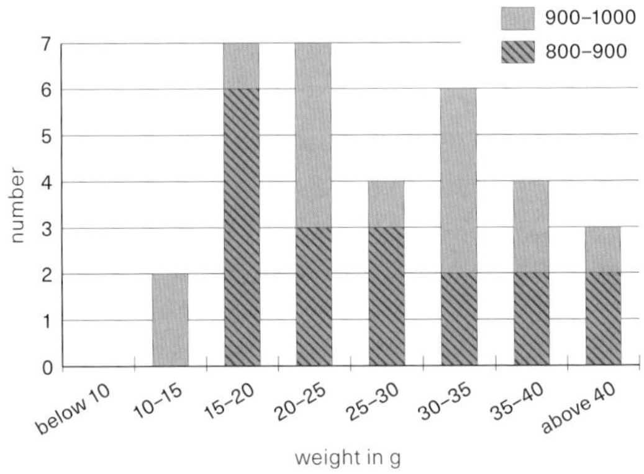

Fig. 5.33: Measurable spindle whorls as for weight from Viking Age in Firdafylki (N:33).

When using spindle whorls as indicators of qualities of yarn and textiles, they may also reflect changes in production over time. Of the 33 measurable whorls, the lightest weight group were able to produce the very finest qualities, while otherwise generally medium-fine and ordinary qualities, to coarser yarns were produced (Table 5.32; Fig. 5.33).

All the seven graves with multiple tools and of general high status had tools for weaving:

- Hilde, Innvik, Stryn, FIII: 1 loom weight, 1 beater of whalebone (R439), length?
- Hilde, Innvik, Stryn, F? + MIII: 2 loom weights, 1 weaving beater (R440), part, length?
- Myklebust, Eid, FIII: 6 loom weights, 1 weaving beater of whalebone (R439), 60 cm long
- Arnestad, Gloppen, FIII: 1 weaving beater (R440), *c.* 80 cm long; 1 R413
- Mardal, Gloppen, FIII: 2 loom weights
- Dale, Fjaler, FIII: 1 loom weight
- Fyling, Gaular, FIII+MIII: 1 weaving beater (R440), *c.* 60 cm long

Compared with the representation in all the graves, five of the seven graves with composite tools including weaving beaters connect to these general high-status graves, while loom weights, 10 in all, are under-represented in this group of graves. The weaving beaters appear in different materials, size, and shape – three of whalebone (R439) – and as noted in the high-status female graves at Hilde and Myklebust. The weaving beaters are notably longer than in the earlier graves. Compared with the 9th-century graves, shears for outdoor sheep shearing are rare in these later contexts. Here, the shears in five or six of these eight graves are now of the type for indoor textile work (R443) and all are related to high-status female graves. Only the double grave at inland Fyling contained sheep shears. The grave at Dale with tools for five processes, also comprised a glass pellet for smoothing, and as a

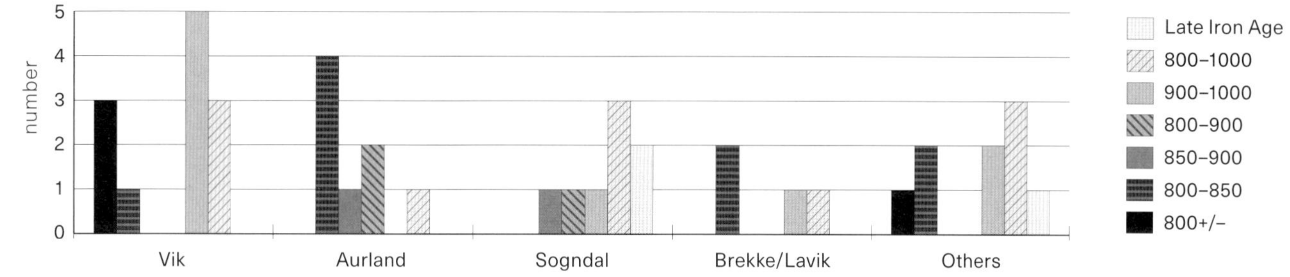

Fig. 5.34: Temporal distribution of graves with textile tools in Sygnafylki (N:41).

Table 5.33: Distribution of textile tools for varying work operations in graves, c. 900–1000 in Sygnafylki (N:10).

Operations	*Combing*	*Spinning*	*Weaving*	*Shearing/cutting*	*Sewing*	*Smoothing*	*Graves*
1	–	–	3	–	–	1	2
2	–	3	2	–	.	1	3
3	1	1	1	1 (R443)		1	1
4	3	3	2	2 (R443)	1	1	3
5	1	1	1	–	1	1	1
N	5	8	9	3	2	5	10

tool of foreign origins a rather rare artefact among the grave goods.

- Hilde, status FIII: shears, R443
- Myklebust, status FIII: shears, R443, a bone needle
- Arnestad, status FIII: shears, R443
- Mardal, status FIII: shears, R443
- Dale, status FIII: shears, R443, 1 smoothing pellet of glass

The spindle whorls from the two centuries show a relatively even spread in weight and with most spindle whorls in the weight groups from 20 to 35 g, for producing rather ordinary qualities. Over time, however, there are differences, with a larger share of light spindle whorls in the weight groups below 30 g in the early part of the Viking period. While nearly all the graves with multiple tool kits from the 9th century related to inland/upland areas, close to the resources and with more tools for varied production, including sheep shearing, nearly all the 10th-century graves of this kind connected to the lowland, at 'new' places, as well as central places. Judging by the tool assemblages and high status, the tools and grave goods indicate roles within production above the household level, as an advanced production, possibly specialised. In contrast to the earlier graves in other environments they indicate shifts of power and with the means to support such a production. The graves with multifunctional tool kits, including keys, were generally few, now appearing only in three high-status graves and all located at central places. Altogether, seven graves with textile tools connected to places that became church centres in the early Middle Ages – Eide/Kirkeide (2 graves), Hilde/Innvik (2 graves), Dale (1 grave) and Grodås (1 grave). Still, other areas and places such as Gutdal with long traditions, carried on the production judging by the tools in the graves and the grave goods.

Sygnafylki

In Sogndal, further in-fjord, the only grave this late in the 10th century connected to Kvåle, was renowned as a central place connected to the early Kingdom and a medieval church site too, *Kirkian á Huale*[101] – and was a grave of similar prominence as the early pre-Viking Age grave referred to earlier in this chapter.

Compared with the graves from the Early Viking Age, there is a reduction in the number of graves, grave goods, and tool assemblages, however, except for Vik with six high-status graves and generally with tools for more processes in the 10th-century graves than earlier (Fig. 5.34). Six of these 10 graves had tools for more than two work operations, covering from three to five processes (Tables 5.33 and 5.34).

In Vik, with five, possibly six graves of this kind, three or possibly four were located at Hopperstad, the other two at Hove and Hønsi (Fig. 5.35).

All the four, possibly five graves were located at the same terrace at Hopperstad as the earliest grave from the late 8th century – and at a central place with an early medieval church, *kirkian á Hoprekstoðum*,[102] and with a still existing medieval stave church, and like in Aurland a site without clear male graves (see Appendix 5.6). The grave at Hove, on

Table 5.34: Tools, and status in graves from the 10th century in Sygnafylki.

Status	*Shears*	*Wool combs*	*Spindle whorls*	*Loom weights*	*Weaving beater*	*R413*	*Needles/case*	*Smoothing plaque*/stone*	*N: Graves*
Status III	7	2	8	22	3	1	2	1*	6
Status II		2	1	1	1	1		1*	3
Status?		–	1	2	–	–	–	1	2
Double		–	1	–	–	–	–	1	1
N	7	4	9 (+2)	23 (+2)	4	2	2	2*, 2 in glass	12 (+)

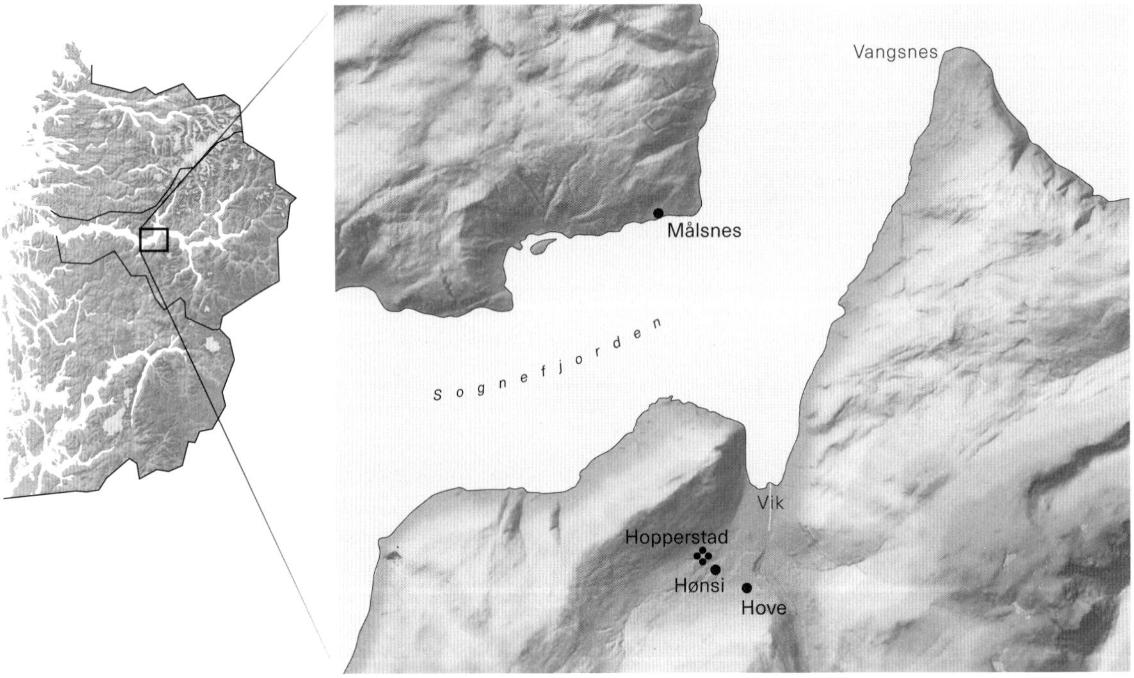

Fig. 5.35: Graves with composite tool kits in Vik from the 10th century (Illustration: Per Bækken).

the other side of the river to Hopperstad, was another place that became a medieval church centre, *kirkian á Howe*,[103] and judging by its name an early pre-Christian cult place. A high-status female burial, at Hønsi, in the main agricultural district too, appeared as a boat grave in a long mound and dated from the late 9th into the 10th century.

Three of the late graves at Hopperstad distinguish themselves not only in number but also for status and with rich assemblages of textile tools and other costly grave goods. One of them was a high-status female in a grave above the very early boat grave. Remains of the teeth showed that she had died at an old age. She was furnished with two oval brooches of Irish origins, and beads for a necklace, mostly of glass and some of amber. The tools in the grave were for five work stages. One of the other graves at the terrace of similar high status contained the same type of brooches, a trefoil brooch, and more than 100 beads. The tool kit included textile tools for three work stages, besides a sickle and hoes and horse gear. The third grave was less distinguished for dress accessories, with only one brooch but of the same type as the former. The tools, however, covered four work stages, shearing, spinning, and weaving and with a smoothing plaque of whalebone, besides food-related equipment. The fourth grave, located close to the former and in an area then with several burials, was an unmarked grave in a boat and contained the largest assemblage of exquisite grave goods of foreign origins besides the somewhat earlier Vinjum grave in Aurland referred to above, and among others a cauldron and ladle of bronze, two glass vessels, besides other costly artefacts and also included a long stick/staff like the Vinjum grave. The textile tools, however, were limited to two spindle whorls, a weaving beater, and some loom weights. Altogether, the grave goods signify high status despite the absence of the traditional brooches. The find of a single weaving beater in an earlier excavated mound at Hopperstad may indicate a possible fifth and secondary grave, however, it was uncertain both for dating and as a separate grave.

*Table 5.35: Graves in Sygnafylki from the late 9th and 10th century with traces of tools for more than two work operations. P = preparation, P** = combing, S = spinning, W = weaving, W* = loom weight(s), W** = weaving beater, W*** = weaving beaters and loom weights, F = finishing (cutting, *sewing**), M = maintenance; M* = pellet of glass, M** = whalebone plaque.*

No.	*Place*	*District*	*Central place for communication and control*	*Tool categories*	*Work operations*	*Keys*	*Foreign artefact*	*Status*
B12131	Målsnes	Balestrand	x	6	4: P**+S+W+F*	1	x?	FIII
B9060	Hopperstad	Vik	x	5	4: P**+S+W**+F*	–	x	FIII
B7761	Hopperstad	Vik	x	4	3: S+W***, F*			FIII
B4637	Hopperstad	Vik	x	5	4: P**+S+W**+M**		x	FII
B4511	Hopperstad	Vik	x	4	4: P**+S+W**+M*		x	FIII
B708–17, B2734+2739	Hønsi	Vik	–	4	3: P**+W**+ M*		x	FII–III
B3456	Kvåle	Sogndal	x	3	3: P+S+W***	3	x	FIII

Another high-status grave further in-valley and probably a boat grave, indicated by iron rivets in a large, long-mound at Hønsi with at least two separate graves, was a male weapon grave of high status too. The female grave of high status contained tools for three processes, combing, weaving, and a smoothing pellet of glass for textile tools and additionally food-related equipment and some special but exquisite dress accessories. The grave at Hove, and another grave without dress accessories, like the Hopperstad grave, contained few textile tools, only wool combs and the strange spatula-shaped tool, besides various food-related equipment. This grave was one of several grave mounds with weapon graves there from the Early and Late Iron Age onwards, thus also differing from Hopperstad without corresponding male graves and graves in monumental mounds.

The concentration of high-status female graves at Hopperstad in the Late Viking Age, all richly furnished with textile tools and other prestigious equipment, clearly distinguished the place both for status and tools, and indirectly as a central place for production too. The textile equipment used as grave goods varied, covering from four to five work processes, besides a wide array of other implements – agricultural tools, sickles and/or hoes, and food-related equipment and even tools for trade. The two other contemporary graves in Vik, however, were more limited for tools. The decline of graves with textile tools in Aurland this late, marks a remarkable change compared with the Early Viking Age and an opposite trend compared with Vik, however, the later grave at Vinjum shows a similar trend as at Hopperstad, possibly triggered by centralisation processes of the time.

The high-status female grave at Kvåle in Sogndal also falls in the pattern of centralisation and concentration, being situated at a place that became the residential place for the King's local representative and with its own church, *kirkian á. Huale*.[104] This grave also appeared at the same site as the very early female grave predating the Viking period and of similar high status (see Appendix 5.0). The new grave contained, like the former, an all-inclusive tool kit, for all processes, from shearing to weaving, besides agricultural tools, equipment for food and storage, three keys, and bridles, and again a long stick/staff. Another prestigious grave related to Målsnes, at a headland on the northern side of the Sognefjord, and with tools for four work stages – shearing, combing, spinning, and sewing.

Compared with the graves from the Early Viking Age in Sygnafylki, the graves from Late Viking Age are generally fewer but with an increase in grave goods and tools, where seven graves contained multiple tool kits (Table 5.35). When assessing the tools related to the various work stages step-by-step, we see that six of them had tools for the preparatory stage of combing, except one of the Hopperstad graves, however, all related to indoor work and for the final stages of production for cutting yarn and textiles (Table 5.35).

- Hopperstad (B7761), status FIII: 2 shears (R443)
- Hopperstad (B4637), status FII: wool combs
- Hopperstad (B9060), status FIII: wool combs
- Hopperstad (B4511), status FIII: wool combs
- Hønsi, status FII–III: wool combs
- Målsnes, status FIII: wool combs, shears (R443
- Kvåle, status FIII: wool combs, shears (R443)

Six of the graves had tools for spinning, with 12 whorls in all and additionally some spindles in one of the Hopperstad graves, and then two with one whorl each and both in the same weight group, 25–30 g. One of the two whorls in the Målsnes grave also matched these whorls, while the other would produce coarser yarn. In the Kvåle grave with four

Table 5.36: Measurable spindle whorls for weight from the 9th- and 10th-century graves from Sygnafylki (N:30).

Weight in g period	*<10*	*10–15*	*16–20*	*21–25*	*26–30*	*31–35*	*36–40*	*>40*	*Unavailable*	*Total N*
800–900	–	2	2	6	4	2	–	1	7	24
900–1000	2	–	1	3	6	–	–	1	2	15
N	2	2	3	9	10	2	–	2	9	39

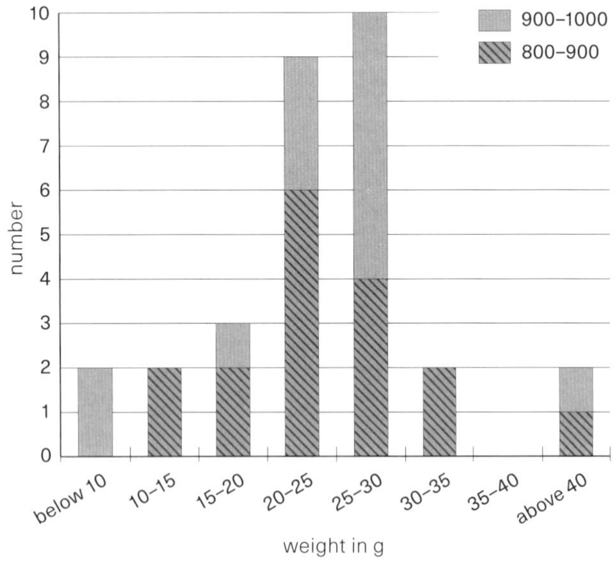

Fig. 5.36: Measurable spindle whorls as for weight from the 9th- and 10th-century graves in Sygnafylki (N: 30).

whorls they vary from very light whorls to spin the very finest yarn to produce more ordinary qualities.

- Hopperstad (B7761), status FIII: 1 spindle whorl, *26–30 g (*estimated weight)
- Hopperstad (B4637), status FII: 1 whorl, *26–30 g
- Hopperstad (B9060), status FIII: parts of 2 spindles
- Hopperstad (B4511), status FIII: 2 whorls (not retrieved)
- Målsnes, status FIII: 2 spindle whorls, 27 g, 47 g
- Kvåle, status FIII: 4 spindle whorls, 5 g, 6 g of fired clay, stone, *30–35 g

Only two other 10th-century graves from Sygnafylki contained spindle whorls, both in the common weight group 25–30 g, and in one case, with two loom weights.[105] Compared with the spindle whorls in the 9th-century graves, the latter represent a smaller assemblage but of wider range of weights, from the very lightest of 5–6 g to above 40 g, with a concentration between 25–30 g. The earlier assemblage, however, spread within all weight groups 10–15 g to 30–35 g and both with a whorl heavier than 40 g, able to produce a variety of yarn qualities (Table 5.36; Fig. 5.36).

All of these graves with composite tool kits, except Målsnes, had tools for weaving, and all except one with weaving beaters, represented by the longer beaters with an obtuse angle to the shaft, and in two graves, combined with loom weights.

- Hopperstad (B7761), status III: 1 weaving beater (R440), 83 cm long, 1 loom weight, 1 R413
- Hopperstad (B4511), status III: 1 weaving beater (R440), 9 loom weights
- Hopperstad (B4637), status II: 1 weaving beater (R440)
- Hopperstad (B9060), status III: 12 loom weights, 1 R413
- Hønsi, status III: 1 weaving beater (R440)
- Kvåle, status III: 1 weaving beater (R440), 67 cm long

Two other 10th-century graves in Sygnafylki have left traces of weaving: the grave at Brekke at the entrance of the Sognefjord but from an unclear context as a secondary grave in a mound as a boat burial, however, with only a single loom weight. The other, a high-status female burial in an upland grave at a dispersed place in the mid-zone had a weaving beater as the only grave gift.

Relatively few 10th-century graves, four in all, contained weaving beaters, and not exclusively related to high-status graves; one only with loom weights and another both a weaving beater and loom weights. One of the Hopperstad graves comprised both a beater and the spatula-shaped tool R413. That the tool appeared in many high-status graves with textile tools such as the graves at Hopperstad and Hove in Sygnafylki and Firdafylki throughout the period corroborates its role as a tool in textile production.

Smoothing plaques of whalebone related to the Hopperstad graves too – as the only graves in Sygnafylki with these tools from northerly areas. The glass pellets in two of the graves, in one of the Hopperstad graves and in the Hønsi grave, are two of three graves with such tools in Sygnafylki: the third, however, was in a less prestigious double grave further out the fjord in Gulen. When including tools for the finishing processes of sewing, cutting textiles, yarn, and smoothing textiles of plant fibres, the graves with composite tools distinguish themselves from the others: three with shears for cutting threads and textiles, and five graves with tools for sewing and/or smoothing:

- Hopperstad (B7761), status FIII: 2 shears (R443), 16 cm, 22cm long

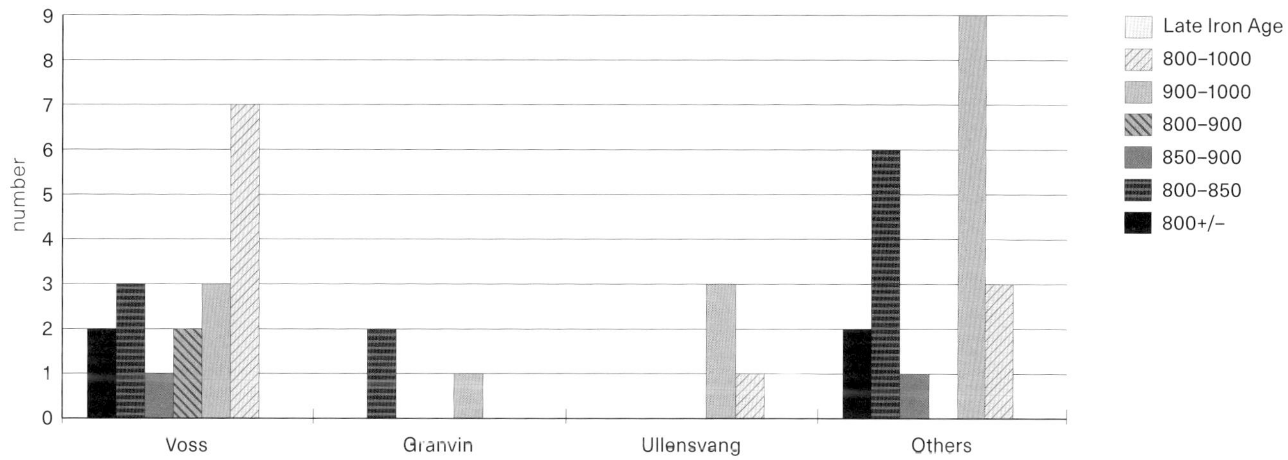

Fig. 5.37: Temporal distribution of graves with textile tools in Hordafylki (N:46).

- Hopperstad (B4637), status FII: 1 smoothing plaque of whalebone
- Hopperstad (B9060), status FIII: needle case of bronze, 1 whalebone plaque
- Hopperstad (B4511), status FIII: smoothing pellet of glass
- Målsnes, status FIII: shears (R443), needle case of iron
- Hønsi, status FIII: smoothing pellet of glass
- Kvåle, status FIII: shears (R443)

These graves from the Late Viking Age show both continuity and new patterns related to location and representation with a concentration of high-status graves at central and strategic places, such as at Hopperstad in Vik, Kvåle in Sogndal as well as Vinjum in Aurland – they all were places that became medieval church places. This is also a similar trend as in Firdafylki, with fewer graves in this category at dispersed places compared with the 9th century and instead concentrated to one or a few graves at central places and nodes for communication, like the development in Firdafylki. The tool assemblages represented in prestigious graves at central places and at special places and environments in both Sygnafylki and Firdafylki, may indicate a production above household level. Their connection to central places and probably as nodes within larger estates may further indicate an advanced production. At such places, the high-status graves reveal tools able to produce a variety of qualities and with grave goods that indicate means and resources to supply a trained work force and with capacity to fulfil specialised tasks in production.

Hordafylki, with 14 graves with textile tools from the Late Viking Age and with nearly one-third of all dated to the 10th century, represents a relatively higher share of such graves than the counties to the north (Fig. 5.37). However, 15 other contexts had, as we have seen, broader dating frames: five to the transitional decades into the 9th century, or late 9th to the early 10th century, and six roughly dated to the Late Iron Age, but most likely belonging to the Viking period, possibly to the 9th century, although a later dating cannot be excluded.

Of the 10th-century graves, only one related to the coast, the Toranger grave, presented in the beginning of this chapter, six to the mid-zone, and seven to the inner zone – and many of these in the category of double graves, such as at Ytre Arna or possible double graves and altogether unclear contexts. These unclear graves related to both upland areas and central agricultural areas in the lowland. One was a grave in the Voss area at a junction in the western part between waterways and land route. This was a boat burial, modestly equipped with two spindle whorls of weight categories 10–15 g and 35–40 g, able to produce yarn to textiles from the very finest qualities to the coarse. Two related to mountain farms in an upland valley at the fringes of the *fylki*, leading eastwards to Vik. Two high-status double graves signalling equal high status between the genders, were the earlier presented graves at Ytre Arna in the mid-zone and at Hyrt in upland Voss, nearly 400 m.a.s.l. Three of the six graves related to the inland communities at places such as Veka, Trå, and Bu, in respectively Voss, Granvin, and Ullensvang by the Hardangerfjord.

A relatively high share, seven in all and nearly half of the 15 late Viking Age graves in Hordafylki, now had tools for more than two work processes – for three to five work stages, and with the most frequent tools for weaving (see Appendix 5.6). These graves with composite tool kits follow the same geographical pattern as the earlier but of generally more blurred contexts – one from the coast, two from the mid-zone, and four from the inner zone, Voss and inner Hardanger, but without the same connection to central places, as in Firdafylki and partly Sygnafylki, and represent graves in both upland/inland and central lowland areas.

Table 5.37: Representation of tool categories in grave contexts from Hordafylki c. 900–1000 (N:14).

Operations	*Shearing*	*Combing*	*Spinning*	*Weaving*	*Sewing*	*Smoothing*	*Graves*
1	–	–	1	1	–	–	2
2	2	1	1	3	1	–	4
3	3	3	2	4	–	–	4
4	2	3	3	3	1	1	3
5	1	1	1	1	–	1	1
N	8	8	8	12	2	2	14

*Table 5.38: Hordafylki, 10th century status III with traces of tools for more than two work operations. P = preparation, P** = combing, S = spinning, W = weaving, W* = loom weight(s), W** = weaving beater, W*** = weaving beaters and loom weights, F = finishing (cutting, *sewing**), M = maintenance; M* = pellet of glass, M** = whalebone plaque.*

No.	*Name*	*District*	*Central place*	*Fjord*	*Coast*	*Valley/ Upland/ Inland*	*Tool categories*	*Keys*	*Foreign artefacts*	*Status*
B5874	Toranger	Austevoll	–		x	–	P*+S+W***+M**	1	–	FII–III
B5627	Vangdal	Kvam	–	x			P***?W**	–	–	FIII
B6657	Trå	Granvin		x			P***+S+W*** +M*	–	x	FIII
B6228	Veka	Voss	–			x	P**+S+W***+F		x	FIII
B4864	Hyrt	Voss	–			x	P**+S+W***	1		FIII/ MIII

The tools in the graves: new modes of production?

The most common tools in these graves related to weaving, and with an equal representation of shearing, combing, and spinning (Table 5.37).

The graves with more composite tool kits, five in all (Table 5.38), constitute then a relatively higher share than in the 9th century and of general higher status too, both as separate female graves and as double graves. Two contained keys, four foreign artefacts, and all from the inner zone, Voss and Granvin with the earlier presented high-status graves at Veka and Trå and the double grave at upland Hyrt.

All of these graves/contexts had tools for the preparatory stages, either shearing, combing, or both. The shears were all the type for sheep shearing, and in three graves combined with wool combs, while two graves from Voss only had wool combs:

- Toranger, Austevoll, status F II–III: shears (R442)
- Vangdal, Kvam, status III: shears (R442) + wool combs
- Trå, Granvin, status FIII: parts of shears and wool combs
- Veka, Voss, status FIII: 1 wool comb
- Hyrt, Voss, FIII/MIII: wool combs

Five grave contexts contained spindle whorls, and all but one with a single whorl, and seven whorls in all – and of no more than nine in all the 10th-century graves, and with generally lighter whorls than from the earlier graves (see Table 5.27):

- Toranger, Austevoll, status FII–III: 1 spindle whorl, 10 g
- Trå, Granvin, status FIII: 1 spindle whorl, 10 g
- Veka, Voss, status FIII: 2 spindle whorls, 9 g, 17 g
- Hyrt, Voss, MIII/FIII: 1 spindle whorl, 22 g

The three high status graves at Toranger, Trå, and Veka contained the lightest spindle whorls, 9 and 10 g. Still, the highest share of light whorls related to graves from the 9th century and with a broader representation of varied weight groups too, up to above 40 g.

Tools for weaving, as the most common category in the 10th-century graves, were represented in all the seven graves either by loom weights, three with weaving beaters of the traditional type, R440, or both categories, and four only represented with loom weights, from two to 20:

- Toranger, Austevoll, status F II–III: 4 loom weights, 1 weaving beater (R440), 61 cm
- Vangdal, Kvam, status III: 1 weaving beater (R440), 74.5+ cm
- Trå, Granvin, status FIII: 11 loom weights, 1 weaving beater (R440), 79 cm
- Veka, Voss, status FIII: 11 loom weights, 1 R413
- Hyrt, Voss, MIII/FIII: 20 loom weights

Three of these graves comprised weaving beaters, varying from 61 to 80 cm in length, and may seem less socially distinguishing as grave goods than earlier. The prestigious Veka grave with the most complete tool kits contained, for

Table 5.39: Distribution of spindle whorls in Hordafylki within weight groups of 5 g in graves c. *800–*c. *1000 (including wider dating frames) (N:36 whorls).*

Weight in g period	*5–10*	*10–15*	*16–20*	*20–25*	*26–30*	*31–35*	*36–40*	*40+*	*N: unavailable/ measurable*	*Total N*	*N: Graves*
800+/–	–	4	–	–	3	.	–	–	1	8	3
800–900	–	4	1		1	1	–	1	–	8	7
900–1000	2	2	1	1	1	1	–		2	10	13
Viking Age	–	2	–	–	1	–	1	–	1	5	3
Late Iron Age	–	1	1	2		–	–		1	5	5
N	2	13	3	3	6	2	1	1	5	36	31

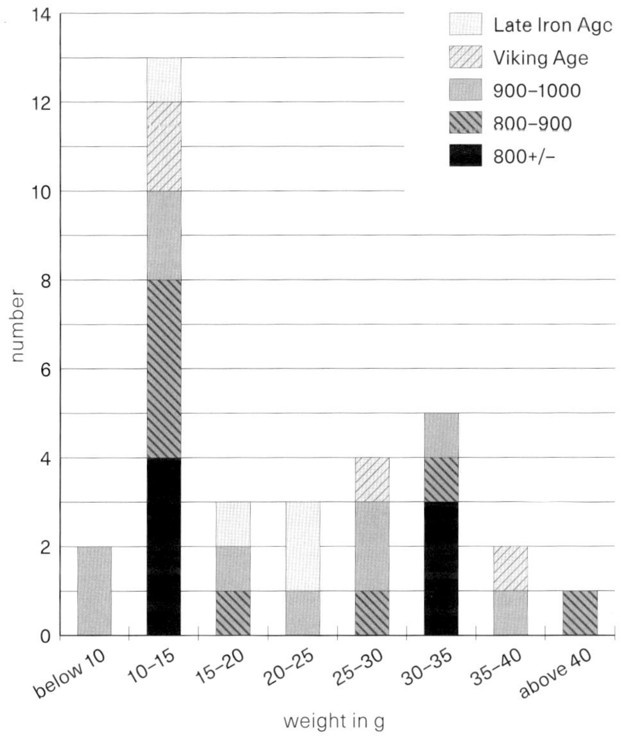

Fig. 5.38: Distribution of spindle whorls in Hordafylki in weight groups of 5 g in graves c. *800–c.1000 (including wider dating frames) (N:36 whorls, of which 31 measurable).*

example, only loom weights, like the upland grave at Hyrt. The loom weights as grave goods could vary from four to 20. Tools for finishing process connected to three high-status graves, and related to sewing or smoothing:

- Veka, Voss, status FIII: 1 needle case of copper alloy, with sewing needles
- Trå, Granvin, status FIII: 1 glass pellet for smoothing
- Toranger, Austevoll, status FII–III: 1 whalebone plaque for smoothing

A needle case of the same kind as in the Veka grave appeared in the high-status double grave at Ytre Arne too, and in three graves with wider dating frames, one in Voss, one in Etne the mid-zone, and a third made of bone in the inner zone.

These graves with composite tool kits share both common and different traits with graves in Firdafylki and Sygnafylki, and to a larger extent included keys and foreign artefacts, and broader assemblages of food-related equipment and farming tools, such as at Trå and Veka. However, the late graves in Hordafylki connected to a lesser degree to central places like in the counties to the north. Three to four graves in the Voss area, especially Veka, Trå, and Hyrt, seem to indicate special working milieus in rather different landscapes but all in the inland area, two in the lowland and the Hyrt in an upland area. They share many common traits as for tools, types of artefacts, and social standing and from about the same time, within reachable distances and possibly interconnected somehow. Still, the tools and grave goods indicate a production above the ordinary and on an advanced level compared with the others, and judging by the wealth in the some of the graves, were able to support such a production.

When comparing all the available spindle whorls from Hordafylki and from the whole period, including those within broader dating frames, they count 37 whorls from 33 contexts, of which 32 spindle whorls are measurable (Table 5.38; Fig. 5.38). The distribution in weight groups reveal a different pattern than earlier graves and a higher share of light spindle whorls below 15 g than the whorls in the later graves. This is a parallel pattern as for the graves further north, with the highest representation in the 9th-century graves. However, in Hordafylki, the two lightest date to the 10th-century graves (Table 5.39).

The graves with textile tools from Hordafylki this late largely connect to 'new' places but in many of the same districts as the earlier graves, such as Voss and Granvin in the inner zone, and generally in good agricultural areas with access to grazing land but not least to upland areas, close to the outland resources. The graves in the mid-zone also connected to districts with favourable conditions for farming, but very few to coastal areas and to central places. In this respect they are rather unlike the patterns revealed in Sygnafylki and Firdafylki. In these districts largely based

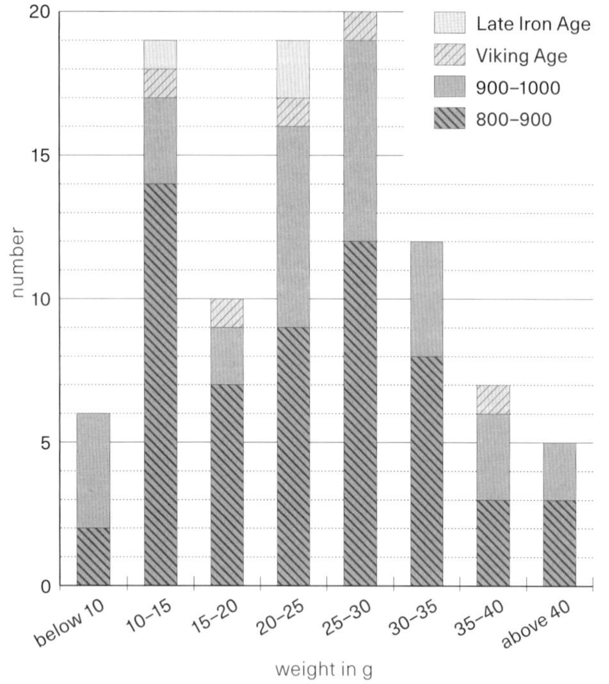

Fig. 5.39: Distribution of all measurable spindle whorls within the study area in weight groups of 5 g (N: 98) (Graphic design: Per Bækken).

on freehold, the production seems to be spread out on the household level. Still, it may not necessarily represent an ordinary household production. The tools at some of these places indicate the competency to produce a variety of qualities, from common use areas to the luxurious segment.

Having assessed the graves with textile tools on meso levels, within the three counties over time and at places in varied environments inland/upland and lowland with access to varied resources, dispersed or concentrated to special places, different patterns emerge and change over time. Altogether, they indicate differences in production and modes of production as may be reflected in the textile remains, and the wealth of grave goods also indicating means to supply a workforce above the ordinary.

Agents, environments, and modes of production: changes in space and over time?

Has this examination of graves and grave goods with textile tools made it possible to come closer to agents within production in these varied landscapes, or only the top of society? After all, the graves represent a small segment of society, and according to the burial customs, generally people of the highest social levels, and as for gender an even smaller segment. The 'inverted pyramid', as presented earlier in this chapter (cf. Fig. 5.13), clearly hides a large and archaeologically, largely invisible work force doing the manual work. They are, however, reflected by the tools they used and the finished products, displaying their working capacity and competency. The source value, therefore, exceeds the relatively small numbers and uneven social representation of tools and asymmetric grave goods. The tools and their contexts provide insights at many levels – not only to the social aspects of production, but also technologically, exposed by the tools and in various geographical and social settings and landscapes and over time.

The expansion of Viking Age graves with textile tools represented a boom in the burial customs by the turn of the 9th century, to expand significantly through the century, but following different directions through the 10th century before these burial customs faded out. Throughout the Viking period, the tools dominantly related to female high-status graves, reflecting their role in textile production basically as organisers and administrators. Although the graves of this type are relatively few, they are represented in the whole area of study related to western Norway, but shifting over time. The boom in the early 9th century largely coincides in time with the westbound expansions of raids and early settlements in the North Atlantic, also characterising the Viking period. They again depended on textile production, needing both sails and other gear at sea, as well as sociality distinguishing outfits. The spread of these graves varied, however, over time between the three counties, also internally within the geographical zones, which seem to have a bearing too. In broad outlines, the largest expansion relates to the first part of the period, especially in the early 9th century and connected to the inner zone – upland and inland areas and northern parts of Firdafylki but also along the coast and inland further south in Hordafylki. The people at these places also seem to connect somehow.

The graves in Sygnafylki from the 9th century, fewer than further north, relate to other environments and without similar connections to upland in the inner zone. They often connected to focal or strategic places in fertile lowland landscapes by the fjords but with access to upland areas too. The 9th-century graves in Hordafylki differ, connecting to coastal as well as inland areas and by the fjord. In all three counties, high-status graves with textile tools had a parallel affiliation to central places and junctions of communication, along the coast in Hordafylki and to some extent to inland/upland areas too. In the latter part of the period, this type of graves to a larger degree connected to central places, which later became medieval church centres and/or at nodes in larger estates. The graves in such contexts to a larger extent contained multiple tool kits, and not only for textile production, indicating a role as managers of wide work domains. When assessing graves with composite tools for textile production in Hordafylki, they generally represent high status, from both inland/upland areas and central places, and of varied qualities, as indicated by 'small things', such as spindle whorls as indirect signs of yarn qualities and achieved technological competencies of producing textiles of varied qualities (Fig. 5.39).

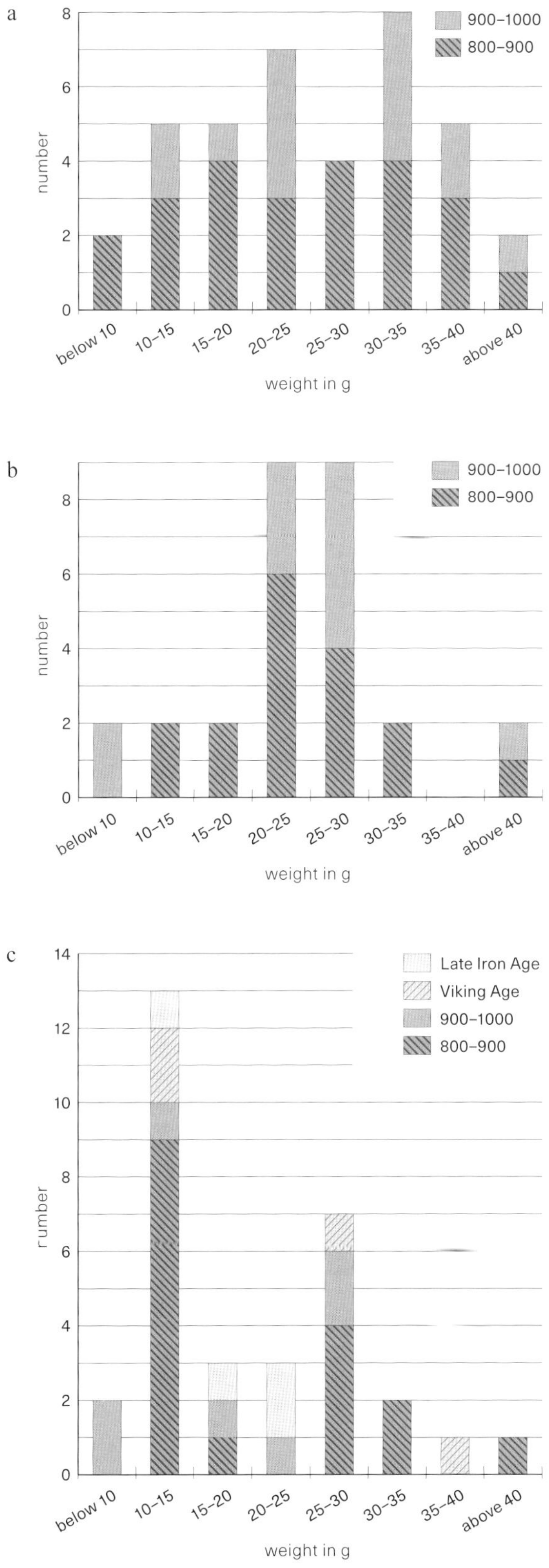

Fig. 5.40: a) Distribution of spindle whorls in weight groups of 5 g in Firdafylki (N:38), b) distribution of spindle whorls in weight groups of 5 g in Sygnafylki (N:28), c) distribution of spindle whorls in weight groups of 5 g from Hordafylki (N:32).

Figure 5.39 shows that most of the measurable exemplars of these spindle whorls, 58 in all, or 59%, date to the 9th century, and 28 or 29% to the 10th century. A few have wider dating frames, where four date broadly to the Viking period and a few to the Late Iron Age. They show a whole range of weights, from weighing less than 10 g to more than 40 g, and cover the whole spectrum needed for producing yarn of various qualities and use areas. The largest concentration of spindle whorls, counting 70 whorls, with 10 or more whorls in all these weight groups, weigh between 10–15 g and 30–35 g. The diagram gives thus an impression of general development and variations but conceals regional differences and is on a smaller spatial scale.

When comparing the distribution of weights on regional levels, the diagrams indicate both qualitative and quantifiable differences over time in the three counties. Light spindle whorls, below 10 to 15 g, appear earliest in Firdafylki, with a relatively high number from the 9th century, counting 24 measurable spindle whorls, or 63% of all, and with a rather even spread within the weight groups up to *c.* 40 g (Fig. 5.40 a and b).

The spindle whorls from Sygnafylki, with the lowest number of measurable whorls, the highest representation stems from the 9th century, 61%, with a concentration in the common weight groups from 20 to 30 g, or 64% of all measurable whorls.

The largest number of light spindle whorls, below 15 g, relate to Hordafylki, 12 whorls, and represent 47% of all the measurable whorls, and another but smaller peak in the weight group 25–30 g, represented in 22% of all 32 contexts (Fig. 5.40c). Here, 53% of all measurable whorls relate to the 9th century and only 16% to the 10th century and with a higher share from graves with wider dating frames.

The tools have a clear social bias, related to women from the upper echelons of society and seem to reflect the luxurious segment of production. However, several graves contain more than one whorl, reflecting production of both fine and coarser yarn and textiles.

The graves with multiple tool kits largely concur with high status throughout the period, often connecting to special places and textile production environments. Some places show long traditions over centuries, such as Myklebust at Eid in the lowlands and Gutdal in the inland/upland Stryn. Others indicate intensification of production, marked by monumentalisation of graves within shorter time spans, such as in Aurland, ending abruptly by the late 9th century, however, with one of the most prominent graves with a central location at the entrance of the valley. On the larger spatial level of counties, there is a notable shift, such as in Firdafylkit, with many graves in the 9th century while in the Late Viking Age they tended to concentrate to central places, fewer in number but of high prominence and with richer tool assemblages. The graves in Sygnafylki reflect female actors in the higher and highest strata too, again at central places, such as at Kvåle in Sogndal, Vinjum in Aurland and Hopperstad in Vik, but at more dispersed places too, such as Målsnes, however, with a strategic location by the Sognefjord. At places like these, the production and women in charge must have leaned on a pool of subordinates to carry out the manual work in all the required processes and to produce a wide assortment of textiles.

The graves in Hordafylki reveal another pattern, marked by expansion by the late 9th and 10th centuries and different spatial distribution, probably influenced by different types of landownership and organisation in the outer and inner zones. The tools indicate variations and changes in time and space, basically following these trends. At most of these places, the graves tend to include other types of tools too – cooking utensils and agricultural tools, hoes, and sickles, indicating broader domains. In some districts in Sygnafylki, such as Vik and Sogndal, and in Firdafylki, Gutdal in Stryn and Myklebust at Eid, graves with textile tools are represented throughout the period, from around 800 or earlier, into the 10th century indicating special places for production of *longue durée*. The later graves in Hordafylki, marked by high prominence in the inner zone, especially Voss, connect to an area traditionally held by freeholders of high status. They differ then from the coastal graves related to central and strategic places as for communication. The distribution of graves connected to central places indicate their roles at nodes within estates, as also reflected in later sources connected to tenanted farms. Many high-status graves were marked by other status symbols too, such as keys and not least foreign artefacts, indicating wider networks. The graves with textile tools seem thus to reflect larger societal structures and centralisation processes and both prosaic and conspicuous needs for textiles for varied use areas.

As single tools or parts of multiple tool kits and socially distinguishing artefacts, tools as grave goods contain many layers of information related to agency and actions in various environments over time – technologically, as for work domains, gender roles, and status.

Judging by the graves and grave goods containing textile equipment, women were the main agents within textile production and with high-status women as organisers, probably also taking part in large-scale enterprises of the period, directly or indirectly. The inverted pyramid of social representation in Viking Age graves with textile tools from western Norway (cf. Fig. 5.13), shows that as many as 59% of these graves were of the highest status III, compared with 16% in that position in Viking Age female graves as a total.

Indirectly, then, the women buried with their fine dress accessories and textile tools had a special position, at the same time hiding a large and archaeologically largely invisible work force, necessary to carry out all the laborious work processes involved in such environments. The tools in the graves as symbolic grave goods cannot therefore be taken as a 1:1 representation of the textile production as such, which to a large degree must have been carried by a subordinate work force judging by their status. Altogether, these graves with prestigious grave goods and tools indicate a production on an advanced level and with means, status, and resources to support this kind of production needing varied competency and a large input of skilled labour. Still, many of the more anonymous graves also indicate an ordinary household production.

Notes

1 Shetelig 1904, 13–14
2 Härke 2014, 1.
3 Friðriksson & Vésteinsson 2011.
4 Eldjárn 1956, 246–50; Pétursdóttir 2007; 2009.
5 Hayeur Smith 2004.
6 Kaland 1973.
7 Sigrdrifumál, stanzas, 33, 34.
8 Hárbarðardarljoð 45, Hervararsaga III 12.
9 Birkelid 1938; Steinsland 2005.
10 *G* 29.
11 *L* VI, 16.
12 *G* 270–74.
13 Mundal 2020, 227–231.
14 *G* 23.
15 *G* 150.
16 Shetelig 1912.
17 Pétursdóttir 2009.
18 Iversen 1997, 47.
19 Shetelig 1947 with references; Jesch 1991, 25, 27.
20 Iversen 1997.
21 Shetelig 1947, with references.
22 Härke 2014, with references.
23 Iversen 1999; 2008; cf. Chapter 3.
24 Shetelig 1940; Kaland 1973.
25 Eldjárn 1956.
26 Gellein 2007, 66.
27 Skipstad 2009, with references.
28 Dommasnes 1982, 73; Skipstad 2009.
29 Stylegar 2010.
30 *e.g.*, Shetelig 1912.
31 See Chapter 7, Table 7.13.
32 Shetelig 1912, 219–228.
33 Eldjárn 1956.
34 Eldjárn 1956, 204, 214, 226; Kaland 1973; Owen & Dalland 1994; 1999.
35 Solberg 1985, 61–76; 2010, 269.
36 Solberg 1985.
37 In such cases, indicated as *e.g.*, status FII–III; FI–II in Appendix 5.1.
38 *L* VII, 54.
39 Meulengracht Sørensen 1983.
40 *G* 275.
41 Lund 1994; 1996.
42 *G* 308.
43 Þorláksson 1991.
44 Eldjárn & Friðriksson 2000.

45 Eldjárn & Friðriksson 2000, 590; Friðriksson 2009; Friðriksson & Vésteinsson 2011.
46 Goody *et al.* 1976, 10–18.
47 Williams 2007, 107–123.
48 Müller 2013.
49 Eldjárn 1956, 97; see also Chapter 9.
50 Eldjárn 1956, 73–77; Kaland 1973; Owen & Dalland 1999.
51 Kaland 2006; Øye 2015a.
52 Shetelig 1912, 176–177; Petersen 1951, 421–24.
53 Steinsland 2005.
54 *a.o.* Price 2002, 177, Gardeła 2009; 2016; Lunde 2018.
55 Heide 2006.
56 Shetelig 1912, 211; cf. Fig. 5.12.
57 Kaland 1973.
58 Solberg Andreassen 2001, 199 with references.
59 Eldjárn 1956, 95–97.
60 Eldjárn 1956, 180; Guckelsberger 2017; Hayeur Smith *et al.* 2019.
61 Eldjárn 1956, 340–341.
62 Eldjárn 1956, 185–186, 347; Álaugarey.
63 *KLNM* XII, 383 with references.
64 Eldjárn 1956, 117; Pétursdóttir 2009.
65 Rognø 2019. Meling 2000.
66 Eldjárn 1956, 193.
67 Mundal 2020.
68 Shetelig 1947, with references.
69 Eldjárn 1956, 218
70 Eldjárn 1956, 168–169
71 Eldjárn 1956, 75–76
72 Eldjárn 1956, 62–63.
73 Meulengracht Sørensen 1983.
74 Price *et al.* 2019.
75 Eldjárn 1956; Þorláksson 1991.
76 Indrebø 1935.
77 *KLNM* V, 39; XV, 546–551; XVI, 378–379.
78 *e.g.* Glørstad 2010; 2012.
79 a. o. Baug et al. 2019.
80 Bakka 1963.
81 *a.o.* Tveiten 2012; 2015.
82 Shetelig 1905.
83 Shetelig 1905.
84 B6618/B6688, Bø, Breim, status MIII; B11479, status MIII.
85 Magnus 1978.
86 cf. Shetelig 1912, 203, note 3.
87 B5146, Grov, Oppstryn, Lorange 1875, 238.
88 Hellesøe 2006, 59.
89 Moberg 1992, 181.
90 A male grave, B8320, from the head of another side fjord of the Aurlandsfjord, is represented with a single weaving beater but its dating and status is uncertain.
91 BK, 46
92 Solberg 1986, 198.
93 B3972-86, Tokvam II, status F?, cf. Solberg 1986, 212.
94 *BK*, 9.
95 Gabrielsen 2007, 189–190.
96 *BK*, 12.
97 *BK*, 13
98 Shetelig 1905.
99 Iversen 1999, 56.
100 *BK*, 26.
101 *BK*, 35.
102 *BK*, 50.
103 *BK*, 50.
104 *BK*, 35.
105 B14183, Osland, Høyanger, status FI; B8272, Gryten, Gulen, status MII/ F?

6

Tools and textiles in Viking Age graves

In the spectacular burial ceremony at Veka in Voss, sometime in the 10th century, the deceased woman was dressed in a high-quality blue gown equipped with exquisite dress accessories with all her grave goods and tools placed in a wooden coffin before it was covered by a large cairn (cf. Chapter 5, Fig. 5.15). When the grave was excavated nearly a thousand years later, only fragments were left, representing different techniques and qualities of varying fineness: several remains of 2/2 twills and broken diamond twill and enough to be analysed and identified as parts of all her garments. The textile finds relation to the two oval brooches and a third brooch indicate that the remains of 2/2 twill belonged to a cloak or caftan, fastened to a third brooch, and worn over a strapped gown of diamond twill, blue in colour indicated by traces of indigotin,[1] and held up by straps fastened by two oval brooches, A fragment of 2/2 twill with a long selvage shows that the cloak was decorated by a narrow band of patterned weave, probably of silk *samitum*. The dress accessories – the oval brooches of gilt copper alloy, decorated with silver filigree and partly covered by silver, two bracelets of copper alloy and 118 beads for a necklace[2] – also underlined her high status.

Archeologically, only fragments of the female dress are left in these graves, however, not in all. In the best cases, the preserved fragments have been in direct contact with dress accessories, such as brooches, chains, armlets, and other jewellery of copper alloy, sometimes of silver and even gold, depending on status and type of dress and ways of expressing or legitimising identity and social affiliations.

Remains of textiles have been found in 38 of the Viking Age graves from the study area of western Norway,[3] and of these a majority of 29 graves contained both textile tools and textiles – 19 from the 9th century and 10 from the 10th century. As grave contexts, the textile remains relate primarily to the funeral clothes of high-status women. Even the tiniest fragments are invaluable sources, able to shed light on social, technological, and economic aspects of the production. Besides the Veka grave, among others, two of the other high-status graves in the inner region of Hordafylki with larger tool assemblages and prestigious grave goods have left remains of textiles of the same qualities as in the Veka grave – at Trå in Granvin and at Hyrt in upland Voss (see Fig. 6.6).

Of these, the woman in the double grave at Hyrt was buried with two oval brooches, without a third brooch but with a rich assortment of beads for a string between the brooches, supplied with various grave gifts – textile equipment, a key, iron knife, and remains of a wooden casket. Several types of textiles were preserved both above and underneath the brooches – both exclusive and more common and coarser fabrics – three of plant material, probably linen, the others of wool, all z/z-spun, woven in three weaving techniques: tabby, 2/2 twill, and broken diamond/lozenge twill (Fig. 6.1). Hana Lukešová's analysis of the textiles has made it possible to connect them to different garments and qualities. Six fragments of plant material, dyed blue, were tabby woven with a thread count of 18/10 in the warp/weft. Representing the first layer, they seem to represent the undergown, worn innermost. Beneath the brooch, the fragments of 2/2 twill with the same structure and with a thread count of 12/10, represent a second layer, presumably the tunic, the strapped gown. The biggest fragment, dark blue-green in colour, connected to a plaited string as a strap. The third layer and of the finest quality – a piece woven in broken diamond/lozenge twill – most likely represents the cloak. This was black-blue in colour, woven with a repeat pattern of 20/10 and a density of 50 warp threads and 20 weft threads per cm, and of very high quality. Some wool fragments woven in tabby techniques had a tighter weave than those of plant material and with a thread count of 32/20.[4]

The funeral clothes in the Veka grave were rather similar with a blue gown woven in broken diamond twill technique and even finer quality and a kaftan or cloak, woven in 2/2 twill fastened by a third brooch, an oblong brooch possibly of foreign origin.[5] The grave at Trå in Hardanger was also buried with similar clothes, a dress woven in broken diamond quality and a cloak in 2/2 twill (Fig. 6.2).[6]

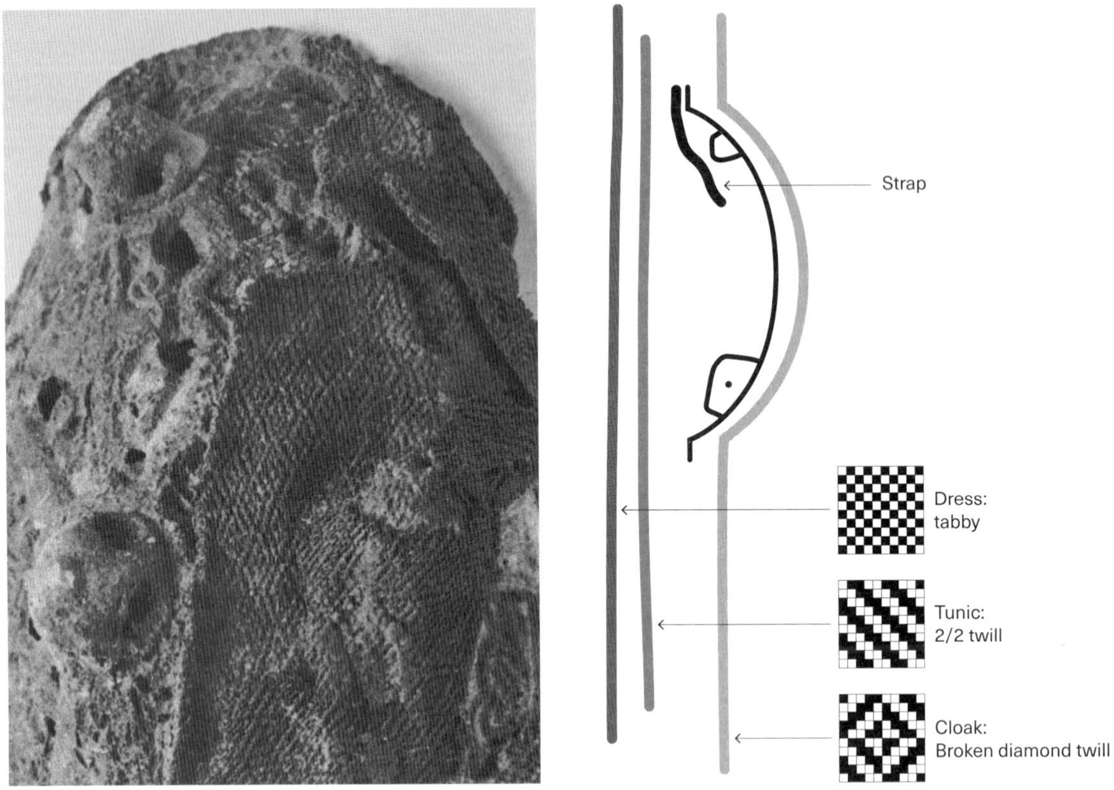

Fig. 6.1: Close up of one of the brooches in the grave at Hyrt, with preserved textile of broken diamond twill (Photo: © University Museum of Bergen) and with a sketch of the various layers woven in different qualities and techniques (Illustration: Per Bækken).

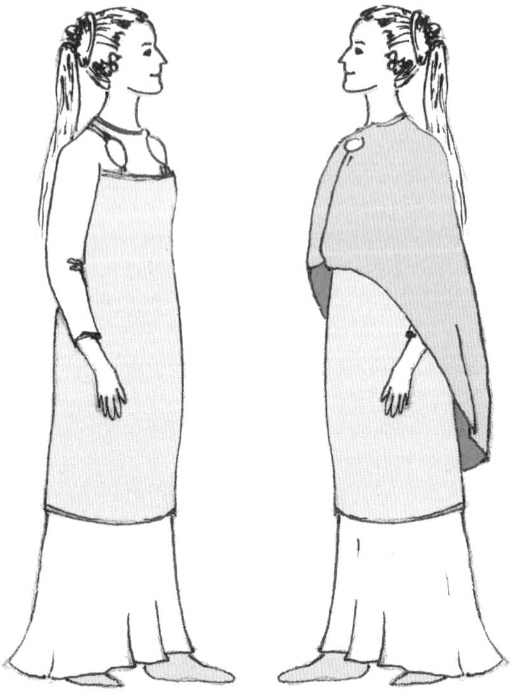

Fig. 6.2: Sketch of the Viking Age garments related to high-status women: a strapped gown and undergown, and with and without a cloak (Illustration: Per Bækken).

About the same time, a young woman of high rank at Ketilstaðir in eastern Iceland was buried in a similar outfit – and when excavated about a millennium later, still with remains of textiles under the brooches of guilt copper alloy and a small trefoil brooch and a necklace of more than 40 beads. The fragments were of linen and wool: a tabby woven linen shift and remains of a blue-dyed gown and supplied with 42 smaller and larger beads for a necklace between the brooches. The blue dress was woven in diamond twill of z/z spun yarn, 11/10 threads per cm, and with remains of straps woven in 2/2 twill, z/s 6/5 per cm and dyed blue too.[7] This is a rather unique find in Iceland sharing many traits with the west Norwegian counterparts, however, in a coarser quality (Fig. 6.3).

The clothes in Viking Age graves, their colours and quality, as visible signs of status and identity, were further accentuated through the associated dress accessories and other grave goods and not least, tools that reflected roles and work domains, as shown in the foregoing chapter. In some cases, as in the graves referred to above, the pin of the brooches penetrated and preserved several layers of the garments – underdress, the strapped dress/tunic and a shawl or cloak of various qualities and weaving techniques. Others had simpler garments and probably fewer dress accessories.

Fig. 6.3: Reconstructed garments from Ketilstaðir based on the textile remains in the grave: underdress, dress with shoulder straps, and with copies of the same type of dress accessories as in the grave. Reconstruction by Marianne Guckelsberger and woven by Marled Mader (with kind permission).

In all these cases, the textiles had belonged to women of high status, and with remains of textiles woven in diamond twill techniques of a very fine quality and most frequently represented the strapped gown. Whether these qualities represent high-level domestic production or imported goods has been an issue of debate (cf. Chapter 1). Earlier research regarded textiles of these qualities of broken diamond twill as Frisian import,[8] or possibly from the British Isles.[9] However, later studies argue for a domestic and advanced production in western Norway, woven on the warp-weighted loom, pointing out that the technique has been overestimated as for complexity and moreover, known in the area prior to the Viking Period.[10]

So far, textiles from graves have largely been analysed as a separate category and only to a limited extent assessed as integrated elements of their contexts or related to their wider social and physical environments.[11] Tools and other grave goods intentionally placed in the graves were hardly arbitrary but according to norms and practices of the period, visualising the deceased's social status, fully dressed, and equipped with tools representing roles ritually and symbolically. As practical tools, their range and variations additionally reflect levels, modes, and technological capacity. Remains of the dress, the dress accessories, and other artefacts, moreover, signify social, technological, and economic aspects of production (cf. Chapter 3). It could be small-scale within common households or more advanced production and possibly as attached specialist production. As a regular provision for a certain group of people of local or regional elites, the production could be differentiated and regularised to produce special textiles, such as status-enhancing clothes for the elite or at the other end of production, for practical purposes on a larger scale, such as sails. Within a hierarchical society, the work most likely leaned on a subordinate work force, as substantiated by the earlier presented inverted social pyramid and the variety of textile tools in the graves (cf. Chapter 5, Fig. 5.13).

How graves with status-indicating dress and dress accessories, tools and other artefacts link to various places and environments in different geographical zones and settings is therefore a central issue in this chapter. By assessing tools, textiles and social emblems as integral elements in the funeral customs, the aim is to trace milieus that were capable of producing the more advanced textile techniques. The strategy is to assess the 'small things' – spindle whorls and other artefacts such as dress accessories as parts of larger societal structures and systems of production, and with a focus on the most advanced techniques.

As Holm-Olsen has argued, the textiles produced in the few urban centres of the period, such as Hedeby, Birka, and Kaupang, would appear as more homogenous than in decentralised rural environments, however, not necessarily of lower quality. If produced in various rural environments, an even spread of varied textile qualities should be expected, while related to centres with a specialised workshop would give clusters of certain qualities.[12] How qualities diverge or cluster in various settings thus serves as an indicator of possible specialisation when assessing the textiles and related to tools in the same contexts. Do they reflect domestic production at various levels, or more likely represent import? Although the tools are rather simple, size and weights of spindle whorls and loom weights indirectly indicate differences in quality and fabrics. Presence or absence of tools for fibre preparation and weight differences of spindle whorls and loom weights may, then, indicate different types of production too. Special types of cloth needed various raw materials: wool, flax, or hemp, and with sorting and combing of wool or hackling as for plant fibres, and with adjusted tools for producing different textile qualities. Standard cloth of defined qualities, in contrast, would require a homogenous wool type as well as standardised or adapted tools. The quality of the dress, tools, and dress accessories as symbols of social identity and hierarchies may additionally indicate relations and communication.

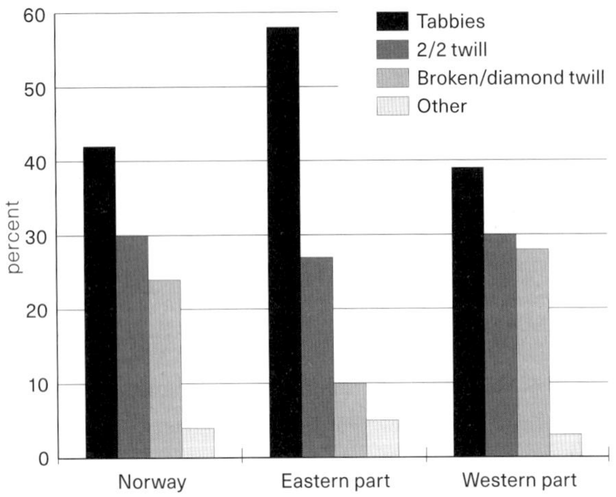

Fig. 6.4: Relative representation of weave qualities in Viking Age Norway (Source: Bender Jørgensen 1986, Graphic design: Per Bækken).

Weaves – cloth types: a backcloth

A wide range of Early Iron Age textile types, mostly from the Roman and Migration periods (AD *c.* 0–550) have been identified in Scandinavia, although the most frequently found textile type is a z/z spun 2/2 twill. The earliest traces of textiles in broken diamond twill from Norway appear in the Late Merovingian period and in increasing quantity by the Early Viking period, not least in the western part of the country.[13] They represent different qualities and techniques – some exclusive, tightly woven with very fine threads, and the most exquisite, broken diamond twill, with a high thread count in the warp, such as at Veka, among others. The textile remains from grave contexts within the study area from western Norway, analysed by Inger Marie Holm-Olsen, Lise Bender Jørgensen, and Hana Lukešová, have revealed various techniques and qualities of tabby, 2/2 twill, and broken diamond twill as main techniques (cf. Fig. 6.1). Basically, they comprise the same types and qualities as documented from Birka, Hedeby, and Kaupang, however, where they appeared as standard high-quality textiles produced by specialised textile workers within urban workshops.

Bender Jørgensen has shown that the dominant weaves among the preserved Viking Age textiles from Norway are tabby and constitute about 40% of the textiles, 2/2 twill about 30%, and 25% in broken diamond twill, and the rest, only denoted as 'other weaving'. Of the graves with preserved textiles within the present study area of western Norway, 18 graves contained textiles woven in broken diamond twill, implying that 68% of the graves with tools and preserved textiles contained textiles of the most advanced techniques and well above the average but still of various qualities (Fig. 6.4).

Both Holm-Olsen and Bender Jørgensen emphasise how the quality within this weave system varies as for the density of the threads and within the thread systems. Holm-Olsen's analysis, based on a statistical approach, divides the material according to three elements: (1) density in the densest thread system, (2) density in the more open thread system, and (3) the relation between densities in the two thread systems. These groups comprise finer textiles in the more advanced weaving techniques, broken diamond twills and 2/2 twills, both with a higher density of threads per cm in both directions than tabby weaves. The weaving is also more regular and homogenous, and generally of higher quality. The group of broken diamond twill shows differences in density and quality and forms two main groups. The first comprises material where the arithmetic mean quantity for the number of threads in the densest thread system is maximum 37 threads and 15 threads in the more open system, normally representing warp/weft. The second and finest group comprises weaves, the arithmetic mean quantity for the number of threads in the densest thread system is minimum 38 threads and 16 threads in the more open system. The textiles of category II, then, are more homogenous and regular in both directions and thus more standardised than those of category I and indicate specialised work. However, she emphasises that the divide between the two is not sharp and with a relatively even transition – and moreover, to achieve this high density of threads demands good sorting of wool from different parts and layers from the fleece, an even and good spinning and similarly, high skills in weaving. According to Holm-Olsen, diamond twills of quality I could very well represent an advanced domestic production, as well as all tabbies and most 2/2 twills. She finds the differences between categories I and II to be too small to characterise the weaves as the result of a specialised and professional craft.[14] Technological competence *per se* does not indicate specialisation, which is even more a matter of social organisation, as also emphasised by Cathy Costin (cf. Chapter 2).

Bender Jørgensen uses largely the same analytical methods as Holm-Olsen but without the same statistical approach. She divides the textiles woven in broken diamond twill in three qualities: quality I, the coarse quality, with up to 30 warp threads per cm in the tightest thread system, quality II, the medium quality of 30–40 threads per cm, and quality III, the finest and densest with more than 40 threads per cm. In western Norway, all three qualities were represented in the Viking period according to Bender Jørgensen's analyses, and in contrast to the Birka textiles, which only contained the two finest and densest qualities II and III.[15]

In the following, Bender Jørgensen's tripartite classification serves as an analytical tool when assessing the textiles contextually on a macro level. However, Holm-Olsen's statistical approach, revealing more gradual transitions, gives a more nuanced impression of production and technological capacity, as diamond twills also appear in the same qualities as the traditional and common qualities of tabbies of 1/2

Table 6.1: Graves with textiles of broken diamond twill (BDT) and textile tools from the late 8th to the late 9th century (N:10), with their tools, and when with more than two marked in bold. H = Hordafylki, S = Sygnafylki, F = Firdafylki (N:10) (Source: Bender Jørgensen 1986).

Ref. no.	*Museum no.*	*Site/ Community*	*County*	*Shears*	*Wool combs*	*Spindle whorls*	*Loom weight*	*Weaving beater/R413*	*Needle (case)*	*Weaves*	*Gender/ Status*
1	B7639	Store Kongsvik, Tysnes	H	1	–	–	6	–	–	BDT 2/2 twill	MIII/ FII–III
2	**B9014**	**Sandal, Breim**	F	1	1 pair	1	6	–	–	BDT	FIII
3	**B9501**	**Bø, Breim**	F	1	1	2	9	1	–	BDT 2/2 twill	F/III/ MIII
4	B9626	Naustdal, Førde	**F**	–	–	**1**	–	–	–	BDT Tabby	FIII
5	**B9064**	**Nes, Sogndal**	S	1		3	–	1	–	BDT (imprint), Tabby	FIII
6	B9063	Terum, Aurland:2 burials	S	1		3	–	1 R413	–	BDT 2/2 twill Tabby	FIII
7	**B7731**	**Vinjum, Aurland**	S	2	1 pair	1	48	1 R413		BDT Tabby	FIII
8	**B9065**	**Nes, Sogndal**	S	1	1	–	15	1 R413	–	BDT/ Tabby, 2/2 twill	FIII
9	B8000	Grodås, Hornindal	F	–	–	–	–	–	(1)	BDT Tabby 2/2 twills	FIII
10	B7893	Holmedal, Fjaler	**F**	–	–		7	–	–	BDT	F/M?

and 2/2 twills, and not unlikely, then, produced in domestic settings. Only the very lightest, thinnest, and homogeneous bindings in both directions would have demanded a highly competent and possibly specialised work force.[16] Within settings representing an elite economy and higher strata with the means to support, provide, and organise all stages in the work, this is therefore also a viable option.

The quality of some of these fabrics – fine worsted cloth of high uniform standard –implies both special environments and a certain level of organisation. It needed access to grazing land, large flocks of sheep, and a competent work force to achieve high-quality wool, and acreages to cultivate and prepare plants into textile fibres. Not least, it demanded coordination, logistics, and administration of skilled labour for sorting and combing wool, and for spinning in various gauges into different but homogenous thicknesses of yarn for the warp and weft when weaving various textile qualities. Both tools and textiles, as well as social status, then, may indicate special production places, milieus, and possibly also links between such places. Composite tool kits for textile production may in this way reflect active textile production milieus above the ordinary. A domestic production limited to the household's own needs would of course also need basic tools, however, probably less differentiated as for producing qualities above the ordinary.

Tools and textiles in graves from the early Viking Age

Of the 29 graves from western Norway with both textile tools and preserved textiles, 19 date to the 9th century and with an over-representation in the group of high-status graves with textile tools. Of these early graves, 10 graves comprised broken diamond twill qualities and, in most cases, also with remains of textiles/garments in other qualities, of 2/2 twill and tabby (Table 6.1).

Often, the textile appeared on the underside of the brooches belonging to the strapped gown in female graves, and in a few cases only as mineralised imprints (No. 5). In some cases, such as in the Hyrt grave among others, various layers of textiles were preserved by the pins of the brooches, and with varied qualities belonging to different garments.[17] However, in most cases only smaller parts have been preserved, and of varied qualities according to use areas and status. All the 9th-century graves with these high-quality

Table 6.2: Representation of graves from the late 8th into the late 9th century with both tools and textiles of tabby, 2/2 twill, and chevron twill qualities. Graves with composite tool kits marked in bold (N:9) (Source: Bender Jørgensen 1986).

Ref. no.	*Museum no.*	*Site/Community*	*County*	*Shears*	*Wool combs*	*Spindle whorls*	*Loom weights*	*Weaving beater*	*Needle (case)*	*Smoother*	*Weaves*	*Gender/ Status*
11	B9061	Hopperstad, Vik	S	–	–	1		1	–	–	2/2 twill	FIII
12	B6500	Skjervum, Vik	S		–	–		1	–	–	2/2 twill,	FIII
13	**B8669**	**Tokvam no 1**	**S**	**1**	**1**	**3**	**28**	**1**	**–**	**–**	**Tabby**	**FIII**
14	B5807	Myklebust, Eid, grave I	F	–	1	–	–	1	–	–	2/2 twill: fine and coarse Tabby	FII–III/ MIII
15	B7184	Beinnes, Olden, Stryn	F	–	1	–		1	–		Type?	FIII
16	B10975	Sårheim, Breim	F	–	–	–	–	1			Tabby	FIII
17	B6483	Svingeset, Stryn	F						1		Chevron twill	FIII
18	**B11413**	**Sandal, Jølster**	**F**	**1?**	**1**	**4**	**10**	**1**	**–**	**1 whale-bone plaque**	Tabby	**FIII**
19	B7732	Skjervheim, Voss	**H**								2/2 twill	**FIII**
	N: 9			1?	4	8	38	7	1	1		

textiles represented then general high status – two, possibly three as double graves, and nine or 10 female graves, and all belonging to the elite. Geographically, one related to Hordafylki from a coastal area, four to Sygnafylki from the inner zone, and five to Firdafylki, of which three are from the mid-zone (Nos. 8–10) and two are from the inner zone (Nos. 2 and 3) (Table 6.1).

The nine other graves with tools and textiles but in other qualities, 2/2 twill and tabby and chevron twill (Table 6.2: Nos. 11–19), represented similar high status. Of these, three were from Sygnafylki (Nos. 11, 12, 13) in Vik and Aurland, with many graves and five from Firdafylki (Nos. 14–18), and only one from Hordafylki (No. 19). Of these, two related to central places in the mid-zone, Hopperstad in Vik and at Myklebust in Eid (Nos. 11 and 14). All but two (Nos. 15 and 19), related to lowland areas in rich agricultural areas (Table 6.2).

Four of the graves with textiles in diamond twill qualities connected to central places, Vinjum in Sygnafylki (No. 7), and Naustdal, Grodås, and Holmedal (Nos. 4, 9 and 10) in Firdafylki, and as mentioned, probably as nodes within smaller or larger estates of older heritage (cf. Table 6.1 and Fig. 6.5). The other graves connected to larger farms, often strategically located at harbours or at the entrance of valleys (Nos. 1, 6, 7, 8). Only one of these graves related to Hordafylki and the earliest from the end of the 8th century (No. 1), while the others stretched over the 9th century as such. In Firdafylki, two of the five graves connected to the inland/upland district Breim, at Sandal and Bø (Nos. 2 and 4), in the transition zone between the mid- and the inner

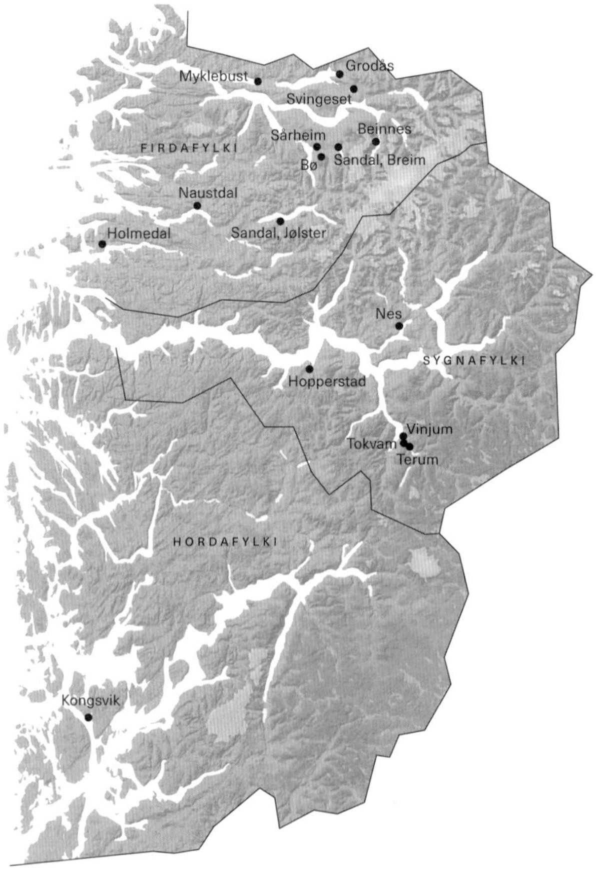

Fig. 6.5: Graves with tools and textiles from the late 8th to the late 9th century from western Norway (Illustration: Per Bækken).

Table 6.3: Representation of textile qualities in space and time from the late 8th century to the end of the 9th century (Source Bender Jørgensen 1986, 257, 263–267).

Time	*Representation/ Place*	*District/ County*	*Grave*	*Diamond twill: Quality I*	*Diamond twill: Quality II*	*Diamond twill: Quality III*	*Other textiles*
750–800	No 1 Store Kongsvik	Tysnes, H	Double grave in a mound		38/12	–	2/2 twill 16/9
	No 2 Sandal	Breim, F	Mound with several graves	25/15	–	–	–
800–850	No 3 Bø	Breim, F	Double grave, boat burial in a mound	24/11 24/16 25/15 22/12	–	–	2/2-twill 18/13
	No 4 Naustdal	Førde, F	Flat grave	–	–	48/16	Tabby 14/14; pattern woven fabric 20/12
	No 5 Nes	Sogndal, S	Flat grave	Imprint?	Imprint?	Imprint?	Imprint, 2/2 twill 16/8, tabby: 14/12
	No 6 Terum	Aurland, S	mound?	Imprint?	Imprint?	Imprint?	Imprint, 2/2 twill, tabby
850–900	No 7 Vinjum	Aurland, S	Boat burial, mound?	?	38/26 (linen)	62/25	Tabby 28/7: wool Tabby 18/15: linen Rib weaves, strap
	No 8 Nes	Sogndal, S	burial in a cairn	28/14	–	–	2/2 twill 16//9; tabby 21/?
	No 9 Grodås	Hornindal, F	Flat grave?	24/16, 24/13	–	42/20	Tabby/rib weave, 18/9
	No 10 Holmedal	Fjaler, F	Boat burial in a mound, Gender?	–	–	50/18	Small unidentified pieces
N:	10			3 (7)	1 (2)	5 (5)	6 (+ 2 with imprints of BDT

areas, and with three other contemporaneous graves with textile tools in the district. In Sygnafylki, all the four graves with remains of diamond twills connected to the mid- and inner zones (Nos. 5–8), and to the communities of Sogndal and Aurland – and at Nes in Sogndal with two such graves (Nos. 5 and 8), and additionally three other graves with textile tools only; Aurland with two, at Terum and Vinjum (Nos. 5 and 7), besides eight other graves with textile tools but without textiles.

The qualities

Bender Jørgensen's meticulous analyses of the textiles make it possible to give a more precise assessment of the qualities, which is important when taking Holm-Olsen's observations into account too – that there are often gliding transitions between the categories, and not as standardised as expected from specialised workshops. Based on Bender Jørgensen's analyses, Table 6.3 gives an overview of the graves with diamond twill qualities and other textiles in the 10 graves over time, and represented by their thread count in the warp/weft, normally measured by per cm.

Table 6.3 shows that the 10 graves with textiles woven in diamond twill could be classified in differentiated qualities, I–III. Three of the four graves with textiles of the coarser quality I (Nos. 2, 3, 8) represent an interesting group, being all prominent graves, richly equipped with textile tools. Two of them, moreover, connected to neighbouring places, Sandal and Bø in Breim (Nos. 2 and 3), while the third in Firdafylki connected to a central place at Naustdal in the southern part of the county (No. 4). Two of the graves in Sygnafylki this early (Nos. 5 and 6), in, respectively, Aurland and Sogndal, mid-valley at the place Terum and at Nes, a headland by the Sogndalsfjord and in both cases only represented by mineralised imprints of undefined quality. The four graves with the finest quality III all connected to central places, however, all but one, the grave at Vinjum in Aurland (No. 7) with diamond twill in quality II and III and one of unidentified quality and with a comprehensive tool kit for all work stages. The three

Table 6.4: Viking Age graves from the late 700s into the late 9th century with tools and textiles of diamond twill qualities I–III (N:10); graves with composite tool kits marked in bold (Source: Bender Jørgensen 1986)

No.	*Museum no.*	*Locations*	*Shears*	*Wool combs*	*Spindle whorls*	*Loom weights*	*Weaving beater*	*Other textile related tools*	*Quality*
1	B7639	Store Kongsvik, Tysnes	1			6			II
2	**B9014**	**Sandal, Breim**	**1**	**1**	**1**	**6**	–		**I**
3	**B9501**	**Bø, Breim**	**1**	**1**	**2**	**9**	**1**		**I**
4	B9626	Naustdal, Førde	–	–	1	–	–	–	III
5	**B9064**	**Nes, Sogndal**	**1**	–	**3**	–	**1**		**?**
6	**B9063**	**Terum, Aurland**	**1**	–	**3**	–	–	**R413**	**?**
7	**B7731**	**Vinjum, Aurland**	**2**	**1**	**2**	**48**	–	**R413**	**III, II. ?**
8	**B9065**	**Nes, Sogndal**	**1**	**1**	–	**14**	**1**	**R413**	**I**
9	B8000	Grodås, Hornindal	–	–	–	–	–	Needle case, 3 needles	III
10	B7893	Holmedal, Fjaler	–	–	–	7	–	–	III
		N: 10	**8**	**4**	**12**	**90**	**3**	**3 + 3 R413**	

other graves (Nos. 4, 9 and 10) with quality III textiles, in contrast, all had few tools.

All the graves except the early grave at Sandal in Breim (No. 2) contained moreover textiles in other qualities, 2/2 twills and tabbies, belonging to other garments, such as undergown or cloak. Four of them contained both 2/2 twills and tabby-woven textiles (Nos. 5, 6 and 8), and two only with 2/2 twills (Nos. 1 and 3). Three graves with the finest quality III (Nos. 4, 7 and 9) contained remains of tabby woven textiles with a general high thread count, in both linen and wool. We see then that also the tabbies and 2/2 twills have a general high thread count, between 16 and 28 threads in the warp per cm and 7 and 15 in the weft.

When differentiating the various qualities of broken diamond twill according to Bender Jørgensen's tripartite division as for fineness and density, quality I–III, with quality III as the finest, it provides an extra facet related to the issue of domestic production versus import. Could these fine textiles just as well have been produced locally, connected to special places within textile production environments?

Tools and qualities

Of all the graves with textile remains of diamond twills that had tools, six of them had tools for three to five work processes (Nos. 2, 3, 5, 6, 7, 8), one for two processes (No. 1), and three for only one work stage (Nos. 4, 9 and 10). Altogether, shears, spindle whorls and weaving tools represent the most frequent tool categories, including the spatula-shaped tool, R413 (Table 6.4).

Only two of the five graves with diamond twills of qualities III and II comprised more than one tool category: tools for shearing and weaving and at Vinjum in Aurland (No. 7), with textiles of qualities II and III, and with tools for nearly all processes. Two of the three graves with remains of quality I, were both early graves furnished with multiple tools and from the same district of Breim, as noted at Sandal and Bø (Nos. 2 and 3), and rather like one of the graves at Nes as for tools and textiles (No. 8). Two of the other graves with textiles woven in diamond twill technique, however of unidentified qualities (Nos. 5 and 6), had tools for three work processes. Nearly all these graves connected to central agricultural districts, and four to central places (Nos. 4, 7, 9, 10) or as nodes in communication (Nos. 1, 5, 8). Judging by the tools, however, some of them were hardly distinguished or reflected as central places for production judging by the textile tools (Nos. 1, 4, 9 and 10).

Changes over time?

When assessing the graves represented by textiles of diamond twills in a closer temporal perspective, we see that three represent some of the earliest graves stemming from around AD 800 or some decades earlier (Nos. 1, 2, 3) (Table 6.5).

The earliest grave represented by textiles woven in diamond twill, at the coastal grave in Hordafylki, was in fact of quality II (No.1), while the textiles in the upland grave at Sandal (No. 2) and the double grave at Bø (No. 3), both in Breim were woven in diamond twill of quality I – and rather homogenous, with almost the same thread counts in the weft. Both contained composite tool kits too, with shears, wool combs, spindle whorls and loom weights (cf. Table 6.4). However, the tools vary and with whorls of different weight, at Sandal with a light spindle whorl of 11 g, and at Bø with two spindle whorls for different use areas, 19 and 46 g. The grave at the central place Naustdal in Firdafylki from the first half of the 800s, represents the earliest grave with diamond twill of quality III, and as noted with few tools. The third and somewhat later grave at Nes in Sogndal (No. 8) and most prestigious grave as for other grave goods, but with a single weaving beater as for tools. In two of the early 9th-century graves in respectively

Table 6.5: Chronological representation of graves with textiles of broken diamond twill qualities.

Period	Quality I	Quality II	Quality III	Undefined quality	Representation nos.
c. 750–800	2	1			1, 2, 3
c. 800–850	1		1	2	4, 5, 6
c. 850–900	1	1	3		7, 8, 9
800– 900			1		10
N	4	2	5	2	10

Sogndal and Aurland (Nos. 5 and 6), the textiles were only preserved as mineralised imprints and unidentifiable as for a finer classification of quality.

The three graves from the latter part of the century (Nos. 7, 8 and 9), reveal another pattern comprising the finest quality III of diamond twill and all connected to central places. Here, only the Vinjum grave (No. 7) with the most composite tool kit had remains of textiles woven in both qualities II and III. The grave with textiles of quality I, at the headland, Nes by the fjord in Sogndal (No. 8), connected to the same place as the earlier grave (No. 5), represented by an imprint (Table 6.3), and with remains of a third possible grave context[18] and all with composite tool kits. The representation of tools in these three contexts varied from covering one work process to include tools for all stages in production. In contrast, the graves located at central places in Firdafylki contained few tools. Only one of them (No. 2) contained more than one tool category, for spinning or weaving. The two others, with one tool category, was either for weaving (No. 10), spinning (No. 4), or sewing (No. 9). Two of the graves with textiles of diamond twill of quality I (Nos. 2 and 3), in contrast, had tools for four processes, for shearing, combing, spinning, and weaving, and in two cases (Nos. 5 and 8), and at Vinjum for all stages of production except sewing (No. 7), and one grave (No. 10) only for weaving (cf. Tables 6.3 and 6.4).

The correspondence between high-quality textiles, high-status graves with various textile tools and their location at central places, and seemingly textile producing environments in the mid- and inner zones, does not seem accidental, either for status or locations. The early high-status graves in Breim, at Sandal and Bø, share many common traits – both with composite textile tools and textile remains of diamond twills and of about the same quality (cf. Table 6.2). Together with other contemporary graves with textile tools in this area, they indicate a special environment for textile production and seemingly as an advanced domestic production.

The varied representation of tools and textile qualities may indicate different modes of production at different places and over time. The highest representation of the finest qualities appeared in the second part of the 9th century and mainly at central places, which may indicate an advanced household production, and probably attached specialist production, at least at Vinjum.

The grave at Vinjum (No. 7) with its broad spectrum of tools and high-quality fabrics, in contrast, left textiles of broken diamond twill of quality II and III and one unidentified quality. The high-quality textiles of diamond twill of the dress in the Vinjum grave match, in fact, qualities from Birka and Hedeby. Compared with the early representation of quality III in the three graves from Firdafylki (Nos. 4, 9–10), the qualities in the Vinjum grave were even finer and tighter, especially in the warp with a thread count of 62/25, and with from 12 to 20 threads more per cm in warp than in the three counterparts. Here, the diamond twill of quality II differed from the others, being woven of plant fibres and with a tighter weave in the weft, with a thread count of 38/26 in the warp/weft and much tighter than the earliest representation in the coastal grave in Hordafylki (No. 1). The textiles in the Vinjum grave covered many use areas, even represented by coarser felted pieces with a thread count of 8/7. This grave included moreover some coarser tabby-woven textiles (rib-weaves) of wool and linen, probably as parts of the undergown. A strap made of horsehair and a fragment of broken diamond twill with a plaited border with an unidentifiable thread count,[19] add to this versatile collection of textiles.

The other three graves from central places in Firdafylki with remains of diamond twill quality III-textiles (Nos. 4, 9–10), differ from Vinjum with a far smaller range of preserved textiles and with a limited assortment of textile tools. The grave at Grodås in the inland district Hornindal (No. 9), however, included fragments of both quality I and quality III with a thread count of 24/16 and 42/20 per cm. The traces of verdigris at the brooches indicate that it was a strapped gown, and blue in colour. The grave goods of a needle case with some small and thin needles used for finer textiles, all connect to female graves of high status. The grave at the central place Naustdal further south from the early part of the century (No. 4), with textiles of quality III had a thread count of 48/16 with tighter warp threads and a looser weft. The single spindle whorl of 26 g in this grave represents, however, spinning rather ordinary yarn. The last of these graves was an 'ungendered' boat grave, without female dress accessories and possibly an asymmetrical double grave (No. 10). The preserved textile of broken diamond twill of quality III was made of plant fibres and a fine and tight warp with a thread count of 50/18 per cm, and with a looser weft than the quality II textile from Vinjum.[20] Still, being woven in linen, it is not quite compatible. The tools in the grave were limited to a few rather light weights. The remains of diamond twill quality III in these three graves vary, then, from 42 to 50 threads per cm in the warp and from 16 to 25 threads in the weft.

These 10 graves represent qualities that vary in both time and space during the 9th century, where the finest textiles of qualities II and III appear rather early, around the turn of the 8th century represented by quality II, while the earliest representation of quality III appear in the grave at Naustdal in the first part of the century. Still, textiles woven in diamond twill of quality I, seem to be more frequent this early, represented in four graves and with two from a limited area in Breim. They are rather homogenous in quality as for the warp but with a somewhat tighter or looser weft but generally balanced. Comparing textiles of quality I from the graves in Breim and Sogndal, the warp in the later contexts is somewhat tighter, and the weft relatively looser, however, these are rather insignificant differences. Quality II, as represented in the very earliest of the graves and in the later grave at Vinjum and woven in both wool and linen with similar thread counts in the warp but differed in the weft; 38/12 and 38/26 respectively, and more balanced in linen. However, the Vinjum grave contained both quality II and III, and the finest of them all.

Although being categorised within three quality groups, they do appear as homogenous and standardised products, although repeat patterns and the common pattern of 20/10, can be recognised over time and in space: in Breim from graves from the early part of the century and very earliest at upland Sandal (No. 2) and the latest from Vinjum in Sygnafylki (No. 7) in Sygnafylki and Grodås (No. 9) in the northern part of Firdafylki. The even spread of qualities with only smaller variations within and between the quality groups speaks in favour of an advanced production with traditions in weave patterns.[21] A general trend seems to be those graves with textile tools and high-quality textiles connected to central places (Nos. 7, 9 and 10), however not all of them in environments for textile production judging by the tools in the graves.

Some of the graves with the largest tool assemblages stem from Breim, Sogndal, and Aurland (Nos. 2, 3, 5, 7 and 8). Here, the connection between high-status graves and high-quality textiles in both upland/inland areas and at central places as for status and communication corroborate their role as special places for textile production, possibly as an attached specialist production. However, when high quality textiles appear at central places and with few textile tools, such as at Naustdal, Grodås and possibly Holmedal and few other such graves in their neighbourhood, traded textiles cannot be excluded.

Other graves with textile tools and other textile qualities

Seven of the graves with both tools and textiles of the more common weave techniques of 2/2 twills and tabbies, were all from female contexts and in graves of status III too (Table 6.6). Two related to areas with many graves with textile tools, in the district Vik in Sygnafylki as the earliest (No. 11 and 12) from around the turn of the 9th century or somewhat earlier, and seemingly some decades later in Aurland (No. 14). Of the four other graves with textiles from Firdafylki, the earliest connect to the central place Myklebust by the Nordfjord (No. 13), with other graves with textile tools (cf. Chapter 5). Two graves relate to the fringes of Stryn, one at upland Svingeset, the other in the lowland related to the fjord (Nos. 15, 16), while one connects to the place Sandal the inland district of Jølster further south from the latter part of the century (No. 17) Of these the grave from Svingeset in Stryn (No. 15) stands out with textiles woven in chevron twill of high quality (Table 6.6).

Compared with the 10 contemporary graves with textiles of diamond twill qualities (Tables 6.1 and 6.2, Nos. 1–7), they differ with more common textile qualities and their smaller range of textile tools – two with tools for combing and weaving, one for spinning and sewing, and one only for weaving. Only one grave from the latter part of the century, the grave at Sandal in Jølster (No. 17) had tools for all processes from shearing to smoothing (cf. Chapter 5). Still, all were high-status female burials, furnished with brooches belonging to the strapped gown, however, only with remains of the tabby weaves, often connected to the undergown, like at Vinjum. The preserved textiles seem then to represent some other garments than the strapped gown. As noted, qualities not only varied as for status but also functionally, and depended not least on preservation conditions.

One of the two female graves in the Tokvam mound at Aurland (No. 14), one with an all-inclusive tool kit and the other with a single spindle whorl, demonstrates another intricate pattern as for representation of social status and agency (cf. Chapter 5). Here, in the central grave of the high-status woman, only a small fragment of tightly woven tabby was preserved, however, hardly representative for her garments. Judging by the accessories, two oval brooches and a third brooch, they indicate that she had worn a cloak or a shawl but most likely an undergown, and with only one piece preserved. The same underrepresentation of textiles applies to the high-status female grave at Sandal in Jølster, again with a fragment of ordinary tabby, which hardly represents the dress that was furnished with exquisite brooches and a silver chain, and moreover supplied with an all-inclusive tool kit.

When comparing textile qualities of 2/2 twill and tabby in graves with broken diamond twill (Table 6.5, Nos. 1–7) with those without diamond twills, the 2/2 twill weaves in the latter group appear in several qualities too, fine, and coarse, from 11 to 25 threads in the warp and from 6 to 24

Table 6.6: Graves from c. *750 into the late 9th century with combinations of tools and textile without diamond twill qualities (N:7); graves with composite tool kits marked in bold (Source: Bender Jørgensen 1986).*

Ref.	*No.*	*Place*	*Shears*	*Wool combs*	*Spindle whorls*	*Loom weight*	*Weaving beater/R413*	*Needle (case)*	*Smoother*	*Weaves*	*Status*
No. 11	B9061	Hopperstad Vik	–	–	–					2/2 twill:10/9	FIII
12	B6500	Skjervum, Vik	–	–	–	–	1	–	–	2/2 twill: 18/10	FIII
13	B5807	Myklebust Eid	–	1	–	–	1	–	–	2/2 twill: 25/12; 11/6, Tabby/reps: 16/6	FIII
14	**B8669 I&II**	**Tokvam, Aurland**	1	1	3	28	1	–	–	Tabby: 18/24+	FIII
15	B6483	Svingeset, Stryn	–	–	1	–	–	1	–	Chevron twill: 36/13, Tabby:16/13, Strap of tabby 24/12	FIII
16	B7184	Beinnes, Olden, Stryn	–	1	–	–	1	–	–	Type?	FIII
17	**B11413**	**Sandal, Jølster**	**1?**	1 pair	4	10	1	–	1 whale bone plaque	Tabby: 20/15 Tabby ?/13	FIII
	N: 7		2?	4	8	38	5	1	1		

threads in the weft, and rather in the same qualities as in the graves represented by 2/2 twills and tabbies too. However, a textile of chevron twill with thread counts 36/13 (No. 15), corresponds to the quality of diamond twill quality II. The similar qualities in graves combined with broken diamond twill qualities and a few pieces of 2/2 twills: two of the same quality, 16/9, and looser then, and with tabbies with thread counts of 16/6, 16/13, 24/12, 20/15 (Table 6.1). The pieces of tabbies found in the graves with diamond twill qualities generally had a tighter and finer warp, with from 14 to 28 threads per cm, but with a looser weft from 7 to 15, and only in one case more than 24 threads per cm. The differences in qualities do not only relate to presence and absence of diamond twill qualities and may signify different textile production milieus. They were, however, limited to what was used as funeral clothes and dress accessories, with a preserving effect on their connecting textiles. Still, it depended on the general preservation conditions, and all the layers of the garments were rather seldom preserved. This is important to have in mind to avoid conclusions based on *e silentio*-reasoning. Generally, all are woven of z/z yarn, however, in two cases, the 2/2 twill woven textiles in z/s spun yarn, the early grave at Hopperstad (No. 11), and tabby woven textile in the Terum grave (No. 6).

Graves from the 9th century with textiles without textile tools

Five of the 11 graves from the 9th century with textiles but without textile tools comprised only the ordinary 2/2 twills and tabbies, like the six graves with tools and textiles (Nos. 11–17) presented above. The graves differ in other ways too, both for quality and gender (Table 6.7).

Only one of these five graves, an upland grave from Hordafylki on the border to Sygnafylki (No. *i*), represents a high-status female burial. Two are double graves of unequal status, both from Gloppen (Nos. *ii*, *iii*). The last two, both from the same community, Fjaler in Firdafylki (Nos. *iv, v*), but seem to represent high-status male graves.[22] One connected to the central place Dale in Firdafylki (No. *v*), together with another grave with textile remains. The textiles in the two male graves were both of 2/2 twill and tabby and of coarser quality than in the female graves, represented with other types of garments.

Graves with tools and textiles from the Late Viking Age

The 10 graves with tools and textiles from the 10th century reveal a different spatial pattern than in the earlier graves with textile remains. They include also a relatively higher share of high-quality textiles of diamond twills, preserved in eight of the 10 graves, of which seven had composite tool kits (Table 6.8). Nearly all related to the inner zone, with respectively three from each county: Hordafylki (Nos. 18–20), Sygnafylki (Nos. 21–23), and four from Firdafylki (Nos. 24–26). Only one related to the mid-zone, and at the central place Dale (No. 24) in Fjaler, in the south-western part of Firdafylki. Nearly half of these graves connected

Table 6.7: Graves from the 9th century with textiles without textile tools (N:5) (Source: Bender Jørgensen 1986).

Museum no. Reference	*Place*	*District/County*	*Diamond twill*	*2/2 twill*	*Tabby*	*Gender/Status*
B7732 *(i)*	Skjervheim	Voss, H	–	12/5 12/9	20/15	F III (R647)
B6026 *(ii)*	Sande	Gloppen, F	–	7/6 9/6	–	M/F II/III (R647, Irish ring pin)
B6685 *(iii)*	Eide	Gloppen, F	–	–	13/12 (pattern) 13/13 Reps 13/7	M/F III/II (R647)
B9765 *(iv)*	Korsvoll	Fjaler, F	–	–	16/?	M III
B3106 *(v)* N: 5	Dale	Fjaler F	–	8/6, pattern, 3 colours	–	M III

Table 6.8: Graves with tools and textiles from the 10th century (N:10) (BDT = Broken diamond twill); graves with composite tool kits marked in bold (Source: Bender Jørgensen 1986; Lukešová 2011).

Ref.	*Museum no.*	*Place/District/ County*	*Shears*	*Wool combs*	*Spindle whorls*	*Beater*	*Loom weights*	*Needle (case)*	*Smoother*	*Weave*	*Gender/ Status*
18	**B6657**	**Trå, Granvin, H**	1	1	1	–	11	–	1 pellet	BDT	FIII
19	**B6228**	**Veka, Voss**	–	1	2	–	11	1 needle case	.–	BDT	FIII
20	**B4864**	**Hyrt, Voss**	–	1	1	–	20	–		BDT	MIII/FIII
21	**B7761**	**Hopperstad, Vik**	2	–	1	1 R413	1	–	–	BDT	FIII
22	**B9060**	**Hopperstad, Vik**	–	1	–		12	–	1 whale bone plaque	BDT	FIII
23	**B12131**	**Målsnes, Balestrand**	1	1	2	–	–	1	–	BDT	FIII
24	**B5910**	**Dale, Fjaler**	1	1	4	1		–	1 pellet	BDT, 2/2 twill, tabby	FIII
25	**B8953**	**Kirkeide Stryn**	–		1					BDT Tabby	FII–III
(26)	B8818	Loen, Stryn	1	–	–	1		–	–	Tabby	FIII
(27)	B6433	Brekke, Oppstryn	–	1	–		3	–	–	Tabby	FII
	N: 10		6 (5)	7	12 (7)	3	58 (6)	2			

to central and strategic places, at gateways or points of control and were probably nodes within early estates (Nos. 21, 22, 24, 25).

Except for the three graves in the northern part of Firdafylki, all had composite tool kits as grave goods. Still, they contained more ordinary qualities of tabby too. In Sygnafylki, two again connected to the central place Hopperstad in Vik (Nos. 21, 22). A third grave with high-quality textiles (No. 23) stems from the high-status grave at Målsnes, located at a promontory at the northern side of Sognefjord, opposite to Vik. The graves from inland Hordafylki – Veka, Trå, and Hyrt – also stand out with their exquisite textiles of diamond twill and their composite tool kits, and other artefacts as grave goods.

The textile equipment in these graves cover from one work process to include tools for all stages of production in seven of them (Tables 6.8–6.10). While the two graves without textiles of diamond twill of quality I (Nos. 23, 25), one had a tool for spinning, the other grave with remains of textiles woven in diamond twill of both qualities I and II (No. 22) had tools for four processes – shearing, combing, spinning, and sewing. Generally, the graves with diamond twill of quality III had more composite tool kits than the others. The two graves without diamond twill qualities in contrast had tools only for respectively shearing and weaving or combing and weaving (Table 6.8).

Eight of these graves contained textiles of diamond twill qualities. Only two of the graves represented with diamond twill, included quality I: Målsnes (No. 23) by the Sognefjord and the other at Kirkeide (No. 25) in Stryn besides the finer quality II, as also found in the Hopperstad grave (No. 22). The textile remains in the Målsnes grave, however, consisted of seven layers of textiles connected to the brooches, woven

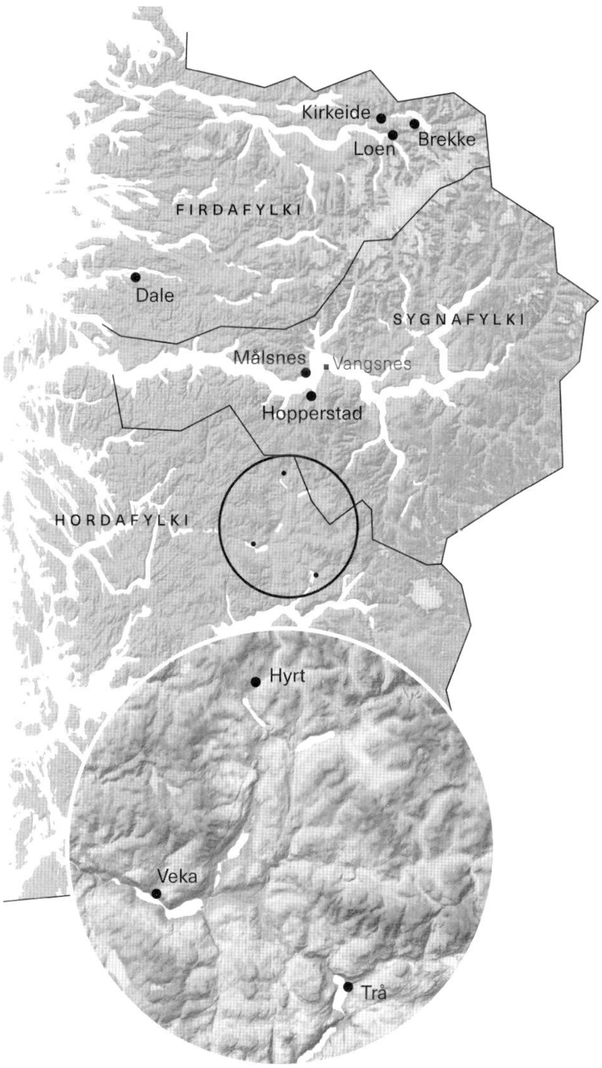

Fig. 6.6: Graves with tools and textiles of varied qualities from the 10th century within the three counties (Illustration: Per Bækken).

in different techniques and qualities, as parts of different garments. Two fragments of 2/2 twills at the upper side of the brooch and in different qualities fine and coarser, with a thread count of respectively 14/11 and 8/4 per cm, both probably representing some outer garments, kaftan, cloak, or shawl. The third and fourth layer and under the brooch was woven in diamond twill of quality II, with a thread count of 30/11 and 36/14 z/z spun yarn. All the other fragments woven in diamond quality II, of about the same quality seem to represent the strapped gown and possibly undergown.[23]

Of all these graves, five included then the finest quality III (Nos. 18–21): Trå, Hyrt, Veka, Hopperstad (No. 21), and Dale (No. 24), and all of high status (Table 6.9).

Of these, three of the graves with the finest broken diamond quality III, combined with other qualities, also

Table 6.9: Representation of textiles of broken diamond twill qualities in 10th-century graves.

Period	*Quality I*	*Quality II*	*Quality III*	*Undefined quality*
900–1000	2	2	5	–
Representation Nos.	Nos 23 & 25	Nos 22 &23	18–21 & 24	

contained the most complete tool kits – at Veka, Trå, and Hyrt – and all from the inner zone of Hordafylki (Fig. 6.6, cf. Table 6.8, Nos. 18–20). Here, the rich tool assemblages offer a special opportunity for a closer assessment of tools, textiles, social emblems as closed contexts and for comparison with other contexts that are less complete (cf. Fig. 6.6). The fragments of the dress in the grave at Veka (No. 19), referred to earlier in this chapter, was woven in broken diamond twill of quality III in a repeat pattern of 28/10 threads, and a density 42/20 and 50/22 threads per cm. The cloak, woven in 2/2 twill, varies in density in both the warp and the weft: 17/14, 16/20, and 23/17 per cm with the warp threads as normally tighter and finer than in the weft, as also is the case in the tabby weaves with 30/16 threads, and as noted, also included a pattern-woven band of samitum silk.

The Trå grave revealed remains of a similar outfit of varied qualities. The two largest fragments, woven in 2/2 twill and dark blue-green in colour, had a thread count of 26/16, where the selvage showed tiny remnants of a blue sewing thread, possibly of linen. Remnants of a tabby loop on the third brooch were probably fastened to the cloak from the 2/2 twill. The new analyses by Lukešová, referred to in the beginning of this chapter are summarised together with the other contemporary graves with textiles (Table 6.10).

The textiles from the Trå grave included the finest quality with the highest thread count, woven in pattern 20/10 and with the higher density of 64/26 in the warp/weft and in z/z spun fine yarn.[24] The textiles in the graves at Hyrt and Veka, with a wider assortment of qualities, the diamond twills were woven in the same repeat pattern and with a high thread count too, with a thread count per cm, varying from 48 to 50 in the warp and 26 in the weft at Hyrt and in the Veka grave varying from 42 to 52 threads in the warp and 20 to 22 in the weft. Like the graves at Trå and Vinjum, the Hyrt grave also contained textiles of plant fibres too, probably linen.

The textiles of diamond twill from the two Hopperstad graves (Nos. 21 and 22), both of quality III and with the same repeat pattern and of similar quality with a thread count of 45/20 and 48/16 respectively, were about the same quality as the in the grave at upland Hyrt. One of them (No. 22) had remains of diamond twill quality II too, with a thread count of 34/20. Remnants of three upper straps indicate two different dresses, both woven in diamond twill of different qualities, one with a selvedge and traces of the seam for

Table 6.10: Representation of textile qualities from the 10th-century female graves (N:10) (Source: Holm-Olsen 1972; Bender Jørgensen 1986; Lukešová 2011).

Reference no.	*Museum no.*	*Place/Community/ County*	*Grave*	*Gender/ Status*	*Broken diamond twill, Quality I*	*Broken diamond twill, Quality II*	*Broken diamond twill, Quality III*	*2/2 twill*	*Tabby*
18	B6657	Trå, Granvin, H	Secondary grave in mound, wooden coffin?	FIII	–	–	64/26	26/16	24/14, 20/18
19	B 6228	Veka, Voss, H	Mound	FIII			44/22 50/20 42/20 50/22 52/22	17/14 16/20 23/17	–
20	B4864	Hyrt, Vossestrand, H	Mound, double grave	MIII/FIII	–	–	50/20 48/20	(1) 32/20; (2) linen 18/10, 16/11 (3) 24/16	32/20, Linen; 18 /10, 16/11, 24/16 pleated band
21	B7761	Hopperstad, Vik, S	Mound with several burials	FIII	–	–	45/20	18/10 16/11	Reps 20/7 Tabby linen: 24/17 pattern woven band
22	B9060	Hopperstad, Vik, S	Mound with several burials	FIII	–	34/20	48/16	17/11	Pleated band strap, 2 tabby woven: 23/12; the other as imprint
23	B 12131	Målsnes, Balestrand, S	Mound	FIII	28/14, 26/13	30/11, 36/14 30/14 32/12, 32/12 30–36/10–14	–	14/11; 8/4	Tabby 14/14, rests of a string
24	B5910	Dale, Fjaler, F	Grave chamber in a mound	FIII	–	–	44/17 40/17 40/17 46/17	7/9 10/8 17/15 12/8 16/17 18/16	Tabby 22/16 Tablet weaving: 16/18, handwoven strap
25	B8953	Kirkeide, Stryn, F	Flat grave, in a wooden coffin, covered by flat stones	FIII	26/16	–	–	–	linen: Tabby 11/7, 13/17, 9/? felted
26	B8818	Loen, Stryn, F	Flat grave	FIII			–	–	Tabby 28/14, Tabby 20/10 (linen) Strap: 1:tabby/reps14/16, Strap 2: tabby 14/18–20
27	B6433	Brekke, Stryn, F	Flat grave	FII					25/12

Table 6.11: Graves from the 10th century with preserved textiles but without tools (N:7) (Source: Bender Jørgensen 1986).

Museum no.	*Place*	*District/County*	*Gender/ Status*	*Grave*	*Broken twills, Quality I*	*Broken twill, Quality II*	*Broken diamond, Quality III*	*2/2 twills and tabby*
B5625 (vi) (R652)	Vangsnes	Vangsnes, Balestrand, S	FII	Mound (incomplete find)	–	37/15	–	Tabby: 30/15 pleated, Reps/tabby 30/13, 16/8 Pattern woven band I: rib weave, band II: 20/16,
B7812 (vii) (R652)	Fjellsende	Lindås, H	FIII	Flat grave	24/10	–	–	2/2 twill: 15/13 10/?
B7535 (viii) (R654)	Nes.	Kvinnherad, H	MIII//FIII	Flat grave, mixed find?	–	–	–	2/2 twill: 16/7
B7664 (ix) (R652)	Sørheim	Dale, Luster S	FIII	Flat grave Complete?	?	?	?	?
B7655 (x) R652	Arnestad	Gloppen, F	FII	Small mound	?	?	?	?
B5939 (xi) R652	Skrøppa	Breim, F	FIII	Flat grave	–	–	–	Tabby 18/? Tabby/reps 22/15
B10539 (xii) JP 51	Skoden	Innvik, F	FII	Flat grave	–	–	–	Tabby 20/13

braiding at the edge.[25] The last of these five graves with quality III textiles, at the central place Dale in Firdafylki (No. 24), the fragment of diamond twill preserved under the brooch, represents the layer closest to the dress. It was woven in the same pattern as the others, with a thread count 44/17. Fragments of coarser fabric and of a selvedge made in patterned woven twill, and a tablet woven cord were also found.[26] While four of the graves from the 9th century comprised textiles of diamond twill of quality I, only two of the later graves (Nos. 23 and 25), contained this quality, while five of the 10 graves from the 10th century with tools and textiles, comprised diamond twill textiles of quality III.

The eight graves represented with textiles in diamond twill and the other represented by quality I, came from a central place, Eide/Kirkeide, in Stryn further north in Firdafylki (No. 25). The textile fragments, like many others, were woven in a repeat pattern of 20/10, and of about the same quality as noted in one of the layers in the Målsnes grave, although somewhat finer, with a thread count of 26/16 and similar fineness to the 2/2 twills from the Trå textiles. Two of the graves *without* diamond twill-woven textiles were in the Stryn area too, at Loen, another central place by the fjord (No. 26), and the other in an upland area in Oppstryn (No. 27). Both revealed only tabbies, one with a thread count of 28/14, another of looser linen with a thread count of 20/10, and two tabby woven straps with thread counts of 14/16 and 14/18–20, and the other grave with a tabby woven textile with a thread count of 25/12 per cm. While many of the graves from the 9th century only comprised a few textiles in 2/2 twill, the 10th-century graves to a larger extent contained qualities of tabbies and 2/2 twill, representing different parts of the garments.

Additionally, seven 10th-century female graves within the present study area but without textile tools (Nos. *vi–xii*) had preserved textiles beneath their brooches (Table 6.11). Two of them stem from Sygnafylki, of which one at Vangsnes (No. *vi*), close to Vik. The other was located at the central place Dale in Luster further in-fjord on its northern side of the Sognefjord (No. *ix*). Three of the graves related to Firdafylki and again in the same districts as graves as with textile tools, in Gloppen, Breim, and Innvik (Nos. *xi, xii, xiii*), and of the two graves from Hordafylki, one related to the coast (No. *vii*) the other to the mid-zone (No. *viii*). Of these graves only one had traces of broken twills of quality II, the female grave from Hordafylki (No. vii), and the two in, respectively, Sygnafylki and Firdafylki with remains of undefined qualities (Nos. *ix, x*) (Table 6.11).

Of these seven graves, the grave at Vangsnes (*vi*) contained the most varied types of textiles: from ordinary tabbies to 2/2 twills, including fabrics woven in broken herringbone twill, two bands, of which one a pattern woven, multicoloured band in rib weave and with a rather unique exemplar of a pleated fabric of wool, z/z spun tabby woven fabric with a thread count of 28/15 threads per cm (*vi*). It may be remains of an undergown, like the finer pleated fabrics of linen, as identified as among the Birka textiles, although coarser (Fig. 6.7).[27]

Vangsnes had other prestigious graves too, and one of them contained textiles with remains of gold threads, and a twined lace of gold, as known from Birka,[28] probably fastened to some fabric, besides red coloured wool and supplied with some beads, of which a larger one was amber.[29] The other graves with textiles without textile tools contained more common qualities and mainly tabby-woven fabrics, in contrast

Fig. 6.7: Fragment of a pleated tabby-woven fabric of wool, from the grave at Vangsnes in Sygnafylki (Photo: © University Museum of Bergen).

to the graves with both tools and textiles (cf. Table 6.10). The graves with textiles but without textile tools relate then to the same districts as the other graves of general high status and in some cases with textiles of the same qualities as the graves with both tools and textiles.

Tools and textiles as sources

That as many as 29 graves with textiles and textile tools, and nearly one-third of all the graves in this category and additionally at least seven graves without such tools, show that women of the higher ranks in western Norway were buried in their best clothes – and it was a burial custom lasting from the very beginning of the Viking period and throughout the period. However, of about 140 graves with textile tools – no more than *c.* 20% left preserved remains of the funeral clothes the deceased had once been buried in. This implies that the corpus of textiles related to graves is limited in quantity and narrow in terms of functions and is generally related to the funeral dress, belonging to women from the upper echelons – in one or two gowns – an inner and outer of different qualities, and/or a chemise and cloak, and often with textile tools besides other grave goods, largely depending on status. Being on display as the funeral garments of the deceased, in some cases resting on undercovers as in the Vinjum grave, the textiles in these contexts exclude all the other trivial but important use areas within textile production. They reflect, however, both the luxurious segment of production as well as clothes of coarse and more ordinary qualities. Still, the tools selected as grave goods were hardly chosen to match the funeral clothes and with a potential to reflect broader use areas too, but as selected grave goods they also serve as indicators of female or gendered work domains.

Even so, the remains of both tools and textiles are meaningful representations that indicate textile production on a high level, with traces of diamond twills of qualities II and III from the earliest period to well into the 10th century. Similar weave patterns and techniques of *long dureé* indicate milieus with technological capacity, transmitted over generations, however, seemingly within special social spheres, and a segment with the means and opportunities of displaying a conspicuous lifestyle. The fabrics, however, were not uniform and varied in quality, woven in different patterns and techniques, not least related to functions of the various garments and the finest often the strapped gown, worn under a cloak, often woven in 2/2 twill and broken twills of varied patterns, but also with a tabby-woven undergown either in wool or linen or a cloak as reflected in in the Hyrt grave, among others.

The find distribution in 2/2 twill not least demonstrates a relatively even distribution in qualities, indicating a domestic production, as Holm-Olsen has also pointed out. This was a technique that most women mastered, based on traditions and experience achieved through generations. However, some high-standing milieus also produced more complicated patterns and of finer qualities as social attributes.

Table 6.12: Graves with textiles and composite tool kits from the 9th and 10th centuries (N:14).

Reference	*Museum no.*	*Locations/Districts*	*Shears*	*Wool combs*	*Spindle whorls*	*Loom weights*	*Weaving beater*	*Other tools*	*Quality*
2	B9014	Sandal, Breim	1	1	1	6	–	–	I
3	B 9501	Bø, Breim	1	1	2	9	1	–	I
5	B9064	Nes, Sogndal	1	–	3	–	1	–	? imprint
7	B7731	Vinjum, Aurland	2 R443	1	2	48	–	R413	III & II
8	B9065	Nes, Sogndal	1	1	–	14	1	R413	I
14	B8669	Tokvam 1&2	**1**	1	3	28	1	R413	Tabby
17	B11413	Sandal, Jølster	**1?**	1	4	10	1	–	Tabby
18	B6657	Trå, Granvin	1	1	1	11	–	Pellet of glass	III
19	B6228	Veka, Voss		1	2	11	–	Needle case, R413	III
20	B4864	Hyrt, Vossestrand	1	1	1	20		–	III
21	B7761	Hopperstad, Vik	2 R443	1	1	1	1	R413, Needle case	III
22	B9060	Hopperstad, Vik	–	1 (linen?)	–	12	–	Whalebone plaque	III
23	B12131	Målsnes; Balestrand	1 R443	1	2	–	–	Needles, needle case	III
24	B5910	Dale, Fjaler	1 R443	1	4	1	1	Pellet of glass	III
		14	14	13 +1	26	171	7		

The finest qualities in the earliest decades of the 9th century connect to central places with access to both natural and human resources with the means and competency, such as in Vik, Sogndal, and Aurland in Sygnafylki and the districts of Breim and Stryn in Firdafylki – areas with access to wide pastures for larger flocks of sheep. From the last part of the century, the grave at Vinjum signals a special place both as for status and for textile production, while other central places with graves with tools and high-quality textiles of another character, foreshadow the development in the 10th century at central places but without surrounding settlements with such graves. This seems to reflect ongoing centralisation processes. By the late period, other areas, such as Voss in Hordafylki, came more into the foreground, while earlier textile production milieus, such as Aurland and Breim, were no longer marked by the same distinguishing graves and with textile tools as grave goods.

The varied spectrum of textile tools reflects the importance of textile production, and it is interesting to observe that the most composite tool kits concur with graves with high-quality fabrics and other indicators of high status. By the late Viking Age, graves often appear at nodes in later royal and ecclesiastical administration centres, which most likely had become manors for local or regional elites or chieftains, and with a subordinate work force and tenanted farmsteads without or only with few contemporary graves. However, by this time, graves with high quality textiles and a rich assortment of tools appeared in other environments too and without central functions judging by their later status, such as the graves at Veka, Hyrt, and Trå, all places of high-quality production at farms with wide resources in the lowland and with access to wide upland resource areas.

Places or nodes of production?

To what degree the tools in the graves from western Norway could produce the high-quality textiles or for other use areas, and whether they form patterns and links in time and space, are the next issues to consider. Tools and textiles in the same contexts offer an opportunity for closer comparison, especially related to the finer qualities and the more luxurious segment of production. The graves with composite tool kits provide the best point of departure, especially the tools for combing, spinning, and weaving that can distinguish qualities. Fourteen of the graves from the 9th and 10th centuries contained both multiple tool kits and textiles remains – seven from each century – and in 11 cases combined with remains of the finer qualities of broken diamond twill of qualities I–III (Tables 6.8 and 6.10).

The most complete tool kit of all came from the 10th-century grave at Trå in inner Hardanger (No. 18) with tools for all stages in the production: wool combs, a light spindle whorl, two weaving beaters for making a tight weft, and loom weights of medium weight for stretching the warp, as well as a glass pellet of foreign origin for smoothing fabrics of linen (cf. Chapter 5). In principle, the tools could have been used to produce some of the high-quality textiles in the grave. Altogether, 14 graves from the shift to the 8th century and increasingly into the 10th century, are represented with varied tool kits and preserved textiles. This makes it possible then to assess tools and textiles from the same contexts in more detail, not least based on the weights of spindle whorls and loom weights (Fig. 6.8; Table 6.12).

The Trå grave with a spindle whorl weighing only 9 g may well have been used to spin threads of the finest quality,

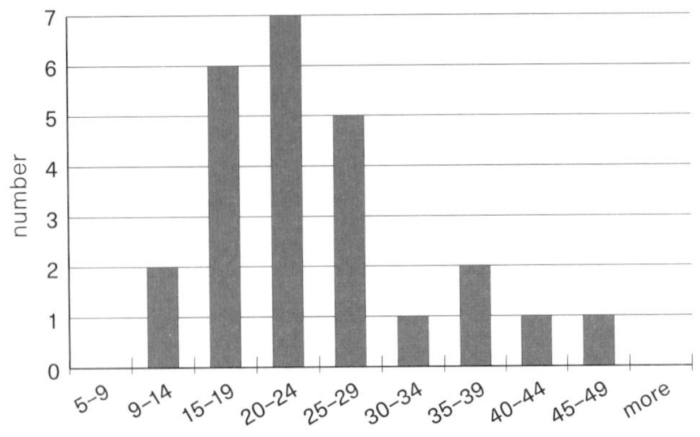

Fig. 6.8: Distribution of measurable spindle whorls (N:25) in the graves from the 9th- and 10th-century graves (Illustration: Per Bækken).

represented by the preserved textiles and loom weights within the weight group of 450–500 g. Altogether, the weights indicate somewhat coarser yarn in the weave, and they may have been used in production of various qualities, as identified among the textiles. The two weaving beaters, both of iron and rather long, originally around 80 cm, the other more than 85 cm, could produce stronger and coarser textiles too. The 12 loom weights selected as grave goods vary in weight from *c.* 450 to 600 g, which indicates somewhat stronger warp threads. Theoretically, at least, the tools found could have been used in the processes of producing the different fabrics found. The thin sewing threads from plant material as documented in the finest textile quality and the glass pellet too show that both linen and wool formed parts of production. As one of 28 general high-status graves with tools and textiles, the combination of tools, textiles, and social emblems substantiate that production in rural settings of high status could produce high-quality fabrics, which have earlier been assessed mainly as imports or related to urban settings.[30]

The finds in graves with the same types of dress accessories may appear with the same types of dress too. In fact, all the graves related to high-status women, identified as for gender by their dress accessories, imply a dress with straps fastened by brooches, however, of various qualities. When the graves included a third brooch, which was the case for some 30 of these graves, they imply a cloak or possibly an undergown or shawl, depending on the size and shape of the brooch. In some cases, layers of various clothes have been preserved, such as in a grave from Hopperstad and, not least, the Målsnes grave. This implies that the remains in most cases appear as arbitrary and incomplete representations of what had once been the funeral outfits. For instance, when only tabby woven textiles are left it does not necessarily imply a low-status dress but rather an incomplete representation due to preservation conditions. Altogether, *e silentio* reasoning is risky without assessing the overall dress customs and dress accessories attached to various types of garments.

Many textile fragments represent only one layer of the garments and not necessarily the strapped gown that was often woven in the broken diamond twill or 2/2 twill qualities when related to high-status women, or other qualities belonging to other parts of the dress. However, when 12 of the graves with textile remains under the brooches are represented by other qualities but without textile tools, they may cover other roles than those supplied with tools in the grave (cf. Tables 6.7 and 6.11).

Textile tools and tool assemblages may indicate status and textile qualities too. As for shears, 12 of the 14 graves with preserved textiles had shears in their tool kits, both for the initial and final stages of production – of the type for cutting textiles (R443) but mostly for sheep shearing. Four of the renowned graves with textiles and textile remains of highest quality – Vinjum, Hopperstad, Målsnes, and Dale (Nos. 7, 21, 23, 24) – had shears related to the final stages of production, as in 13 other high-status graves, among others, two of the graves at Myklebust, at Kvåle, Gutdal and Vik in Stryn, and Arnestad in Gloppen, but without traces of textiles. This type of shears appears in 17 of the graves with textile tools (cf. Chapter 5). All the 10 graves with textiles in broken diamond twill contained wool combs too, implying combing of long cover wool, but also represented in graves with ordinary qualities of tabby and twills (Nos. 14, 17), and in many graves without textile remains. Still, two of the high-status graves in Sogndal and Vik, at Nes and Hopperstad (Nos. 8 and 21), had none. As combing must have been required to produce most of the high-quality textiles, the presence of wool combs as grave goods may not discriminate between qualities represented in the graves. When tools for combing increased from 13 from the 9th century to 15 graves from the 10th century and 60% more and with more varied tools and a higher representation of finer qualities, the tools seem to reflect a greater emphasis on textile production in all its aspects.

Since weights of the spindle whorls indicate ability to spin yarn of various gauges, and 18 of these 27 graves included spindle whorls, they indicate various use areas, both the finer segment of textile production and other qualities. As noted, even small differences of *c.* 5 g affect the thickness and thread quality, and subsequently the woven fabric (Fig. 6.9).

Of the whorls from 13 graves, seven relate to the 9th century, and six to the 10th. Of the earliest seven, five comprised textiles of broken diamond twill, three of quality I or of undefined quality and one of quality III, while the later six represent the finest qualities. Of these, four graves had spindle whorls within the weight group 15–20 g and were represented with more than one whorl. The two early graves from Breim, Sandal, and Bø (Nos. 2 and 3) had diamond twill of quality I; one had two whorls (No. 3); one heavy

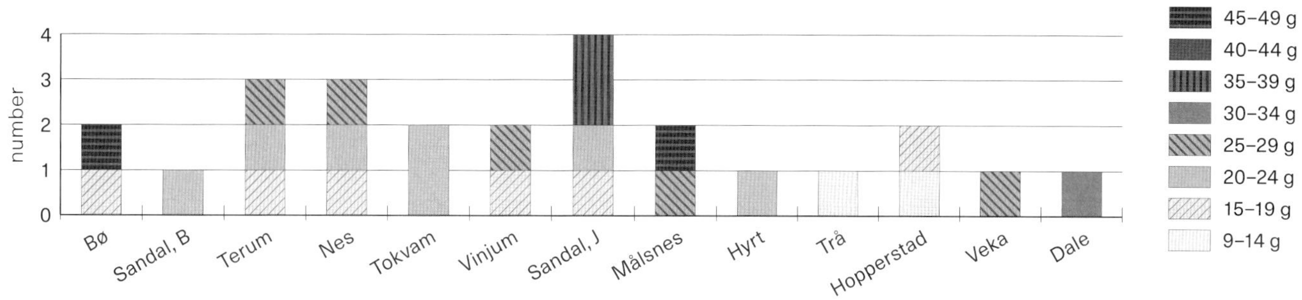

Fig. 6.9: Spindle whorls from graves with tools and textiles, according to weight (N:25).

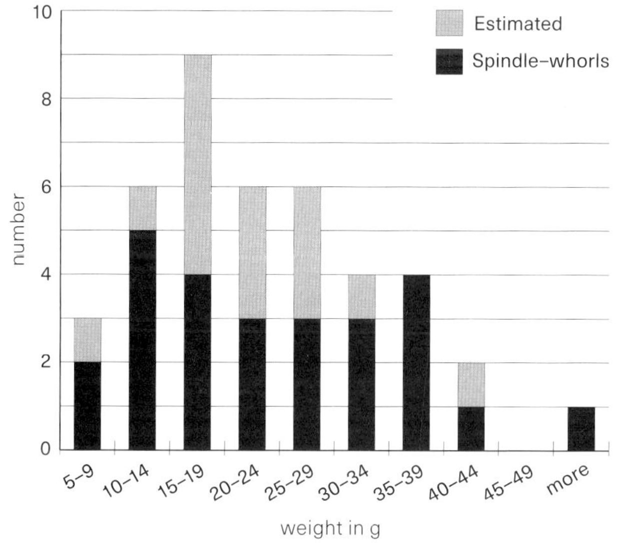

Fig. 6.10: Distribution of spindle whorls, according to weight groups of 5 g in graves without textile remains (N:41).

and the other 19 g for spinning finer yarn. The whorl in the upland Sandal grave (No. 2) was in the weight group 21–25 g. The one in the Hopperstad grave (No. 21) with textiles of quality III too, was heavier in weight group 25–29 g, as in the grave at Dale (No. 24) of 33 g (Fig. 6.9). The two graves in Aurland and Sogndal, at Terum (No. 4) and Nes (No. 5), respectively, each had three whorls covering the whole weight spectrum from 15 to 30 g. The two whorls in the Vinjum grave (No. 7) weighed 19 and 29 g respectively, suitable for spinning medium fine yarn, however, hardly the finest qualities. The two spindle whorls from the graves in the Tokvam mound had almost identical weights, 24 and 25 g, and here represented by an ordinary tabby quality. The grave at Sandal in Jølster had four spindle whorls of 18 g, 21 g, 38 g, and 40 g, which could produce various thread qualities, however, only tabby woven textiles were preserved.

The lightest whorls appeared in two of the 10th-century graves, as noted from Trå (No. 18) and Veka (No. 19), the latter with two whorls, 9 g and 17 g, suitable for spinning very fine, and fine threads matching the remains of fine textile qualities. The grave at Hyrt (No. 20) with remains of diamond twill of quality III contained a heavier spindle whorl of 22 g. The contemporary grave at Målsnes with two whorls were even heavier, 27 g and 47 g: the lighter matching the weight of the one in the Hopperstad grave (No. 22), and both graves containing diamond twill weaves, of, respectively, qualities II and III.

Altogether, the weight spectrum of the whorls represents different yarn qualities, from the thinnest to the coarse, like represented in the graves with preserved textiles, however, not always matching the textile qualities represented in the same contexts, mainly relating to the dress. Nevertheless, the spindle whorls indicate spinning of various gauges, adapted to various types of textiles, but in most cases not as fine as the very finest warp threads as in the textiles of quality III, woven in diamond twill techniques (Fig. 6.9).

In grave contexts with tools but without textiles, an assemblage of 41 spindle whorls covers an even wider weight spectrum, from 5 g to more than 50 g. Of these, 44% weigh less than 20 g, 29% between 20 g and 30 g, 24% between 30 g and 40 g, and 7% more than 40 g (Fig. 6.10). Compared with the spindle whorls as part of broader tool assemblages and with textile remains, they include more light whorls under 20 g and above 30 g too, and with a peak in the weight group 15 g to 19 g. They indicate a higher variety of qualities than evidenced in the graves with multiple tool kits and preserved textiles and that production of finer qualities could be more common than indicated by the preserved textiles.

Altogether 59 grave contexts include loom weights, more than 380 loom weights, mostly from the Viking period, and represent the largest assemblage of textile tools and with a concentration in the inner zone, with more than three times more graves than along the coast and with the mid-fjord area in a mid-position. The average representation of loom weights is also much higher in the mid- and inner zones with six and eight weights per grave respectively than at the coast (Table 6.13). This also corresponds with the representation of textiles – mostly from the inner zone and only few from the mid-zone, and then concentrated to central places.

Table 6.13: Number of graves with loom weights within the study area according to climatic and geographic zone division. Number of weights is in brackets.

County	*Coastal areas*	*Mid-fjord areas*	*Inner fjord and inland areas*	*Total finds*	*Total number*
Hordaland: *Hordafylki*	6 (18+)	7 (42)	11 (68)	24	128+
Sogn & Fjordane: *Sygnafylki & Firdafylki*	3 (5)	10 (68)	22 (180)	35	253
Total	9 (23+)	17 (110)	33 (248)	59	381+

Table 6.14: Graves with weaving beaters in the three geographical zones within the two counties, and in graves combined with loom weights, marked in brackets.

County	*Coastal areas*	*Mid-fjord areas*	*Inner fjord and inland areas*	*Total graves*	*Total number*
Hordaland: *Hordafylki*	4 (3)	3 (1)	13 (1)	20 (5)	20
Sogn & Fjordane: *Sygna- & Firdafylki*	1	11 (5)	30 (11)	42 (16)	42
Total	5 (3)	14 (6)	43 (12)	62 (21)	62

Table 6.15: Graves (N:10) with composite tool kits: loom weights (measurable weights in brackets), other textile tools and textiles.

Ref.	*Grave/Place*	*Loom weights*	*Beater*	*R413*	*Spindle whorls*	*Wool combs*	*Shears*	*Plaques/ Pellets*	*Needle case*	*Quality*	*Date*
2	Sandal, Breim	6 (4)	–	–	1	x	x	–		I	750–800
7	Vinjum, Aurland	48 (18)	–	x	2	x	x (2)	–		III	850–900
8	Nes, Sogndal	15 (14)	x	x	–	x	x	–		I	850–900
14	Tokvam, Aurland	28 (25)	x	x	2	x	x	–		Tabby	800–900
17	Sandal, Jølster	10 (10)	x		4	x	–	x		Tabby	850–900
18	Trå, Granvin	12 (11)	x	–	1	x	x	x		III	900–1000
19	Veka, Voss	14 (11)	–	x	2	x	–	–		III	900–1000
20	Hyrt, Voss	20 (8)	–	–	1	x	–	–		III	900–1000
21	Hopperstad, Vik	1	x	x		–	x (2)		x	III	900–1000
22	Hopperstad, Vik	12	–	x	spindles	x	–	x		III	900–1000
N:	10 graves	166 (101)	5	6	13+ spindles	9	8	3	|		

The remains of the tools for weaving – loom weights as indicators of the tension of threads in the warp weighted loom and their ability to keep an even weave – are likewise important sources when assessing level and quality of production. However, being long neglected as an archaeological source explains why many no longer remain in the museum. The representation of weaving beaters shows a similar geographical pattern but not always combined with loom weights (Table 6.14).

Weaving beaters related to five of the 10 graves with composite tool kits and textiles. Why some graves contained a weaving beater and others not, is not always evident, although it is largely connected to high status but not always. All these graves had tools for the main work processes of spinning and weaving. However, only half of them contained weaving beaters (Nos. 8, 14, 17, 18, 21) and then only two in contexts with remains of broken diamond twill of the finest quality (Table 6.15). Six of these graves contained moreover the spatula shaped tool R413, which indicates a strong connection with production of high-quality textiles. Three of the graves without weaving beaters, at Hopperstad, Vinjum, and Tokvam, contained these spatula-shaped tools. In three graves, however, and another grave at Hopperstad, and Nes and at Tokvam, these tools supplemented weaving beaters. All the 12 spatula-shaped tools in the present corpus relate to graves within textile producing environments – Vik, Aurland, Breim, and Stryn. The upland grave at Sandal, however, contained neither spatula-shaped tools nor weaving beaters. Still, weaving beaters could also have been of perishable materials, such as bone and wood. As symbolic grave goods, they may not echo the production as a 1:1 representation but still give hints (Table 6.15).

All 10 graves with multiple tools contained loom weights in varying numbers, from one to 48. Of the originally 167

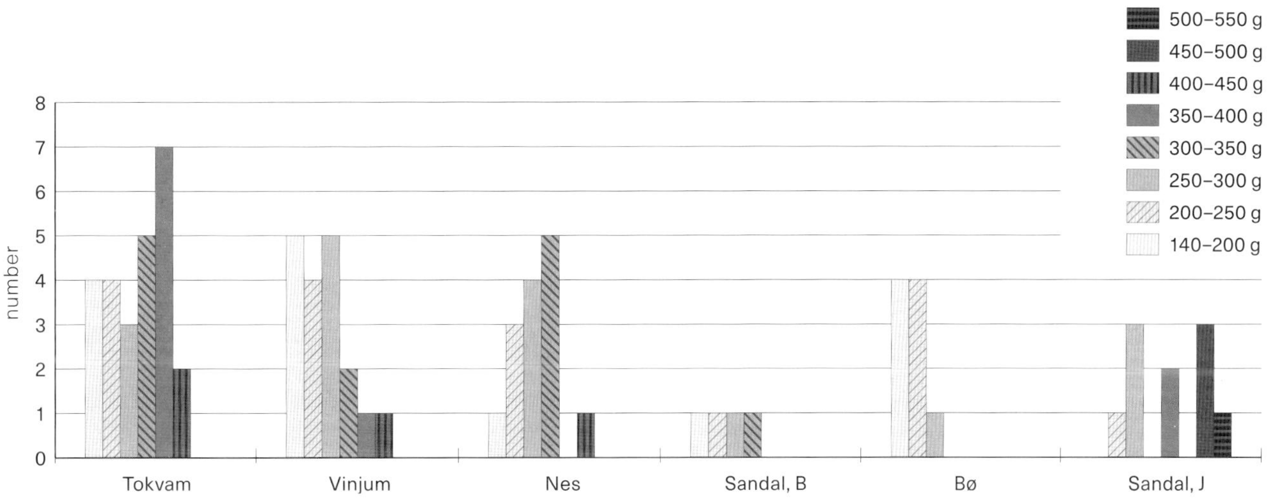

Fig. 6.11: Weight patterns for measurable loom weights from six graves from Sygnafylki and Firdafylki from the 9th century: Tokvam, Aurland (N:25); Vinjum, Aurland (N:18); Nes, Sogndal (N:14); Sandal, Breim (N:4); Bø, Breim (N:9); Sandal, Jølster (N:10). Total number: 80.

weights recorded, no more than 103, or 62%, have been retrieved and are in the museum. The two graves from Aurland, Tokvam (No. 14) and Vinjum (No. 7), may have represented complete looms, with 28 and 48 loom weights respectively, which are the highest representation of loom weights within the present corpus from western Norway, but unfortunately, not all of them remained in the museum (Fig. 6.11). Of these, the weights from Tokvam (No. 14) count 25 preserved loom weights, ranging from 140 g to 450 g but mostly within the array of 300–400 g with a mean of 298 g. About half of the complete or measurable loom weights weigh between 300 g and 400 g, where some seem to form pairs within a smaller weight group of about 200 g.

Presupposing a tension of 20 g per thread, a weight of *c.* 200 g could stretch 10 warp-threads, a weight of 400 g, 20 threads, and so on. A dense thread count with many threads per cm would require many light weights and preferably relatively equal weights, which should be possible in this case. The two spindle whorls in the Tokvam mound, both within the same weight group, 24 g to 25 g, imply only that they could spin yarn of medium fineness. The balanced tabby-woven textiles with a thread count of 18/24+, could very well have been spun and woven by the tools in the secondary Tokvam grave.

As for Vinjum, as noted, only 18, or 38%, of the once 48 loom weights remain. They weigh from 145 to 427 g, mostly within the range of *c.* 200–300 g, with a mean of 253 g. The preserved weights from the Nes grave have more or less the same weight range, from *c.* 160 g to 415 g with a mean of 283 g. The loom weights from Sandal in Jølster (No. 17) are generally heavier, from 230 to 516 g with a mean of 371 g. The loom weights in these five assemblages have a top point around 150 g to 300 g at Vinjum, 100 g to 250 g at Bø with an even representation, and between 100 g and 300 g at Sandal, and clearly higher at Nes from 200 g to 450 g, and Tokvam from *c.* 140 g to above 400 g (cf. Fig. 6.11).

The two graves at Trå and Veka from the 10th century and with the lightest spindle whorls of all, comprised 12 and 11 loom weights each, within a weight range of *c.* 450 g to 600 g at Trå and *c.* 230 g to 400 g, with a mean of 300 g at Veka. The assemblage from Trå, however, ranging from *c.* 350 g to over 800 g stands out with a mean weight, *c.* 230 g to 280 g heavier than from Veka, Tokvam, Nes, and Vinjum, and 160 g heavier than from the graves from Hyrt and Sandal. The third of these graves from Hordafylki from the upland double grave at Hyrt had textiles of quality III and composite tools had 20 loom weights, possibly representing a whole loom. The 10 retrieved exemplars weigh between 261 g to 413 g, with a mean just under 300 g and significantly lighter than those from Trå, and more like weight patterns from the Veka grave. They both share nearly the same weight range and mean weight of around 300 g, while the weights from Trå are significantly heavier within a frame from *c.* 450 g to 600 g with a mean weight.

These assemblages demonstrate rather similar although not completely homogenous weight patterns. In three of the graves, Veka, Hyrt, and Tokvam, they cover nearly the same weight range, mostly between 300 g and 400 g and mean weights of around 300 g, while the assemblages from Nes and Vinjum are *c.* 20 g to 50 g lighter. The weights from Sandal in Jølster are *c.* 70 g heavier in the mean, with about the same range when taking the missing weights from Hyrt into account. These cases indicate then that the loom weights within the assemblages did not necessarily have to be completely homogenous as for weight. The number of threads attached to them could probably also be regulated. Generally, they reflect weaving with relatively thin to medium thick warp threads. The heavier stones in the

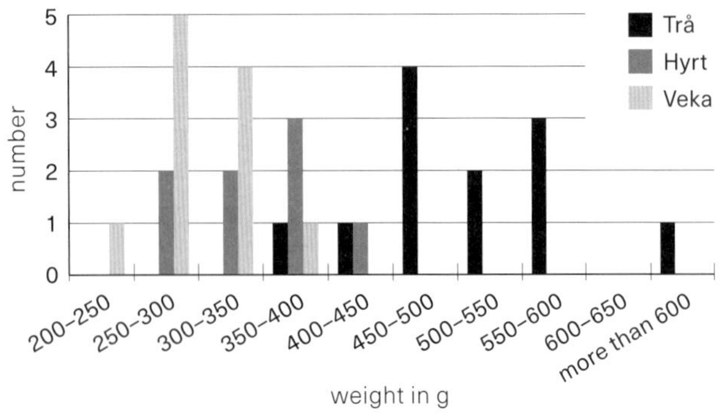

Fig. 6.12: Weight patterns for measurable loom weights from three graves from Hordafylki from the 10th century: Trå, Granvin (N:12); Hyrt, Vossestrand (N:8); Veka, Voss (N:11). Total number: 31 (Illustration: Per Bækken).

Trå assemblage were for stronger and coarser warp threads than the others, nearly 240 g heavier than those at Veka and Hyrt (Fig. 6.12).

Four of the assemblages include both loom weights and weaving beaters of the traditional Viking Age type with an obtuse angle, as well as other textile tools: the graves at Tokvam, Nes, Sandal, Jølster, and Trå assemblages, with different weight representation. Of these, the loom weights from Trå, and to some degree Sandal, indicate weaving of thicker and stronger warp threads, while Tokvam and Nes represent lighter assemblages (cf. Figs 6.11 and 6.12). This may indicate that weaving beaters were used when weaving fabrics of both coarser and finer warp threads.

Does the combination of tools for the various stages of production and textiles support or weaken the comprehension of domestic textile production milieus able to produce the textiles of the fabrics of high quality? Here, the tools and textiles from the extraordinarily rich and well-preserved Vinjum grave from the late 9th century provide a strong case, with ample evidence of tools and textiles and when combined also able to assess the level of textile production, as represented within the elite and by the funeral dress and the textile tools. With an almost complete tool kit and preserved textiles of the finest qualities of broken diamond twill in wool and plant fibres, and with thread counts of 62/25 and 38/26, it altogether indicates it as a special place and milieu for textile production. Still, as we have seen, the Vinjum was not the only grave at these levels; five other graves from the 9th century comprised broken diamond twill of quality III too, and all located at central places (cf. Table 6.4), and another five such graves from the 10th century (cf. Table 6.10). Of these, Hopperstad in Vik is one of the places that signifies continuity, from around the turn of the 9th century (No. 11) and with two other graves of this kind from the 10th century (Nos. 21, 22). The textile qualities in two of these three graves are nearly identical with thread counts of 45/20, and of 48/16 in the third, which indicate continuity and traditions over several generations. Unfortunately, the textile tools among others for spinning and weaving from these graves, are no longer available for measurements. Of the other graves, only one, at Holmedal (No. 10), had a density above the quality in the Hopperstad graves, with a thread count of 50/18 but in this case with few tools, only some loom weights.

The textile remains from the graves at Trå and Veka from the 10th century match the qualities from the somewhat earlier Vinjum grave. They appear in other environments and with a high representation of tools as grave goods too, which again makes it possible to compare tools and textiles. The high share of wool combs as grave goods, and not least the textile remains, indicate a conscious and highly developed wool sorting technique and careful combing of long fibres into high quality fibres and yarn.[31] The weight of spindle whorls varies, as noted, from 9 g to more than 45 g in the contexts with tools and textiles (Fig. 6.8) and even more in contexts with tools without textiles, weighing from 5 g to more than 50 g (cf. Fig. 6.9).

They signify then spinning in various gauges, from the very finest to coarser yarns, but generally within the finer but not the finest segment. The weight spread of loom weights shows a similar pattern (Figs 6.11–6.12). The lightest loom weights relate to the double grave from Breim (No. 5), all complete and weighing from 148 to 260 g with a mean of 197 g. The two other graves represented with the coarser quality I, in contrast, had tools covering all main processes from shearing to weaving (Nos. 1 and 3). The weaves appeared rather homogenous, with a thread count of 22/12 and 24/11, and only of slightly coarser quality than the upland grave with a weave of 25/15 threads per cm in the warp/weft. The third grave at Nes (No. 7), also represented by broken diamond twill of quality I, had a slightly denser warp but looser weft, 28/14. Here, the weights were heavier, from 200 to 450 g but concentrated between 200 g and 350 g, and with a mean of 283 g. Theoretically at least, the quality I textiles could be made within rural domestic settings of high social status and with access to resources, both human and natural. The production of the finer quality would be more demanding for fibre preparation, as well as for spinning and weaving.

In the Vinjum grave (No. 7), with remains of varied textile qualities, including exquisite textiles, the loom weights had a top point around 150 g to 300 g. As half of this assemblage weigh between 150 g and 250 g, and the other half between 250 g and 450 g, they may rather represent separate mountings – weights used for warping looms when weaving either very fine or medium-fine qualities. The number of warp threads fastened to the weights could also regulate the density but not too much to keep an even weave and selvedge. When weaving textiles of varied qualities thus needing various tensions, a larger storage and assortment of weights would, as noted, have been needed at hand. When loom weights appear as grave goods, they do not necessarily represent the mounted loom. In principle though, the tools in the Vinjum grave could produce many of the qualities found, although perhaps not the very finest. The fine and tight weft with weft threads up to 25–26 threads per centimetre would probably have needed a weaving beater but perhaps of softer material than of iron to press the thin threads. The spindle whorls, 19 g and 28 g respectively and with a difference of 9 g, should be able to spin yarn of various gauges of fine and medium-fine qualities (Figs 6.8 and 6.9). At least theoretically, the tools could produce many of the qualities represented in the Vinjum grave and within the setting of a high-status farm at highest level, and indirectly as a superior and administrative role, possibly including wider areas in Aurland too with its high representation of graves with textile tools; altogether eight graves from the 9th century (cf. Chapter 5). In one of these graves further up the valley (No. 5), and from some decades earlier than the Vinjum grave, the traces of broken diamond twill demonstrate an early-established mode of production and a special textile production milieu. This speaks in favour of a domestic production within the luxurious segment of production within an elite economy and probably provided by a subordinate highly competent work force of spinners and weavers but also with the means to get hold of textiles in other ways too.

The other three graves had textiles of about the same qualities as at Vinjum: Trå (No. 18), Veka (No. 19), and Hyrt (No. 20), all with assemblages of loom weights, and two of them, like Vinjum, without weaving beaters as grave goods, and included spindle whorls of light and medium/light weight. The two whorls from Veka, 9 g and 17 g, indicate their ability to spin both very fine and fine threads. Here, the loom weights also indicate weaving with rather fine warp threads (cf. Fig. 6.9). In principle, the whorls could have been used to spin threads for both thread systems in the loom. The spindle whorl from Hyrt, 22 g, would also have been able to spin threads that would match the weights of the loom weights. The same applies to the two whorls from Vinjum of 19 g and 29 g. There is, then, a match between loom weights and spindle whorls, used as grave goods at Veka, Hyrt, Vinjum, as well as Tokvam in Aurland and Sandal in Jølster, but not so at Trå with a very light spindle whorl and heavier loom weights. The weights for spinning and weaving both indicate the production and use of medium thick yarn that would have functioned well in a more open tabby between the warp-threads, or as weft-threads. The weaving beater was an effective tool to make the weft denser, while the finest qualities perhaps would need a tool for softer strokes. Still, the Trå grave too seems, by the varied tool kit, to represent a production environment of high standards.

Some of the prominent graves at special places too and with the more seldom found tools as grave goods for cutting textiles and sewing, such as at Hopperstad, Myklebust, Grodås, and the not faraway upland graves at Maurset and Svingeset (cf. Chapter 5), may indicate special places for production related to the indoor work of finishing the clothes. The grave at Målsnes is another case with a toolkit for shearing, combing, spinning, and sewing, and represented with broken diamond twill quality I and II.

In most cases, these graves with tools and textiles represent women in the higher echelons of society at farms above the average, and four of them were located at central places in the 9th century and further extended with three by the 10th century. Within a hierarchical society, people of lower rank, including unfree labour, would be those who carried out the manual work. As substantiated by the inverted social pyramid, presented in the foregoing chapter, the tools in the graves rather allude to the person in charge of production, organising and supervising a smaller or larger work force. Not least, the production of more prestigious and high-quality textiles would need skilled labour and an organisation based on carefully planning, logistics and coordination of a differentiated work force for all the processes involved to achieve the qualities in demand, coordinating all stages from sorting the wool in various qualities, combing, spinning, and weaving. Moreover, such production needed access to grazing land and arable land for textile plants besides the acres for food production (cf. Chapter 4). This would have needed a highly competent person – most likely the women buried in the graves – with the means, competency, and authority to produce textiles above the level of the common manufacture of daily wear and tear. Still, the whorls and weights furnished as grave goods may not have been the tools used to spin and weave burial dress, which according to experiments would have needed even lighter spindle whorls and weights to achieve the highest thread counts. The tools in the graves still are important signifiers of status, textile production, and level of production without implying a 1:1 relation between them.

Common threads?

In comparison, textiles found in 10th-century graves from Iceland, rather few graves were furnished with textile tools and remains of textiles above the ordinary homespun vaðmal. One of the few graves with textile tools was a high-status female grave at Kornsá in north-western Iceland, with a weaving beater of whalebone and shears of the type for sheep shearing, besides an iron cauldron and part of a bronze scale, however, without textile remains. Another female grave of high status at Daðastaðir contained shears, a spindle whorl, and a hoe, like grave goods in west Norwegian graves too. Yet another grave at Álaugarey contained some shears and remains of textiles too and according to Eldjárn of a vaðmal quality. The Ketilstaðir grave also contained a spindle whorl and shears for sheep shearing too and two small hones besides the textile fragments,

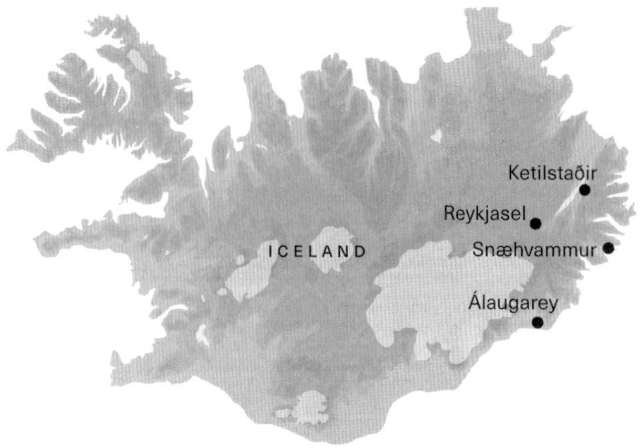

Fig. 6.13: Places with textiles of diamond twills in Icelandic contexts (Illustration: Per Bækken).

classified as vaðmal by Eldjárn, however, later recognised as diamond twills.[32]

Two other high-status graves at, respectively, Snæhvammur and Reykjasel revealed high-quality textiles preserved by the brooches, woven in diamond twill pattern and chevron twill qualities.[33] They stand out with their fine and densely woven qualities: z/z spun threads and with a thread count of 30/12 per cm from Snæhvammur,[34] and even finer and denser 44/22 in the Reykjasel grave, both located in the south-eastern part of Iceland as also the Ketilstaðir grave (Fig. 6.13). These qualities match then broken diamond twills of qualities III and II in the west Norwegian corpus. The textiles in the grave at Ketilstaðir, also woven in diamond twill quality with a thread count of 11/10[35] is coarser then than in the corpus from western Norway (cf. Table 6.3; cf. Fig. 6.3).

Generally, the textiles preserved from Icelandic burials vary from 4 to 25 threads in the warp and between 4 to 15 threads in the weft and largely using single yarns and normally z/z spun in both the warp and weft, as was the norm for textiles produced in western Norway. A large share has also tested positive for indigotin, and dyed blue,[36] like in Norway too. New strontium-isotope analysis of the dress remains from Ketilstaðir has also shown that the wool could have been locally sourced in Iceland[37] and consequently was not an import and was possibly based on traditions and acquired competency achieved through their ancestral roots in their 'homelands'. In contrast to the textiles from the graves, textiles from settlement sites altogether show a more uniform pattern of textiles of 2/2 twills of wadmal for varied use areas and generally of coarse and standardised qualities (see Chapter 9).[38]

The grave with the finest quality of chevron belonged to a middle-aged woman buried in a mound at Reykjasel,[39] the thread counts of 44/22 per cm,[40] they match several qualities from graves in western Norway, but still it has been claimed that these fine qualities represented imports, possibly from the Far East,[41] an old debate that has been modified by, among others, Lise Bender Jørgensen.

Still, the fine qualities in graves from Iceland again bring up the question of provenance and place of production since there are few tools for assessment. However, the wool analysis of the textiles woven in diamond twill from the Ketilstaðir grave has verified a domestic production,[42] but not of the same fine quality. As for the finer qualities, such as at Snæhvammur and Reykjasel, similar fine qualities are found in western Norway too, and in textile producing environments and in female graves of high status.

Many of the graves both in western Norway with textiles and with textile tools among the grave goods included artefacts of foreign origin, which may have included imports of textiles too. Altogether 10 of the graves from western Norway with tools and textiles of the finest qualities of broken diamond twill III, contained grave goods of foreign origins (Nos. 2, 6, 7, 9, 12 and 19–25), of which at least three comprised composite tool kits for textile production. Dress accessories, mass produced in the Viking Age towns of Kaupang, Ribe, Hedeby, Åhus, and Birka, represented in the graves from the early period onwards, moreover, indicate networks, regionally and across borders. Such contacts, interaction and networks are therefore also relevant when assessing the tools and textiles in the graves and indirectly the agents behind the production – and a theme to be further assessed in the following chapter.

Notes

1 Walton 1988, 153.
2 Shetelig 1912, 206–207.
3 Bender Jørgensen 1986, 263–267
4 Hana Lukešová, pers. comm. 30 September 2013.
5 B62228b, JP 140, Petersen 1928, 129.
6 Øye 2015a with references, based on information from Hana Lukešová.
7 Marianne Guckelsberger, pers. comm. May 2021; see also Guckelsberger & Mader 2017; Hayeur Smith *et al.* 2019.
8 Geijer 1972; Hoffmann 1964.
9 Ingstad 1979, 88; see Bender Jørgensen 1986, 358–360.
10 *a.o.* Holm-Olsen 1972; Bender Jørgensen 1986.
11 Øye 2015b; 2015c; 2016b.
12 Holm-Olsen 1972.
13 Holm-Olsen 1972; Bender Jørgensen 1986; Lukešová 2011.
14 Holm-Olsen 1972.
15 Bender Jørgensen 1986.
16 Holm-Olsen 1972.
17 Bender Jørgensen 1986, 267–268.
18 B3971-72/86, Nornes, Sogndal.
19 Bender Jørgensen 1986, 265.
20 Bender Jørgensen 1986; Lukešová 2011.
21 Holm-Olsen 1972.
22 B5910, Dale; B9769, Korsvoll.
23 Bender Jørgensen 1986: 264.
24 I thank Hana Lukešová for letting me use her new results.

25 Lukešová 2011.
26 Bender Jørgensen 1986, 266.
27 Holm-Olsen 1975; see also Hägg 1969; Bender Jørgensen 1986, 264.
28 Hägg 1971, 151.
29 B690-706, Vangsnes, Lorange 1875, 177–178.
30 Hoffmann 1964; Geijer 1965; 1979.
31 Holm-Olsen 1972, 98.
32 Eldjárn 1956, 179–181; cf. Guckelsberger 2015; Hayeur Smith *et al.* 2019.
33 Reykjasel, Eldjárn 1956, 169–170, 184–185; Hoffmann 1966, 87–95
34 Snæhvammur, cf. Eldjárn 1956, 181–185, Hoffmann 1966, 87–95.
35 Guckelsberger & Mader 2017.
36 Hayeur Smith 2018.
37 Hayeur Smith *et al.* 2019, 108.
38 Hayeur Smith 2015.
39 Eldjárn 1956, 169–170.
40 Solberg Andreassen 2001, with references.
41 Rosenquist 1966, 96–108; Hoffmann 1966.
42 Hayeur Smith *et al.* 2019.

7

Nodes and networks in textile production?

This chapter explores the social aspects of textile production and interaction across local and regional boundaries, as far as can be revealed through the burial evidence, expressing status and normative group identities. The key issue is how tools and textiles, combined with socially distinguishing dress accessories and other artefacts, indirectly reflect actors, group identities, and possible networks at shorter and longer distances. As earlier demonstrated (cf. Chapters 4 and 5), textile production as a long chain of work operations involved many people at different levels, and a work force of varying scale and competency. Central places, many probably as nodes in early estates with subordinate farmsteads, would involve a wider group of people than of the household at the main farm itself – both unfree labour and based on obligatory duties in kind and/or work. The economy of the elites in the Viking Age, obviously relied on co-operation but also competition within and between elite networks at various levels – for exchange, raiding and trading, and other enterprises, which required a large workforce and co-operating followers. Both raiding and the growing trade included internal and external slaves as a portable 'commodity'.[1] Towns, trade, and exchange interacted with the rural economy, which stimulated production and innovations, such as quarrying, iron production, forging, boat and shipbuilding, and others, when transforming natural resources into specialised products. It also included textiles, covering large use areas – from keeping the body warm and for shelter to expressing status, and in the Viking period onwards, also sails. All this required an expanding group of workers of diverse competency within the many work stages, which again depended on logistics and organisation within and above the household sphere (Fig. 7.1).

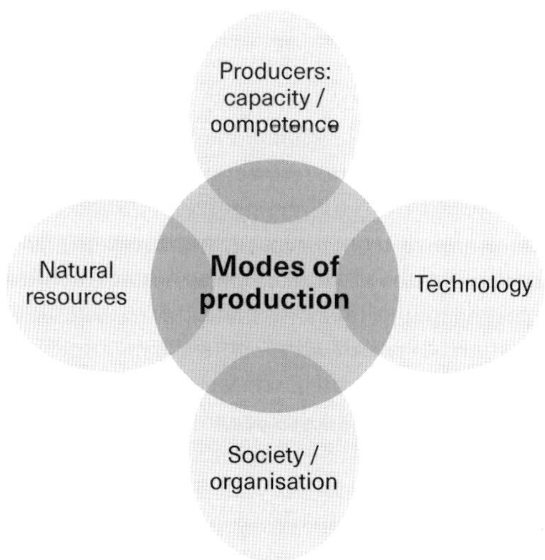

Fig. 7.1: Model of factors with influence upon modes of textile production (Design: Per Bækken).

How the graves connected to and concentrated at special places and environments, and how the funeral and burial practices displayed status and identity through social emblems and grave goods, such as tools and foreign artefacts, reflect both individual life histories and wider social relations. Indirectly, the grave goods may also reveal networks and large-scale enterprises within overarching socio-political structures. Artefacts of Insular origin, fragments of different types of ecclesiastical artefacts, alters and books, reliquaries, chalices, crosses, and mounts from crosses, often reworked into the brooches and occasionally the pendants that appear in graves from around AD 800, are generally understood as originating from goods looted during raids. According to Egon Wamers' research, more than 80% of around 150 such finds in Scandinavia stem from graves and as much as 85% stem from the jewellery found in women's graves as probable 'souvenirs' given to wives and relatives from husbands, fathers, brothers, or friends.[2] Only 4% stem from hoards and the rest from stray finds.[3] Many of these finds relate to the present study area, as shown in Chapter 5. As many as about three-quarters of the finds can be dated to the period AD 800–850 and related to the first wave of Viking attacks *c.* 800–830, while a further quarter can be dated to *c.* 850–950 and related to the expulsion from Dublin in AD 902, with a new wave a decade later to re-establish dominion in both Dublin and York.[4] The largest

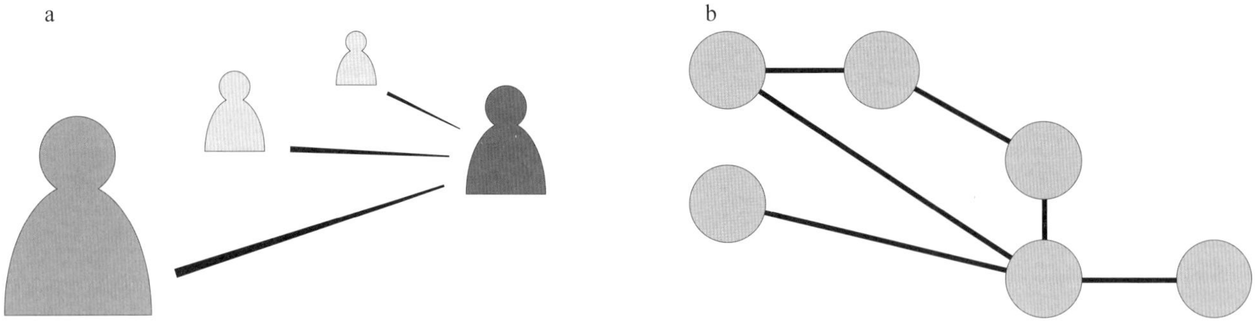

Fig. 7.2: a) Sketch of organisation within a hierarchical structure, b) networks between partners at shorter and longer distance (Graphic design: Per Bækken).

concentration of such finds relates to Firdafylki and Sygnafylki – far fewer in Hordafylki and rather few interpreted as Insular loot.[5] That many of these graves appear as boat burials (cf. Table 5.1; see Table 7.15) may also reflect and symbolise the role and prestige of commuting by sea and waterways in this expansive period.

These foreign objects indirectly indicate joint enterprises organised by 'big men', with the means, authority, and manpower to carry out such expeditions, also taking new land in both inhabited and established settlements. The female graves with such artefacts indirectly reflect roles and participation in networks too, not least related to textile production. The funeral ceremony as a rite of passage, displaying and honouring the deceased with grave goods as relicts and visual manifestations of status, roles, and relations, also confirmed status and positions of the heir(s), thus extending, and enhancing social relations and alliances.

Places and modes of production

The physical, social, and cultural aspects of production are crucial for understanding technologies, innovations, technological abilities, and levels of production – organisation, actors, and work operations within varying social and natural environments (Figs 7.2 a, b). Textile production with its many work operations – as indicated by the tools as grave goods – relied on various resource areas and labour during different periods of the year in lowland districts and upland grazing areas in the summer season, at shielings far or near. Some of the areas of vast grazing land for sheep and wool production also included woodland, hunting areas, as well as access to minerals and iron ores, able to produce quern stones, hones, and all kinds of tools and products of iron, stone, bone, and antler to provide a surplus and commodities for trade. Being extracted and produced locally or regionally, such products found their way to far off Viking Age towns in Denmark and into the Baltic, as has been documented in recent research.[6] These new and expanding industries required organisation and co-operation between people, places, and managers, bringing other wares in return through exchange networks.[7] The production and transport depended not least on textiles for all kinds of gear and wear on board ship. Altogether, the production represented combined efforts of both genders, utilising raw materials from various resource areas, reflected directly at sites of production (see Chapter 8) and only indirectly in the burial practices.

The double representation of female and male graves with various tools reflects such mutual dependencies. When the grave goods contain foreign artefacts, whether in graves at production places or at central places as nodes in early estates, they seem to signal a production above household level and often within reciprocal work domains. The location of the graves (cf. Chapter 5) thus supports a strong connection to the social aspects of production and agency – how the graves represent diverse social groups, although overwhelmingly the upper strata. Indirectly through the material messages, they reflect social and cultural identities that imply social and immaterial relations too. By exploring the contents and contexts of the graves, their spread across landscapes from coast to inland/upland and from south to the north, the aim is to trace such relations as for gender and status – whether on the same social level or of different rank – and as for elite women with the capacity to support people of special competency and as organisers of production, or simply a production on basic household level.

Dress accessories, as visual signs of social status and as normative 'constructed' social group identities, may in this way indicate connections and affiliations between people both near and far off. Not least are specific types of dress accessories, the form and style of oval brooches belonging to the Norse dress, often understood as symbols of married women[8] and as signs of social affiliations, such as bridal gifts, dowries, or heirlooms. They may, therefore, indicate relations and strategic alliances within the elite, achieved through various types of networks, locally and regionally.[9] When the graves contain artefacts of foreign origin, they may indicate such gifts too, based on long distance exchange and trade networks or participation of the crew when raiding, especially in the early part of the period. When textile tools appear in individual graves, or as double or separate graves in the same mound or cairn and with tools for different types of gendered work, they may reflect family bonds,

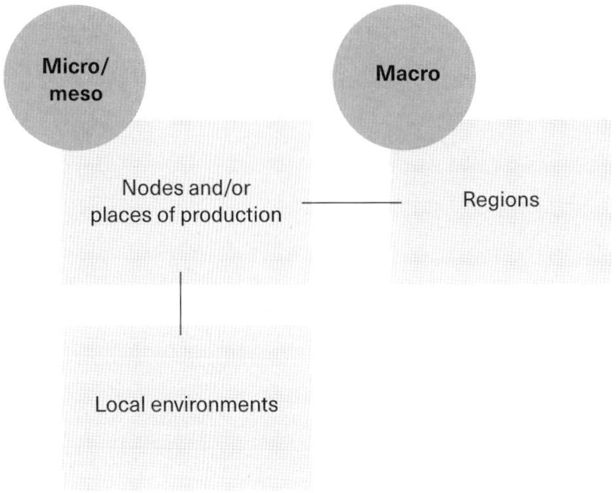

Fig. 7.3: Schematic model of interaction at various spatial levels, integrated in larger societal structures at different spatial levels (Illustration: Per Bækken).

complementary areas of gendered responsibilities or shared work domains.

Dress accessories, being developed from the late 8th into the 10th century and produced in large numbers in the Scandinavian Viking Age towns and to some degree locally, may indirectly indicate such wider networks. When artefacts of Insular or other foreign origin appear in female graves, they reflect both the male donor and the female donee. Penannular brooches of Insular origin, both of plain copper alloy and finer tin and silver coated, have been interpreted as symbols of possible exogamous marriages,[10] although not unambiguous as for meaning. Within the study area, foreign artefacts appear more often in female graves with textile tools than in female graves generally and overall, with strong affiliation to the elite. Indirectly, they also reflect an elite economy organised within wider hierarchical structures and networks. Standardised types of jewellery, however, can hardly count as direct relations, since their distribution far exceeds the number of graves with textile tools and such emblems. When appearing together with other and more special grave goods, such as foreign artefacts and in special contexts – central places, junctions, and transport routes related to special production environments that form patterns of material communication and exchange – they may indicate connections and networks within both shorter and longer distances

As referred to in the introduction, this study integrates several spatial levels and larger societal structures – places, districts, regions, and types of climatic zones and at different latitudes, lowland and highland, and nodal points for communication – related to production, distribution, and access to wares in demand, and to people of similar social standing or as asymmetric relations in the hierarchical structures in the Viking Age, compressed in the schematic model shown in Figure 7.3.

Textiles, not only of utilitarian value but costly objects too, displayed wealth and prestige. Time-consuming and large-scale projects, such as the production of sails were especially demanding – as for access to and control of resources, raw materials, and a work force of diverse skills.[11] These are all relevant aspects of production above a subsistence economy. Context in a wide sense of the term also includes relations between consumers and producers, and the overall socio-political and economic component of supply and demand. The tools provide concrete evidence of production and technological capacity connected to various types of products, either for the luxurious segment or for other and wider use areas. Other kinds of non-locally produced tools and raw materials, such as of whalebone, glass, jet, amber, and other products of distant resources, may indicate status and roles in production within networks and at places such as ports for trade or exchange.

Brooches and other dress accessories, often of copper alloy but also silver, beads of glass, amber and varied minerals and gemstones and others of largely of foreign origin but produced in specialised workshops in the early towns of Ribe, Hedeby, Birka, and Kaupang, most likely spread through social and/or trading networks further on.[12] As centres of production, they imported raw materials through long distance trade – such as carnelian, amethyst, fluorspar, garnet, amber and jet, and precious metals such as gold and silver, tin, brass, copper and lead – to be made into precious dress accessories.[13] They were distributed within networks, representing common enterprises and trading networks at an early stage.

Dress accessories in the graves made of exotic materials from the Far East, the Baltic, or England, could therefore be achieved through exchange networks or other alliances also at closer distances. From western Norway, the closest of these production and proto-urban places were Kaupang and Ribe. By the 10th century, brooches and other items of copper alloy were to some degree subject to domestic production too in special environments but may still reveal far-reaching connections and relations. The same type of brooches as found in graves from western Norway, appeared in 10th-century Icelandic graves too, where 23 of all the gender-identified female graves had brooches and 16 without.[14] Such artefacts could, then, connect to earlier established internal networks related to western Norway. They would involve both men and women as organisers and producers for providing boats, sails, crew, clothes, covers, weapons, and other provisions. Similar types of dress accessories, special textiles as parts of the funeral clothes, tools, and other symbolic grave goods, exposed at the funeral, either as gifts, heirlooms, or other relics, signal roles and positions and indirectly familial relations and lineages, and above all, the position of the closest heir.

Dress accessories and other artefacts thus reflect both social standing and social relations: marital, familial, and other types of partnerships. When they include foreign artefacts from Britain, Ireland, or of Frankish origin, they may, as noted, represent gifts from spouses, acquired through raiding, trading, or exchange networks, or other types of alliances. A recent study by the archaeologist Zanette Glørstad has interpreted the special Insular artefacts of penannular brooches of mainly Irish origin in female graves, as signs of rank and relations to foreign elites in the new colonies to the west and as associated with elites and milieus within superregional networks. They may indicate kinship or other alliances, including exogamous marriages.[15] However, reworked ecclesiastical artefacts of foreign origin appearing in female graves, represent another type of gift, acquired through raiding or secondarily by trading. Insular grave finds show that 6.5% of all the Viking Age graves from the present study area contained Insular artefacts and the majority, as much as 60%, from female graves.[16] By tracing where and how Insular artefacts ended up in graves with textile tools may therefore reveal signs of such networks.

Signs of networks?

The cultural communication of technologies and codes of practice spread through personal networks or long-distance trade between special sites and regions, and as grave goods, artefacts of foreign origin may appear as distinctions within and between otherwise similar types of places.[17] Dress accessories and other personal adornment in the graves, signalling social and cultural identities, are still not uniform in material, décor, and in their various combinations. Generally though, they were connected to a certain stratum of society – the highest – and also to graves in monumental mounds or cairns and as boat graves, often only represented by preserved rivets and nails, and usually more than a hundred to be recognised as a boat.[18] By combining social emblems, textiles, and tools as integral elements in the burial customs at special places and in various environments over time, the aim is to trace possible relations and affiliation between elite milieus – how they connected spatially at lines of communications, at nodes and/or at places of production and in production environments. Not least, foreign artefacts may signal wider networks, gained as parts of local or regional enterprises, linked to the westward expansion and colonisation in the North Atlantic, which again reflect burial customs and affiliations to special districts in their 'ancestral' homeland.

Tools, textiles, and socials emblems in the early Viking Age graves

When combining tools, types of dress accessories, preserved textiles, and foreign artefacts, 12 of 42 of the graves with textile tools from the 9th century from the area of study showed traces of high-quality fabrics, woven in diamond twill (cf. Tables 6.3 and 6.4). Moreover, three of the earliest of these graves from the late 700s contained artefacts of foreign origins, such as at Sandal in Breim. The spread of various types of social emblems, such as brooches, beads, and other dress accessories of foreign origins in the graves may reveal spatial patterns and social relations – vertical or horizontal – at nodes, hubs, and places in such networks – whether stable or fluid over time. As exclusive artefacts produced in the early Viking Age towns, such as Ribe, they represented common enterprises and trading networks to achieve such artefacts but could also represent other types of exchange networks.

The earliest types of brooches from the transition from the Merovingian into the Viking period were plain undecorated thin-shelled brooches of copper alloy (R640) and a finer and ornamented type of single-shelled brooches too (R643; cf. Fig. 7.4). Eleven of the graves with textile tools comprised these early brooches, although they can be difficult to date more accurately when found without other datable artefacts.[19] Five of them comprised remains of textiles too (Table 7.1), and four had grave goods of foreign origin.

One of these graves with these plain brooches, textile tools, and textiles was the double grave at the island Tysnes at Øvre Kongsvik in Hordafylki. Two graves connected to the district Vik by the Sognefjord, and the other two graves furthest north related to Gutdal in inland/upland Stryn and furthest east in Firdafylki – a place with several graves with textile tools.

Of these, the female in the double grave at Tysnes by the coast in Hordafylki (Table 7.1, No. 1), was furnished with a pair of the plain thin-shelled brooches, supplied with a fine rectangular ornamented brooch of gilt bronze as a third brooch, a single bead of jet, and with remains of the dress by the brooches, woven in diamond twill quality II (cf. Table 6.3). About the same time, two high-status women were buried in Vik in Sygnafylki with the same type of brooches but used in different combinations. One was the prestigious boat grave at Hopperstad (No. 2), furnished with one such brooch, supplied with two small rectangular brooches of gilt bronze, and as many as 60 beads for a necklace, mostly of multicoloured glass, eight of bronze and one of amber. Here, the remains of tabby-woven fabrics therefore hardly represent the dress and rather an underdress judging by the quality. Three hooked keys moreover added to the tools for spinning and weaving. Another grave in the lower part of Vik, at Skjervum, close by (No. 13, cf. Table 7.1), contained a single plain brooch, combined with two brooches of the ornamented type R643, and an Irish reworked bronze brooch,[20] and five glass beads. Four plain arm rings of gilt copper alloy were found separately. Three of these early graves related to Gutdal in Oppstryn in the northern and inland part of Firdafylki, with close connection to the county of Oppland and as noted an important area for iron production. Two of the three Gutdal graves with the plain brooches, both cremations, were located rather close to each other, one in a mound, the other as a flat grave. Of these, the grave in the mound was the more prestigious, furnished with glass beads and rock crystal and a key, besides the textile tools. This mound contained a high-status male grave with weapons too and remains of various artefacts of iron, among others, a cauldron and a knife, besides some unidentified objects and a hone.[21]

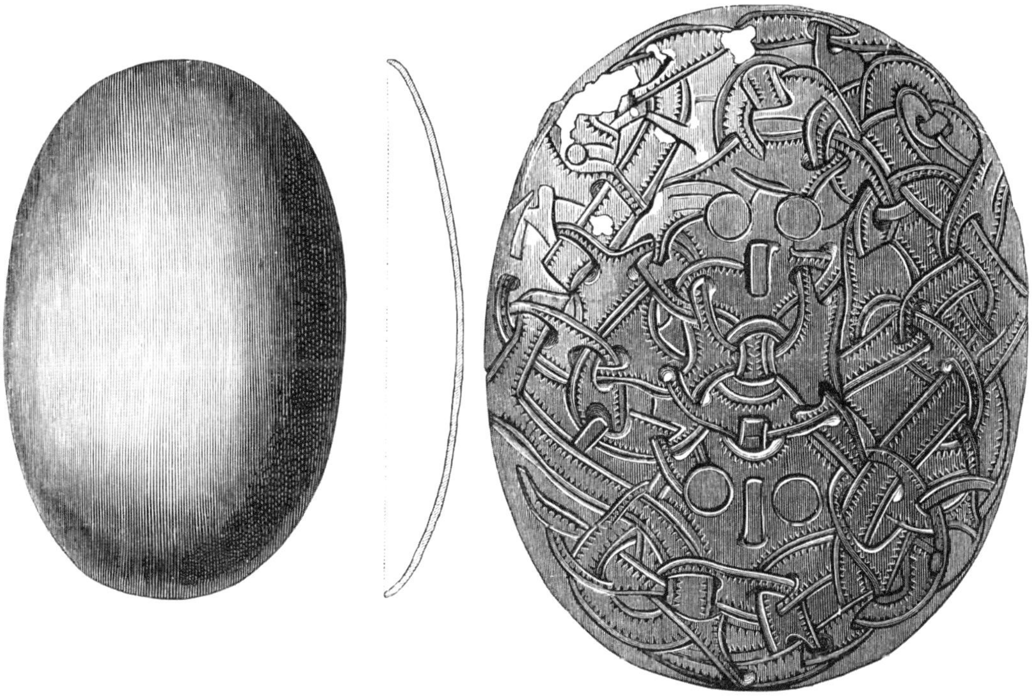

Fig. 7.4: Brooches of type R640 (left); R643 (right), dated to c. *750–800.*

The brooches in two of the Gutdal graves differ from all the others, being made of iron, and may possibly represent underparts to support the thin-shelled brooches of R643.[22] The brooches in the third female grave at Gutdal, were made of copper alloy and with the same shape as type R643, but without ornaments. This was a chamber grave in a small mound and again shared the space with a weapon grave. In contrast to the other graves with tools for three and four work processes, from shearing, combing, spinning to weaving, this grave contained only a single loom weight, remains of wool combs, and a casket.

Compared with the five other early graves with textile tools and with the ornamented thin-shelled brooches (type R643), they echo the same spatial pattern, stretching from the coast of Hordafylki to in fjord inland Stryn furthest north-east. One was again a coastal grave to the south in Hordafylki and a prominent female grave at Fosse, strategically located by the main inner sailing route (see Table 7.2). The dress accessories comprised two brooches of the kind, nearly 60 beads for a necklace – 20 monochrome glass, five red glass-metal, seven foliated, 16 mosaic, one melon shaped, and one of bronze – two plain bronze armlets, and a circular gilt reworked bronze mounting of Insular origin. The female grave at Skjervum in Vik contained two of these brooches, in this case, R640 and R643 combined, and again an Insular reworked brooch. The third grave was the prestigious female grave at upland Sandal, presented in the previous chapters. Here, the woman was furnished with two brooches of type R643 and some 80 glass beads of different shape, size, and material: one silver coated glass, 10 gold coated glass, three blue, and one green, and 66 blue micro beads and two of amethyst. It also contained some exquisite and rare foreign artefacts of both Frankish and Insular origin – a small glass vessel and an enamel decorated bronze mounting and, as we have seen, remains of the funeral dress of broken diamond twill quality I (cf. Chapters 5 and 6). The fourth grave of this kind connected to the central place Myklebust (No. 14) at Eid by the Nordfjord. This was one of the four graves in the large mound and where the woman was placed outside the chest of the male in full armament (cf. Fig. 5.19). The woman, furnished with a thin-shelled plain brooch with unclear traces of ornamentation of an early type and with textile remains. It contained moreover a bead, parts of a casket, and hackles and a weaving beater as textile tools[23] (cf. Chapter 6, Table 6.2, No. 14). The last of these early graves connected to an upland area in the western part of the Stryn district at Langeset along the land route and in the border zone and between Stryn and Hornindal. This was a boat grave, judging by the many rivets over an area of 5 m.[24] Here, the deceased had one brooch only, however, supplied with 32 beads for a necklace, 16 monochrome of different shape, green, blue, and yellow and two gold coated, and the rest mosaic. Besides tools for combing and weaving and food related utensils, there was a long-shafted ladle and a frying pant, among others.

Two of these five graves contained the thin-shelled brooches, R643: the prestigious graves at coastal Fosse in Hordafylki and at upland Sandal in Breim in Firdafylki distinguish themselves with their multiple tools, dress

Table 7.1: Graves with textile tools and special types of brooches from the 9th century (R = Rygh (1885); JP = Petersen 1928). Graves with textiles are marked in bold, and numbers refer to numbers in Tables 6.1 and 6.2 in Chapter 6).

R 640 (=JP3)	*R 643 (=JP4)*	*R644 (=JP15), Berdal type, group D*	*JP Fig. 11 & 12, Berdal type, group A*	*R648*	*R647 (=P51)*	*R649*	*R657 (=JP42)*	*Rectangular brooches*	*Equal armed brooches*
***c.* 750– 800**	***c.* 750– 800**	**800–850**	**800–850**	**800–850**	**800–850**	**800–850**	**850–900**	***c.* 750–800**	**800–850**
No. 1 Kongsvik Tysnes	Fosse, Meland	Vatne/Votno, Alversund	Myklebust III	Fosnes, Stryn	Seim, Granvin	Svingeset Stryn	**No. 7 Vinjum Aurland**	**No. 1 Kongsvik, Tysnes**	Myklebust Eid (R661)
No. 11 Hopperstad, Vik	**No. 13 Skjervum, Vik**	Gimmeland Fana	Mindre sunde, Stryn	Maurset Hornindal	Stalheim. Vossestrand	Lunde, Voss	**No. 8 Nes, Sogndal**	**No. 11 Hopper stad, Vik**	**No. 4 Naustdal, Førde (R661)**
No. 13 SkjervumVik	**No. 2 Sandal, Breim**	1 Hauge, Lærdal	**No. 4 Naustdal Førde**	Hauge, Lærdal	Hauge, Lærdal	Brekke, Lavik	**No. 9 Grodås Hornindal**	Vatne/Votno, Lindås	Maurset, Hornindal (R660)
Gutdal, Stryn (iron)	**No. 14 Myklebust Eid**	Tveiti, Vangsnes		Hodneland, Lindås	Terum, Aurland		**No. 17 Sandal, Jølster**		Hodneland Lindås
Gutdal, Stryn (iron)	Langeset, Stryn	Berdal, Feios			**No. 3 Bø, Breim**[1]				Gutdal, Stryn
Gutdal, Stryn, copper alloy		**No. 6 Nes, Sogndal,** Type C			Sårheim, Breim				
					Eide, Gloppen				
					Myklebust III Eid				
					Langenes Selje				
					Beinnes, Innvik				
					Vik, Nedstryn				
					Folva, Oppstryn				
					Gutdal, Oppstryn				
6	5	6	3	4	13 (11)	3	4	3	5

1 Slomann 1986, 26–28, a mixture of R647 and the older Oseberg style, R650, dated to *c.* 800.

Table 7.2: Representation of foreign artefacts in graves with textile tools from c. 800 +/- (N:5) Bold: with textiles. H = Hordafylki, S = Sygnafylki, F = Firdafylki.

No.	*Place*	*County*	*Brooches, types*	*Other accessories*	*Foreign artefacts*	*Gender/Status*
B12012	Fosse, Meland	H	2 R643	2 bronze armlets (R709; JP191) 58 beads	Mounting, secondary use as brooch	FIII
B6500	**Skjervum, Vik**	S	2 R643, 1 R640	4 arm rings of copper alloy, 5 glass beads: 4 white, 1 blue/green	Irish 3rd brooch: circular ornamental plaque of gilt bronze	FIII
B9014	**Sandal, Breim**	F	2 R643	80 beads, some silver- and gilt coated, amethyst	Glass vessel, enamel decorated bronze mounting of Insular origin	FIII
B5807 I	**Myklebust Eid**	F	1 R643	1 bead, small multichrome	–	MIII/FII
B6035/B6036	Langeset, Stryn	F	1 R643	32 glass beads, two gold foliated, some mosaic	–	FII

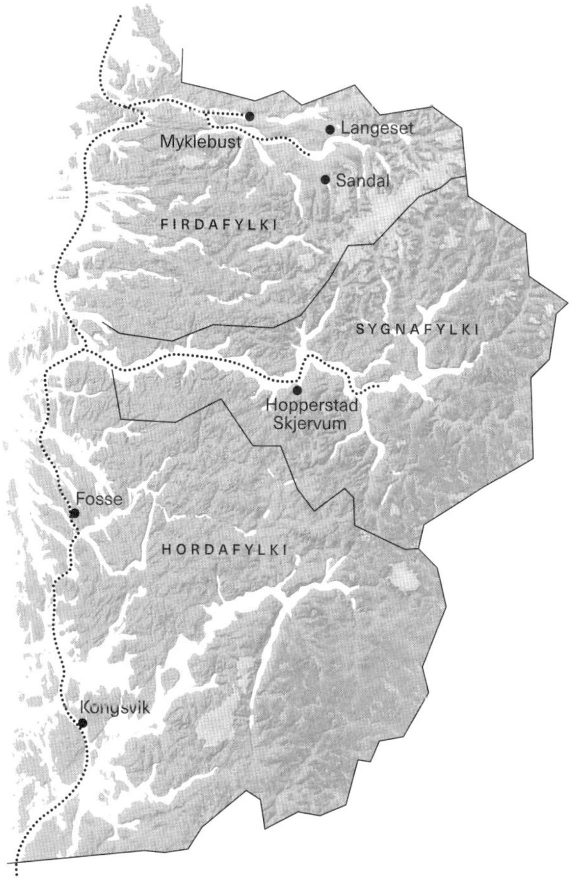

Fig. 7.5: Graves with textile tools and the early brooches of types R640 and R643 at places and along seabound routes for communication (Illustration: Per Bækken).

accessories, most likely produced in Ribe in this early phase and distributed through networks, besides other foreign artefacts, such as glass vessels, bronze cauldrons, and dress accessories of reworked Insular artefacts (Table 7.2). Three of these five early graves connected to sailing routes – one at the coast at Fosse by the inner sailing route in Hordafylki and with seabound connections to central places further north, mid-fjord by the long Sognefjord and Nordfjord, at, respectively, Vik and as noted at Myklebust at Eid. The inland and upland grave at Sandal in Breim and the spread of the graves thus largely repeats the same geographical and social pattern as for the graves with the plain brooch type (Fig. 7.5).

Of these, the grave goods in the graves at Gutdal and Langeset in inland Stryn, as also at Myklebust, were simpler than those in the coastal grave at Fosse and the graves in Vik and Breim with more differentiated tools and artefacts of foreign origin too. They may thus indicate places along routes that served as traffic arteries – routes that had become routinised, to use Sindbæk's terminology, to characterise how networks were established along functional and commonly and repeatedly used routes (Fig. 7.5).[25]

Another type of brooch, the so-called Berdal type from the early 9th century, connected to nine of the graves with textile tools. These brooches, however, are more heterogeneous with several variants.[26] In seven of the nine graves with textile tools and Berdal type brooches, the brooches are of type R644, and Berdal type D or JP 14–15, which appears as an early West Norse type and relatively few representations as grave goods from the area of study (Fig. 7.6). Of these, one grave related to an inland/upland area in the mid-zone of Hordafylki, and four each to Sygnafylki and Firdafylki (Table 7.3).

All four graves in Sygnafylki with variants of the Berdal brooches related to the inner and mid-zones, and three combined with foreign artefacts, again demonstrating westbound connections. Two of these graves at the margins of Vik mid-fjord in Sogn, respectively Tveiti, and the other at the very place Berdal giving its name to this type

Fig. 7.6: The Berdal brooch from the grave at Mindresunde belonged to group A, and a rather small group with this variant (JP 11) (Petersen 1928).

of brooches. This was boat burial in a large cairn in an upland area, *c.* 200 m.a.s.l. with two such brooches and a cruciform brooch of Irish origin and remains of a mounting of gilt bronze, thus demonstrating westward connections. As for tools, however, it only contained a spindle whorl and some loom weights. The grave at Tveiti located by the fjord arm heading into Vik, was less prestigious, containing only a single spindle whorl as for grave goods. The third grave in contrast, was one of the high-status graves at the headland Nes in Sogndal and with another variant of the brooches, supplied with a penannular brooch in gilt silver of Irish origin,[27] two bronze armlets without parallels within the study area, and 134 glass beads of various shapes and colours for a necklace. The remains of the funeral dress (No. 7; 6 cf. Chapter 6), woven in diamond twill of quality II further underline high rank. The fourth grave, a double grave, was located at Hauge, mid-valley in the long Lærdal valley in the inner zone and a central and strategic place for communication and with access to vast upland areas. Besides the two brooches, one of the Berdal type and the other of the common type, R647, it also comprised a penannular brooch of Insular origin,[28] supplied with a few beads.

In Firdafylki, the northernmost grave with Berdal brooches, related to Stryn, at Mindresunde, connecting to the narrow strait by the main river from the large Lake Oppstrynvatn (Fig. 7.6). Besides the brooches, the deceased was supplied with about 50 glass beads of different shape and colours, some gilt, and mosaic, and with a special stone pendant, and a penannular brooch of Insular origin.[29] In the grave at Myklebust at Eid in the mid-zone, in the same large mound at Myklebust as the earlier 'double grave' (cf. Fig. 5.19), the deceased woman before she was cremated was furnished with two types of brooches, with a similar representation of brooch types as the grave in Lærdal, plus a special horse-shaped brooch as a third brooch, without parallels in the present corpus, and a cauldron of copper alloy of foreign origin.[30] The third grave in Firdafylki with Berdal brooches, however, as a single brooch and of another variant, related to an upland area at Bjørnarheim in Breim, and was only supplied with tools for spinning and weaving. In contrast, the fourth grave at the central place Naustdal in the mid-zone to the south, contained two brooches of yet another variant – a bronze armlet and an equal-armed brooch,[31] however, with only a spindle whorl for tools.

The only grave with Berdal brooches from Hordafylki connected to an upland area at Gimmeland in the mid-zone in the district Fana, and with two such brooches and in this case with a penannular brooch of Irish origin. It may possibly connect to the other and more prestigious grave at the central place Fana, however, without brooches but a silver chain and with exquisite grave goods of foreign origins (cf. Chapter 5).

These nine graves with Berdal brooches or in combination with other brooches connected, then, to graves in different landscapes and environments – three to central places that later became medieval church centres – Hauge in inner Sygnafylki, Naustdal in the southern part of Firdafylki, and Myklebust at Eid further north – and related to other strategic places for access and control too – such as Nes by the Sognefjord and Mindresunde by the narrow strait in Stryn. Three of the graves related to upland areas with vast resources – Bjørnarheim in Breim, Berdal in Sygnafylki, and Gimmeland in Fana in Hordafylki.

As many as six of these nine graves reveal westbound connections with their artefacts of Insular origin, and three of them with traces of textiles. One was the grave in the large Myklebust mound, with varied twill qualities preserved by the brooches, while the graves at Nes and Naustdal had traces of diamond twill, of qualities I and II respectively (cf. Chapter 6).

Two other types of contemporary brooches, and rather alike, R647 and R649 (Fig. 7.7), relate to altogether 12 graves with textile tools and sometimes together with Berdal brooches and represented in more than half of all the 9th-century graves from the present study area. Of these, 10 graves connected to Firdafylki, two to Sygnafylki – Aurland in the inner zone and at Brekke by the inlet to the Sognefjord, and in three graves in inland Hordafylki – Voss and Granvin (Tables 7.4 and 7.5).

Table 7.3: Graves with textile tools from the early 9th century with special types of Berdal brooches (N:9) and foreign artefacts (N:6); in bold: with textiles. H = Hordafylki, S = Sygnafylki, F = Firdafylki.

No.	Place/District	County	Brooches, types	Other accessories	Foreign artefacts	Status
B594, 596–603	Berdal, Feios	S	2 R644	Insular (Hiberno–Saxon) cruciform mounting of gilt bronze, used as a 3rd brooch	x	FIII
B5702	Tveiti, Vik	S	2 R644	–	–	FIII
B9065	**Nes, Sogndal**	S	2 Berdal brooches JP13, Group C	Irish penannular brooch in gilt silver (Glørstad Type 1C), 134 glass beads of different colours, the majority blue in different shades, 40 yellow, 7 green and some white	x	FIII
B2160	Hauge, Lærdal	S	1 Berdal Type D/1 R648	Penannular brooch, 2 glass beads	x	MII/ FIII
B4505	Mindresunde, Stryn	F	2 R644 Berdal type A	Irish penannular brooch with wide terminals with a piece of amber or red stone	x	FIII
B5807 III	Myklebust, Eid	F	1 R647 1 R644	1 third horse-shaped brooch	Bronze cauldron	FIII
B9502	Bjørnarheim, Breim, Gloppen	F	1 JP16	1 bead, mosaic	–	FI–II
B9626	**Naustdal, Førde**	F	2 JP12	1 equal armed brooch R661	–	FIII
B5884	Gimmeland, Fana	H	2 R644	1 ringed pin	x	FIII
∑ 9	H: 1 S: 4 F: 4				6	

Although the graves with these brooches were spread over wide distances in different environments, they share other common traits too, not least foreign artefacts that may indicate relations within closer or looser networks (see Table 7.4). This may especially be the case in Firdafylki. In Stryn, the four graves with R647 brooches, all had foreign artefacts. One was the prestigious double grave at Vik, by the fjord and the main entrance to the Stryn valley – a boat grave, also marked above ground as a ship-shaped stone setting. Supplied, amongst others, a spearhead of Carolingian origin,[32] it indicates foreign relations. While the textile tools were few, it contained a whole range of other tools for forging and carpentry, like graves from Oppstryn (cf. Chapter 5).

Of these, the Gutdal grave was the most prestigious, equipped with equal-armed brooches and 13 beads of glass, jet, amber, some silver-coated, and two rare small earrings of gold or electrum – artefacts that point both west and eastwards. The dress accessories in the neighbouring Folva grave with a reworked brooch from an Irish mounting of gilt bronze points westwards.[33] The fourth grave in the wider Stryn area with these brooches connected to its fringes, at Beinnes in Innvik by the fjord and with a ring-headed pin of bronze too but more uncertain as for origin. The last of these graves in the northern part of the county with these common brooches, were located at each end of the Stryn valley and at each end of the long Nordfjord and at a strategic place Langenes for seawards communication, facing the North Sea.

In the community of Gloppen to the south, the three graves with R647 brooches related to the main agricultural landscapes. One was, the earlier presented double grave and boat grave at the central place Bø in Breim,[34] with three brooches of a special mixture of styles R647/R650-type, indicating a somewhat earlier date to the early 800s,[35] and with remains of textiles of diamond twill preserved by the brooches. It also contained a pair of bronze scales, pointing westwards, possibly used for weighing hack silver when sharing loot, as suggested by Wamers.[36] The high-status female grave at Sårheim close by, was represented by R647 brooches too, and with a stray find at the place, a silver thistle-shaped brooch of Irish origin,[37] which again indicates westbound connections. Another grave with a R647 was combined with a R649 brooch and was a double grave located on a tongue of land at Eide, close to the Gloppenfjord, however, without textile tools but tools for trade – a balance scale of Insular origin and tools for wood working and forging, as found in the graves in Breim and inland

Fig. 7.7: Brooches of types R647 (to the left) and R649 from the first part of the 9th century (Petersen 1928).

Oppstryn. This was an early place of power, judging by its at least four Viking Age weapon graves, of which three were boat graves.

The grave in Sygnafylki with this common type of brooch in one of the graves at Terum in to Aurland, connects, as noted to an area with several graves with textile tools, and with remains of high-quality textiles (cf. Chapter 6, No. 5). The two graves from Hordafylki with these brooches relate to rather contrasting environments, one in an upland place in Voss at Stalheim in Hordafylki, bordering Sygnafylki and the community of Vik. One of the two graves from Hordafylki contained these types of brooches from an upland area at Stalheim on the border between Hordafylki and Sygnafylki and included a penannular brooch too.[38]

The other, a high-status grave at Seim from the central agricultural area in Granvin in the Hardanger region[39] to the south, was richly equipped with tools for both textile production and food processing, but without traces of foreign connections. Still, a whalebone plaque[40] may indicate wider relations to the north, possibly distributed within wider networks too, including people in the close neighbourhood at Folkedal, judging by the grave goods.

Altogether, five of the 12 graves with textile tools and the common R647 brooches contained foreign artefacts but relatively fewer than in the graves with Berdal brooches. This spread of graves with special dress accessories and common types of brooches, foreign goods, and overwhelmingly at central locations as for communications and production does not seem accidental. It largely repeats the same spatial pattern as indicated by the other contemporary graves with other types of brooches, social attributes, and foreign artefacts. They connected to the coast, in-fjord areas, as well as lowland and inland/upland areas – often at strategic places and nodal points by land and sea routes, and from inland production areas to the outer coast and with the largest representation in Firdafylki.

It is moreover noteworthy that the earlier mentioned high-status female grave at Westness in Orkney was furnished with the R647 brooches, multicoloured beads for a necklace, and an older Celtic ring brooch of silver from the 8th century, besides a cauldron of copper alloy, as well as a mounting to a girdle of bronze.[41] The Orcadian grave largely matches, then, the graves from western Norway with these brooches as for status, tools, and grave goods.

The three graves with the R649 brooches, one from each county, are too few to reveal any pattern. They relate to rather different landscapes; an upland grave in Stryn at Svingeset at the border to Hornindal, and another at Brekke by the entrance of the Sognefjord but from an uncertain context. Of these, only the double grave at Lunde in the central Voss area,[42] contained a foreign artefact – a third

Table 7.4. Graves with tools and textiles and brooches of type R647.

No.	*Place/District*	*County*	*Brooches*	*Other dress accessories*	*Foreign artefacts*	*Gender*/Status
B8635/B9091	Seim, Granvin	H	2 R647	–	–	FIII
B4736	Stalheim. Vossestrand	H	1 R647	2 bronze armlets (R721)	–	F II–III
B9063	Terum, Aurland	S	1 R647	1armlet of bronze (Fig. JP 186–88), 6 beads	–	FII–III
B9501	Bø, Breim	F	3 R647/ 650	6 beads, 4 of amber, 2 of glass	A pair of bronze scales	M III/FIII
B10975	Sårheim, Breim	F		18 beads, uncoloured glass, mosaic	Insular silver thistle brooch (B6675: Stray find)	FIII
B14583	Eide, Gloppen	F	2 R647		–	FIII
B5807*	Myklebust III, Eid	F	2 R647+ R644	1 horse-shaped brooch	1 bronze cauldron (R725)	FIII
B4756	Vik, Nedstryn	F	2 R647	Ringed pin, 1 bead of green glass-metal	Spearhead, Carolingian style	MIII/ FIII
B5030	Gutdal, Oppstryn	F	3 R647	2 equal-armed brooches (JP fig. 83), 13 beads: 11 glass (blue, green, 1 silver coated and one gold-foliated, and 1 amber and 1 jet), 2 gold earrings	–	FIII
B4752	Folva, Oppstryn	F	2 R647	Irish brooch, 3 small blue beads	Decorated Hiberno–Saxon buckle sheet-bronze buckle-plate	FIII
B7184	Beinnes, Innvik, Stryn	F	2 R647	Ring brooch	Insular	FIII
B5806	Langenes, Selje	F	1 R647	20 beads: 11glass, blue and green, 3 mosaic, I silver- & 1 gold coated, 1 amber, 1 jet	–	FII
∑ 12			3 with other brooches	9 with more than brooches	6 graves	

Table 7.5: Graves with textile tools and brooches of R649 and foreign artefacts (N:3).

No.	*Place*	*Brooch*	*Other*	*Foreign goods*	*Gender/ Status*
B6483	Svingeset, Stryn	2 R649, variant F	57 beads, mostly multi-coloured glass beads	1 miniature bronze axe (Irish parallel)	FIII
B6062-6-0 63	Brekke, Lavik	1 R649		–	FI–II
B2233	Lunde, Voss	2 R649	1 third round brooch (Hiberno Saxon), 3 beads	x	MIII/ FIII

brooch, reworked from a bronze mounting of Anglo-Saxon workmanship (Table 7.5).

The majority of these 15 graves with brooches are of types R647–649, 11 connected to Firdafylki and four each to Sygnafylki and Hordafylki. Of these, three were located on the coast, four in upland areas – the others in central agricultural areas, of which seven connected to strategic places in Stryn, Breim, Gloppen, Sogndal and Lærdal. More than half of them contained foreign artefacts and many of them connected to upland areas, two in

Table 7.6: Graves with textile tools from the early 9th century with brooches, type R648 (N:4).

No.	Place/District	County	Brooches	Other dress accessories	Gender/Status
B4704	Hodneland, Lindås	Hordafylki	2 R648	1 equal armed (JP fig 59). R658	FIII
B2151*|	Hauge, Lærdal	Sygnafylki	1 R648 + 1Berdal, type D	2 beads of glass, 1 penannular brooch	MII/FIII
B7524, B7611	Maurset, Hornindal	Firdafylki	2 R648	1 equal armed brooch R660, 57 beads + 3 (glass, mosaic, jet and amber	FIII
B12695	Fosnes, Stryn	Firdafylki	1 R648	1 armlet of bronze R721	FII

Fig. 7.8: Brooch of type R657 from the late 9th century (Rygh 1885).

Stryn, and two in Hordafylki at the border to Sygnafylki (Tables 7.3, 7.4, 7.5).

When adding four of the female graves with textile tools that contained the less common type (R648), otherwise only found in five graves from the study area,[43] their spread does not look accidental. This was an overall Norwegian type from the first part of the 9th century only are represented in high-status graves with textile tools as grave goods (Table 7.6). Although few, they were widely spread and connected to graves at Hauge in Lærdal, Maurset in inland Hornindal on the fringes of Stryn, at Fosnes by the large lake in Oppstryn, and coastal grave at Hodneland by the inner and sheltered coastal sailing route, and in the same area as the other earlier coastal graves in Hordafylki. The graves then relate to the same areas as represented with this type of grave from the very beginning of the 9th century. Except one, they lacked foreign goods in contrast to the other 15 graves with other types of brooches. The grave at Hauge in Lærdal, as noted, being represented by both a Berdal brooch and an R648 type, contained a penannular brooch of foreign origin (cf. Table 7.3).

Still, there are other corresponding similarities between all these 19 graves, when including these four graves (Tables 7.4–7.6). Two of them, one by the coast and the inland grave at Maurset contained equal-armed brooches as third brooches, although not identical, and the Maurset grave equipped with beads for a necklace of jet, among others, and with equal-armed brooches as in the contemporary graves at Gutdal, Myklebust, and Naustdal in the same county (cf. Table 7.1). The tools in the Maurset grave with a needle-case of bronze with iron needles, a rather rare find in these grave contexts, had its nearest parallel from an upland grave on the other side of the border to Stryn, and in this case furnished with brooches of R649. The graves with brooches of type R648, although few, repeat the spatial pattern as for the other types of brooches, but generally with fewer tools. Altogether, they add up to 19 graves related to the first part of the 9th century.

Only four graves with similar brooches date specifically to the latter part of the 9th century with a more rare and exclusive type, R657. The type has been characterised as the most beautiful of all Viking Age bronze brooches with inlaid silver threads and knobs of highly skilled workmanship (Fig. 7.8),[44] and all related to prominent graves and with a limited distribution. These brooches, although few, all connected to graves with textile tools located at central places in the inner zone in Sygnafylki and Firdafylki (Table 7.7).

The prestigious Vinjum grave in Aurland was then one of four such graves, however, it distinguishes itself with loose inlaid silver-coated knobs also along its sides and ends and framed by silver threads and with lines of ridged silver alloy beneath. Three silver threads form a frame and with plates in the same material and with the animal heads at all the flat parts and beneath the knobs. According to Petersen, they appear as the finest type of workmanship and with distinct animal figures. Besides the

Table 7.7: Graves with textile tools and brooches of R657 dress accessories (N:4) In bold: with textiles of diamond twill qualities.

No.	Place/District	County	Brooch type	Other dress accessories	Foreign goods	Gender/Status
B7731	**Vinjum, Aurland**	Sygnafylki	2 R657	four-sided silver brooch Irish, 37 beads, + silver beads, 2 thin bronze armlets.	mounting for a drinking horn (Irish), 1 cup and lid of bronze (Hiberno–Saxon art)	FIII
B9064	**Nes, Sogndal**	Sygnafylki	1 R657	–	–	FIII
B8000	**Grodås, Hornindal**	Firdafylki	2 R657	–	–	FIII
B11413	Sandal, Jølster	Firdafylki	2 R657	silver chain	–	FIII

two brooches, a special brooch of Irish origin with quadrangular gilt silver ornament, and with a piece of amber, served as a third brooch. It also contained nearly 40 beads, 1 large of amber, 15 of glass mosaic of different shapes, 21 monochrome glass of different colours, red, white, blue, and yellow, and some silver coated. Two thin armlets of copper alloy added to her jewellery. The grave included treasures of Irish origin – a mounting for a drinking horn and a cup and lid of bronze (Fig. 7.9).[45]

The other graves with these brooches were from Nes in Sogndal and Grodås in Hornindal, all three with remains of high-quality textiles of diamond twill. The fourth grave, at Sandal in Jølster, included a rare silver chain. Like the grave at Vinjum, this was a boat burial and similarly equipped with all-inclusive tools (cf. Chapters 5 and 6), however, the Sandal grave in Jølster had a weaving beater of whalebone. Still, of these four graves, only the Vinjum grave comprised foreign grave goods (Table 7.6).

When summing up, 42 graves with textile tools and various types of brooches stem from the turn of the 9th century into the latter part of the century. They generally represent women from the upper echelons of society and connect to varied landscapes from coast to inland, both central agricultural and inland upland areas – some also co-located with male weapon graves. Frequently, they relate to strategic places for communications and control – at crossroads between land routes and waterways, by larger lakes, fjords, and harbours, connecting to the sea. Of these graves, 17 (40%) also comprised foreign grave goods, and altogether is thu a significant correlation (Table 7.8).

The graves with these grave goods then largely connected to the same areas and routes throughout the century, from the coast of Hordafylki to the south, in-fjord and inland in Sygnafylki and Firdafylki. The representation of Insular artefacts indicates crossings over the North Sea, as common enterprises on local or regional levels.

Foreign artefacts connected largely to high-status graves at central places but also in upland and inland areas – and some of the most prominent at Sandal in upland Breim by the late 700s and about a hundred years later, in the monumental grave at Vinjum at the entrance of the Aurland valley. The combination of high-status female graves with the same types of brooches, textile tools, grave goods of foreign origin, and with an exceptionally high representation in graves with textile tools, may indicate relations and probably networks, related to the westward expansion, primarily raiding in the early phase, and trading too in the latter part. The graves thus indicate a pattern, where the earliest graves from around AD 800 or possibly some decades earlier, furnished with grave goods of foreign origin, appeared by the coast of Hordafylki, and in high-status female graves in Sygnafylki and Firdafylki, in respectively Vik and Breim, and inland/upland Stryn, and some of them with textile remains too, such as in the grave at Sandal (Fig. 7.10). At the same time, other graves with common types of brooches and textile tools appeared along the coast, at central places and in upland areas from Hordafylki, and the inner zones of Sygnafylki and Firdafylki. As for status and location, the female grave in the large Myklebust mound stands out as a nodal point in the outer part of the fjord system in Firdafylki, connecting to both land routes and waterways (cf. Fig. 7.5).

The graves reveal, moreover, a remarkable correlation with females buried in special high-quality garments, woven in diamond twill, represented in 11 of 12 graves. Five of these graves contained foreign artefacts, which may indicate special inter-connected milieus, related to textile production of an advanced or specialised character, as well as foreign connections that may indicate joint enterprises. The graves with the most varied textile tools for three and four work processes connected to graves with fabrics of quality I, and only the Vinjum grave with preserved textiles of diamond twill quality III (Table 7.9) – all, then, seemingly places of advanced production or even attached specialist production.

Seeing the prestigious artefacts as socially emblematic distinctions and tools as indicators of administrative roles within production, it is tempting to interpret the specific and common dress accessories as indicators of alliances within closer or looser networks within the elite and with textile tools as a common denominator for the female component. However, the number of tools for the various processes – with six of the 10 graves with tools for more than two

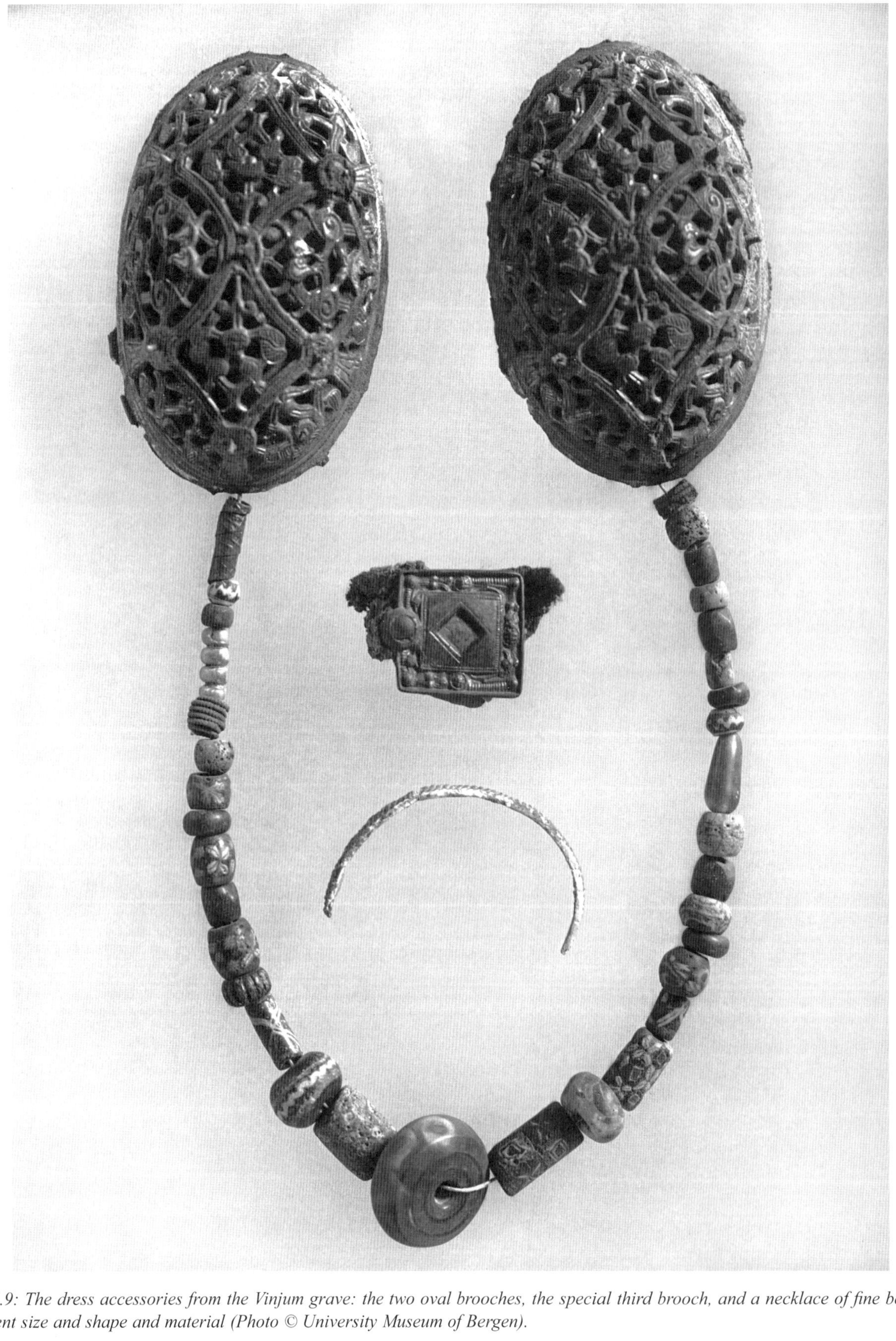

Fig. 7.9: The dress accessories from the Vinjum grave: the two oval brooches, the special third brooch, and a necklace of fine beads of different size and shape and material (Photo © University Museum of Bergen).

Table 7.8: Graves from the late 700s and the 9th century furnished with oval brooches and artefacts of foreign origins; with textiles (N:17).

Museum no.	*Place/District*	*County*	*Brooches*	*Various artefacts of foreign origins*
B12012	Fosse, Meland	H	R643	circular gilt reworked bronze mounting
B6500	Skjervum, Vik	S	R643/R640	circular ornamental plaque of gilt bronze
B9014	Sandal, Breim	F	R643	glass vessel, enamel decorated bronze mounting of Insular origins
B3308	Vatne/Votno, Alversund	H	Berdal variant D	Irish rectangular bronze brooch
B5884	Gimmeland, Fana	H	Berdal: R644	1 penannular brooch
B2233	Lunde, Voss	H	R649	1 third round brooch (Irish)
B594, 596–603	Berdal, Feios	S	Berdal: R644	Insular cruciform mounting of gilt bronze used as a 3rd brooch
B5807	Myklebust III Eid	F	Berdal R644/ R647	1 bronze cauldron
B4756	Vik, Nedstryn	F	R647	ring brooch
B7184	Beinnes, Innvik	F	R647	1 penannular brooch
B4505	Mindresunde, Stryn	F	Berdal Variant A	1 Irish penannular brooch
B4752	Folva, Oppstryn	F	R647	1 Irish 3rd brooch
B5030	Gutdal, Oppstryn	F	R647/ Berdal variant D	penannular brooch, gold earrings
B6483	Svingeset, Stryn	F	R649	miniature axe, Irish?
B9065	Nes, Sogndal	S	Berdal brooches variant C	Irish penannular brooch in gilt silver
B8669	Tokvam I, Aurland	S	R650	cruciform Irish brooch
B7731	Vinjum, Aurland	S	R657	Irish cross shaped mounting of gilt bronze used as brooch, quadrangular silver and drinking horn

processes – is to a lesser extent reflected in the quality of the preserved textiles, except for the Vinjum grave.

Graves without the traditional oval brooches: signs of lower status?

Not all the women buried with textile tools, however, had oval brooches but could be equipped with other types of dress accessories, social emblems, and other grave goods. The graves, grave goods and location may still indicate high status. At least 16 other graves, from the first part of the 9th century, or within a broader time framework, belong to this category of social emblems but of varied status (Table 7.10). Many of them came from the same areas as graves distinguished and identified by the brooches and other social emblems. At least four, and 25% of these 16 graves contained grave goods of foreign origin. The graves could be cremations or inhumation burials, primary and secondary graves in cairns or mounds, but also flat graves and in three cases, as boat graves. All these graves count as female being without weapons. The accessories vary – the simplest, represented by a few beads, while in others contain up to nearly 50 beads, which indicate armlets or necklaces. In a few cases, they contain chains, pendants, and armlets of silver or bronze, and other types of brooches, and some without parallels in the present corpus (Table 7.9). Five of these 16 graves connected to Firdafylki, also represented by the highest number of graves with textile tools and the traditional types of oval brooches. Of these, three related to Stryn in the northern part of the county, one to Eid in the mid-zone and one in its southern part. Six graves within this category connected to Sygnafylki – at Brekke, Sogndal, and Aurland, districts also represented by other contemporary graves with textile tools, and four of them with brooches of other types than the oval types. Five graves in this category related to Hordafylki, from the coast to the inner zones, and two from Voss.

The textile tools in these graves are generally few, mostly for one or two work processes, most frequently for spinning and weaving, and only three graves with tools for more than two work stages: the early boat grave at Folkedal in Hardanger with tools for five work stages, and one of the two graves in a mound in Aurland, with tools for four processes. As noted, four of these 16 graves also contained foreign artefacts and the richest assemblage in the large mound in an upland valley in Fana, besides three other graves with brooches of Insular origin. Compared with the 42 graves with the traditional oval types of brooches where 40% of them also contained foreign artefacts (cf. Table 7.8), these graves contained then a relatively smaller share.

Fig. 7.10: Tentative reconstruction of the dress, dress accessories, portable textile tools in the Sandal grave, also supplied with foreign artefacts (Illustration by Wamers 1991, with permission).

Why graves without the oval brooches connected to the Norse dress were equipped with foreign goods and female-related artefacts has been an issue of discussion.[46] One interpretation is that they may represent foreign women in exogamic marriage alliances, or other types of relations, although not necessarily of lower status when buried with other costly artefacts and in monumental barrows, cairns, or as boat burials. They could also have been given as tokens from seafaring spouses not least when supplementing other types of brooches. Compared with the other graves with brooches related to the Norse dress, possibly marking marital status, these graves largely connected to the same environments and with the strongest affiliation to Sygnafylki. At Brekke, located by the entrance of the Sognefjord, as an example, the two early graves, both cremation burials in large cairns,[47] one contained a single common Norse oval brooch (R649) and a few tools, the other, only with a single circular brooch of Irish origins, connected to another type of monuments at the place, mainly weapon graves.

In Sogndal, a cairn by the entrance of the Sogndalsfjord, yet another grave with textile tools lacked the traditional oval brooches but contained a single equal-armed brooch, like represented in the Myklebust grave, however, used as a third brooch. Containing only a few loom weights as grave goods it does not signal high status but was still given a monumental grave.

In Aurland in contrast, as many as three of the seven female graves with textile tools lacked the traditional Norse brooches. Two of them, located furthest in-valley at Belle at the north-eastern side of the large lake Vassheradsvatnet, both appeared as cremations in boat graves, and both supplied with special dress accessories: one, with two rare S-shaped snake-like brooches, a small ring-headed pin, and a bronze button copying an older style, the other with a reworked brooch from an ornamental mounting of gilt bronze with traces of Irish style ornamentation and a ring-headed pin. This was a secondary grave in a mound with remains of an older grave,[48] where the brooches seem to underline old heritage. The other grave at Belle, in contrast, with a reworked brooch from an ornamental mounting of gilt bronze with traces of Irish style ornamentation and a ring headed pin, rather seems to emphasise westbound relations. Still, both graves as monumental boat graves followed Norse funeral traditions of the elite and neither indicate inferior status. The two separate graves at a site mid-valley at Terum, in, respectively, a mound and a cairn, also differ as for symbols of status and range of tools. The grave in a mound comprised multiple tools for textile production, cooking, as well as farming – wool combing and possibly band weaving, weeding and mowing, and was supplied with three keys. It also contained a metal pin that possibly represents remains of a brooch. Altogether the tools found suggest it was a high-status female grave. The other grave close by in contrast, contained a brooch of a common Norse type (R647), an armlet of bronze and some beads, remains of the dress, woven in diamond twill, besides textile tools

Table 7.9: Graves from c. 750 into the late 9th century with special types of oval brooches, tools, textiles of broken diamond twill qualities, and five with foreign artefacts (N:9).

Museum no.	*Place/District*	*County*	*Type of brooches*	*Foreign artefacts*	*Tools for*	*No. of work processes*	*Quality*
B7639	Øvre Kongsvik	H	R640	Bead of jet	shearing, weaving	2	II
B9014	Sandal, Breim	F	R643	Glass vessel, enamel, bronze mounting	shearing, combing, spinning, weaving	4	I
B 9501	Bø, Breim	F	R647/6	x	shearing, combing, spinning, weaving	4	I
B9625	Naustdal	F	Berdal, group B		spinning	1	III
B9063	Terum, Aurland	S	R647	–	shearing, spinning, weaving?	3	imprint
B9064	Nes, Sogndal	S	R657	–	shearing, spinning, weaving	3	imprint
B7731	Vinjum, Aurland	S	R657	Irish four-sided silver brooch, mounting for a drinking horn (Irish), 1 cup and lid of bronze	shearing, combing, spinning, weaving	4	III & II
B9065	Nes, Sogndal	S	Berdal Type C	Irish penannular brooch in gilt silver	shearing, combing, spinning, weaving	4	I
B8000	Grodås, Hornindal	F	R557	–	sewing	1	III

for shearing, spinning, and band weaving, as well as a lock but without keys.[49] Still, as monumental graves with varied grave goods, both graves signal status above the average. The third site in Aurland with these double sets of graves, at Tokvam, had both graves connected to a large mound, one in its centre, the other as a secondary grave (cf. Chapter 5). Of these, the woman in the central grave as the more prominent was equipped with two oval brooches (R650), a common type in the northern parts of the country,[50] supplied with a cruciform Irish brooch but only with a single spindle whorl and a sickle as grave goods. The secondary grave was less prestigious as for jewellery, only furnished with 40 beads of glass and one of amber for a necklace but in contrast to the other contained an all-inclusive tool kit for all processes, from shearing to weaving, besides a casket and a ladle of iron. They seem thus to represent an asymmetrical relation, however, it is somewhat ambiguous.

At two of these three places in Aurland, then, one of the graves without the traditional brooches contained artefacts of Insular origin. However, when furnished with brooches connected to the Norse dress, they appear rather as additional status markers than as markers of ethnicity and probably acquired as gifts from seafaring spouses. Of these graves, the Vinjum grave as the latest and the most prominent both as for social emblems and tools. In contrast to the many female graves in Aurland, contemporary male graves were few with only two and at each end of the valley. The one innermost place at Belle[51] connected to two female boat graves. The other at the entrance of the valley was a rich weapon grave in a mound at the site of the vicarage in Aurland,[52] and close to Vinjum. It included also various tools – among others, a scythe, a sickle, and tools for woodworking and forging, besides a ring-headed pin of copper alloy.[53] This double representation of gendered graves and tools is a pattern observed in Vik and Sogndal too,[54] as well as in inland districts in Firdafylki, however, in Aurland with notably few male graves compared with the female graves.

In Firdafylki, three of the four graves without the traditional oval brooches connected to Stryn and one to Myklebust in Davik at the southern side and outer zone of the long Nordfjord, and a flat grave, modestly equipped with both dress accessories and tools. Only one grave – a female burial by the entrance of the main Stryn valley at Tonning – contained a foreign artefact: a penannular brooch of Insular origin, besides two armlets of copper alloy and 28 beads for a necklace. The other grave in Stryn at a place mid-valley at Ytre Eide, was furnished with a single armlet of copper alloy[55] and 11 beads. The third grave in a mound at Lyslo close to the fjord in Innvik contained a special rhombic bronze brooch without an equal in the present corpus, and with a single spindle whorl (see Table 7.10). Judging by the dress accessories, tools, location, and relation to other graves, these graves seem to represent female graves 'in the middle' as for status.

The five graves from Hordafylki with textile tools and without the traditional oval brooches also vary in status, when based on grave monuments and grave goods and as for relations to other graves – close by or at larger distances.

Table 7.10: Graves with textile tools from the 9th century with various dress accessories and ornamentation but without the traditional Norse oval brooches (N:16).

Museum no.	*Place/District*	*County*	*Accessories*	*Foreign artefacts*	*Textile tools for*	*Burial type*
B283/496/504	Lyslo, Innvik, Stryn	F	1 rhombic shaped bronze brooch (JP141), c 50 beads from a necklace	–	spinning	Mound
B584–93	Tonning, Stryn	F	1penannular brooch, 2 armlets of copper alloy R709?), 28 beads, mosaic	x	spinning, weaving	Type? (old finding)
B5804	(Ytre) Eide, Stryn	F	1armlet of copper alloy R719, 11 beads	–	weaving	Flat grave
B7005	Myklebust, Davik	F	1 equal-armed brooch, R661	–	weaving	Inhumation, flat grave
B15813	Osen, Gaular	F	13 beads monochrome glass, blue and red.	–	spinning, weaving	Flat grave, demolished?
B6637–B6649	Brekke, Lavik	S	Irish round, convex ornamental plaque, used as brooch	x	spinning, weaving	Cremation. in a mound
B9864	Lomelde, Sogndal	S	1 equal-armed brooch	–	Weaving	Cairn
B5773	Belle, Aurland	S	1 Irish brooch, reused ornamental mounting	x	Weaving	Cremation in a boat
B5815	Belle, Aurland	S	2 S-shaped brooches, 1 ring-headed pin		combing, weaving?	Cremation in a boat
B8669	Tokvam II	S	38 beads, one of them of amber	–	shearing, combing, spinning, weaving	Secondary grave in a mound
B8991	Terum, Aurland	S	1 mosaic bead, iron pin from a brooch	–	combing, weaving (?)	Mound
B6185	Lygra, Lindås	H	1 armlet of bronze R721		spinning, weaving	Flat grave
B322	Fana	H	1 silver chain, 2 beads	Bronze cauldron and bowl of Insular origin, melted glass	Spinning	Cremation in a large mound
B7836	Rykkje, Voss	H	1 bronze pendant, 4 glass beads	–	spinning, weaving	Cremation burial, grave chamber in a mound, possibly two graves
B1544	Saue, Voss	H	several beads for a necklace?	–	Spinning	Uncertain grave
B7114+ B7149, 7867	Folkedal, Granvin	H	8 beads	–	shearing, combing, spinning weaving, smoothing	Boat grave
N	16			4		

One of the most prominent graves in this category, connected to a large mound at the central place Fana in the mid-zone of the county. It was furnished with an exquisite silver chain, besides special grave goods of foreign origin[56] – a bronze cauldron and remains of glass. It thus resembles grave goods in high-status female graves at Hopperstad, Sandal in Breim, and Myklebust, but in contrast, was only supplied with two spindle whorls and without the traditional brooches. Although the dress accessories do not qualify for highest status according to Solberg's criteria, the special grave goods and the monumental grave still indicate high prominence, as also corroborated by its central position in its district, and as a central place and a later medieval church centre.

Three of the other four other graves in Hordafylki without the traditional oval brooches are all sparse both for dress accessories and tools. Of these, a female grave without superstructure in an in-fjord island in the mid-zone at Koppen contained tools for spinning and weaving – a single spindle whorl and 14 loom weights – and a penannular brooch of an Eastern type,[57] and thus is ambiguous for status. This is also the case for the two graves from inland Voss, at, respectively, Rykkje and Saue. One connected to a large mound with more than one grave, and again it included a single spindle whorl and some loom weights, besides a bronze pendant and a few glass beads.[58] The other, appearing as a demolished grave, and represented with a few mosaic beads for a necklace and a single spindle whorl.[59] These graves without brooches do not match either in quality or quantity the grave in Granvin at Folkedal and without dress accessories too, but still as a prestigious burial in a boat and with an all-inclusive tool kit and keys. The tools included moreover a smoothing plaque of whalebone of northernly origin, a find that also links to other high-status graves (cf. Chapter 5; Table 7.10).

This outline shows then that not all high-status women wore brooches as part of the funeral dress of a strapped gown but used other types of dress accessories, often numerous beads for necklaces, armlets and/or gender-related artefacts. They largely relate to the same districts as the female graves furnished with the traditional oval brooches, but do not necessarily indicate inferior status. As shown, four of these 16 graves included Insular or other foreign artefacts, signifying both rank and possible networks – one from the inner zone of Firdafylki, two from Sygnafylki (Brekke and Aurland), and one from Fana in the mid-zone of Hordafylki.

Three of the 16 graves without the traditional dress accessories coincide moreover with central places that later became church centres – Eide/Kirkeide in Stryn in Firdafylki, Brekke in Lavik in Sygnafylki, and Fana in Hordafylki. They add then to five other graves connected to central places from the 9th century and later church centres: three in Sygnafylki: Hopperstad in Vik, Vinjum in Aurland and Hauge in Lærdal, and three in Firdafylki: Naustdal, Grodås, and Fosnes in Stryn furthest north-east,[60] but these in contrast contained other high-status dress accessories too.

This overview shows then that status is not always straightforward and can be ambiguous – and even a larger share of the burials represents higher status than defined by the dress accessories alone and when taking their broader context into account, and changes in the burial practices over time.

The late Viking period: common accessories, common relations – stronger networks?

The number of graves with textile tools are considerably fewer in the last century of the Viking Age: 23 of 38 identified by their dress accessories, and nine with preserved textile remains. The dress accessories appear more uniform than in the earlier graves, where one type of brooch, type R652 (Fig. 7.12), dominated, represented in 19 graves with textile tools. In total, the type connects to at least some 50 graves from the study area and most commonly in Hordafylki.[61] The 19 graves with textile tools thus represented *c.* 38% of graves with this type of brooch. Since nine, or about half, of the graves comprised foreign goods, these brooches are clearly overrepresented in graves with textile tools.

Nine graves with this type of brooches related to Firdafylki, while Sygnafylki and Hordafylki each had seven such graves, and mainly in the same districts as earlier, although a few in 'new' environments (cf. Chapter 5). Four of the nine graves in Firdafylki connected to central places that became medieval church centres – Loen, Hilde, Myklebust, but only Dale in the southern part of the county. All but one of these 23 graves, a double grave, were high-status female graves. Five of the seven late graves in Sygnafylki also connected to central places and with earlier graves of this kind – one at Brekke in the outer area, three at Hopperstad in Vik, and one at Kvåle in Sogndal and places that later became church centres (see p. 164). The two other graves represent different environments – the prestigious Målsnes grave located at a dominating headland opposite to Vik; the other of lower status was in an upland area *c.* 300 m.a.s.l, in the mid-zone. Like in Sygnafylki, only one of the seven 10th-century graves in Hordafylki with the same type of brooches connected to a coastal area, the earlier mentioned Toranger grave (cf. Chapter 5) and only one grave from the mid-zone. All the others related to inland areas. They also lacked the same connection to central places. Four of the graves were moreover double graves but of the same general high status as 15 of the female graves further north (Table 7.11).

Altogether, 19 of the 23 graves had brooches of the common type R652, and generally with two such brooches but in three cases with only one brooch. Five graves, possibly six, comprised a brooch of type R654: two graves with only type R654, either one or two, the others with both types. Another type, R655, which Petersen characterises as a magnificent type that succeeded R657 from the 9th century,[62] was found in a dispersed grave at Gjelsvik, by the outer fjord in the community of Førde together with a bronze chain and tools for weaving.

Nine of all these graves with common types of dress accessories, or 39% of the 10th-century graves with textile tools, contained foreign artefacts, generally of Insular origin, representing about the same level as in the preceding century. Nearly all connected to the inner zone; three from Firdafylki, two from Sygnafylki, and four from Hordafylki. In many cases they combined with other dress accessories, nearly all with a third brooch, beads for either a necklace or armlet, and chains, many of silver, in contrast to the dress accessories from the 9th century that were mostly made of copper alloy. One of the three graves in Firdafylki

Fig. 7.11: Oval brooches (R652), the most common type in the 10th century (Petersen 1928).

with foreign grave goods related to the central place Hilde in Innvik, close to the double grave with textile tools but without dress accessories.[63] Here, the high-status female was buried in a boat grave, covered by a mound, and furnished with a pair of R652-brooches, a few beads of glass and one rather special and larger than the others of silver, and tokens of nine small Thor's hammers of iron, which may suggest an early cult place. The grave goods comprised moreover a lamp, cooking equipment and agricultural tools, besides a composite tool kit for textile production. It included, among others, a weaving beater of whalebone and as earlier noted a long iron spit or staff (cf. Chapter 5, Fig. 5.11). The grave also contained a bronze cauldron of Insular origin, like the one found in the grave at Fana, another central place in its district.

The two other graves that were located at central places were monumental graves too, one at Myklebust at Eid but now in another mound than the two earlier with the two high-status female graves. The other was a chamber grave in a mound at Dale in the district Fjaler in the southern part of the county. Both stand out with their costly grave goods with artefacts of silver, with a silver chain, like found in the earlier graves at Fana and Sandal in Jølster, and at Dale in Fjaler, also with a multiple tool kit, which included a glass pellet of foreign origin too. Another grave in the southern part of Firdafylki with textile tools and foreign artefacts was a double grave at Fyling in the community of Gaular. The female was again furnished with the same type of brooches, besides two trefoil brooches and beads, and the male with an ornamented Hiberno-Saxon bronze buckle and buckle plate. In the third grave at Mardal by the Gloppenfjord further north, the high-status woman had a pair of R654 brooches, as well as a penannular brooch, besides its many tools.

Of the seven 10th-century graves with textile tools from Sygnafylki, two related to the outer zone, six graves mid-fjord, of which four at Hopperstad in Vik, one at Målsnes and another at Kvåle in Sogndal. Of these, five graves contained the most common type of brooches, R652, while two of the graves at Hopperstad had R654 brooches, and one grave only one brooch but together with a brooch of Irish origin. This was also the only grave with a foreign artefact, besides the grave at Kvåle with a cauldron of tin-coated copper alloy of Insular origin. This was a high-status woman adorned with two brooches of the common type R652, two bronze fibulae and a trefoil brooch, and supplied with tools emphasising broad work domains. The high-status female grave in a long mound at Målsnes, with a similar pair of brooches,

Table 7.1i: Graves with common dress accessories from the 10th century (N:23) In bold: with textiles.

No.	*Place/District*	*County*	*R 652*	*R 654*	*R 655*	*Other dress accessories*	*Foreign artefacts*	*Gender/Status*
B6433	Brekke, Stryn	F	–	1	–	–	–	FII
B8818	**Loen, Stryn**	F	2	1	–	–	–	FIII
B5717, B5755	**Hilde, Innvik/Stryn**	F	2	–	–	5 beads of glass, 1 silver, 9 small Thor's hammers	bronze cauldron	FIII
B5764 IV	**Myklebust Eid**	F	2	–	–	30 beads, 1 of silver filigree and spiral ornaments, silver chain	–	FIII
B4605–06?	Mardal, Gloppen	F	–	2	–	1 penannular brooch	X	FIII
B11246	Gjelsvik, Førde	F	–	–	2	1 bronze chain	–	FIII
B9769	Korsvoll, Gaular	F	2	x?			–	FII–III
B11470	Fyling, Gaular	F	2	–	–	1 trefoil brooch (JP. Fig.97) 2 beads	Hiberno–Saxon bronze buckle and buckle plate, ornamented	FIII/MIII
B5910	**Dale, Fjaler**	F	2	–	–	3 small, ornamented brooches of silver, armlet of silver 2 silver finger rings, 107 beads, 1 trefoil brooch	–	FIII
B7761	**Hopperstad, Vik**	S	2	1	–		–	FIII
B9060	**Hopperstad, Vik**	S	–	2	–	Irish 3rd brooch, 21 beads	X	FIII
B4637	Hopperstad, Vik	S	1	–	–			FI–II
B3456	Kvåle, Sogndal	S	2	–	–	2 bronze fibulae, 1 trefoil brooch (R671)	Cauldron of tin coated bronze	FIII
B12131	**Målsnes, Balestrand**	S	2	–	–	1 penannular brooch (Norw, copy), 1 bronze chain, 3 beads	–	FIII
B7366	Mølmesdal, Høyanger	S	2	–	–	–	–	FII–III
B8117–B8818	Brekke, Lavik	S	1	–	–	Rectangular silver brooch, beads	–	FÏI–III
B5874	Toranger, Austevoll	H	1	–	–	–	–	FII–III
B5800	Ytre Arna, Haus	H	2	–	–	2 round brooches of copper alloy (R665, R683), 1 pin with a ring. 7 beads of carnelian and glass		FIII/MIII
B5627	Vangdal, Kvam	H	2	–	–	–	–	FIII
B4432	Fladen/Jamno, Evanger/Voss	H	1	–	–	2 ring-headed pins of bronze, tin and silver-coated (Glørstad type III b), 2 beads	X	FIII/MIII
B6228	**Veka, Voss**	H	2	–	–	Carolingian oblong bronze brooch (JP. Fig. 140), 1 silver and 1 bronze armlet 118 beads, mostly glass	Anglo Saxon silver coin (c792/6)	FIII
B4864	**Hyrt, Voss**	H	2	–	–	20+75 beads	Mounting of a copper alloy to a drinking horn	FIIIM/III
B6657	**Trå, Granvin**	H	2	–	–	silver chain with silver rings, 15 beads, 1 trefoil brooch	Finds of Insular origins (ladle, brooch)	FIII
∑	23		19	5	1		9	

also comprised a penannular bronze brooch, however, a Norwegian copy,[64] besides a bronze chain and beads and again a composite tool kit.

In Hordafylki, the seven graves with these common 10th-century brooches form another pattern than the graves further north. They relate mainly to agricultural districts in the inner zone, by the coast and in the mid-zone, only with one each, three as double graves, the others, mainly high-status female graves. All but one, the Toranger grave by the coast and a double grave at the crossroad between water and land routes to Voss, had two brooches. Four of the seven graves contained artefacts of foreign origin: the double grave and the graves Veka and Hyrt with composite tool kits and high-quality textiles (cf. Chapters 5 and 6). Of these, the double grave contained two tin and silver-coated ring-headed pins of bronze,[65] and two beads. The dress accessories in the Veka grave were even more prestigious; besides the two brooches, it contained a relief brooch, a Carolingian oblong brooch of iron, and two armlets of silver and bronze, respectively. Additionally, there were 118 beads for a necklace, mostly of glass, a few of amber, carnelian of foreign origin, and of rock crystal, and an Anglo-Saxon coin from the late 700s, secondarily used as jewellery. The double grave at Hyrt in the upland area further east at the fringes of Voss, had almost a hundred beads for a necklace besides the brooches. However, the mounting of copper alloy for a drinking horn, also of foreign origin, most likely belonged to the male. In the fourth grave, the earlier mentioned grave at Trå in Granvin, the dress accessories were even more impressive. Besides the pair of brooches, it included a silver chain with silver rings, 15 beads and a trefoil brooch, and artefacts of foreign origin – a ladle, a bronze cauldron, and a goblet of green glass. As shown in the preceding chapter, these graves had comprehensive tool kits too and revealed remains of broken diamond twill from the dress by the brooches.

A large share of these 23 10th-century graves with textile tools, nine or 39%, comprised foreign goods, and even higher, 53% when based on 17 graves of the highest status III, and clearly overrepresented in such graves which were often located close to the earlier female graves and the two other graves from the late 9th century, again in a boat (cf. Table 7.9).

Graves with textile tools and other types of dress accessories

Seven other graves from the 10th century or some decades earlier comprised other types of brooches: oval, rounded, or circular, oblong/tongue-shaped, or trefoil brooches and in four of the graves with brooches appearing in pairs and in all but one in high-status graves (Table 7.12; see Fig. 7.12). Five of the graves connected to Firdafylki, the other two to respectively Sygnafylki and Hordafylki. In Firdafylki, two of the graves related to the northern part of the county, one at Kirkeide, mid-valley in Stryn, the other furthest north-east at Gutdal – both places with several graves with textile tools.

Table 7.12: Graves with other types of brooches from the late 9th and 10th centuries (N:7).

No.	Place/District	Other brooches	Other dress accessories	Foreign artefacts	Gender/Status
B5525	Gutdal, Stryn	1 oblong, tongue shaped brooch (R664), remains of a second, 1 pendant of copper alloy (JP167) 3 other pendants, Borre style	Bead of glass, carnelian, crystal	–	FIII
B8953	Kirkeide, Stryn	2 oval brooches, JP.57, Jelling style	–	–	FIII
B7653	Arnestad, Gloppen	2 brooches, Borre style (JP47), circular brooch in Jelling style	5 round pendants of bronze (Jelling style), beads	–	FIII
B12043	Gjerland, Førde	2 circular brooches 1 bronze (JP 116) and 1 silver (1 R665/9?)	7 beads	–	FIII
B10878	Gjerland, Førde	1 trefoil brooch of copper alloy (JP109)	–	–	FI
B708–17; B2734+2739	Hønsi, Vik	Round silver brooch Jelling style	1 silver pendant (JP 157)	Textiles- gold threads	FIII
B4599	Bu, Eidfjord	2 oblong tongue shaped brooches. JP137 Jelling style	2 beads, chain of iron		FIII
Σ	7			1	

Here, the woman in the Gutdal grave had been furnished with a rather special tongue-shaped brooch and remains of another, as well as pendants in Borre style. The brooches in the grave at Kirkeide in Jelling style are without equivalents in the present corpus from western Norway but share traits with the high-status grave at Snæhvammur in Breiðdal,[66] furnished with the fine textiles woven in diamond twill. Two of the graves, the brooches of Jelling style appear as rather rare artefacts in Norway – oblong tongue-shaped brooches and of the animal ornamented type (Fig. 7.12). The Kirkeide grave connected moreover to an earlier grave nearby with tools for weaving but without the traditional oval brooches, only an armlet.[67]

One of the other 10th-century high status graves in Firdafylki, from Arnestad in Gloppen by the fjord, contained a rare type of oval brooch of Borre style too, combined with pendants of Jelling style and in this case with a composite tool kit (cf. Chapter 5). The grave by the Gloppenfjord thus differs from the other grave in the district as for social emblems but both seemingly on the same social level (cf. Table 7.11) and as for tools. The two other graves with textile tools and special dress accessories both related to the southern part in the mid-zone of Firdafylki at an inland and upland place, Gjelsvik, with the same type of brooches as in the Ketilstaðir grave. While one contained two circular brooches of silver and bronze, respectively,[68] the other grave had only a single trefoil brooch,[69] however, both were rather modest for textile tools.

Another grave with more rare brooches and prestigious dress accessories of Borre/Jelling style related to Hønsi in Vik in Sygnafylki – and was an early discovered grave and possibly a female boat grave. It contained a round silver brooch (R663), two silver pendants, and one of bronze with remains of fabrics with traces of gold threads. The textile tools included wool combs, a weaving beater, and a smoothing pellet of glass, also signalling foreign connections.

The last of these seven graves was from Bu in the inner parts of Hardanger and the other once again from Gutdal in Oppstryn, furthest north-east.[70] Both appear as impressive female graves – at Gutdal, with many graves with textile tools from the very beginning of the Viking Age onwards, and the latest, the most prominent but only represented by two light spindle whorls as for tools. This late grave contained a complete tongue-shaped brooch too and remains of a second, and like represented in the Bu grave, it included several pendants of copper alloy too. These are the only two graves with such brooches in the present corpus from Norway but found in a grave at Kornsá too, located in north-western Iceland, and as the only grave in Iceland with these brooches, and otherwise represented in Hedeby among others.[71] This grave was also a grave with textile tools and of higher status than those with the more common oval brooches R652–654.[72] This may also be the case in western Norway, where brooches of Borre and Jelling style represented general high status, nearly all of status III, and of which four connected to central places later to become church centres and all to districts with other graves furnished with textile tools.

Five other 10th century graves with textile tools and furnished with social emblems, however, without the traditional Norse brooches, call special attention related to the issue of status and heritage (Table 7.13). Another late grave in Vik, and the fourth grave of this kind at Hopperstad, represents an intriguing case, as it stands out as an exceptionally rich grave but lacked the traditional emblems related to the dress, only some beads for a necklace, however costly, one large of silver, agate, chalcedony, and several of glass and mosaic. As shown in Chapter 5, the grave goods comprised moreover several prestigious and exotic artefacts, as well as of Insular origin. They included, among others, several rather unique containers – a large cauldron of bronze with three bird-shaped escutcheons and mountings shaped like

Table 7.13: Graves with textile tools from the 10th century with dress accessories but without the traditional Norse brooches (N:5).

Museum no.	*Place/District*	*Accessories and status objects*	*Foreign artefacts*	*Textile tools for*	*Burial type*
B8060	Kirkeide, Stryn	2 pendants of bronze, JP155, 30 beads (Borre style)	–	shearing, spinning	Flat grave
B4511	Hopperstad, Vik	Various beads, *e.g.* a large of silver, 1 filigree, agate, calcedony, several of glass, and mosaic 1 staff/stick with a handle of copper alloy	Bronze cauldron, bowl, bronze-bound pail, 2 glass vessels, scales, weights, Kufic dirhams	shearing, spinning, weaving	Boat grave? Flat, inhumation grave
B4841	Hove, Vik	Bronze chain	–	combing/ band-weaving	Mound
B6089	Koppen, Haus	1penannular bronze brooch (Type I, Eastern variant), 3 beads	–	spinning, weaving	Cremation burial, flat grave
B13951	Bolstad, Evanger, Voss	Small bronze ring, guilt bronze ring, 2 beads	–	spinning	Cremation, boat grave, cairn
N: 5			5		

truncated cones and fragments of another with a low foot, and an egg-shaped enamel multicoloured plate, a small bowl, a saucepan, and a wooden pail with bronze mounting. The special finds of two balance weights of lead with gilt bronze plaques and a pair of tin-coated bronze scales together with two Irish weights and three others, point westwards. Two glass goblets, one small bottle-shaped and the larger ornamented with ribbons, of Frankish origin[73] and an old Kufic dirham dated AD 738/39 with origin in the Middle East – altogether rare and prestigious grave goods. Remains of glass vessels are, as noted, rare finds in Viking Age graves[74] but as we have seen are represented in the early Sandal grave, in the Myklebust mound and the high status Trå grave and in the grave in the large mound at Fana, another grave without the traditional oval brooches. The textile tools in the Hopperstad grave served three work processes – shearing, spinning, and weaving – and comprised other domestic tools too, a push key, a casket, a strike-a-light, a hoe, and bridles.

Noteworthy as for status and social relations, is a long spit/staff/distaff of iron (cf. Chapter 5, Fig. 5.12) of the same type as found in high-status 10th-century graves at Gutdal in Stryn, Hilde in Innvik, Stryn, Myklebust at Eid, and Arnestad in Gloppen, all in Firdafylki, and in Sygnafylki in the grave at Kvåle in Sogndal, besides Vinjum in Aurland, and as noted also from Hordafylki, the graves at Veka and Trå, distinguished graves, and all with the common denominator of tools related to textile productions.

The stick/staff in the Hopperstad grave is of the same character as the others, and has a handle and end ring of copper alloy, respectively 16 cm and 18 cm and the stick 74 cm long and altogether 108 cm in length. It also resembles the sticks/staffs found at in two graves in Iceland. Of these, one was from the grave of a high-status woman at Álaugarey with brooches of R652 and 654 and with remains of fabrics of both wool and linen and furnished with textile tools too, while the other grave had left few remains.[75] Altogether, this representation of high-status women with these special staffs and textile tools and connected to central places, may indicate wider networks. The representation of exquisite beads, of which one is a large finely decorated silver bead, in graves with these staffs, such as in the Hopperstad grave, as also in the Myklebust and Arnestad graves may be another link, which Shetelig observed and commented upon.[76] Altogether, this grave at Hopperstad was the best-equipped grave with the most complete tool kit for textile production and with prestigious foreign artefacts and appeared at the same site as a boat grave, also like the earliest and prominent boat grave from the late 8th century.

Of all these places, Hopperstad stands out, with four female graves with textile tools, of which two had costly dress accessories and generous grave goods, including tools[77]– one of them with four brooches – a pair of R654, plus a single brooch of R652 and one of Irish origin. Being represented in high-status graves – at Gutdal, Hilde, Myklebust and Arnestad in Firdafylki, and at Vinjum, Hopperstad and Kvåle in Sygnafylki, besides Trå and Veka in Hordafylki – they may indicate relations within high-status networks, stretching to Iceland. Still, Insular finds were relatively more frequent in the 9th-century graves in Sygnafylki generally, with 11 such graves, and then in 38% of the graves with textile tools.

However, only one of the other graves, the grave in Vik, indicates foreign wares, thus differing from the graves with the common brooches of type 652–654, where six contained foreign artefacts and remains of high-quality textiles of diamond twill (Table 7.14). While the brooches and grave goods in the earlier Hopperstad graves substantiate connection between elite milieus in Vik, Breim and at Myklebust at Eid that point both west and eastwards, the woman buried in a boat without the traditional Norse dress accessories with many foreign artefacts may indicate a special standing, possibly gained through an exogamic marriage, and as a confirmation of elite alliances across the North Sea, as has been suggested. Still, it appears as a traditional Norse boat grave

Table 7.14: Graves from the 10th century with tools, textiles of diamond twill, and foreign artefacts (N:6).

Museum no.	*Place/District*	*Brooches*	*Foreign artefacts*	*Tool categories*	*Work stages*	*Quality*
B6657	Trå, Granvin	R652	Ladle, brooch, pellet of glass	shearing, combing, spinning, weaving, smoothing	5	III
B6228	Veka, Voss	R652	Anglo Saxon silver coin	combing, spinning, weaving, sewing (R413)	4	III
B4864	Hyrt, Voss	R652	Mounting of a copper alloy for a drinking horn	combing, spinning, weaving	3	III
B9060	Hopperstad, Vik	R652	Irish 3rd brooch	combing, spinning, weaving, sewing, smoothing (R413)	5	III
B12131	Målsnes; Balestrand	R652	Penannular brooch	shearing combing, spinning, sewing	4	III
B5910	Dale, Fjaler	R652	Pellet of glass	shearing, combing, spinning, weaving, sewing, smoothing	6	III

of prominence, and with a seemingly prestigious stick/distaff that may indicate contacts and possibly local and regional networks. The tools for trade and other grave goods indicate relations to both internal and external elites. Another female grave in Vik buried without traditional oval brooches in a monumental grave at the central place Hove with a chain of copper alloy for personal adornment and a few textile tools for wool combing and band weaving, besides bridles, an iron cauldron, and a casket.[78] Like Hopperstad this was a place that became a medieval church centre and a place marked by monumental but mostly male graves, which must have dominated the landscape both in the 9th and 10th centuries, however, with a dominantly male supremacy in contrast to the graves at Hopperstad.

The graves as signs of nodes in production, communication, and cooperation

By the late Viking Age, the foreign artefacts seem generally to consist of grave goods that could have been traded or acquired as gifts. This seems to be the case at Trå, Veka, and Hyrt, with, among others, a large cauldron, a ladle of copper alloy, and two glass goblets, a brooch of a reworked Anglo-Saxon silver coin, and a mounting for a drinking horn, a cauldron and a saucepan of bronze, and a drinking glass of Insular origin.[79]

In the six graves with remains of textiles of diamond twill, they were all the finest quality III. The tool kits with tools for three to six processes, however, indicate special places for production. Seen as possible indicators of networks, these staffs/distaffs as grave goods generally connected to people from the upper echelons of society in central agricultural areas and at nodal points and strategic places, at crossroads between land routes and waterways, again connecting to the coast and sea routes. Artefacts of foreign origin seem largely to have followed such routes, appearing in three of these graves, one at Breim at the head of the fjord, and two in the upland area of Stryn, also with a common passage leading eastwards, and all in graves of high status.

The long-distance distribution of jewellery, dress accessories and artefacts produced in the new Viking Age towns and traced, appear in graves as accessories on the funeral dress in generally high-status graves in various parts of the three counties under study. They often appear at central places and along routes of communication, which also indicate social relations and networks. That the graves by then contained a much higher share of foreign artefacts, pointing westwards, may further substantiate cooperation, networking, raiding, or trading within the elite on a larger scale and at longer distances. The reversed social pyramid reflected in the graves with textile tools, as presented in Chapter 5, is another indicator of the social aspects of production, their role as production places, hubs, and nodes in various types of networks. The graves with textile tools were most visible and widespread in the early 9th century, to become more concentrated and seemingly centralised by the latter part and into the 10th century. In this way, they reflect socio-political structures on a macro level.

It is not only the graves at central places and later church sites that indicate nodes for communication, cooperation, and production. Inland and upland areas also seem to be integrated in the production, and as for gender with reciprocal work domains, judging by the graves and grave goods, including both male and female graves, related to woodworking, forging, animal husbandry, and textile production. In Firdafylki, and both Oppstryn and Nedstryn with many such graves, developed as central places, later to become church sites – at Fosnes by the large Lake Oppstrynvatnet, at Eide/Kirkeide mid-valley, and by the inner Nordfjord at Loen and Innvik – serving as nodes for communication east–west and lowland–upland and by the fjord and larger lakes.

That many of the graves with textile tools – both with and without social emblems of brooches and other jewellery – appear as boat graves – both female graves and as double graves – is another characteristic of the period, from its very beginning around AD 800 into the 10th century, and most frequently from the early Viking period, and 24 in all from the area of study from western Norway, as far as they can be dated (Table 7.15).

Of all these boat graves, 10 related to Firdafylki (Nos. 1–10), another nine to Sygnafylki (Nos. 11–19) and four to Hordafylki (Nos. 20–25). The largest concentration in Firdafylki connected to Stryn, where two of them related to Oppstryn by the large lake (Nos. 5, 6), and another two connected to the inner fjord, Nordfjord (Nos. 4, 7), in, respectively, Nedstryn and Innvik, while another related to an upland area close to the border to Hornindal with its large lake Hornindalsvatnet (No. 3). Yet another boat grave was located at Myklebust (No. 9), mid-fjord in Eid, while one connected to inland Jølster to the south by the large Lake Jølstravatnet (No. 8), and at Holmedal a central place by the fjord, Dalsfjorden in the district Fjaler furthest south in Firdafylki. Of the two graves in the community of Gloppen, one was located by the Gloppenfjord (No. 1) and the other in inland Breim (No. 2), not far from the large Lake Breimsvatnet. All these boat graves were located rather close to waterways, either larger lakes or fjords. In Sygnafylki, one of the nine boat graves related to the entrance of the Sognefjord (No. 11), four to Vik mid-fjord in the long Sognefjord (Nos. 12–15), one to upland Berdal further in-fjord (No. 16) and three to Aurland further east by the Aurlandsfjord, a side arm of the Sognefjord, and with two of these graves in the inner valley by the large lake Vassheradsvatnet (Nos. 17, 18) and the last the prestigious grave at Vinjum closest to the fjord (No. 19). In Hordafylki, four of the six boat graves related to the inner zone – Voss and Hardanger (Nos. 20–25). Of these, one

Table 7.15: Boat graves with textile tools (N:25).

County	District	Ref. no	Place	Museum no.	Female	Double	Date
Firdafylki	Gloppen	1	Ytre Hauge	B5150	1?	?	750–800
	Stryn/Innvik	2	Langeset	B6035–6036		1?	750–800
	Breim	3	Bø	B9501		1	800–850
	Stryn (Nedstryn)	4	Vik	B4756		1	800–850
	Stryn/Oppstryn	5	Fosnes?	B12695			800–850
	Stryn/Oppstryn	6	Flo	B5865/B10176		1	800–850
	Stryn/Innvik	7	Hilde	B5717, B5755			900–1000
	Jølster	8	Sandal	B11413	1		850–900
	Fjaler	9	Holmedal	B7893		1?	800–900
	Eid	10	Myklebust	B5764	1		900–950
Sygnafylki	Brekke	11	Brekke	B8117–B8818	1		Late Iron Age
	Vik	12	Hopperstad	B9061	1		750–800
	Vik	13	Hopperstad	B4511	1		900–1000
	Vik	14	Skjervum	B6500 a–c	1		750 –800
	Vik	15	Hønsi	B708–17, B2734+2739	1		850–950
	Feios	16	Berdal	B594, 596–603	1		800–850
	Aurland	17	Belle	B5773	1		Late Iron Age
	Aurland	18	Belle	B5815	1		800–850
	Aurland	19	Vinjum	B7731	1		850–900
Hordafylki	Granvin	20	Folkedal	B 7114+ B7149	1		800–850
	Eidfjord	21	Vardberg	B7867	1		800–850
	Voss	22	Bolstad	B13951	–	1?	900–1000
	Vossestrand	23	Bygd	B8598	–		Late Iron Age
	Kvinnherad	24	Naterstad	B5927–28		1?	900–1000
	Etne	25	Sørheim	B12058	1		Late Iron Age

related to the long Hardangerfjord (No. 20), another to the very inner part (No. 21), while the two in the Voss district were situated in each end to the west and east: the former to a junction between the fjord leading west and land routes to Voss (No. 22), the other to the upland/inland district to the east (No. 23), not far from the large lake Myrkdalsvatnet. The last two graves related to the mid-zone (Nos. 24 and 25), in agricultural districts and quite close to waterways (Table 7.15).

Although the burial customs of using boats as a casing for the dead has many symbolic connotations, they are also concrete manifestations connected to the living and related to status. Many of these general high-status graves connected to areas where people depended on waterborne transport, and that the boats serving as necropolises for people of high rank but were smaller boats for smoother waters, *c.* 5–7 m long. The same tradition of burying people in boats was also transferred to the North Atlantic, and many are known from Icelandic contexts.[80] Still, larger ships, were required when crossing the Norse Sea and the Norwegian Sea sailing to the islands in the North Atlantic and were produced on home ground for practical reasons but seldom ended up in graves, and then the very top of society.

Across the Norwegian Sea

As we have seen, the burial customs in Iceland largely mirror the burial traditions in western Norway, burying the deceased with the same types of social emblems according to rank, and when of high status and with brooches of R652–654 as the most common but also other types of brooches and dress accessories. These were also of the same types as represented in western Norway, also the more rare round and tongue-shaped brooches and pendants in Borre or Jelling style, and with beads, chains and sometimes armlets, but overall, with fewer tools. One of the best equipped with tools was in fact a grave at Kornsá in north-western Iceland, with textile tools and other prestigious grave goods, a large cauldron of iron, and remains of scales of bronze with tongue shaped brooches

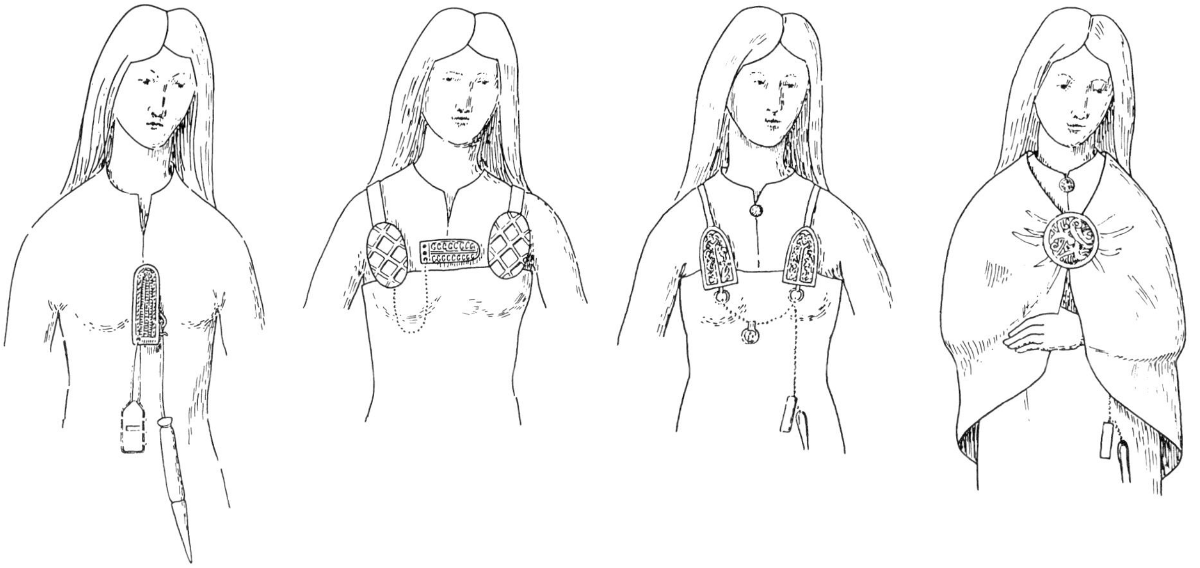

Fig. 7.12: The oblong tongue-shaped brooches could be worn in different ways, judging by burials in Hedeby (Illustration: Wamers 1984, with permission).

and was also the only grave with these tongue-shaped oblong brooches (Fig. 7.12).

When two of the graves from western Norway, at Bu in Hardanger and the other at Gutdal in Stryn, had the same rare type of brooches as in the Kornsá grave, it may indicate closer contacts (Fig. 7.13). The grave at Álaugarey contained, moreover, the same type of long stave/staff/distaff[81] as found in several prestigious graves with textile tools in western Norway, among them one of the richly equipped graves at Hopperstad without brooches, the graves at Vinjum, Kvåle, Myklebust, Hilde, Arnestad and again in the late Gutdal grave of high status it may again indicate relations across the sea. Two of the most prominent graves in Hordafylki, at Veka and Trå in Voss and Granvin, follow the same pattern – and with an attribute that does not seem accidental, and clearly a marker of status that connected to textile production, probably both practically and symbolically, as a distaff (cf. Chapter 4, Fig. 4.6).

Some of the graves with preserved textiles included artefacts of foreign origin of a special kind, such as found in a high-status male grave with remains of gold spun threads beside one gold and three silver buttons, and a lead weight of 6 g, and rather like finds from western Norway, and notably from a grave at Hønsi in Vik and yet another at Vangsnes, both with textile with gold threads.[82] This may indicate trade and/or wide-ranging networks, since rare dress accessories and other artefacts are of basically the same types as found in graves from western Norway and again acquired through wider networks related to raiding and trading.

The woman buried at Ketilstaðir in a blue gown was furnished with oval brooches of type R655/JP55 and this is also a rare type of brooch in both Iceland[83] and Norway, and in the present corpus from western Norway only represented one grave with equals at Gjelsvik, in Firdafylki (cf. Chapter 6, Fig. 6.3). The grave at Snœhvammur in Breiðdal[84] contained, moreover, the same type of brooches in Jelling style as in one of the graves with textile tools from western Norway, at the central place Kirkeide also in Stryn but otherwise without representation in the present corpus

The fine diamond qualities in graves from Iceland again bring up the question of provenance and place of production. As for the textiles woven in diamond twill from the Ketilstaðir, the wool analysis substantiated a domestic production,[85] and indirectly probably traditions and competency transferred from the 'mother-country'.

Concluding remarks

By exploring common characteristics in the burial practices related to graves with textile tools as grave goods – their locations within varying social and natural environments, their status assessed by their dress accessories, tools, and other grave goods, and not least of foreign origins – they altogether, at least indirectly, reflect how people acted and cooperated within hierarchical structures and elite networks at various levels.

The graves and grave goods imply and reflect large-scale enterprises within over-arching socio-political structures, related to both raiding and trading, as well as colonisation, which affected and included both genders. Like other expanding industries of the period, the textile production must have leaned on a larger work force that

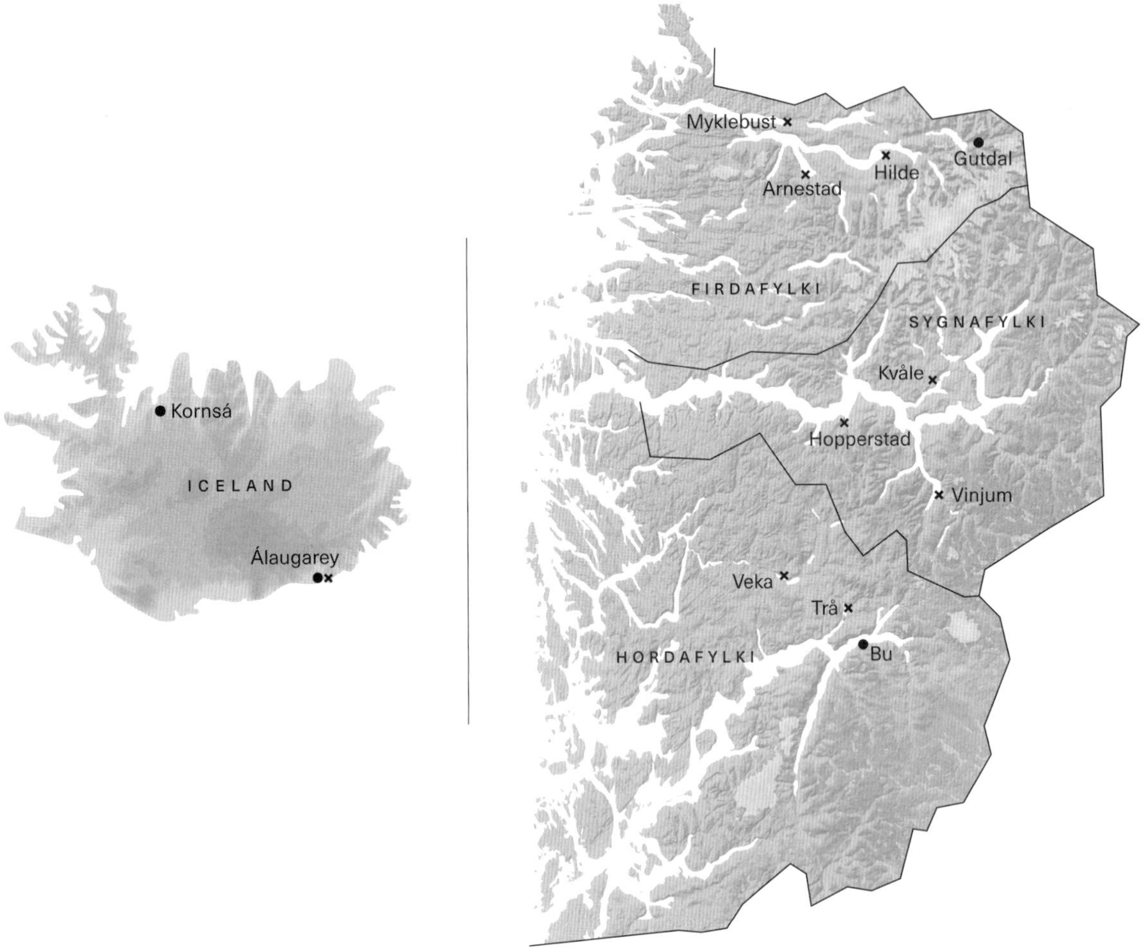

Fig. 7.13: Bu in Hardanger and Gutdal in Stryn and Kornsá in north-western Iceland with the rare type of tongue-shaped brooches. The crosses on the maps indicate places with graves with sticks/distaffs (Illustration: Per Bækken).

left few or no archaeological imprints. Judging by the tools and grave goods, the production was organised by high-status women and within networks that can be traced at shorter and larger distances over time. The social emblems and dress accessories in these graves, produced in centres such as Kaupang, Hedeby, and Birka reflect such networks directly and indirectly, probably in the form of exchange networks.

Differentiated tools and textile remains indicate a versatile production of varied qualities and concentrated to special places and environments, and by the Late Viking Age tending to locate at central places in their districts, such as Hopperstad and Vinjum, probably as nodes of production, being represented by many graves with tools and textiles. Myklebust is another place marked with this kind of grave that may signal central places as nodes in production over a long period from the Early Viking period and throughout the Viking period, although without leaving textile remains of these qualities. Veka, Trå, and Hyrt, which share many common traits for tools and textiles, are other examples of places that seem to have related to each other at shorter distances.

That the seemingly symbolic distaff connecting to so many of the most distinguished graves (among others, Veka and Trå in Hordafylki, Hopperstad, Kvåle and Vinjum in Sygnafylki, and Hilde, Arnestad, Myklebust, and Gutdal in Firdafylki), many also represented with high-quality textiles, exquisite grave goods and tools for textile production within the luxurious segment, does not seem accidental and may possibly indicate wider networks, even stretching to Iceland. Still, textiles of such qualities were produced in the same centres as the dress accessories and not necessarily locally. After all, the grave goods are only indirect traces of products and production. The next step is therefore to explore the direct traces at production places as far as they have been revealed archaeologically in both rural and urban settings.

Notes

1 Holm 1986; Downham 2009; 2017.
2 *e.g.*, Wamers 1998.
3 Wamers 1998.
4 Wamers 1998, 51.
5 Wamers 1998, 57.
6 Baug 2002; 2015; Ashby *et al.* 2015; Hansen & Storemyr 2017; Baug *et al.* 2019; Ashby & Sindbæk 2020.
7 Sindbæk 2017.
8 *e.g.*, Sindbæk 2012; Hayeur Smith *et al.* 2019.
9 Sindbæk 2012.
10 Glørstad 2010.
11 Andersen *et al.* 1989; Andersen 1995; Andersson Strand 2016.
12 Sindbæk 2012.
13 Plather 2011; Gjøstein Resi 2011.
14 Hayeur Smith 2004, 233.
15 Glørstad 2010, 120–125.
16 Jåtten 2006.
17 Sindbæk 2011.
18 Hellesøe 2006, with references.
19 Petersen 1928, 7; Gjessing 1934, 144–145.
20 Wamers 1985, Catalogue 62.
21 B5143, Gutdal, Stryn, male grave status MIII.
22 Helgen 1982.
23 Shetelig 1905, 20–22.
24 B6035-36, Langeset, status FII, a boat grave.
25 Sindbæk 2005.
26 Petersen 1928, 16–17.
27 Wamers 1985, Catalogue 67.
28 Glørstad 2012.
29 Glørstad Type 1A.
30 B5807, Myklebust, status FIII.
31 B9626, Naustdal, status II. Armlets of the same kind stem, moreover, from four other graves with textile equipment – three from Hordafylki, of which two are from coastal areas and one is from an upland area.
32 Petersen 1940, 139.
33 Petersen 1940, 51.
34 B6675, Bø. Breim, Gloppen status MIII/FIII, Slomann 1986, cf. note 22.
35 Slomann 1986, 25–28.
36 Wamers 1998.
37 Petersen 1940, 162, 187.
38 Glørstad 2010.
39 B8635, Seim, Granvin, status FIII.
40 Petersen 1940, 182.
41 Kaland 1973; Jesch 1991, 9-10.
42 B2233, Lunde, Voss; not in the upland area of Voss as *e.g.* referred to in Sørheim 2018, 254.
43 Petersen 1928, 29.
44 Petersen 1928, 44–51.
45 Sørheim 2011.
46 *e.g.*, Dommasnes 1982; Bødal 1998.
47 B6637, B6649, Brekke, status FI–II; B6062-6063, Brekke. Status FII.
48 B5815, Belle, Aurland, status FIII. Petersen dates the grave to the late 9th century (Petersen 1951, 501), while Helgen (1982, 26) to the 7th century, based on two brooches of a type dating back to the late Migration period, though Shetelig claims that it may be a local reinvention of old forms, and a bronze button (B5815d) also indicates a younger date.
49 B9063, Terum. FI–III
50 Petersen 1928, 26.
51 B6741, Belle, Aurland.
52 B9008, Aurland vicarage, status MIII.
53 Type JP237.
54 Bødal 1998, with references; Barndon & Olsen 2018.
55 Petersen 1928, 151.
56 B322, Fana prg. Fana, status FII–III, later one of the King's 12 royal chapels.
57 B6089, Koppen, Haus; Glørstad 2010.
58 B7836, Rykkje, Voss
59 B1544-1559, Saue, Voss
60 Referred to in the cadestre *Bjorgynjar Kalvskinn* (BK).
61 Petersen 1928, 59.
62 Petersen 1928, 71.
63 B5588, Hilde, Innvik, Stryn, status MIII/FI.
64 Glørstad 2010.
65 Glørstad type III b.
66 Eldjárn 1956, fig. 121, 306.
67 B5804, (Ytre) Eide, Stryn; cf. Table 7.10.
68 B12043 Gjerland, Førde, status FIII: one of bronze (JP 116) and one of silver (R665).
69 B10878. Gjerland, Førde, status FI.
70 B4425, a & b; cf. Wamers 1984, no 25, 121.
71 Wamers 1984, no. 19, 73.
72 Hayeur Smith 2004.
73 Petersen 1940, 51, 101–104, 162; Sørheim 2011 dates the finds to the 9th century, disregarding the latest artefacts.
74 Gaut 2011, 173: B9014, Sandal, Breim; B322-26, Fana prg.; B6657, Trå, Granvin; Bø, Karmøy.
75 Eldjarn 1956, 185–186: Stærri. Árskógi 73 and Álaugarey. No.119.
76 Shetelig 1905, 46-47
77 B9060, Hopperstad, status FIII, B7761, Hopperstad, status FIII.
78 B4841, Hove, Vik, status FI–II.
79 Petersen 1940, 98.
80 Eldjárn 1956, 212–214.
81 Eldjarn 1956, 95–97. Álaugarey No. 119.
82 Eldjárn 1956, 28–31, cf. the grave at Hønsi in Vik, also with textiles with gold threads, as also the grave at Vangsnes.
83 Eldjárn 1956, 180.
84 Eldjárn 1956, fig. 121, 306.
85 Hayeur Smith *et al.* 2019.

8

Textile production in Viking Age contexts

One of the few excavated deserted Viking Age farms from western Norway connects to the community Årdal, in the inner zone of Sygnafylki at Ytre Moa, located on a terrace about 100 m.a.s.l. in Øvre Årdal. Situated in the inner and upper valley, on the eastern side of the long lake Årdalsvatnet and the fjord further out, it may seem remote. It was, however, central in terms of access to wide upland resource areas. The site, excavated in the 1960s, revealed six buildings dated from the 9th to the 11th centuries, and denoted A–F in field reports and later publications.[1] The buildings appeared as small horseshoe-shaped one-room buildings with outer walls of stone, varying in size from 22.5 to 37.5 m^2 internally and with a wide entrance on the gable side. In contrast to the traditional multi-functional longhouses, these one-room buildings represent a structural innovation with differentiated functions. Five of the six buildings showed traces of textile production, four of them with central fireplaces and two without – buildings C and E – the latter tentatively interpreted as a stable. Except building A, which burnt down *c.* 900, they represent the latest phase of the Viking period. Originally the arable land at Ytre Moa included about 17 acres, with several unexcavated grave mounds and clearing cairns (Fig. 8.1).

Buildings A, D, and F also contained cooking pits for food processing. Buildings A, B, and F had remains of inner walls of wood, and with flagstones at the entrances. In all the four buildings with central fireplaces, the debris indicates combined living and work areas. The tools and other artefacts scattered in the buildings shed light on both production and social status, and with artefacts of the same kind as deposited in the Viking Age graves from the study area. The tools for textile production, as well as farming, forging, woodworking, and hunting, reflect both male and female agents. However, the highest number of tools related to textile production – spindle whorls, loom weights, shears, and even the less common spatula-shaped tool R413. The tools for cooking and agriculture included remains of a steatite vessel, quern stones, a sickle, a hoe, and an *ard* share for ploughing, a hammer, a chisel, and a drill for woodworking, and arrowheads for hunting, alluding to work and tasks close by and at larger distances. One of the buildings, building B, also left remains of dress accessories of types found in several 10th-century female graves of high status: beads and brooches of copper alloy – a trefoil brooch, an unfinished pendant in Borre style, and a fragment of an oval brooch (R652). These artefacts indicate, then, that the fireplace was used for advanced metalwork and production of dress accessories of the same kind as produced in the new Viking Age towns. Remains of two caskets and a strike-a-light in the building are other items also used as grave goods in high-status graves. Remains of a glass beaker of foreign origin is not least noteworthy, since otherwise only represented in a few graves and of highest status (cf. Chapters 5 and 7). The arable land, however, was limited and does not represent a farm above the average size. On the contrary, the tools in the buildings indicate that it was rather the outlying resources in upland areas that made it possible to provide a surplus and to achieve and produce such status markers.

Tools and artefacts found in working contexts differ from intentionally furnished graves and represent often damaged, discarded, or lost objects, tramped into the ground inside or outside buildings within their working environments. Some abandoned settlements may also leave somewhat better-preserved material, such as at Ytre Moa. This site stands out in that it represents well-preserved and well-documented *in-situ* finds of both buildings and tools. Reflecting a site with differentiated workplaces it is able to illuminate both types and levels of production. Such workplaces could be spatially integrated in the farm, or dislocated, closer to the outland resources, which seems to be the case at Ytre Moa. This may also explain its nucleated structure and with buildings of the same type as lately have been discovered at higher altitudes in the Årdal mountains and, at yet another contemporary production place, at upland Bjørkum in the Lærdal valley. Such places and finds shed light on people involved in the different work processes – and as for textile production, most frequently spinning and weaving judging

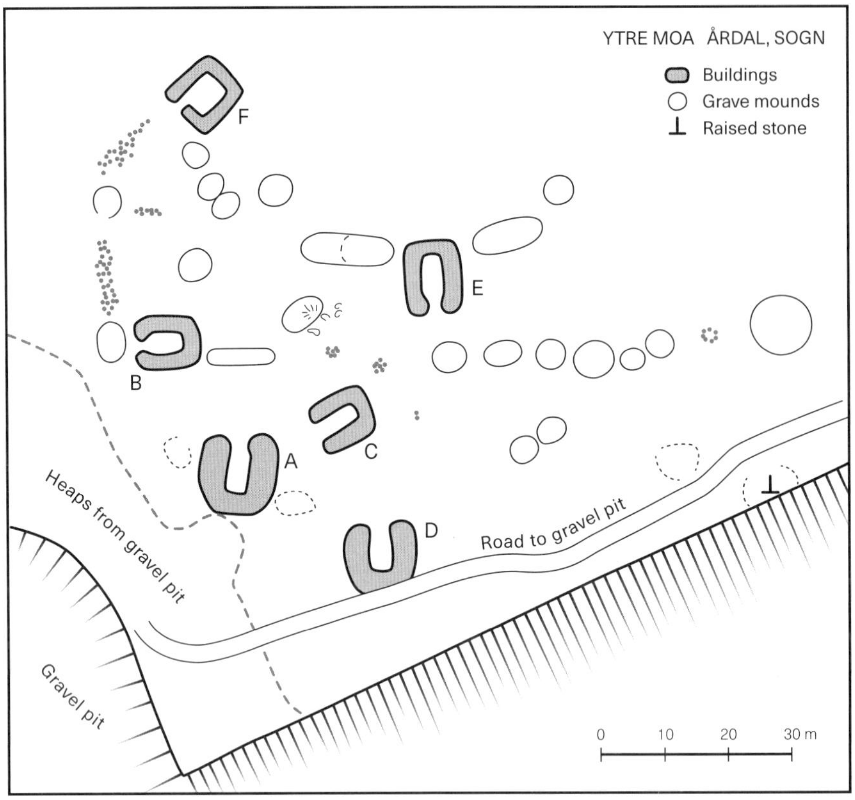

Fig. 8.1: The cluster of buildings A–F at Ytre Moa, Årdal (Illustration: Per Bækken, based on Bakka 1971).

by the preserved remains. Whether the tools indicate household production, combined with other kinds of production, or on an advanced level possibly as a specialised craft, is a central issue when assessing tools from *in-situ* working environments in both lowland and upland areas. In contrast to the grave finds, such traces of tools and textiles from various production places reveal concrete working environments and production practices.

The overall aim in this chapter is to trace and assess tools as indicators of primarily textile production in varied contexts – such as *in-situ* finds of tools from Viking Age production places in both western Norway and in the North Atlantic from the late part of the 9th century onwards. Do they differ from the tools selected as grave goods or tools from the early urban workplaces at Kaupang?

The case of Ytre Moa

The finds of textile equipment in five of the six buildings at Ytre Moa, and in all the buildings with hearths, demonstrate the importance of a production that seems to have exceeded ordinary household production and possibly on a specialised level. All the four buildings with fireplaces had traces of spinning and weaving, which indicate at least one loom in each of the buildings, A, B, D, and F, and with a workspace varying from 30 to 44 m². The spindle whorls

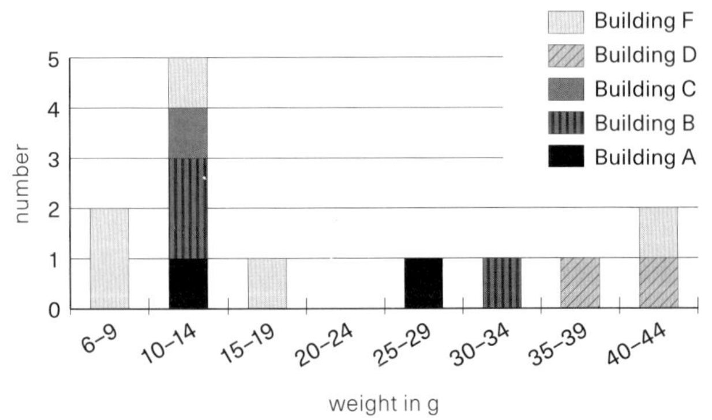

Fig. 8.2: Spindle whorls (N:13 measurable) in groups of 5 g in the five buildings A–D and F at Ytre Moa, Årdal.

left behind, 14 in all and all of stone, mostly steatite, weigh from *c.* 10 g to 41 g, with half less than 15 g (Fig. 8.2). They signify then a production of varied qualities – both fine and coarser.

The largest number of spindle whorls, five in all, connected to the largest building, F, with both a hearth and a cooking pit, and represent a broad weight spectrum: two in the very lightest weight group, respectively 6.4 g and 7.6 g, another two, *c.* 12 g and 15 g, but also the heaviest, *c.* 40 g. The whorls thus reflect spinning yarn of varied qualities in the same working space. Spindle whorls in the weight

Fig. 8.3: The assemblage of spindle whorls from Ytre Moa (N:12) (Fylkesarkivet i Vestland).

group 10–14 g for producing fine yarn were found in buildings A, B, C, and F. The three spindle whorls in building B represent different weight groups too, weighing respectively 10 g, 14 g, and *c.* 30 g in the spectre of light to medium light whorls able to produce both fine and ordinary qualities. Building D, however, contained heavier whorls of 38 g and 40 g for spinning coarser yarn for other use areas. Overall, the spindle whorls signify textile production of varied fabrics for broad use areas, and with internal differences between the buildings (Fig. 8.3).

The finds and weight distribution indicate, then, that many people were engaged in spinning various yarn qualities and that they represent the same weight spectrum as the spindle whorls appearing as grave goods, but in the graves rarely with more than one or a few whorls.

Four buildings had traces of weaving, 10 loom weights in all, of which seven were complete. Traces of the loom, represented by loom weights, related to buildings A and B and in both buildings the weights were spread near the long wall close to the south-western corner, and the two spindle whorls were found near the entrance. Except for a heavier loom weight of 834 g in building F, the others match the most common weight groups represented in the graves, ranging between 224 g and 374 g. Buildings A, B, and D, however, had left only two loom weights each, still representing nearly the same weight range – 224 g and 358 g: 276 g and 374 g, and 317 g and 343 g respectively. If these weights are representative, they also indicate weaving of fine to medium dense qualities. They can be compared with the larger assemblages of weights from some of the graves in Sygnafylki – such as at Nes in Sogndal, and Tokvam and Vinjum in Aurland (cf. Chapter 6, Fig. 6.11), represented by varied weights and in about the same weight range too: between 140 g and 450 g in the Tokvam grave, and from 145 g to 427 g and 160 g to 415 g in the graves from Nes and Vinjum, respectively. The mean weights, however, differ more – at Tokvam *c.* 300 g, Vinjum *c.* 250 g, and Nes with a mid-range mean weight of *c.* 280 g, based on the preserved weights, and rather like those from Ytre Moa.

Shears were found in two of the buildings, A and F, the latter with two, which again indicates a building where several people were engaged in various work processes. The two measurable shears are both medium sized and have a circular loop (R443), which indicates indoor work for cutting yarn and textiles, rather than for sheep shearing.

The largest building F contained other types of artefacts too – a strike-a-light, hones, knives, together with various agricultural tools, a share of an ard, a sickle, and weapons, probably in need of sharpening judging by hones nearby – and all concentrated to areas around the large hearth and along the eastern wall. The find distribution signifies, then, both common spaces and special areas designated for textile production, and with the loom close to the entrance to provide daylight. Food-related

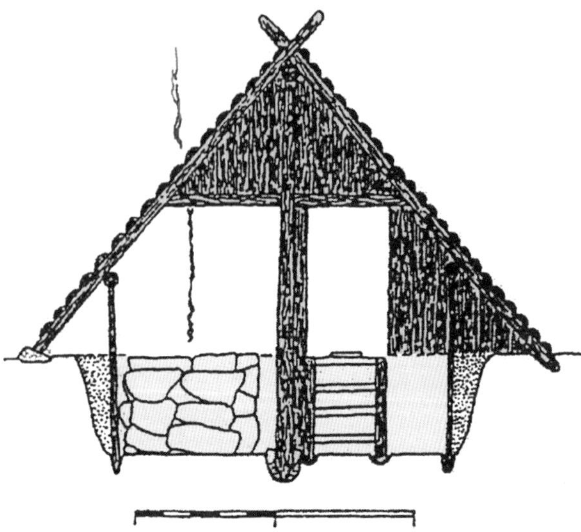

Fig. 8.4: Sketch of a pit house (Summanen 2002; Øye 2006).

Fig. 8.5: Locations of pit houses and working environments from the Viking period within the study area (Illustration: Per Bækken).

equipment, and finds of a hammer, a scythe, and a knife indicate multifunctional use of the buildings for both men and women. A silver foliated bead (now lost) indicates a female presence of some rank.

The building complex at Ytre Moa has long been regarded as an atypical farm structure compared with the traditional three-aisled multifunctional longhouses of the period but has still been interpreted as a Viking Age farm within a basically agrarian framework. The relation and similarities with the newly discovered building structures and complexes of similar construction at higher altitudes that indicate special working environments for various specialised production, seem therefore relevant when assessing tools and environments at Ytre Moa too. This is also the case as for the recent discoveries of a special kind of house construction connected to textile production, so-called sunken-featured buildings, or pit houses, appearing as integrated special buildings at farm level. In contrast to the building complex at Ytre Moa and to the upland pit houses, they tend to connect to central places of high status and often with graves with textile tools too, which further substantiate their role as hubs in production. Such buildings may therefore indicate textile production above the ordinary household production – advanced or specialised.

Sunken-featured buildings: rooms for specialised textile manufacture?

Pit houses, or sunken-featured buildings, are known over larger areas in western Norway and the North Atlantic, both archaeologically and through literary sources. Denoted as ON *dyngja* in literary sources, it refers to the special structure with a sunken floor below ground, and referred to in the sagas,[2] among others, Egil's Saga, where it alludes to events in the 10th century in Aurland and as a place with many women at work, '*ok konur mjæk margar Þar*'.[3] As noted, such houses have recently also been documented archaeologically. They are characterised by their sunken floor, partly dug into the ground, with a light superstructure and a roof often supported by posts and/or walls of stone, often with a hearth, and a wide entrance on the gable side, where daylight could provide light for the loom (Fig. 8.4).[4]

The warm and damp climate inside thus seems to have been especially suitable for working with textile fibres. Archaeologically, pit houses generally connect to various types of production and often textile production, appearing as special workplaces for specialised production in both rural and urban Viking Age contexts. In Norway, relatively few of these constructions have been discovered so far, compared with other areas in Scandinavia and northern Europe, including Iceland.

During recent decades, however, open-air rescue excavations have revealed pit houses in western Norway too, in Sygnafylki and Firdafylki, both in upland areas and in the lowlands and as noted with a tendency to connect to central places – so far discovered at Stedje and Kvåle in Sogndal, Henjum in Leikanger and at Sæbø/Hopperstad in Vik in Sygnafylki and at Osen in Firdafylki and represented with high-status female graves and with textile tools. Bjørkum in the long Lærdal valley and in the border zone between inland and mountainous upland, stands out with an agglomeration of 13 pit houses (Fig. 8.5). Do they share common traits or diverge in terms of types of production?

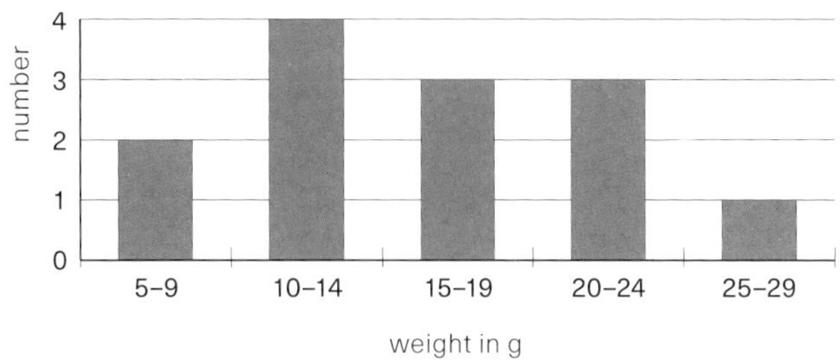

Fig. 8.6: Spindle whorls from the pit house at Stedje, Sogndal, by weight in weight groups of 5 g (N:13) (Graphic design: Per Bækken).

Pit houses at central places: Stedje and Kvåle in Sogndal

Two pit houses have lately been uncovered in Sogndal, at Stedje and Kvåle, both central places of the highest status in the community, and a third in Vik – all in districts with several Viking Age graves with textile tools. Of these, Stedje, with status as a main church centre in the whole Sygnafylki in the Middle Ages, was centrally and strategically located by the Sogndalsfjord on a terrace at the entrance of the Sogndal valley at the western side of the river. Kvåle, also a central place located *c.* 2 km up valley on a terrace, dominated the eastern side of the valley. This was a place that later became a residential place for the King's governor, ON *lendmaðr*, in Sygnafylki and with his private church, and most likely a node in an estate, of older age, and marked by two prestigious female graves with textile tools from the 6th and the 10th centuries (cf. Chapter 5), like also at Stedje with several, large grave mounds and both female and male graves from the 6th century.[5] One of the early female graves echoes the contemporary grave at Kvåle as for types of tools with a short weaving beater and a light spindle whorl of clay.[6] The pit houses, then, reveal working environments at these central places that dominated on each side of the valley, and connected to the landowning elite.

The pit house at Stedje, dated to *c.* 850–1100, represents an interesting textile-producing environment, with its finds of both tools and textiles. The pit house, measuring 5.1 × 6.9 m, or *c.* 35 m^2 internally, contained four hearths – two probably serving as cooking pits.[7] The building comprised tools for all processes – spinning, weaving, sewing, and smoothing, and represented by 16 spindle whorls, of which 13 were complete and measurable and there were additionally four rough-outs. The weight distribution of the complete or measurable spindle whorls varies from 5 g to 28 g (Fig. 8.6), half of them less than 15 g – and generally lighter than those found at Ytre Moa and able to produce both fine and medium fine yarn and textiles. The holes for the spindles vary from 6 mm to 9 mm at the narrowest, signifying spindles of different size and lengths.

The tools also included 67 loom weights, of which some were rough-outs, a possible weaving beater, two small

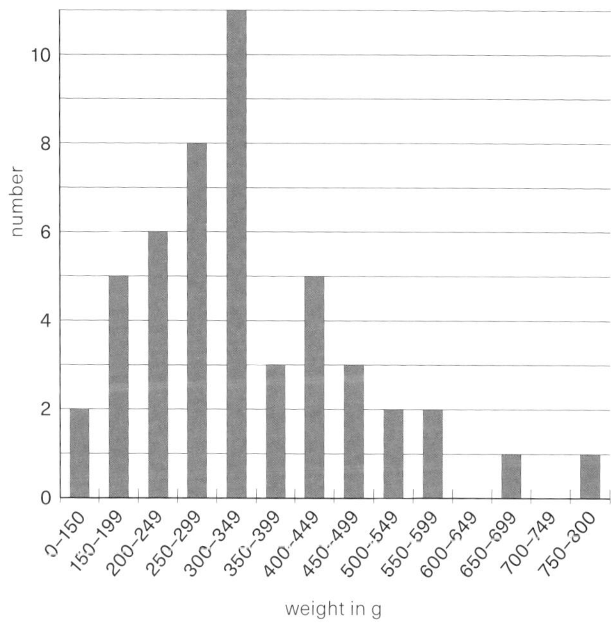

Fig. 8.7: Loom weights from the sunken-feature building at Stedje, by weight in weight groups of 50 g (N:49) (Graphic design: Per Bækken).

needles of bone and bronze, and a possible smoothing stone, together with fragments of burnt textiles. Of the more than 70 loom weights – complete or as rough-outs and fragments, including eight reused parts of soapstone vessels – 49 could be estimated for weight, which varies from 159 g to 754 g. The majority, 33 or 67% of the 49 measurable loom weights weigh between 200 g and 400 g, with a peak between 300 g and 350 g (Fig. 8.7). The weight distribution of spindle whorls and loom weights thus indicates spinning and weaving, primarily of very fine yarn and fabrics of fine and average qualities.

Analyses of the textile fragments found in two carbonised lumps showed that they belonged to two types of textiles of linen, woven as plain tabby and both of z/z and z/s-spun yarn. The thread counts vary within the two types of qualities, from 14–22/10–17 in the warp and weft per cm in the fragments with z/z-spun yarn and in the fragments with z/s spun yarn with thread counts of 10–17/10–19, signifying two qualities within each type. Some of the fragments have a fine gauze-like structure that resembles a thin veil. A similar quality is known from Hedeby and Mammen, denoted as *Schleiergewebe*. Finds of z/s-spun yarn are otherwise rare in Norway in the Viking period – to date only found in a few graves.[8] They seemingly represent another tradition than the common tradition of using z-spun yarn in both the weft and warp, and contrary to textiles produced in Insular areas.[9] Still, the tools found could very well have been used to produce the fine textiles found. The weaving beater needed particularly for producing densely woven fabrics is interesting in this milieu. However, there were no direct traces of textile preparation at the site, such as shearing. The light and different whorl weights and loom weights indirectly indicate both careful sorting and combing of wool to produce the different qualities of yarn and textiles of both linen and wool.

Altogether, the physical environments of the sunken-feature building designated for textile production, the finely woven textiles, the many textile tools, and the production of spindle whorls and loom weights indicate a specialised milieu, where several people must have been working together – spinning, weaving, and sewing. Judging by the size of the building of around 35 m^2, about the same size as the buildings at Ytre Moa, it still appears with two fireplaces and two cooking pits, indicating a working environment for several people. Judging by the many loom weights within different weight spectrums, it had contained more than one loom.

One of the two pit houses, newly discovered at Kvåle was located some 200 m from the clustered medieval settlement and the rich Viking Age grave with textile tools. This structure resembles the sunken-featured building at Stedje with an almost square structure of about the same size as at Stedje, *c.* 36 m^2, and with a sunken floor *c.* 50 cm below the natural surface, however, without tools or artefacts other than a single iron nail. Since not yet ^{14}C-dated, it is dated stratigraphically, possibly to the Early Middle Ages, when connected to some other dated buildings in the area. Whether it served as a work place for textile production specifically, however, cannot be verified by tools. A so-called ON *vefstófa*, denoting special houses designated for weaving, however, is referred to in an official document concerning a division of an inheritance at Kvåle in 1314.[10] It describes the building as part of the clustered hamlet-like settlement and close to the early medieval church at Kvåle. It seems then to build on older traditions.

A pit house at Henjum in Leikanger, Sygnafylki?

Another pit house uncovered in Sygnafylki again connected to a central place and a later medieval church site at Henjum in the agricultural community Leikanger, a neighbouring district to Sogndal. Like at Kvåle it contained neither tools nor other artefacts. The pit house was situated close to the main Sognefjord at a terrace *c.* 40 m.a.s.l. Here, an open-air excavation revealed traces of a long-term settlement with longhouses from the Late Bronze Age, Early Iron Age and into the Viking period. This was a small rounded or oblong pit house of *c.* 15 m^2, without traces of roof-bearing posts, and with a large cooking pit.[11] The sunken floor, *c.* 20 cm deep, had traces of fired clay- and lattice-covered walls, and with a 1.2 m long and 1.3 m wide entrance leading down into the small house with a hearth on its northern side. However, only a piece of iron and remains of an undefined tool were found. Compared with the other pit houses with finds of textile producing implements, it differs in construction and is only about half the size. Since it also lacks tools, its function is rather unclear.

A Viking Age production place in Vik in Sygnafylki

In 2013, the remains of yet another pit house and other constructions were uncovered on the shoreline in Vik at Sæbø – a name meaning habitation by the sea. At an early stage of archaeology, two now destroyed grave mounds were recorded at the place; one of them with a mounting of guilt copper alloy of Insular ornamentation, the other as a weapon grave from the early 9th century.[12] The recent discovery of building structures was the result of a new housing construction project within an area of 890 m^2. However, before the archaeologists were called upon, 50–60% of the area had already been disrupted. The open field excavation that followed uncovered remains of various types of structures – besides the sunken-featured building, cooking pits and garbage pits, related to varied types of production, mainly textiles and metallurgy, but unfortunately by then representing only parts of the past activities at the place.[13]

The small pit house[14] appeared as an almost square submersion in the underground with rounded corners, 3.6 m × 3.8 m, or nearly 14 m^2 and of about the same size as at Henjum and no more than half the size of the pit houses at Stedje and Kvåle. A hearth in the south-western corner was delimited by vertical 40–70 cm high flagstones on three sides. The pit house, looking like a chamber, dates broadly to the Viking period. The ^{14}C-dated layers dated to the 10th century contained three spindle whorls and 10 loom weights of which four as fragments, and a few beads. Of the three spindle whorls recorded,[15] two belong to the weight group 5–10 g, weighing respectively 8 g and 9 g, for spinning very fine yarn, and the third of 31 g for spinning rather ordinary qualities. The loom weights in the same context range between 181 g and 417 g. One of the many pits in the area, possibly a garbage pit, a *c.* 2 m × 1.5 m and 0.5 m deep structure (no. 541), contained, among other things, three loom weights, one of them with a small, engraved cross, shaped like a Thor's hammer, and was possibly used as a mould for melting such small figures as tokens. These discarded loom weights in the pit were generally heavier and of about similar weight as in the pit house (Fig. 8.8).

The discovery of a pit house in Vik with its many rich Viking Age graves with textile tools from the very beginning of the Viking period and well into the 10th century thus provides a context of great interest, not least since the oldest preserved cadaster from the mid-14th century[16] lists Sæbø as a farm that was obliged to pay land rent to Hopperstad Church. This indicates that Sæbø by the fjord was part of an early estate and with Hopperstad as a node and with Sæbø as an interconnected strategic place by the fjord, and archaeologically evidenced as a place for production for both textiles and forging. Compared with the pit house at

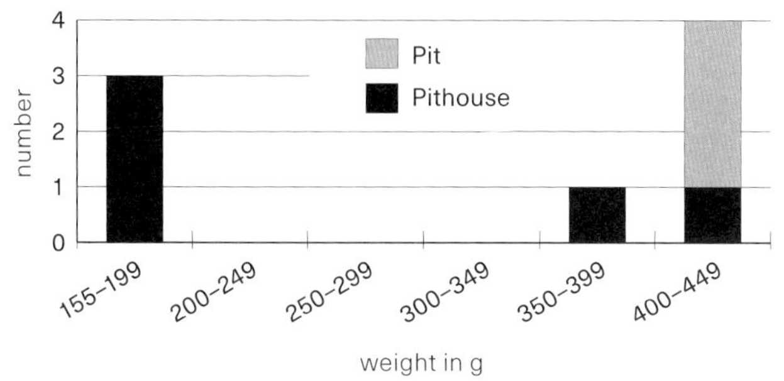

Fig. 8.8: Loom weights from the sunken-feature building at Sæbø, Vik, by weight in weight groups of 50 g (N:8) (Graphic design: Per Bækken).

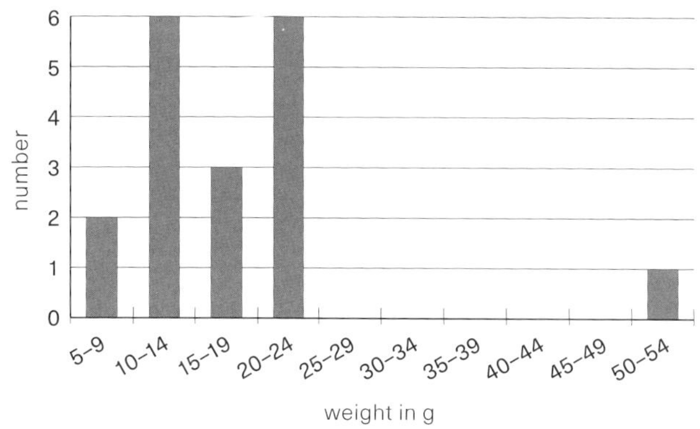

Fig. 8.9: Spindle whorls from Osen, Gaular, by weight in weight groups of 5 g (N:18) (Graphic design: Per Bækken).

Stedje, it was smaller and contained fewer tools. However, the building remains, and pits were only parts of a much larger production place, unfortunately disturbed before the archaeologists arrived at the place. Nevertheless, the few tools recovered all indicate production of textiles of very fine yarn and lighter than those in the graves at Hopperstad (cf. Chapter 5). The production environments at Sæbø thus corroborate Hopperstad's status as a central place and node within a wider estate, functioning as a place for seemingly specialised production

A Viking Age longhouse and pit house at Osen, Gaular in Firdafylki

By the turn of the last millennium, an excavation at the central place Osen in the community of Gaular, also a place with several graves with textile tools (cf. Chapter 5), revealed a pit house, a longhouse, and a female grave with textile tools, all from the Viking period. Osen, situated on a marine terrace at the head of Dalsfjorden, one of the larger fjords in the southern part of Firdafylki,[17] is known through historical records as a central place of transshipment at this junction between waterways and land routes, possibly a thing place too.[18] Here, the remains of a Viking Age three-aisled longhouse *c.* 23 m × *c.* 6 m (Building II) were uncovered. The pit house (A13) 6.25 m × 5 m, or *c.* 31 m^2, was then of about the same size as the other buildings with textile tools. Additionally, 24 smaller pit structures and a Viking Age female grave with textile tools[19] were located on the terrace. There were also traces of an earlier longhouse from the Merovingian period (Building I), *c.* 45 m long and 4.65 m wide at the gables and 6.5 m at the middle and considerably larger than the Viking Age longhouse, which was partly covered by a third building from the Middle Ages (Building III). The Viking Age longhouse, the contemporary pit house, and a pit (A140), all revealed traces of textile production. Altogether 18 spindle whorls were recovered: however, only one in the pit house and all the others in the Viking Age longhouse, mainly in its southern part around the fireplace. Of the 86 complete or fragmented loom weights, as many as 76 connected to the longhouse, and again in the southern part of the building, and evidently a work area. The other 10 connected to the pit house but with considerably fewer tools.

In addition to the spindle whorl and two loom weights in the pit house, several hones, as well as personal belongings, such as seven beads of mosaic, multicoloured glass beads, some faceted, one of amber, with two of the beads fastened by a silver thread, and fragments of a vessel of copper alloy were found, and are of the same type, then, as found as dress accessories and grave goods in high status graves.

All but one of the spindle whorls in these surroundings weigh from *c.* 6 g to 25 g and with a single heavy whorl of 55 g, obviously for plying yarn or spinning very coarse yarn (Fig. 8.9). Of these, the very lightest came from the pit house, while the one in the pit weighed 25 g. The spindle whorl in the female grave from the late 9th century weighed 14 g.[20] The weight distribution of spindle whorls thus resembles the weight spectrum from Ytre Moa.

The loom weights vary in weight from 87 g to 769 g, however, the majority, 35 of the 52 preserved weights, or 62%, weighed between 150 g and 400 g. The two weights in the pit house were in the about same weight range too, respectively 274 g and 329 g (Fig. 8.10). The weight distribution resembles, then, the distribution in the pit house at Stedje, however, most of the textile tools at Osen were found in the longhouse and designated to a special area by the hearth.

The finds of pit houses and places of production at Stedje in Sogndal, Sæbø in Vik, and at Osen in Gaular, share many common traits – they all connected to central places, located close to the fjord and in areas indicating many-sided production, besides textile production, also forging and appearing with cooking pits, both at Sæbø/Hopperstad and Osen. However, at Osen, the traces of textile production are clearest in the longhouse. The sunken-feature building at Stedje, of about the same size as at Kvåle and Osen, although somewhat larger and with two fireplaces and two

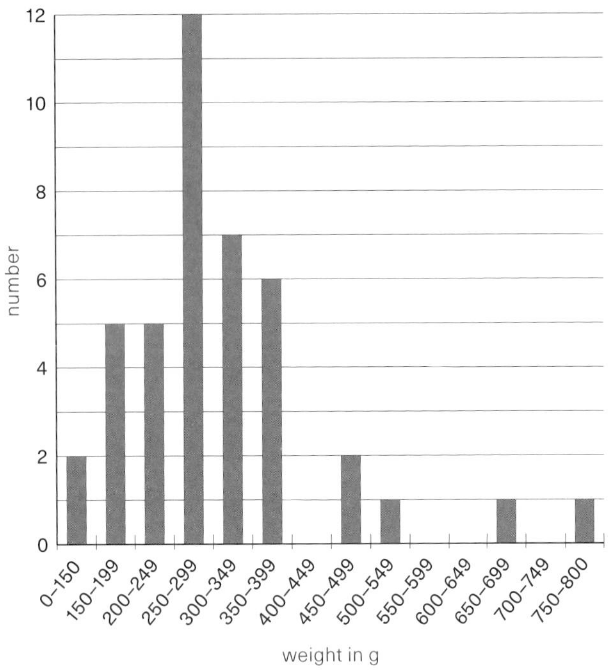

Fig. 8.10: Measurable loom weights from various buildings at Osen, Gaular, by weight in weight groups of 50 g (N:42) (Graphic design: Per Bækken).

cooking pits, stands out with a higher number of textile tools and with better preservation conditions with remains of textiles too. In all these cases, the tools left predominantly indicate production of finer qualities. Altogether, these special environments indicate a specialised production above household level and where the pit houses served as workshops connected to these central places. The buildings at Sæbø and Henjum were smaller and less clear as for function, like also at Kvåle but most likely connected textile production, and at Sæbø, probably connected to Hopperstad as the central place.

Bjørkum in Lærdal: a specialised and seasonal production place

At the margins of the farm Bjørkum in the inner part of the Lærdal valley in Sygnafylki, around 20 km upstream from the river's outlet to the fjord, a large multifunctional site situated 130 m.a.s.l. was uncovered in 2009 (Fig. 8.11). Bjørkum represents an extraordinary site in the Norwegian context, with an agglomeration of 13 pit houses and with traces of varied production. Being situated in the transition zone between the agricultural inland fjord district to the west and high mountains to the east, it appeared as a meeting place for utilisation and processing of different resources from different eco zones – areas for hunting and gathering as well as grazing and animal farming, and wool production.

Of the 13 pit houses and two other two- or three-aisled house constructions, two of them had fireplaces, while all the others lacked hearths. The building complex is ^{14}C dated to the early Viking period, 750–850 and with activities into the latter part of the Viking period. They concur thus in time with the early expansion of graves with textile tools as grave goods. These pit houses were circular structures, while two of the youngest were rectangular but of a lighter construction than those referred to in the lowland, such as at Stedje. To date, the 13 pit houses constitute the largest assemblage recovered in Norway. The excavations revealed moreover about 40 cooking pits and otherwise traces and finds related to different kinds of craft production, among others remains of a smithy, and traces of working with copper alloy, iron, and bone/antler – besides textiles. There were also remains of artefacts and tools for daily use, such as knives as well as other personal belongings, among others, beads, fragments of brooches, keys, horseshoes, and not least, production waste, bones, and metals (Fig. 8.11).

Ten of the pit houses contained finds of textile-production equipment, especially for spinning and weaving.[21] Altogether 23 spindle whorls, 20 more or less complete, were found and of varying weight from 5.4 g to 33 g and a single whorl above 50 g. The whorls had a peak within the weight groups 15–24 g for spinning fine/medium fine yarn (Fig. 8.12).

The largest number was found in two of the buildings, pit house 1 with three whorls, and pit house 33 with five whorls of varying weights. The latter was represented with both the lightest to the heaviest whorls, from 5.7 g to above 50 g, two in weight group 15–19 g and one in weight group 20–24 g. The pit house thus included whorls for producing all types of yarn and textiles, from the very finest, average fine, to coarse, and in this respect, resembling the assemblage of spindle whorls at Stedje and Osen (cf. Figs 8.6 and 8.9). The three whorls in pit house 1 varied in the weight group 10–14 g, and two in the weight group within the range of 24–30 g, able to produce fine and more average yarn. The two pit houses, Nos. 6 and 111, with two whorls each, also represented different weights, the latter for producing very fine and average yarn, and in pit house 6 in the weight groups 10–14 g and 20–24 g, for generally fine yarn. Theoretically, at least, they could have produced yarn for the warp and weft in the same loom.

Five of the buildings, Nos. 1, 2, 5, 6, and 33, all with tools for spinning, contained loom weights too, used then, in combination with weaving and other areas for textile production. Of the 28 weights revealed, 19 connected to eight of the pit houses and with the largest assemblages in pit houses 5 and 1, with eight and three weights, respectively. The others, structures 2, 33, and 111, plus Nos. 4, 7, and 37, contained only one or two whorls each, and the rest in other contexts. The complete and measurable loom weights vary from 94 g to 872 g, with the majority between 200 and 400 g. The eight

Fig. 8.11: The site at Bjørkum, covering an area of 1500 m², carried out as an open field excavation, which revealed remains of 13 pit houses (Photo: Morten Ramstad, with permission).

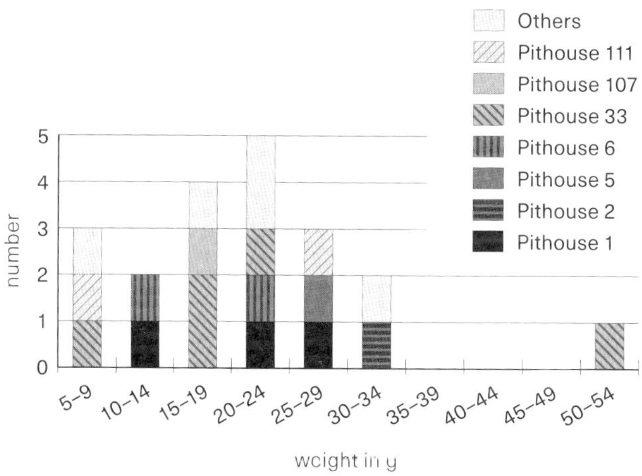

Fig. 8.12: Measurable spindle whorls from the pit houses at Bjørkum, by weight in weight groups of 5 g (N:20) (Graphic design: Per Bækken).

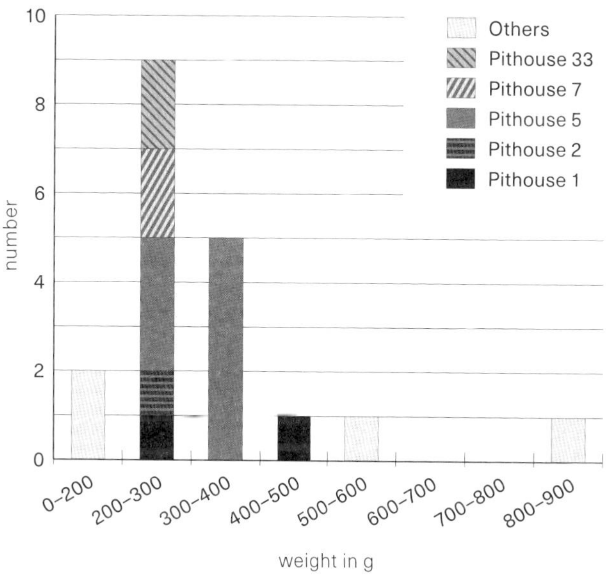

Fig. 8.13: Measurable loom weights from the pit houses at Bjørkum, by weight in weight groups of 100 g (N:19) (Graphic design: Per Bækken).

loom weights in pit house 5 are rather equal in weight, from 233 g to 384 g, but otherwise the numbers are too small to give a representative impression of the fabrics woven on the looms. Still, their presence indicates that weaving was carried out in many of the houses. Four of the weights in other contexts, represented both very light, and light loom weights, including the average and the heaviest loom weights (Fig. 8.13). Of other textile production equipment, 11 smoothers or possible smoothing stones were found in pit houses Nos. 2, 4, 6, 37, and structure 490. Furthermore, 11 needles, or parts of bone and iron, were uncovered, again in the main areas for textile production – five in pit house 1, one each in pit houses 2 and 33, and the rest in other contexts.

Fig. 8.14: Overview of the landscape in Vikadalen at Nyset-Steggje, where four building remains contained textile tools (Photo: © University Museum of Bergen).

Judging by the building structures, tools, and other indicators, such as absence of hearths, the place seems to have been mainly used as a seasonal workplace, and probably also served as a meeting place between west and east, based on outlying resources in the upland and mountainous areas with access to wide grazing land, hunting areas and iron ores at short distances. Although the loom weights are few, their presence indicates weaving in at least five buildings though probably more since the looms must have been dismantled and loom weights reassembled when the annual working season was over.

Textile production in outlying mountainous areas

Seasonal occupation sites in mountainous areas seem to have been rather common in western Norway in the Viking period and Middle Ages,[22] as also later. Since the 1970s, large-scale excavations, carried out in relation to the establishment of new hydroelectric power schemes, uncovered many house sites from these periods in high lying shieling areas. Many of them contained traces of textile production. At Nyset-Steggje in the Årdal mountain area in Sygnafylki (Fig. 8.14), 16 sites with clusters of house sites of 33 Iron Age buildings were located at 950–1300 m.a.s.l. in the climatic forest limit, where the landscape opens with wide areas of land for grazing, hunting, and gathering.

Of these, 24 buildings from nine different areas were excavated.[23] The buildings, often two or three at the same site, revealed small one-room buildings of about the same construction, size, and shape as found at Ytre Moa in lowland Årdal, with an inner floor of 15–30 m^2 and an opening on the gable side. Generally, they have been connected to summer farming, animal husbandry, and dairy production. However, the zoo-osteological material from the Viking period, revealed sheep and goats as the dominant fauna, besides pigs and to a lesser degree, cows. Judging by the textile tools in the buildings, textile production was very important too, as substantiated osteologically in the zoo-osteological material.

Altogether 16 of the 24 excavated buildings cover the Viking period or earlier, with some stretching into the Early Middle Ages. One area, Vikastølen, *c.* 960 m.a.s.l and with a cluster of four buildings from different phases, dated from AD 700–900 into the Early Middle Ages, represents one of the most extensive excavations in the area (Fig.8.14). Here, all the buildings revealed traces of textile production – two with six and seven spindle whorls each and 16 loom weights, plus fragments and rough-outs. The spindle whorls were mostly made of steatite of varying shapes, like those found in the lowland, some also decorated with concentric or vertical lines. Many of the whorls were damaged with only 26 measurable or with estimated weights (Table 8.1; Fig. 8.15). They range from 7 g to 30 g, with a peak in the weight group 20–24 g, and 88% under 25 g.

One of the buildings in Vikadalen. Loc 34, B, had six spindle whorls representing different weights from *c.* 7 g,

*Table 8.1: Buildings at Nyset-Steggje from the Viking period with textile tools, * = estimated.*

Sites	*No.*	*Contexts*	*Dating*	*Whorls*	*Whorl weights in g*	*Loom weights N:*	*Other tools*	*Status indicators*
Loc 34	B14034	Site A Site C	AD 650–890	5 1	13, 20, 21, 22, 24, 24g*	16	1 hone	
Loc 34	B14035	Site B	AD 580–960	6 + 1*	7, 8*, 9, 12, 14, 15, 30	2	2 knives	8 beads, pieces of amber
Loc 34	B14039	Site G	AD 890–1020	1 (half)	20*		Fragments of hones	
Loc 100	B14100	Site A	AD 880–1040	4 + 3 fragments	6, 15, 21, 24	2 (309 g, 388+)	1 awl, 1 chisel, 7 hones, knives, 2 arrowheads	3 beads
Loc 100	B14101	Site B	AD 720–1020	–	–	1 (broken)	2 arrowheads, 2–3 knives	
Loc 82	B14078	Site F	AD 880–1020	2 broken	11g*, 15*	1 (286 g)		4 beads
Loc 122	B14122	Site B	AD 880–1020	**8 (5)**	4, 12, 13, 14, 25	3 (broken & as fragments		
N		7 house sites (one double)		26 + fragments	25 measurable	25 measurable		

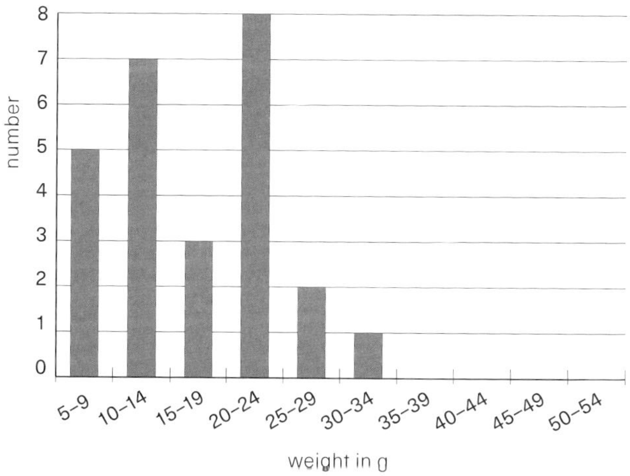

Fig. 8.15: Measurable spindle whorls from Nyset-Steggje, by weight in weight groups of 5 g (N: 26) (Graphic design: Per Bækken).

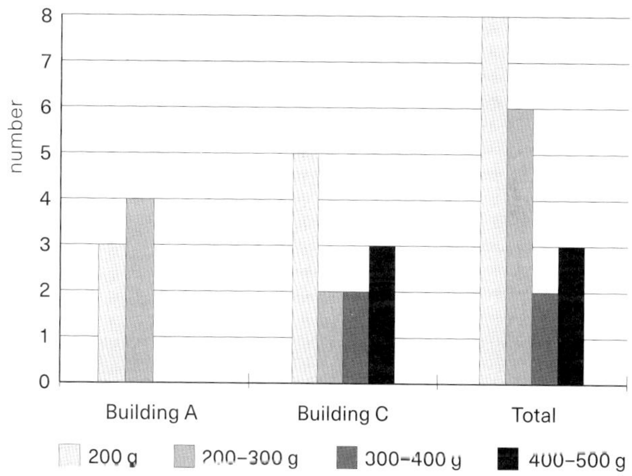

Fig. 8.16: Measurable loom weights from building sites at Nyset-Steggje, by weight in weight groups of 100 g (N:19) (Graphic design: Per Bækken).

9 g, 12 g, 14 g, 15 g, and 30 g for varied yarn qualities. In another area, the valley Fossdalen, one of the house sites, Loc 122, B, from the late 9th and 10th centuries included eight whorls within a similar light weight range (cf. Table 8.1; Fig. 8.15). In contrast to the other areas with textile tools, the easily transportable tools for spinning are far better represented than for weaving.

Compared with the other sites, the loom weights in the shieling contexts are generally rougher in shape, often made of a harder mineral, with a fair number of rough-outs, and many broken or damaged. The few that are complete appear in about the same weight groups as in the lowlands but the damaged loom weights indicate a somewhat heavier assortment, above 400 g (Fig. 8.16).

Differences and similarities between production places and environments

These seven locations, in lowland and upland areas in Sygnafylki and Firdafylki with remains of workplaces for textile production share many common traits but they also reveal

Table 8.2: Representation of textile producing equipment at the six areas.

Sites	*Context*	*Dating*	*Whorls*	*Weights*	*Beater*	*Shears*	*Needle*	*Smoother*
Ytre Moa, Årdal	5 buildings	Viking Age	14	10		3		
Stedje, Sogndal	1 pit house	Viking Age	16	72	1	1	2	1
Kvåle, Sogndal	1 pit house	Viking Age						
Sæbø, Vik	1 pit house	Viking Age	3	10				
Osen, Gaular	1 longhouse 1 pithouse	Viking Age	18	86				
Bjørkum, Lærdal	10 pit houses	700–900	23	28	–	–		
Nyset-Steggje, Årdal	7–8 house sites/buildings	700 –1000	26	22	–	–	–	–
N			100	228	1	4	2	1

differences – geographically, socially, and technologically as for their ability to produce various types of textiles. At some places, the textile production connected to other types of production, forging and metalwork, and work in bone and antler and all related to the Viking Age, especially to its first part, as far as they can be precisely dated. Six of them connected to the inner zone of Sygnafylki, to the communities of Årdal, Sogndal, Vik, Lærdal, and one to the mid-zone of Firdafylki. Four related to central places of high status at Stedje and Kvåle in Sogndal, Sæbø/Hopperstad in Vik, and Osen in Gaular; all strategically located as for communication and control. Four of these places had several buildings with traces of textile production too – Osen in Gaular, Bjørkum in Lærdal, Ytre Moa in Årdal, and Nyset-Steggje in the Årdal mountains. The sites, the varied house sites, and building types reflect social differences with different opportunities and requirements for textile products, reflected in the tools, both in quantity and quality and by their specific environments.

Altogether, the sites have left remains of 328 objects for analysis, 100 spindle whorls and 228 loom weights, including fragments and rough-outs, a weaving beater, four shears, two needles and a pellet for smoothing. The assortment of finds is clearly affected by the conditions for preservation and the way they are deposited through use, or intentionally selected as grave goods. Generally, the find conditions are unfavourable for organic material, which means that tools of stone such as spindle whorls and loom weights and other durable materials are overrepresented. However, of these places, Bjørkum is an exception with favourable find conditions for bone and antler, though only rarely related to textile production. Accordingly, spindle whorls and loom weights are the most common manifestations of the production. Still, as discarded artefacts, they are often broken and left as rough-outs, and not always retrieved by the museum either. Yet, the numbers, much larger than those of selected grave goods, their range and different weight groups should be indicative of the quality and extent of textile production (Table 8.2).

The spindle whorls vary in weight for producing different types of yarn for the weft and the warp and for different qualities and use areas, and they also vary at the different sites. Four of the sites stand out with spindle whorls of the lightest category, around 5 g, able to produce the very thinnest yarn for high quality products – at Stedje, Kvåle, Osen, Bjørkum, and Nyset-Steggje (Table 8.3) – but closely followed by the two other sites – Sæbø and Ytre Moa – with spindle whorls only a few grams heavier. Of all these assemblages, the lightest relate to Stedje with a range from *c.* 5 g to 28 g and peaking in the weight group 10–14 g and with the majority, 87%, under 25 g. The assemblage from Nyset-Steggje represents a wider array of whorls, the heaviest weighing more than 50 g but, like Stedje, with a peak in the weight group of 10–14 g and with a large majority, 92%, under 25 g. The 20 measurable whorls from Bjørkum, with a weight range from *c.* 5 g to 33 g, are generally heavier and more homogenous with a peak in the weight group 20–24 g and represented with 74% under 25 g. The 18 whorls from Osen, ranging from around 5 g to 55 g show two peaks, one in the weight group 10–14 g, the other from 20–25 g and 94% under 24 g. The few spindle whorls from Sæbø, two are light, 8 g and 9 g respectively, and one just above 30 g, and were used, then, for producing rather different textile qualities (Table 8.3).

The loom weights represented in the same contexts also vary both in number and weights but are, as noted, often rather fragmented and with a minor percentage of measurable weights within a wide weight spectrum but generally with a majority weighing less than 400 g. The largest assemblages relate to the high-status places Stedje and Osen with, respectively, 76 and 82 loom weights. However, with only 49 and 42 measurable, they range from light to heavy weights of 754 g and 769 g respectively, both with a peak in the weight group 250–350 g. Generally, all these places comprised a few heavy and a few light weights, and with peaks in loom weights in the mid-range – 200–300 g at Bjørkum, 300–400 g at Ytre Moa, and 400–450 g at Sæbø. These differences, as well as the number of tools within

Table 8.3: Representation of complete and measurable tools in relation to the assemblages of spinning and weaving implements at the six sites. Weight range and relative representation within the range is marked in brackets.

Places/Districts	*Spindle whorls (measurable weights)*	*Weight range (g)*	*Peak: weight group (% under 25 g)*	*Loom weights (measurable weights)*	*Weight range (g)*	*Peak: weight group (% under 400 g)*
Ytre Moa, Årdal	14 (14)	10–41	10–14 (57%)	10 (5)	223–834	300–400 (90%)
Stedje, Sogndal	16 (13)	5–28	10–14 (87%)	72 (49)	159–754	250–350 (61%)
Kvåle, Sogndal	–	–	–	–	–	–
Sæbø, Vik	3 (3)	8–32	5–9 (67%)	10 (8)	189–443	400–450 (50%)
Osen, Gaular	18 (18)	5.7–55	10–14 &20–24 (94%)	86 (42)	87–769	250–300 (88%)
Bjørkum, Lærdal	23 (20)	5–33	20–24 (74%)	28 (24)	94–872	200–300 (67%)
Nyset-Steggje, Årdal	26 (24)	4–49	10–14 (92%)	23	286–800t	?

the structure at the sites, also seem to signify differences in the textile producing environments and the organisation.

How many looms the loom weights represent, is difficult to ascertain as they are often preserved as fragments and leftovers, when dismantled after the season. This may often have been the case at least in seasonal-based buildings, such as at Bjørkum and Nyset-Steggje. These were also both sites sparsely represented with loom weights. At Ytre Moa, their place in the building and size indicates generally one loom each in four of the buildings, while possibly two looms in the sunken feature building at Stedje, as a larger building. At Osen, a loom had been in use in the Viking Age longhouse too. The small pit house at Sæbø could only have housed one loom. Based on the estimate of looms, the ratio of spindle whorls to looms at these sites would be 3.5:1 at Ytre Moa, 2.9:1 at Bjørkum and 3:1 at Sæbø, while 7:1, at Nyset-Steggje, 8:1 at Stedje, and at Osen 9:2, which is more in accordance with the effort of producing enough yarn for the weave and for various gauges for the warp and the weft, and for producing different textile qualities.

In contrast to the workplaces in the lowland and at Bjørkum, the textile producing environments in the Årdal mountains were spread over larger areas at places connected to other activities in these vast shieling and hunting areas. It seems that textile production – wool collecting, sorting, fibre preparation, spinning, and weaving – was combined with at least herding and dairying but also hunting. Spinning also dominates as for tools and the finds of spindle whorls have the widest weight spectrum of all, from under 5 g to nearly 50 g, and able to produce yarn for a wide array of fabrics. The wide range of spindle whorls thus reflects use in a multi-purpose textile production, aimed to supply wool and yarn for different types of fabrics, and not least noteworthy, including the same fine types of fabrics as in the workshops at the central farms, at Stedje, Kvåle, Sæbø, Osen, and at Bjørkum and Nyset-Steggje. At Bjørkum, the production was most likely organised as seasonal manufacture in the mountains in the summer and at connected farms in the winter. At Ytre Moa, deserted in the 11th century, the spindle whorls had a wider weight range than at the other sites – from 10 g to 41 g.

Comparing the weight distribution of loom weights in the pit houses from Bjørkum and Stedje, a rather similar pattern emerges. Although the number of weights differs, they both range from less than 150 g to about 800 g, with the majority concentrated around 200–400 g. Loom weights above 800 g seem to be represented at sites too, dated later than AD 900, which is also in accordance with observations in other areas.[24]

Of all the sites, the building complex at Ytre Moa represents an atypical farm structure compared with the traditional three-aisled and multifunctional longhouse of the period as represented at Osen. At Ytre Moa the structures are more like the buildings and building clusters at Nyset-Steggje, both in shape and size. That five of the buildings, and only two of them had a hearth and used for textile production is also atypical. There are several contradictory factors as for its status and organisation. The many grave mounds close to the buildings at the site at Ytre Moa (cf. Fig. 8.1), and that several of the buildings, A, B, and F., revealed remains of different status objects too, indicate a higher status than would be expected from the modest buildings, the small arable land area, and its early deserted status. One of the buildings, house site A, had remains of a trefoil fibula, while yet another, building F, contained a buckle of copper alloy, a pendant of gilt copper alloy from a necklace, and some beads, together with the textile-producing equipment and other tools. It is therefore difficult to ascertain whether it had a subordinate status to the larger farm Moa in Årdal as perhaps a later special production unit, or whether it was a farm in its own right. In the early modern period, Moa appeared as one of the larger farms in the community of Årdal with a large livestock that, considered as a minimum,[25] would provide an annual yield of *c.* 35–70 kg wool from the sheep in addition to possible fibres from the goats. The strategic location of both Bjørkum and Ytre Moa at the crossroads of three mountain valleys, with access to wider resource areas may have played an important role for production. This may also explain the emphasis on textile production at Ytre Moa, and seemingly an advanced production above household level.

These seven cases represent rather different types of sites and buildings, reflecting diverse forms of textile production, scales, and organisation. The sunken-feature buildings at Stedje and Kvåle, the smaller pit houses in a working area at Sæbø in Vik, probably connected to Hopperstad, at Osen with textile production in a longhouse, the agglomeration of buildings at Bjørkum and Ytre Moa, and spread over larger areas at Nyset-Steggje in the mountains, represent altogether, dissimilar environments, and social contexts. While the sunken-feature building at Stedje and the longhouse and pit house at Osen, with the largest assemblages and a female grave of high status with textile tools too, represent textile production at central places, possibly on a specialised level within the elite, the clustered horseshoe-shaped buildings at Ytre Moa in Årdal and at Nyset-Steggje in the Årdal mountains signify other types of production environments. Ytre Moa as a clustered all-year settlement with traces of varied production split up in several buildings and within an elite economy was probably based on a surplus that depended on outland resources, not unlikely connected to the production in the Årdal mountains. Here, the sites were spread at places over larger areas and based on seasonal activities. The different groups of buildings in the Nyset-Steggje were probably connected to farms or central places in the fjord districts of Årdal, Lærdal, and Aurland. Ytre Moa with the similar house constructions and close for access should, then, probably also be seen in a larger social framework than as a separate farm unit.

The shieling areas with traces and buildings for textile production thus seem to be connected to and integrated in the farm economy, related to farms in the fjord districts of Sogndal, Vik, Årdal, and Lærdal in Sygnafylki and Gaular in Firdafylki – and probably an advanced production above the average, as also indicated in high-status graves from the lowland areas. The structures and tools from both Bjørkum and Nyset-Steggje, seemingly based on seasonal production and exchange of products, raw materials, semi-finished, or finished products or textiles, bone and antler, iron, minerals, and others, is also substantiated by the finds of weights and a scale. Still, the sites represent rather different environments, where Bjørkum indicates a seasonal workplace of communal character, judging by the buildings that lack fireplaces, have many open-air cooking pits and not least by the number of buildings, including a smithy. Altogether, the buildings and finds at Bjørkum and Nyset-Steggje indicate a large-scale industry based on surplus production, of both textiles and other production, such as comb-making, forging, producing various tools of bone, antler, and iron, such as knives and various agricultural tools.

Hjelle in Oppstryn in Firda: an assembly place or a place of production?

The concentration of 11 one-room buildings with entrances facing an inner courtyard at Hjelle in Oppstryn, *c.* 30 km inland from the fjord and so far, interpreted as a thing site,[26] may, however, also relate to large-scale production, as reflected in the Viking Age graves in this upland area with access to both grazing land and with nearby access to iron. These buildings both differ and share traits with the outland buildings in Årdal in Sygnafylki also in the inner and upland areas. The buildings at Hjelle vary in size, *c.* 8–12 m in length and 4–5 m in width, furnished with a long hearth in the middle, with a superstructure and a roof supported by posts and with slightly rounded walls, built of stone and earth, and with entrances facing an inner 'courtyard'. The artefacts left are limited, and as for tools, a pair of shears, knives, a key ring, a fishhook, hones, flints, rivets, and iron nails, which may possibly indicate boat building. A few beads and an Anglo-Saxon coin were also found, the latter a *scyta*, coined by master Eanwich under Ethelred in Northumbria (AD 810–840), which matches the 14C-dating to the period AD 810–880,[27] which again matches many of the graves in this area with tools for varied types of production and represented with Insular artefacts too (cf. Chapter 5).

The concentration of buildings, and area of fire-crooked stones from the more than 60 cooking pits over a large area of 2500 m^2 behind the buildings, may indicate a large concentration of people to be lodged and fed over longer time spans than a few days' meetings. Whether they served as stays during seasonal thing gatherings, as has been suggested by the excavators, or served other functions related to production is not so easy to determine by the findings. There are, for instance, no traces of tools for textile production. Nails and rivets, however, could indicate both forging and boatbuilding, as also revealed in many graves in this area, including a large share of boat graves in the area (cf. Chapter 5). The many graves, both female and male with tools for various work processes, including forging, wood working, boatbuilding as well as textile production, based on a large work force in need of co-operation, would have required large-scale workplaces and accommodation too, as well as access to raw materials. Could Hjelle, at the outskirts of the county, and as a rather peripheral location for a thing assembly, have been such a working place? It concurs in time with the surrounding graves on these northern fringes of Firdafylki, both male and female and supplied with various types of tools, indicating large scale production related to outlying resources

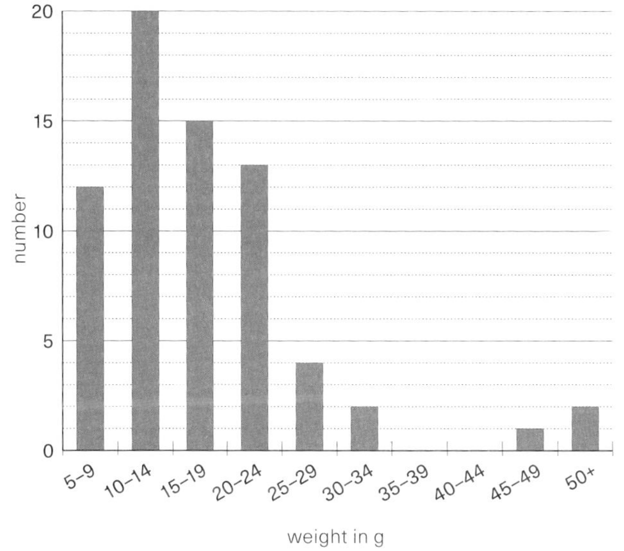

Fig. 8.17: Spindle whorls from the various Viking Age rural workplaces, by weight in weight groups of 5 g (N:69) (Graphic design: Per Bækken)

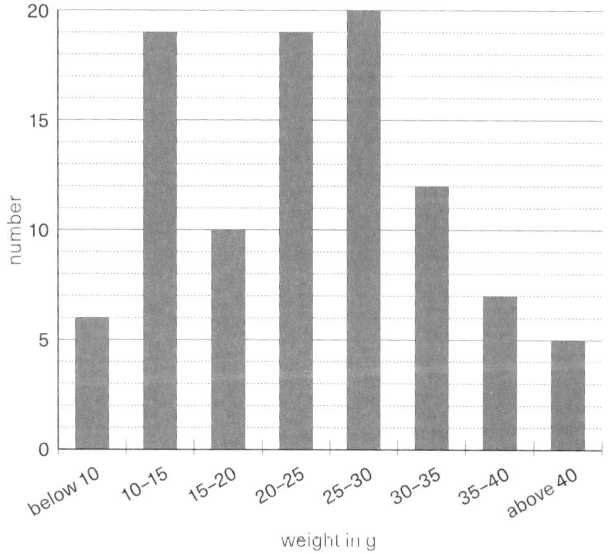

Fig. 8.18: Spindle whorls from Viking Age grave contexts, by weight in weight groups of 5 g (N:98) (Graphic design: Per Bækken).

– wool and wood and close connection to iron producing environments to the east in these upland areas of Stryn. So far, however, this is an open question.

Altogether, textile production seems to have been an important element at all the sites referred to in this chapter. They all indicate an economy based on the extensive use of outlying resources to maximise the output of the farms; beside the manufacture of textiles, there was also the production of iron, extraction of the minerals, and the hunting and gathering of the period. This is not least reflected in the many graves revealed in the same areas and districts as these work areas, thus indirectly reflecting the social aspect of production, as shown in the preceding Chapters 5 and 6.

The tools in varied contexts: signifiers of varied production too?

When comparing the tools as leftovers from Viking Age rural workplaces with tools intentionally given as symbolic laden grave goods in varied grave contexts, socially and geographically, they may give a more nuanced impression of textile production – how they match or mismatch. Here, measurable spindle whorls may serve as indicators of textile qualities – whether in the finer, ordinary of coarser yarn qualities. The measurable spindle whorls from the workplaces referred to count 69 whorls and 98 in all from all the graves with spindle whorls.

Altogether 65 of 69 measurable whorls, or 87%, stemming from various workplaces, weigh overall from *c.* 5–6 g to 25 g, with as many as 12 under 10 g and very fine qualities and with a peak of 20 specimens in the weight group 10–14 g, also within the finer segment and another 15 between 15 g to 19 g and 17 between 20 g and 34 g and more common yarn qualities and only three of more than 45 g for rather coarse yearn (Fig. 8.17).

Although the whorls in the graves represent nearly the same weight range, from under 10 g to above 40 g, they show a more even spread within the weight groups from 10 g to 30 g and with two peaks, one between 10–15 g, as from workplaces, and the other between 20 g and 30 g and for different use areas, spinning fine and medium fine yarn. They are, however, generally heavier than the ones from production places, with 44 whorls weighing more than 24 g, or 45% of all the measurable whorls, in contrast to 13% of the whorls from the various working environments (Fig. 8.18). The spindle whorls from production places are thus generally lighter than represented in the graves.

Figure 8.19, which marks the relative representation of the spindle whorls in these different contexts, makes the differences even clearer. Although fewer, the whorls in working environments are generally lighter than in the individual graves. They indicate production of both finer and of more uniform qualities. Only to a limited extent do they indicate production of coarser fabrics, such as sails – but nor do they indicate standardised cloth.

Traces of textile working areas in the North Atlantic

What is the case in the North Atlantic – does the production follow old traditions or change based on new and other opportunities on new ground? Many of the textile production tools in the North Atlantic are found in specialised work environments – in Iceland, often in small pit houses, representing special work zones for weaving and, to some

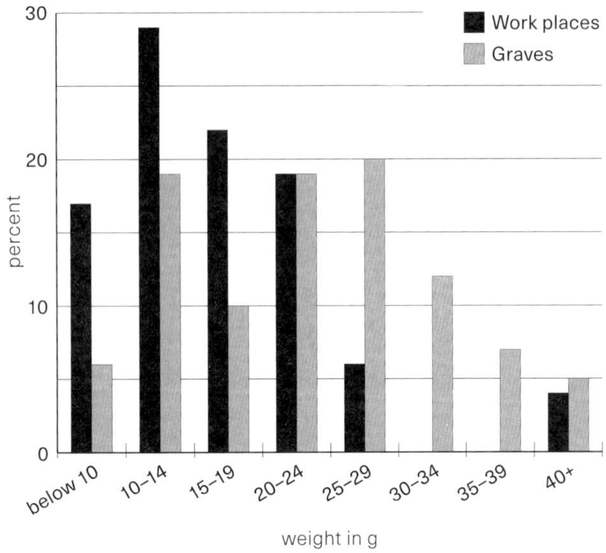

Fig. 8.19: Relative representation of spindle whorls from workplaces (N:69) and graves (N:98) in weight groups of 5 g (Graphic design: Per Bækken).

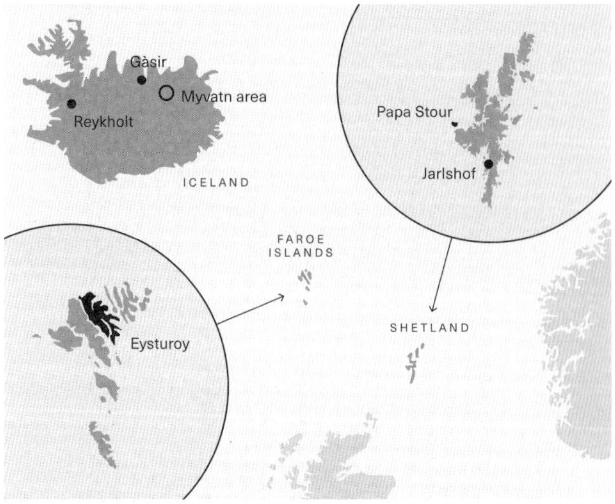

Fig. 8.20: Areas in the North Atlantic mentioned in the text (Illustration: Per Bækken).

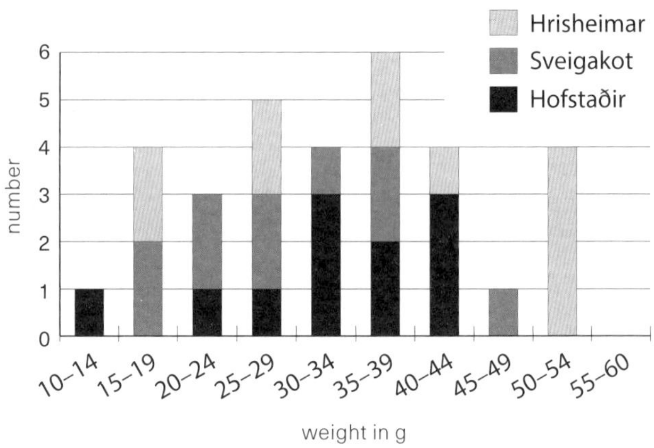

Fig. 8.21: Measurable spindle whorls from Hrisheimar (light grey) (N:11), Sveigakot (grey) (N:10), and Hofstaðir (black) (N:11) (N:32) according to weight groups of 5 g (Graphic design: Per Bækken).

extent, spinning, as also documented in western Norway. A recent study of the Icelandic pit houses has identified 24 such houses from 16 different Viking Age sites, of which the majority, 76%, have traces of textile production.[28] The pit houses appear as rectangular or square structures, measuring between 8–16 m^2 internally, and often with a fireplace in a corner – offering a workplace for about two people for spinning and weaving, resembling, then, the finds from western Norway. These structures seem to have been rather common in Iceland from the late 9th to the 11th century before the loom and textile work was taken into the main longhouse.[29] This change was evidently not accidental, and it is interesting to note that it coincides with the abolition of the 'un-free' workforce of thralls. The building type is known from Shetland too, but not at this time from the Faroe Islands, where textile production also connected to longhouses (Fig. 8.20).

The finds of textile tools and in some cases, textile remains, in such surroundings make it possible to assess work processes and products contextually from three deserted Icelandic settlements in the Mývatn area: Hofstaðir, Hrisheimar (*c.* 930–1050), and the somewhat later settlement at Sveigatkot (*c.* 950–1150) with all representing early textile production environments.

In the Faroe Islands, three Viking Age and Early Medieval sites stand out with larger assemblages of textile tools: the *landnám* farm, Toftanes (*c.* AD 870–1020), í Uppistovubeitinum from the late Viking period and deserted in the 14th century, and the shieling area, Argisbrekka (*c.* AD 850–1150), all on Eusturoy. In Shetland, The Biggings, on a magnate's farm on Papa Stour, and the Jarlshof sites represent other examples.

In Iceland, ample remains of textile work were left in the five pit houses at Hofstaðir: a total of 13 spindle whorls and clusters of loom weights represent the remains of an apparently dismantled loom. The remains of textiles were also found.[30] The measurable whorls from Hofstaðir weigh from 12 g to more than 40 g, suitable for producing threads in different gauges. The spindle whorls from Sveigatkot and Hrisheimar – with 24 and 14 whorls respectively – were generally more fragmented. Here, the measurable whorls weigh from *c.* 15 g to 50 g, although a few are somewhat heavier, with the heaviest at Hrisheimar. They show nearly the same weight pattern as those at Hofstaðir and were able to spin medium fine to coarser threads (Fig. 8.21).

These examples are not unique and seem to represent a general trend in Iceland – where spindle whorls, mostly weighing from *c.* 25 g to 45 g, were intended to produce

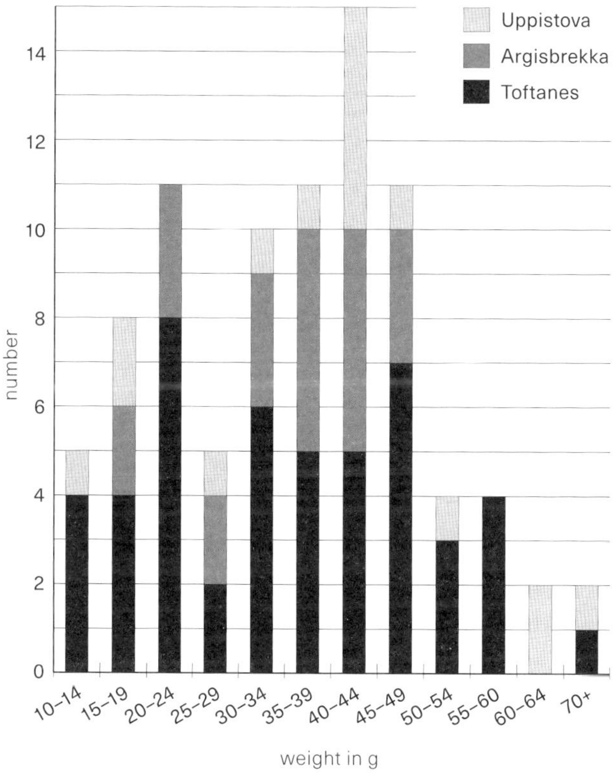

Fig. 8.22: Measurable spindle whorls from Toftanes (N:49), Argisbrekka (N:23) and í Uppistovubeitinum (N:16). Total (N: 87), according to weight groups of 5 g (Graphic design: Per Bækken).

threads of different thickness and strength; they signify the spinning of medium to coarser threads, as well as plied yarn. Most of the artefacts I have been able to inspect visually stem from the later Viking period into the 11th century.

The earliest recorded textiles so far stem from one of the pit houses at Hofstaðir, which dates from the late 900s to the early 11th century. Here, only six fragments could be identified for technique or quality: all in 2/2 twill, z/s spun and with warp threads estimated to be *c.* 8–10 and 10–12 threads per cm and of medium fine quality, with one finer fabric represented with a thread count of *c.* 18 × 10 per cm.[31] The spindle whorls from the site could actually have produced these mostly medium and coarser threads for the warp and weft in the different 2/2 twill qualities. Compared with the textiles found in the graves, mainly represented by remains of the burial dress, such as at Ketilstðir, the textiles found at production places were coarser. However, the spindle whorls in these settings were generally much heavier than those from working places in western Norway. Most of the surviving textiles relate to the Middle Ages (see Chapter 9). At the time of the settlement, they were only occasionally z/s spun, but by the 11th and 12th centuries, twills and tabby woven textiles steadily became the norm, and is explained by other traditions and the mixed population of Norse and Insular, or Celtic, origins.[32]

Although preserved textiles are rare finds in the Faroe Islands, with none yet dated to the Viking period or the Middle Ages, textile tools are common. The number of spindle whorls per site is generally higher than from Iceland, while the looms, strangely enough, have left few traces. At Toftanes, nearly 50 spindle whorls, mostly of soapstone and of different shapes, were found, including rough-outs,[33] while the other two sites, í Uppistovubeitinum close by, and the shieling area, Argisbrekka, included 16 and 23 whorls, respectively (Fig. 8.22).[34] Like the cases from Iceland, they show a wide weight range, from less than 15 g to more than 70 g. More than half of the measurable whorls, however, weigh between 30 g and 50 g, with a peak in the weight group 40–45 g.

Although wide weight ranges are represented at all the sites, the whorls, and mostly of soapstone from Toftanes appear somewhat lighter in general than at the other sites, with a higher percentage (35%) lighter than 30 g. Compared with the Icelandic whorls, the Faroese whorls show a wider weight range and more of them occur in the heavier weight groups, but neither shape nor weight clearly reveal standardised types.

The early traces of textile production in Shetland are fewer and less studied, although the excavations at Jarlshof have revealed ample traces of textile production throughout the Norse period, especially from the earliest centuries after the colonisation. Altogether, 31 spindle whorls and 160 loom weights were recovered, of which several were found in clusters, seeming to represent whole looms.[35] More recent excavations at The Biggings, Papa Stour, have uncovered both textiles and tools from the Early Middle Ages, including 24 spindle whorls of different shapes and material. They weigh from 11 g to *c.* 40 g, capable of spinning yarn of different gauges: fine, medium to rather coarse threads, largely corresponding with the thread qualities in the fabric remains found at the site.[36] The Jarlshof whorls are more homogeneous in shape and size, commonly within the weight group of 25–35 g. Altogether, they seem to indicate a textile production on a larger scale and more standardised.

Textile production in the new Viking Age town at Kaupang

The last issue to consider, then, is how rural tools and textile whorls related to the urban production in the early Viking towns and with a focus on Kaupang as the nearest. Representing specialised environments for varied craft production and thus another economy, the production also related to another social organisation. As a densely built-up settlement area based on crafts and commerce, mainly from *c.* 800 into the 10th century, the textile tools from Kaupang are more numerous and to some degree more comprehensive in composition. The grave fields surrounding the densely built settlement area, revealed about 400 graves with 18 of

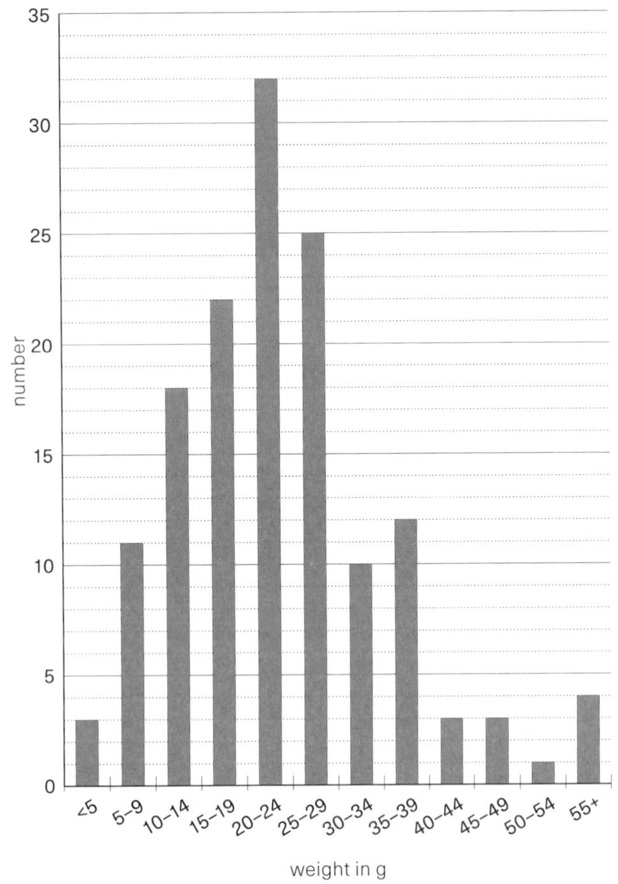

Fig. 8.23: Measurable spindle whorls from the Viking Age town Kaupang in weight groups of 5 g (N:149) (Graphic design: Per Bækken).

the 41 datable female graves from the 9th century, many also with textile tools and some also with textile remains. Do they differ, then, from the tools and textiles from the rural contexts?

Based on my earlier analyses of textile producing implements from Kaupang, which comprise 121 measurable spindle whorls, some 470 loom weights, and one weaving beater, two pairs of shears and some needles, needle-cases and some pellets for smoothing, mostly from the settlement area. Also, here, spindle whorls serve as a primary diagnostic measurement for comparison.[37] In Kaupang, the spindle whorls stem mainly from the settlement area, within buildings at the different plots, in middens, hearths, and ditches and others, as lost, left, and discarded implements. The spindle whorls make up an assemblage of nearly 150 whorls, ranging from *c.* 5 g to around 60 g, of which 27 whorls are from the graves and generally better preserved than from the settlement and work areas. However, they all fall within about the same weight range, from *c.* 5 g to more than 50 g, although few are above 40 g (Fig. 8.23). They represent then a similar trend as the rural assemblages from western Norway. However, the Kaupang whorls demonstrate a wider spectrum of weights and thus larger differences in qualities of fabrics. Both spindle whorls and loom weights, however, are more fragmented than the rural cases represented here, as they largely are made of clay.

The loom weights were mostly made of fired clay and generally very fragmented are therefore best represented in weight rather than in numbers. Sometimes when found in clusters, they may represent a left loom. The loom weights, whether broken or complete, still indicate that weaving was carried out at all the plots. The few measurable or possible to estimate as for weight, 17 in all, weigh between *c.* 150 g to nearly 800 g, but the complete weights tend to be heavier than those of estimated weight. However, only one loom weight appeared in the graves from Kaupang.

The analysis from Kaupang shows that 69% of the measurable spindle whorls, from both the settlement area and the graves, weigh less than 25 g, and 31% between 25 and 35 g, peaking in the weight group 20–24 g (Fig. 8.24). Compared with the corresponding weight distribution of spindle whorls from the rural sites in western Norway, the whorls in fact have a relatively higher share of light whorls than from Kaupang. At all the referred to places, more than half of the recovered whorls are under 25 g. At the central places Stedje, Osen, and in the pit houses from Bjørkum, more than 90% of all the recovered spindle whorls are lighter.

In the Kaupang assemblage, 10% of the spindle whorls weigh less than 10 g, while the corresponding percentage of the Viking Age rural sites constitutes as much as 15%. Compared with the corresponding weight distribution of spindle whorls from the rural sites in western Norway, three of them, Stedje, Bjørkum, and Nyset-Steggje, have relatively more whorls under 25 g than at Kaupang, and when all included, with a peak in the lighter weight groups too.

The diagram (Fig. 8.24) shows then similar weight ranges from *c.* 5 g to above 50 g, represented in all find contexts but with different concentrations within the weight groups in these varied find contexts and the highest representation from Kaupang with a peak in the common weight groups between 15 g and 30 g, both relatively and in numbers. The lightest whorls both relatively and in number are from rural and urban workplaces and with a broader range of whorls of varied weights at Kaupang than the other contexts.

The analysis of the Kaupang assemblage, covering the period from *c.* AD 800 into the early 10th century, indicates a varied and extensive production of mainly fine and average qualities and not a standardised production. It seems rather to have been related to production on demand, not least related to the higher echelons of society, as also substantiated by the tools in the graves at Kaupang.

As for loom weights, the majority of the relatively few measurable loom weights from Kaupang weigh between 300 g and 500 g, peaking between 300 g and 400 g. In this respect, they resemble the weight distribution from

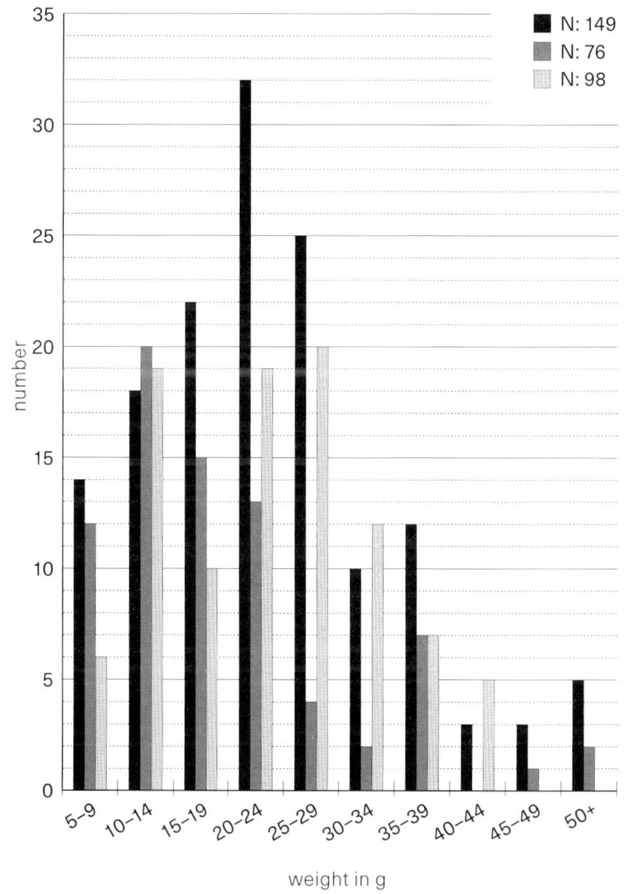

Fig. 8.24: All measurable spindle whorls from Kaupang by weight groups of 5 g (N:149) compared with spindle whorls from rural workplaces in western Norway (N:76) and graves (N:98) (Graphic design: Per Bækken).

Bjørkum and Stedje. By and large, it is still difficult to observe a distinct difference between rural sites and the Viking Age town of Kaupang as for weight and ability to produce various qualities.

Analyses of textile equipment from Hedeby and Birka, carried out by Eva Andersson, Strand,[38] show a similar tendency as in Kaupang, however, based on larger assemblages but also covering longer periods, starting somewhat earlier and lasting into the early 11th century. The range of weight is an interesting indicator of standardisation and specialisation. At all three sites – Kaupang, Birka, and Hedeby – the spectrum of weights is quite wide but widest at Birka, where the weights vary from less than 5 g to 89 g and with a few whorls weighing more than 100 g. In Hedeby, the spindle whorls vary from 4 g to 75 g and at Kaupang, as noted, from less than 5 g to 59 g. At Birka, the whorls between 5 g and 29 g are the most common, with a peak between 10 g and 14 g. At Hedeby, the spindle whorls between 10 g and 25 g are the most frequent, with a peak between 10 g and 20 g[39] and at Kaupang, those weighing 15 g to 29 g are the most common, with a peak in the weight group 20–24 g.

When comparing the range of weights, the assemblages concentrate in nearly the same weight groups, although the assemblages from Birka and Hedeby include a wider range of heavy whorls, weighing up to more than 100 g. At Kaupang, they weigh no more than 60 g. Kaupang also had a larger group of whorls in the weight group 25–30 g than in Birka and Hedeby.[40] The variation in weight thus indicates an almost corresponding range of thread qualities from the thinnest and finest yarn to the coarse. Compared with Birka and Hedeby, however, the Kaupang spindle whorls have a higher concentration of medium heavy whorls, suitable for spinning woollen yarn of rather ordinary medium/thin qualities. The diversity and quantity of spindle whorls and loom weights and their chronological and spatial distribution indicate activities beyond the household level. Charlotte Blindheim, who led the Kaupang excavations carried out in 1950–57, even suggested that Kaupang could have been a centre of textile production, and at that time, she based it on the smaller assemblage of textile producing equipment. Although now founded on a larger assemblage, it still appears as a smaller production centre than the larger centres of Hedeby and Birka where the find assemblages clearly substantiate a specialised textile production in the densely built settlements.

Some of the graves at Kaupang, as in Hedeby and Birka, have left traces of the burial dress too, of which the finest were woven in diamond twill qualities, like qualities found in graves from western Norway (cf. Chapter 6). Anne-Stine Ingstad who analysed the Kaupang textiles,[41] classified the textiles into four rough categories – coarse, medium, fine, and very fine – woven in tabby (39 fragments), 2/2 twill (77 fragments) of two types, z/s-spun and z/z spun yarn respectively, lozenge twill (26 fragments), and chevron twill (8 fragments), with many of them of excellent quality. The majority of the 2/2 twills from Kaupang have z/z-spinning – this type, because of its geographical distribution, denoted the 'Norwegian' type.[42] Ingstad still considered it more likely that most of the finer textiles recorded in the graves were imported, suggesting a western Insular origin. In principle, however, all the textiles found could have been produced using the implements found at the site. Eva Andersson Strand[43] has come to a similar conclusion regarding the Birka tools and textiles, which are generally finer and more exclusive than their counterparts from Kaupang. The very lightest spindle whorls could, according to the experiments at CTR, also produce the very finest threads to achieve the high thread count per cm as in the finest diamond twills.

'Small things' matter

Judging by the textile producing equipment from various Viking Age workplaces in varied production environments and workplaces both in western Norway and in the North Atlantic, they indicate a large-scale and complex production of varied qualities, covering a whole range of use areas, from the average qualities for many purposes to high quality products within the exclusive and even luxurious segment, but also the coarse, among others needed for the large-scale projects of producing sailcloth. The varied *in-situ* workplaces and working environments expose both similar trends but also differences – between central places in the lowland and working environments in uplands in western Norway, and related to the North Atlantic, especially Iceland, and compared with Kaupang as a new town producing for a market.

By using weights of spindle whorls and loom weights as indicators of textile qualities, they reveal a tendency towards the finer qualities in all these environments, as also workplaces connected to high-status central farms, such as Stedje, Osen, and Sæbø/Hopperstad in western Norway, and at Hofstaðir, Toftanes, and The Biggins in the North Atlantic, respectively Iceland, the Faroe Islands, and Shetland. These were all environments connected to the higher strata. The lightest weights for the finest yarn appear to connect to active workplaces, connected to separate buildings as specialised production at special workplaces, in the case of pit houses, but could also be integrated in the longhouses of high status and at central places, or appear as agglomerated building complexes in rural upland context, such as at Bjørkum and the Årdal mountains.

To what extent the production leaned on a subordinate work force, represented by the shift from the pit house to the longhouse at main farms, as suggested in Iceland, is difficult to ascertain. Considering the hierarchical structure, and in the Viking period, based on an unfree work force, it is rather likely, as also indicated by the upturned pyramid as for status in graves from western Norway. At the workplaces in the uplands and in pit houses, this may be reflected by the simple accommodation, tools, and the overall lack of status symbols in these environments. Not all were native judging by the spinning of s/z-spun yarn which probably indicates an Insular origin in some cases, and as indicated, not least in Iceland,[44] but also in western Norway, such as at Stedje and revealed by textiles in these environments. The managers and organisers of production, however, most likely belonged to the high-status stratum, represented in the graves and furnished with dress accessories and with affiliations to others and within networks operating on the same social level.

Notes

1 Lillehammer 1967; 1971; Bakka 1971; Larsen 1995.
2 Fritzner I, 276–277.
3 Egils soga [Egil's Saga] 32.
4 Andersson 1999.
5 Solberg 1986, 161.
6 B6475II, Stedje, Sogndal.
7 Mortensen 1993; 1997.
8 Unpublished report A. Bergli 1993; E. Schjølberg 17.1 1993, University Museum of Bergen.
9 *e.g.* Hayeur Smith 2016.
10 *DN* VI, no. 84; Øye 1986, 410–412.
11 I thank Søren Diinhoff for kindly making his excavation report available to me (Diinhoff 2016).
12 B1240-41; B1622, Sæbø, Vik.
13 Åstveit & Melvær 2013. I thank Leif Inge Åstveit, Bergen University Museum, for giving access to his report.
14 Denoted structure 242 in the report.
15 B17009/3,4,5, Sæbø, Vik.
16 *BK*, 234, 236.
17 Report by Søren Diinhoff; Bergen University Museum. Diinhoff 2002.
18 Sørheim 2018, 260, with references.
19 B15813, Osen, Gaular.
20 B15813/85, Osen, Gaular, status FI?
21 Ramstad 2010; 2018; Ramstad *et al.* 2011. I would like to thank the leader of the Bjørkum project, Morten Ramstad, for making the unpublished material, the artefacts and field documentation, available to me.
22 Øye 2002.
23 Bjørgo *et al.* 1992; Bjørgo 2005.
24 Andersson 1999, 19.
25 Øye 2015b, 58–59.
26 Bruen Olsen 2015
27 Bruen Olsen 2005; 2015.
28 Milek 2012, 93.
29 Milek 2012, 120.
30 Lucas 2009; Sverrisdóttir 2009.
31 Sverrisdóttir 2009, 318–319.
32 Hayeur Smith 2016, 237.
33 Stumann Hansen 1991.
34 Arge 1991; 1887; Mahler 2007.
35 Hamilton 1956.
36 Walton Rogers 2012.
37 Øye 2011.
38 Andersson 2003.
39 Andersson 2003, 118–119.
40 Andersson 2003, 142.
41 Ingstad 1999, 266.
42 Ingstad 1999, 266–267.
43 Andersson 2003, 53, 99.
44 Hayeur Smith 2016.

9

From rural to urban production: changing conditions for textile production in the Middle Ages?

By the Early Middle Ages, in the wake of state formation processes, urbanisation, and expanding trade, textile production became one of the larger industries in northern Europe. New demands, new technology, and new organisation and specialisation were central elements in these processes. The horizontal foot-operated treadle loom played an important role when it reached urban settings in north-western Europe by the early 11th and 12th centuries. The need for standardised and specialised products was an important factor, not least related to the expanding seagoing transport, for sails, tents, waterproof clothing, and packaging for cargoes. Additionally came all the daily requirements for both indoor and outdoor garments from a steadily growing population. As earlier, textiles served as important social status markers and by this time, for both a secular and clerical elite. In consequence, the many use areas required a varied production of different qualities and in larger quantities, which again stimulated new technology – with the horizontal loom as the most important innovation. However, the horizontal loom was primarily adopted in urban centres connected to specialisation within handicrafts, evolving into guilds in northern Europe. The invention of the great spinning wheel came later, in the early 14th century, and earliest in England and Flanders.[1] In the period and area under study, the hand-turned drop spindle was therefore still the basic implement in the time-consuming process of spinning, and different whorls were used for the warp and the weft to achieve the various yarn qualities for different use areas.

To what extent domestic weaving was based on new technology in medieval Norway or continued on the old warp-weighted loom in the period of discussion is not so clear as imported wares eventually became more available commodities from a surplus market-oriented production of textiles in north-western Europe. How and to what extent these changes and large-scale processes affected textile production in these areas at the fringes of northern Europe and how they reacted to these changing conditions, is a central issue in this final chapter.

Centre and periphery in production?

In Norway, Bergen emerged as a town by the late 11th century and soon developed as the political and administrative centre in the Kingdom, which also included the islands in the North Atlantic without towns. Norway's westward orientation in the Middle Ages towards Britain and the North Atlantic continued a tradition going back to the Viking period and is reflected in the development of Bergen as the main commercial centre by the latter part of the 12th century. In the Viking Age, the islands stretching from the Hebrides to the Faroe Islands had periodically paid tribute to the Norwegian king, and Norwegian rule continued over the Orkneys, Shetland, and the Faroe Islands as so-called *skattlands* (tax lands). By the late 12th century, during King Sverri's reign (1177–1202), the political affiliation between Norway and the North Atlantic was further strengthened. Not only the Orkneys but also Shetland, as an integral part of the Orkney earldom, were linked both politically and economically, and in 1195 brought under direct control of the Norwegian Crown.[2] Bergen's westward political and economic orientation reached a peak in the 13th century when Icelanders and Greenlanders placed themselves under the Norwegian Crown in the early 1260s.

Because of the town's strategic location on the west coast, along the main sailing route, halfway from the Continent to the rich fisheries in northern Norway, and with even shorter distances to England and the islands in the North Atlantic, however, in open sea (cf. Introduction, Fig. 0.1), Bergen soon became the main harbour and node for both import and export. The habitation area along the waterfront at Bryggen – *the Wharf* – therefore expanded significantly during the 12th and 13th centuries and onwards. Although export trade was largely based on marine products, primarily dried cod, *stockfish*, in return for imports of grain primarily but also other transportable commodities in demand from both England and the Baltic countries, including textiles of varied qualities, and raw materials – wool, flax, hemp, and other products, and from the North Atlantic, wadmal.

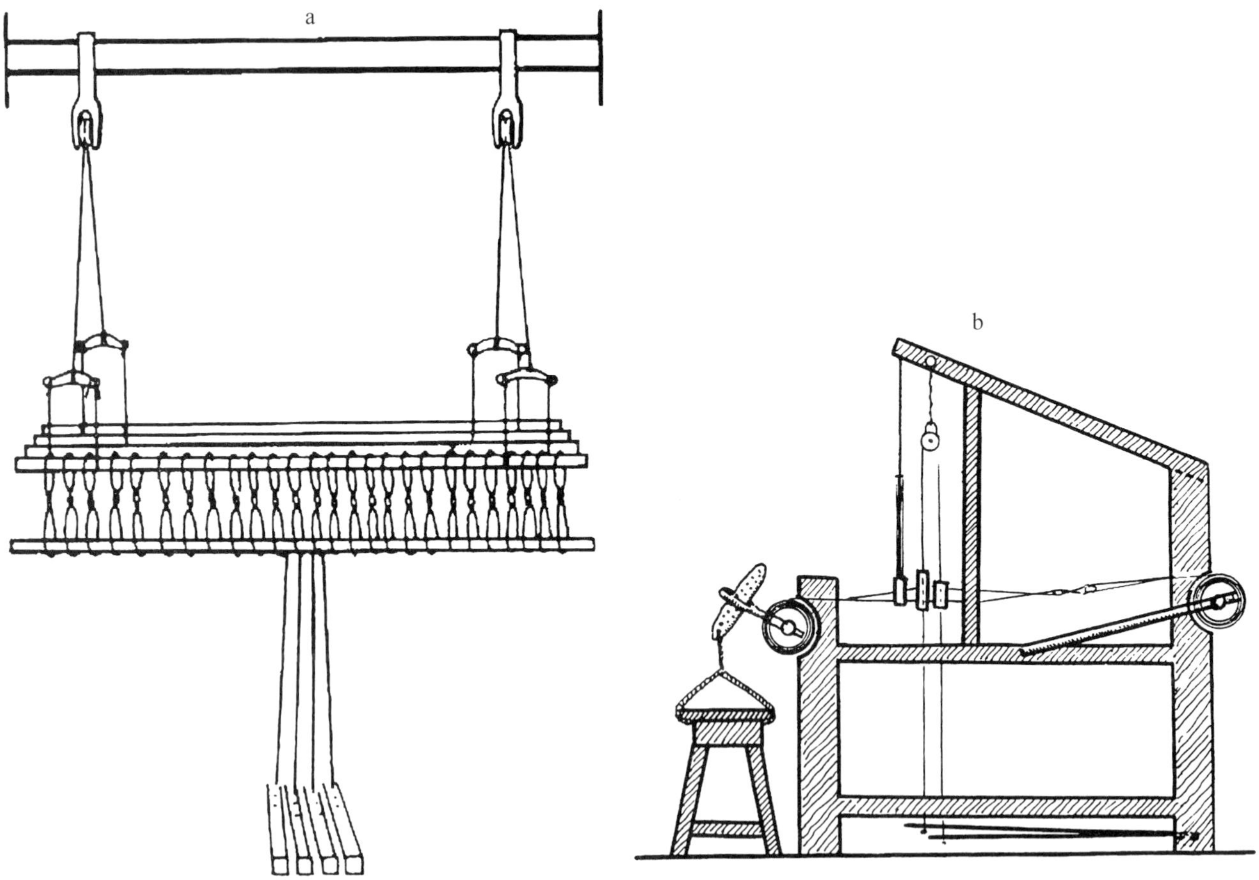

Fig. 9.1 a) Schematic drawing of the horizontal loom operated by treadles (Source: Øye 1988, 74), b) schematic cross section of the loom and the two beams, the seat for the weaver operating the shed by the treadles.

By the late 1200s, the urban handicrafts in the town counted more than 20 crafts, producing vital domestic commodities too. Networks provided access to important raw materials from outlying and upland areas for many of these crafts, including textile production, such as weaving. In Scandinavia, weavers as a group are in fact first mentioned in the Urban Code, specially designed for Bergen (*Bl III, 8*). This is about the same time as the possible archaeological remains of the horizontal loom appear among the finds from the large-scale excavations at the Bryggen site (1955–1968). According to archaeological evidence, however, it is unclear how quickly this would have been adopted and to what extent, and quite when the vertical loom was phased out in different areas in Scandinavia.

In the North Atlantic, production carried on continuously at a decentralised farm level. Here, sheep and wool could provide a surplus production of *vaðmál,* which at an early stage became standardised 'commodity money' when paying the obligatory duties of tithes, land rent, and other obligations, and even provided a surplus for export. Also, in Norway, homespun wadmal was a commodity integrated in the overall value system of interchange in the economy, however, only in the early part of the period and earlier than the Urban Code and Land Law, both from the 1270s.

Assessing tools and textiles from both urban and rural contexts in these areas may provide a clearer conception of these literally interwoven processes; socially, technologically, and economically in varied environments and with actors at many levels.

Two types of looms: two kinds of settings and actors?

A common characteristic of the new horizontal loom is that the warp threads are stretched out horizontally between two cloth beams, one in front of the seated weaver and another at the far end of the loom, and the sheds are steered by foot treadles. These were linked by pulleys, joined by heddles – loops for the horizontal warp threads – and fastened to heddle rods to be lifted and lowered by the treadles so that the weft could go through, creating the shed and the weft inserted by a shuttle (Figs 9.1a and b).

The principle of creating sheds, however, is the same for the vertical and horizontal looms but a major difference between the two looms was the length of the woven cloth,

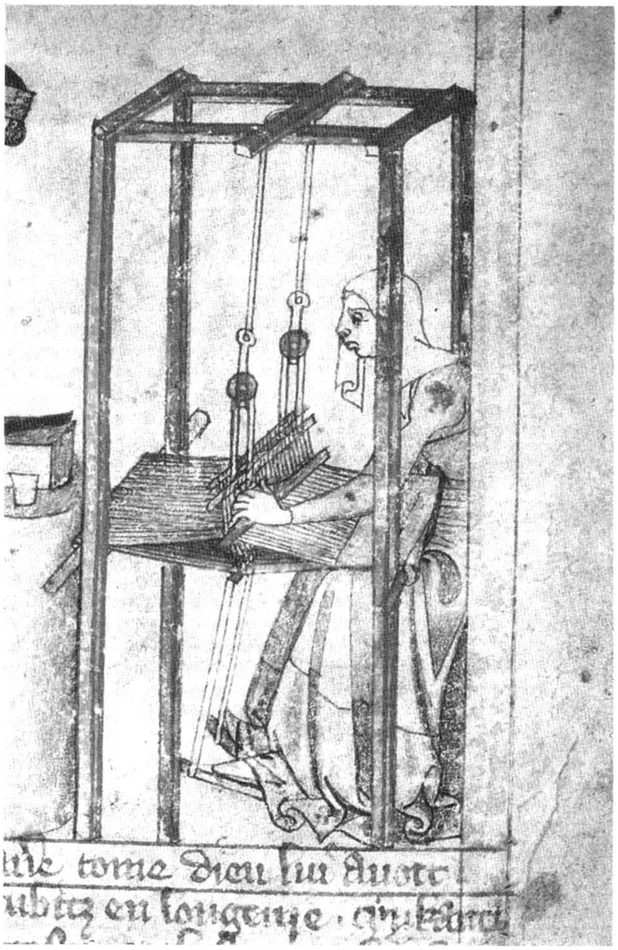

Fig. 9.2: Female by the loom. Egerton Genesis Picture Book c. 1360 (British Library, Egerton 1894, fol. 2v, by permisssion).

with the horizontal loom making it possible to operate 20–30 m long threads and weave a corresponding long cloth.[3] Early depictions of the loom show that it limited the width of the cloth to an arm's length of a single weaver, while those from the latter part of the 14th century also illustrate looms operated by two weavers with four treadles, able to weave broader cloth (Fig. 9.1b).

The advantages of the horizontal loom were the weaver's better control over the sheds, and the ability to weave longer, more uniform and densely woven fabrics, and not least much faster – estimated to represent more than a three-fold increase compared to the warp-weighted loom.[4] Overall, it clearly represented a considerable increase in productivity and trade. The implementation and spread of the horizontal loom and the expansion of the textile industry therefore had wide-reaching consequences, also for all the other foregoing processes involved – sorting, cleaning, combing, and spinning – not only in the new urban textile centres but also in rural areas. In urban settings in these north-western areas, weaving and other areas of textile processing, such as tailoring and shearing achieved status as separate handicrafts, which, as recorded in Bergen and in England, and as on the Continent, developed into guilds and large-scale industries.

That the shift from the vertical to the horizontal loom also represented a shift in gender roles tends to have become a common understanding, although an increasingly contestable preconception. When different textile implements, especially loom weights for the upright loom and spindle whorls for hand spinning, occur in archaeological contexts in urban tenements at an early stage of the towns from the 11th century onwards, it has been presupposed that they related to a household-based production with women as the main actors. Then weavers without further specifications, as noted, were mentioned in the Urban Code and when the first mention of a named weaver in Norway turned up in 1570 in Bergen, it was evidently an immigrant male craftsperson judging by the name.[5] So when textiles that may represent weaving on the horizontal loom appear in urban contexts, they thus tend to be taken as implicit proof of handicraft carried out by men.[6] However, these assumptions are not self-evident and the differences in weave qualities produced by the two types of loom may be overrated in terms of technological complexity and the agents involved.[7] Medieval illustrations show that not only men but also women used the horizontal treadle loom and that the weaving could be more or less advanced (Fig. 9.2).

To what extent and when the new loom was introduced in medieval Scandinavian towns, is not so clear, nor the consequences. Here, the textile producing implements and textiles found in medieval towns, and in this case, Bergen, provide large assemblages of tools that can be compared with corresponding rural technologies thus making it possible to assess similarities and differences as for tools and use areas – also related to the foregoing Viking period.

Textile tools from an urban context, Bryggen in Bergen

The large-scale excavations in the harbour area of Bryggen in Bergen with its good preservation conditions for organic material, revealed a far more varied material of wood and bone than the contemporary rural sites and from the Viking period. Archaeologically, they represent all stages in the textile production except combing: spinning, weaving, sewing, and finishing. The finds included, as noted, remains that possibly belonged to the horizontal loom (cf. Figs 9.1a and b), and of similar shape and functional elements as found in other Scandinavian urban contexts – Hedeby, Sigtuna, and Lund. However, they all appeared as single finds among an overwhelming representation of loom weights belonging to the upright loom – with about 800 loom weights and more than 400 spindle whorls, besides other textile tools in smaller quantities. The tools for spinning and weaving largely concur in time, dated to periods marked by fire layers from

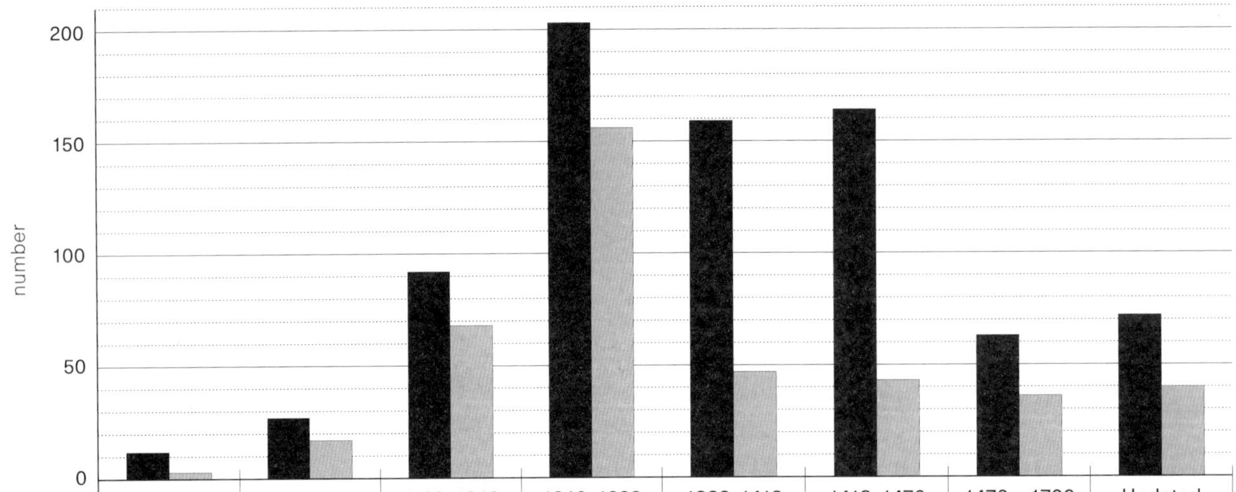

	Before 1170	1170–1198	1198–1248	1248–1332	1332–1413	1413–1476	1476–c1700	Undated
Weights	12	27	92	203	159	164	63	72
Whorls	3	17	68	156	47	43	36	40

Fig. 9.3: Representation of all spindle whorls and loom weights from Bryggen according to periods, marked by fire layers, from c. 1170 to 1702 (Graphic design: Per Bækken).

Table 9.1: Spindle whorls from Bryggen in varied materials (N:410).

Spindle whorls	*Stone*	*Wood*	*Bone/antler*	*Clay/pottery*	*Metal*	*N*
N	249	118	16	20	7	410
%	61%	29%	4%	5%	1%	100%

the major historically documented town fires from AD 1170 onwards, providing chronological frameworks throughout the Middle Ages (Fig. 9.3).

In Bergen, as in other Norwegian medieval towns, the upright warp-weighted loom was in use even longer than on the Continent and was a common tool from the early 12th century well into the 15th century but in very small numbers and evidently on a small scale after around 1400.[8] This was also the case in other excavated medieval towns, Oslo and Trondheim.[9]

The finds of 410 spindle whorls from Bryggen, also recovered in wood and bone/antler due to the favourable preservation conditions indicate a general under-representation of these materials in the rural contexts. Still, the majority, 61%, of the Bryggen whorls are made of stone, mostly steatite but also other minerals, 29% of wood, both softwood and hardwood, less of bone/antler, fired clay, and of metal, all lead (Table 9.1).

The Bryggen spindle whorls of stone, counting 162 measurable specimens out of 249 finds, vary in weight from 6 g to more than 50 g with 62% below 25 g, 11% under 10 g and peaking between 15–20 g (Fig. 9.4). However, only the spindle whorls of stone are compatible with weight distributions of spindle whorls from the rural sites in western Norway.

The concentration of light whorls is also accentuated by whorls in the lighter material of wood, bone/antler, and clay/pottery. The weight distribution of spindle whorls clearly shows that yarn was produced in varied qualities, and many could spin very thin yarn, probably used for textiles with a high thread count. Several of the buildings at Bryggen revealed concentrations of spindle whorls for varied qualities, loom weights for the upright loom and needles, reflecting textile production in the same working environments, while single tools from other contexts, may have been lost, left, or thrown away, in passages, wharves, and other places.

Among the wooden implements were also some scattered rudiments, probably belonging to the horizontal loom, with two pulley blocks (Fig. 9.5), a separate pulley-wheel and a heddle horse, all essential parts of the horizontal loom. Pulley blocks with pulley-wheels, to be fixed to, or suspended from the top-beam of this kind of loom were used to carry the cords connecting to the treadles with heddle rods, and where the heddle-horse had the same functions as a pulley, for making sheds. When the treadles are depressed, the heddle-rod is lifted, and a shed is formed (cf. Figs 9.1a and b). Of these, four finds dated to the late 12th century, and the three others to the latter part of the 13th and the 14th centuries respectively, however, not connected contextually to specific buildings.[10]

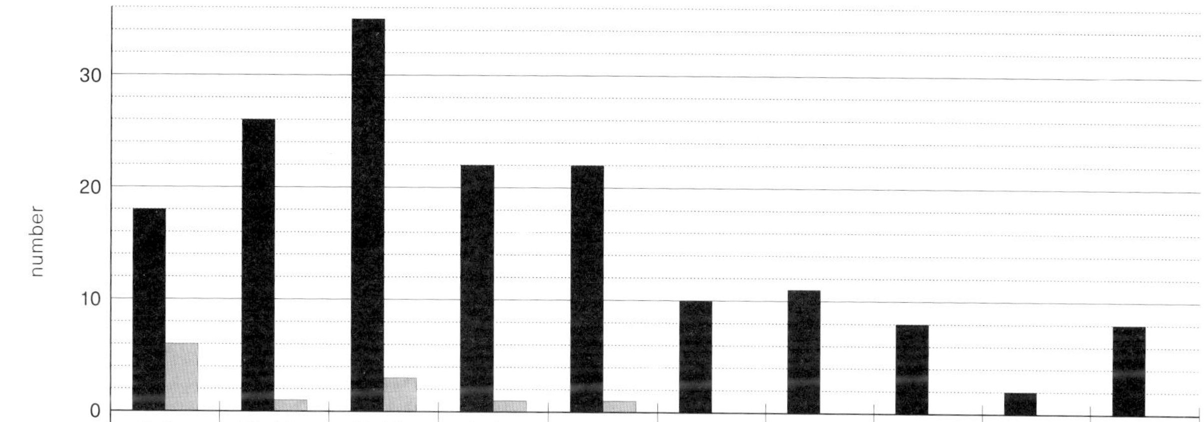

	5–9 g	10–14 g	15–19 g	20–24 g	25–29 g	30–34 g	35–39 g	40–44 g	45–50	50+
Stone	18	26	35	22	22	10	11	8	2	8
Clay/pottery	6	1	3	1	1					

Fig. 9.4: Complete spindle whorls of stone from Bryggen in weight groups of 5 g (N:162) (Graphic design: Per Bækken).

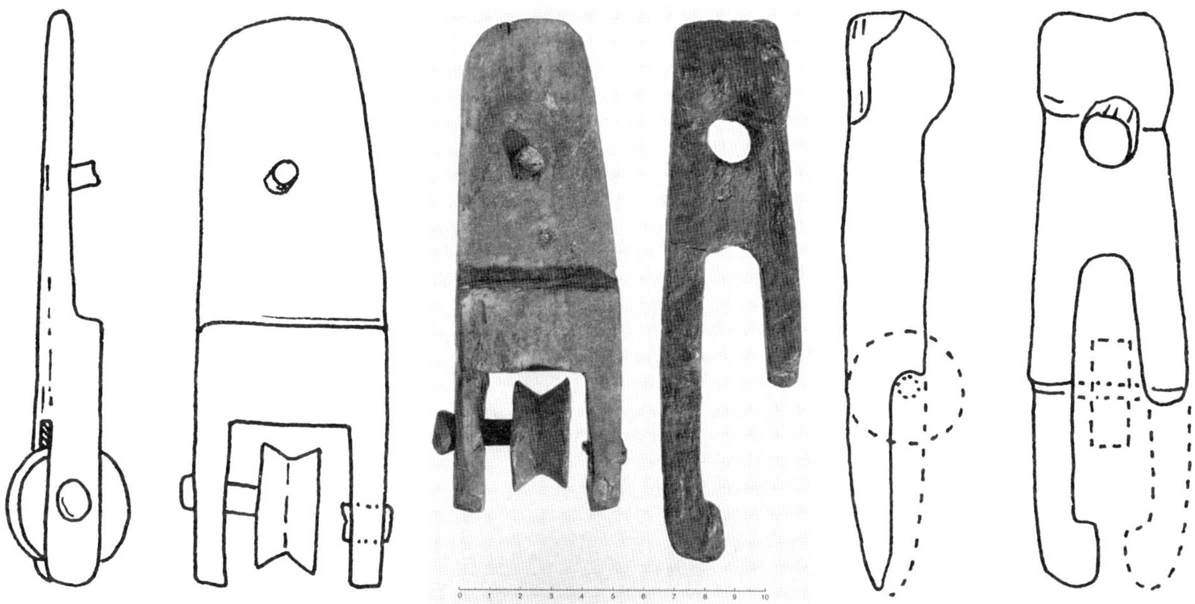

Fig. 9.5: Probable pulley block and part of a casing for a pulley at Bryggen (Source: Øye 1988, 74).

The earliest date of a possible pulley block is found in Viking Age Hedeby and dated to the 11th century; about the same time as the earliest documented from York. From present Sweden, finds dated to the 12th century stem from Sigtuna and Lund,[11] about the same time as the earliest from Bergen.

If the identification of these finds is correct, it means that the horizontal loom was used in Bergen, possibly as early as in the last decades of the 12th century, and from the mid-1300s into the 14th century, however, the earliest date and identification as the most uncertain. Anyhow, the large number of loom weights from the same stretch of periods still indicates that weaving on the upright warp-weighted loom dominated (cf. Figs 4.10 and 4.12), and as documented in at least six buildings in the same period as the finds of the possible remains of the horizontal loom.

The measurable loom weights (N:555) from the Bryggen site vary considerably as for weight, from under 100 g to over a kilo. The majority (80%), however, range between 300 g and 900 g (Fig. 9.6). Many of the loom weights appeared in clusters, reflecting working environments, altogether 238 loom weights from 18 concentrations, of which 12 associated with burnt buildings, and interpreted as *in-situ* finds, representing textile production environments at Bryggen (Fig. 9.7). These clusters are represented in layers from before 1170 to *c.* 1400

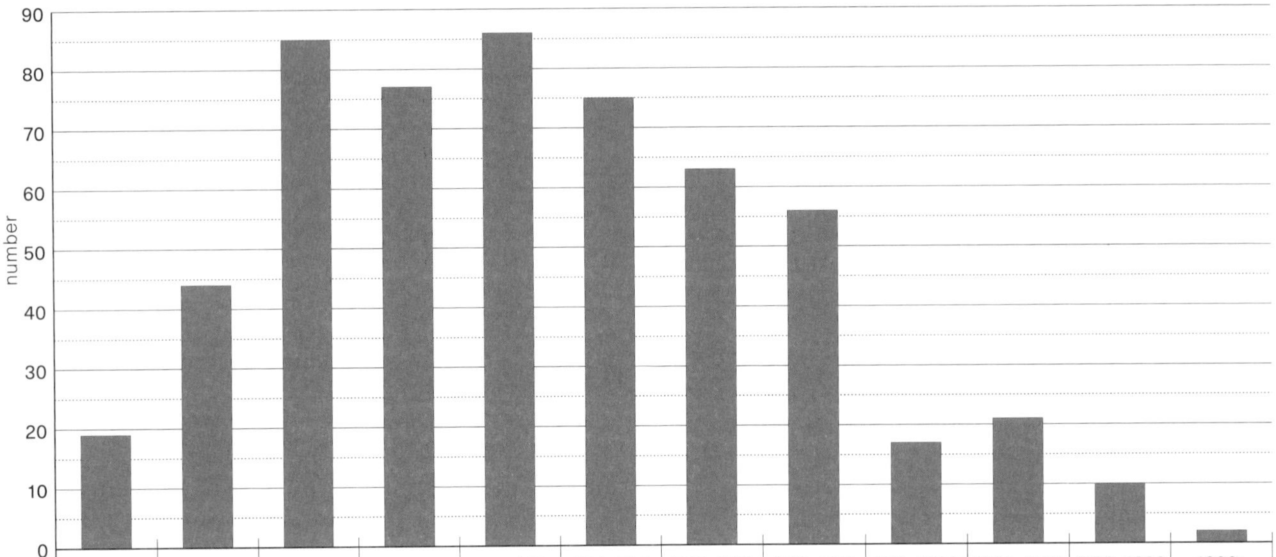

	100–200	200–300	300–400	400–500	500–600	600–700	700–800	800–900	900–1000	1000–1100	1100–1200	1300+
N	19	44	85	77	86	75	63	56	17	21	10	2

Fig. 9.6: Complete loom weights from Bryggen in weight groups of 100 g (N:555) (Graphic design: Per Bækken).

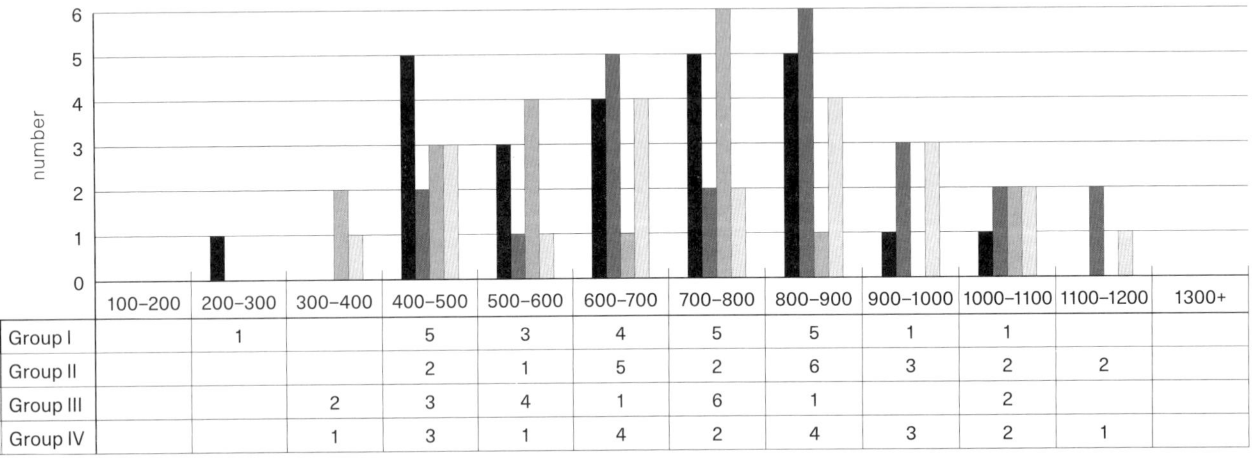

	100–200	200–300	300–400	400–500	500–600	600–700	700–800	800–900	900–1000	1000–1100	1100–1200	1300+
Group I		1		5	3	4	5	5	1	1		
Group II				2	1	5	2	6	3	2	2	
Group III			2	3	4	1	6	1		2		
Group IV			1	3	1	4	2	4	3	2	1	

Fig. 9.7: Groups of loom weights in buildings from Bryggen in groups of 100 g (Graphic design: Per Bækken).

with six from the first part of the 13th century, and six from its latter part and into the early 14th century.

These concentrations show broad weight spectrums too, from 230–300 g to over a kilo, while some are lighter and more homogenous, between *c.* 250 g and 430 g. The mean weight varies from 360 g to 880 g for the clusters and those with more than 10 weights from around 400 g to around 850 g. The weight distribution within the clusters shows a concentration in the weight range from 400 g to 900 g with 71% of all the loom weights within this array.[12] Finds in four of these buildings, denoted groups I–IV, seem largely to represent the whole loom, each with more than 20 loom weights, counting from 23 to 30 weights, and from the late 12th century into the early 14th century (Fig. 9.7). In all the groups, the weights vary considerably but also share common traits. Group I, ranging from 300 g to 1000 g, concentrate in the weight groups 400–800 g, group II from 300 g to 1100 g, group III from 300–1000 g, mostly from 400–800 g, peaking between 700–800 g, and group IV with generally heavier weights ranging from 300 g to 1100 g.[13]

Soapstone was used for both loom weights and spindle whorls along with other minerals found in the surrounding districts as unfinished spindle whorls in stone, but also wood and bone or antler. Several needles made of bone, antler, and wood without an eye should probably also be regarded as unfinished products. Some were found in a building, and were probably produced there, such as a large number of small sewing needles, found in concentrated groups in the

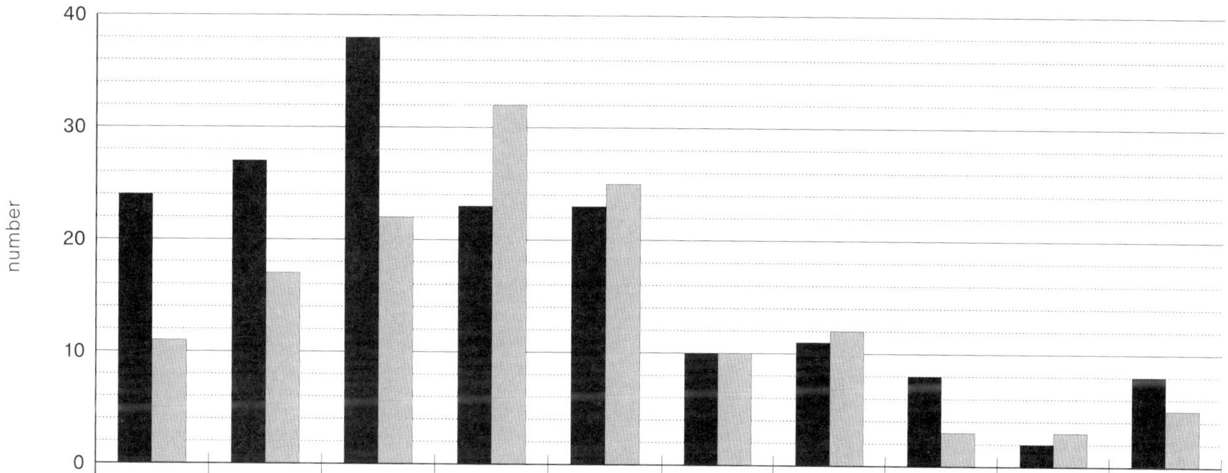

	5–9 g	10–14 g	15–19 g	20–24 g	25–29 g	30–34 g	35–39 g	40–44 g	45–50	50+
Bryggen	24	27	38	23	23	10	11	8	2	8
Kaupang	11	17	22	32	25	10	12	3	3	5

Fig. 9.8: Comparison of weight distribution in the Bryggen (N:174) and Kaupang (N:149) assemblages of spindle whorls, in weight groups of 5 g (Graphic design: Per Bækken).

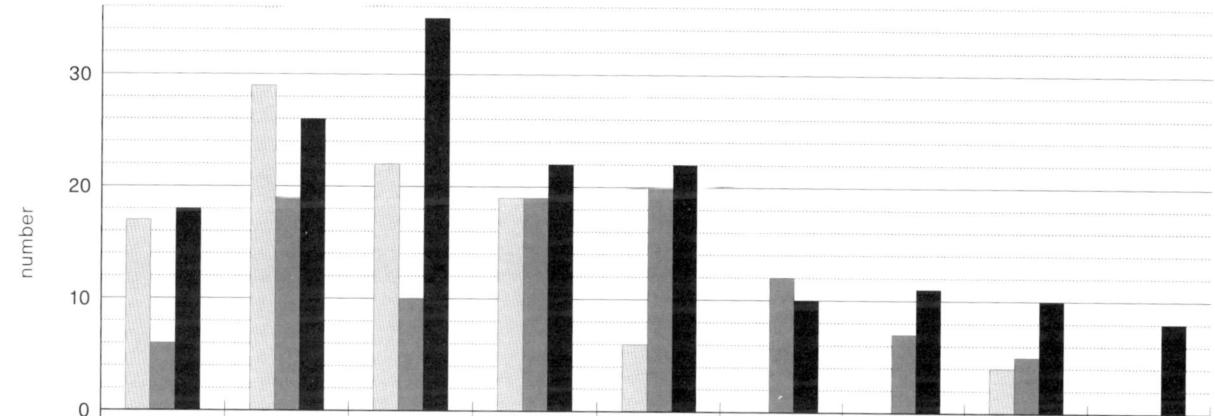

	below 10 g	10–14 g	15–19 g	20–24 g	25–29 g	30–34 g	35–39 g	40+	50+
Work places	17	29	22	19	6	0	0	4	
Graves	6	19	10	19	20	12	7	5	
Bryggen	18	26	35	22	22	10	11	10	8

Fig. 9.9: Spindle whorls from the Bryggen site arranged according to weight groups of 5 g and compared with whorls from Viking Age graves and working contexts (Graphic design: Per Bækken)

front part of one of the tenements, and may reflect tailoring and a workshop. The details of the find situation, however, are too vague to be able to determine this.[14] Anyhow, it is difficult to ascertain whether the production reflects a professional craft or household production.

Compared with measurable spindle whorls from Viking Age Kaupang (cf. Chapter 8), both sites revealed concentrations of spindle whorls in the lighter segment, under 30 g, although with a higher share in the Bryggen assemblage with a peak in the weight group 15–19 g (Fig. 9.8).

When comparing the 98 spindle whorls from Viking Age graves with the Bryggen spindle whorls of stone, with altogether 162 measurable specimens, they largely fall into the same weight range, from 6 g to more than 50 g, with 62% under 25 g, 11% under 10 g, with a peak between 15 g and 20 g. Of the Viking Age whorls from grave contexts, 55% of the whorls weigh less than 25 g, with a peak in the weight group 20–35 g and are thus generally somewhat heavier than the spindle whorls from Bryggen. However, compared with the whorls from Viking Age working environments in western Norway, counting 69 measurable whorls, 86% weigh under 25 g, with a peak in the weight group 10–15 g. The lightest whorls, then, came from various working areas from Viking Age rural sites (Fig. 9.9).

Table 9.2: Representation of complete and measurable tools in relation to the assemblages of spinning and weaving implements at the six sites. Weight range and relative representation within the range is marked in brackets.

Sites	*Spindle whorls (measurable weight) N*	*Weight range (g)*	*Peak: weight group (% under 25 g)*	*Loom weights (measurable weights) N*	*Weight range (g)*	*Peak: weight group (% under 400 g)*
Bryggen, Bergen	206 (162)	6–49	15–19 (62%)	792 (555)	100–1300	300–600 (27%)
Kaupang	149	5–60+	20–24 (69%)	–	150–800	?
Ytre Moa, Årdal	14 (14)	10–41	10–14 (57%)	10 (5)	223–834	300–400 (90%)
Stedje, Sogndal	16 (13)	5–28	10–14 (87%)	72 (49)	159–754	250–350 (61%)
Sæbø, Vik	3 (3)	8–32	5–9 (67%)	10 (8)	189–443	400–450 (50%)
Osen, Gaular	18 (18)	5.7–55	10–14 & 20–24 (94%)	86 (42)	87–769	250–300 (88%)
Bjørkum, Årdal	23 (20)	5–33	20–24 (74%)	28 (24)	94–872	200–300 (67%)
Nyset-Steggje, Årdal	26 +	4–30	20–24 (88%)	40 (<10)	286–800	?

The urban site at Bryggen, covering an area of 7,500 m^2, with larger depositions of cultural layers, building remains, foundations, and shifting waterfronts, relates to the massive expansion in the harbour area from the late 12th century until the late 15th century. When it comes to textile production, it is therefore hardly compatible with the rural sites from the Viking period neither in quantity not in spectrum of tools. However, qualitatively, the tools indicate differences as for textile qualities and use areas based on weight.

Table 9.2 summarises the respective weight spectrums, represented from the urban medieval site at Bryggen and the Viking Age rural workplaces from both lowland and upland contexts, which may indicate modes of production. It shows that the largest assemblage of textile producing implements from Bryggen also represent a broader weight spectrum for spinning and weaving than from the earlier rural contexts, and generally for producing coarser yarn and textiles but finer than those produced at Kaupang, judging by the tools lost and left at these places, while the tools from Viking Age sites, tend to indicate finer qualities.

Tools and textiles

More than 2,000 textile remains have been uncovered at Bryggen, however, only around 670 have so far been analysed, all from the late 12th century,[15] as also some finds of pleated textiles from this early period and stretching into the 1300s.[16] These textiles thus stem from rather different contexts than from the earlier centuries, predominantly from graves with remains of the funeral dress. In contrast, the Bryggen textiles represent worn-out and thrown-away remnants of various qualities – both coarse and finer fabrics, discarded as rubbish and preserved in these waterlogged layers and environments by the harbour.

Nearly all the published textile finds from Bryggen represent twill qualities (95%); about a half (51%) 2/2 twills, 41% 2/1 twills of the fine worsted qualities, and only 5% common and coarser tabbies. Of these, the 2/1-twills form a pattern of distinct diagonal lines of z-spun warp and softer s-spun weft, thus differing from the balanced 2/2 twill and the plain tabbies of z/z spun yarn in both the warp and the weft that were largely produced on the vertical loom. These fine worsted qualities in varied diagonal or chequered patterns found in Bergen, are found in the medieval towns of Oslo and Trondheim too.[17] Being woven very evenly and with a very high thread count, they have been interpreted as signs of specialists at work, and most likely imported goods woven on the horizontal loom.[18] This distinct change in archaeological textiles to fabrics of 2/1 twills coincides with the technological change from the upright loom to the horizontal loom, as pointed out by English researchers concerning English tools and textiles.[19] However, at Bryggen, some of these textiles have left traces of the starting border, typical for weaves woven on the upright loom. This is also the case for the similar fine qualities found in Trondheim and in Oslo, where they cover about 50 and 40% of the textiles respectively from the early Middle Ages, to decrease during the 13th and 14th centuries.[20]

Not only remains of starting borders as used on the warp-weighted loom testify that 2/1 twills were woven on the warp-weighted loom, experimental research has also shown that it is fully possible to weave this pattern on the warp-weighted loom, and in fact especially suitable to produce light and tight weaves.[21] These types of weaves are therefore no longer proof of the horizontal loom in itself. The fine and tightly woven diamond twill qualities from Viking Age graves, moreover, were produced long before the horizontal loom was introduced and pleated textiles are known from such contexts, as, for example, from the 10th-century high-status grave from Vangsnes in Sygnafylki (cf. Chapters 5 and 6). The medieval spindle whorls and loom weights from Bryggen, however, are generally heavier than those from working environments in the Viking period (cf. Table 9.2). It is therefore also of interest to compare the

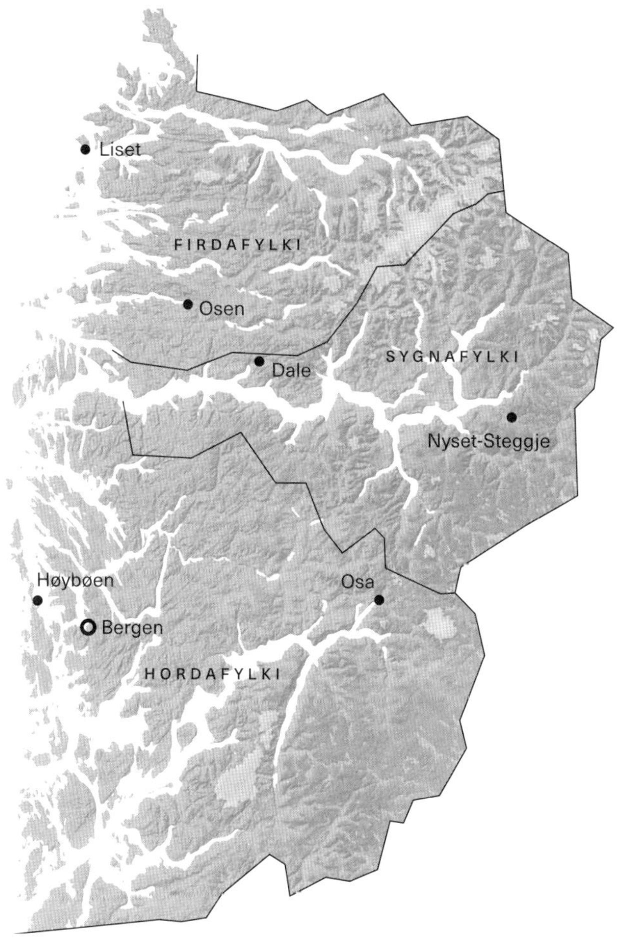

Fig. 9.10: Places with traces of textile producing implements from the Middle Ages (Illustration: Per Bækken).

Bryggen tools with corresponding tools from rural medieval working environments. Could the textiles found at Bryggen just as well have been produced in rural districts, which after all were closely integrated in the urban economy, interacting with the town for both demand and supply of various goods and services? Or, are they more likely imported goods, as suggested by Ellen Schjølberg who analysed the textiles? Supply of wool and yarn from rural districts, which is a prerequisite for urban production, is another option to consider – were wool and yarn possibly transported from rural places in the area to the town?

Textile tools from medieval rural environments: six cases

The archaeological traces of textile production from rural contexts from the Early and High Middle Ages until *c.* 1300, however, are more limited than from the Viking period. So far, they only relate to a few deserted farms in lowland areas but from all the three medieval counties. Two of the six cases referred to here were in Hordafylki – the coastal site Høybøen on the island Sotra, west of Bergen and another site further off connected to Osa in the in-fjord district of Ulvik in inner Hardanger.[22] The two other sites were in Sygnafylki and Firdafylki respectively: Dale in the community of Høyanger, and the aforementioned central place Osen in Gaular with Viking Age house sites too, and the coastal site Liset in Bremanger further north. The mountainous areas at Nyset-Steggje were still in use in the Middle Ages, also for textile production (Fig. 9.10). However, none of these sites have left traces of textiles.

The only medieval rural site from western Norway with textile remains within the study area of western Norway, connected to the medieval churchyard at Guddal, an inland valley in Fjaler in Firdafylki.[23] This was a male grave in the outskirts of the yard in a boggy area, which contributed to preserve remains of the shirt, jacket, and carpet, woven in 2/1 twill of z/s spun coarse yarn, and dated to the Late Middle Ages. Remains of a starting border demonstrate that it was woven on the vertical loom. These coarse textile remains, however, mark a strong contrast to the fine female textiles in the high-status female graves in the Viking period.

The coastal farm Høybøen, Fjell, Hordaland

One of the better-investigated medieval sites with textile tools from western Norway is the marginal coastal farm, Høybøen, abandoned *c.* 1350 and never inhabited and worked again. The site, with two deserted longhouses, was included in a larger farm, Vindenes, and used as an outlying area for grazing. The place itself was a strong textile producing location considering the size of the farmland and its later deserted status, where the arable land formed only a minor part of the area compared to its outlying resources. When including heathland for winter grazing, it has been estimated to have been able to feed a maximum of about 70 sheep, six lambs and probably two to three rams that could have produced a maximum of 120 kg to 175 kg of wool,[24] though probably somewhat less for actual production when considering sorting and combing. Still, it is difficult to assess the number of sheep, based on the potential for grazing and fodder. According to later information, the numbers of sheep and goats at the farm Vindenes of which Høybøen was then a part, were considerably lower.[25]

The two medieval longhouses were divided in two rooms interpreted as living rooms and workshops for textile production, Buildings I b, II a, and II b. Both buildings had a room with traces of textile production, spinning and weaving (Fig. 9.11). Altogether, the textile tools included remains of 72 loom weights and 24 spindle whorls (Figs 9.12 and 9.13). The spindle whorls found represented different weight groups but the majority were under 30 g and could produce a wide yarn spectrum, from fine to coarse yarn, however, except the very thinnest threads. Spindle whorls around and above *c.* 30 g could also produce coarser yarn suitable for spinning coarse yarn for producing, for instance, sail cloth.[26] The spindle whorls

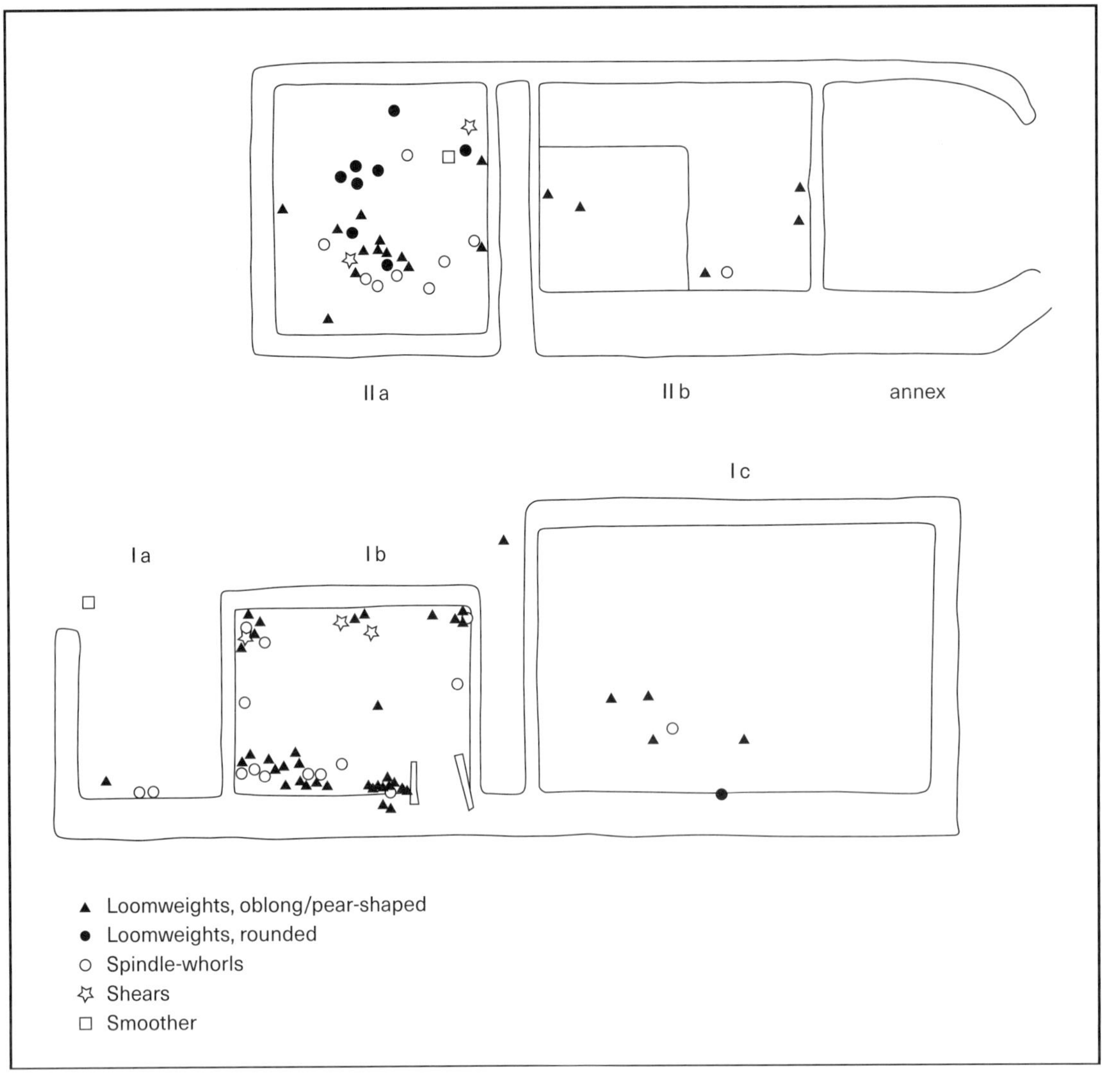

Fig. 9.11: Distribution of textile tools at Høybøen, 72 loom weights and 24 spindle whorls. Buildings I b, II a and II b, interpreted as living rooms and working areas for textile production; I a: workshop; I c: byre (Source: Randers 1981).

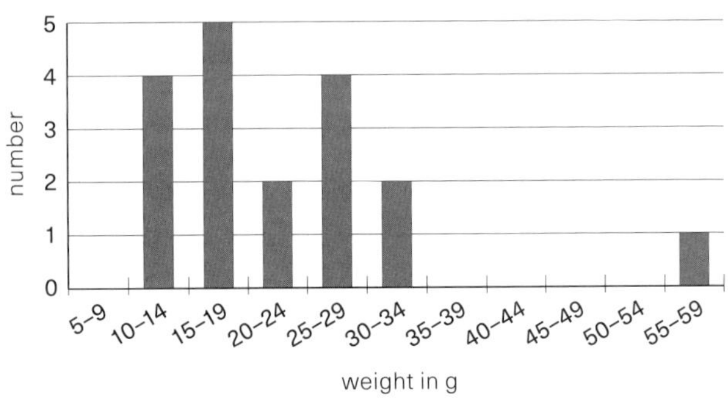

Fig. 9.12: Measurable spindle whorls from Høybøen, by weight in weight groups of 5 g (N:18 measurable) (Graphic design: Per Bækken).

at Høybøen, then, are somewhat heavier than at Bryggen as also at Viking Age rural workplaces.

The loom weights found represent about the same weight pattern in both the two rooms II a and I b within the weight spectrum just under 200 g with most of the loom weights weighted between 200 g and 600 g and with a peak in the weight group 300–400 g. The loom weights in room II b, however, represent a somewhat wider range, from under 200 g to around a kilo with 56% under 400 g. In room I b, the weights spread between *c.* 200 g to around a kilo with 52% under 400 g (Fig. 9.13).

Compared with the loom weights from the Viking Age contexts, the medieval weights thus represent a wider weight range. However, the loom weights and possible loom weights from Bryggen include an even wider weight spectrum, from under 100 g to around 1200 g. At Bryggen, the majority, *c.* 80% weigh between 300 g and 900 g, and only some 10% below 300 g.[27] The medieval loom weights are thus generally heavier than the Viking Age rural assemblages and indicate a trend of weaving coarser fabrics. At both places, the tools seem to have served a multipurpose textile production.

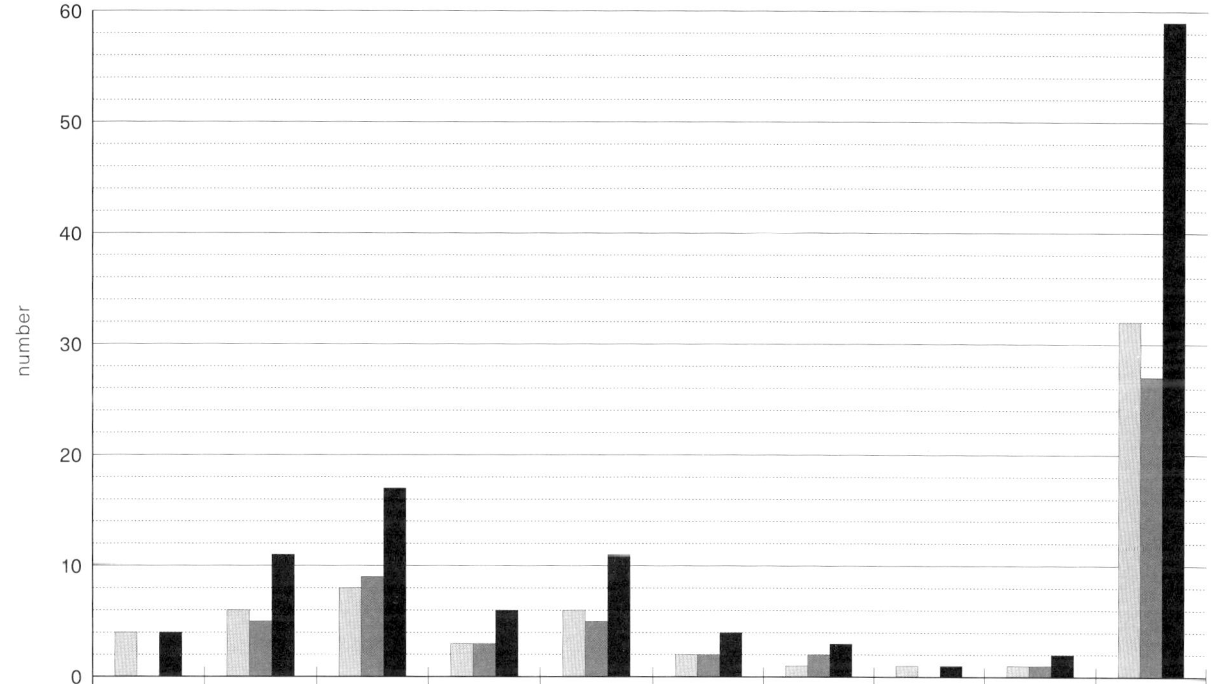

	200 g	200–300	300–400	400–500	500–600	600–700	700–800	800–900	900–1000	N
Room 2a	4	6	8	3	6	2	1	1	1	32
Room 1b		5	9	3	5	2	2		1	27
Total	4	11	17	6	11	4	3	1	2	59

Fig. 9.13: Loom weights from Høybøen, by weight in weight groups of 100 g (N:59) (Graphic design: Per Bækken).

Like the Viking Age site at Ytre Moa (cf. Chapter 8), the medieval longhouses at Høybøen appear as abandoned, with many of the more durable objects *in situ*. The number of loom weights and their spatial distribution indicate altogether two or possibly three looms in two of the rooms in the longhouses while the many spindle whorls of different weights indicate production of fabrics of both finer and coarser yarn, but not of the very finest qualities. The buildings seem to represent two households, and quite a few people seem to have been involved in textile production, probably for more than the household's own needs, although the textiles needed in a coastal environment were many. Finds of imported products within the buildings, among them ceramic drinking vessels and cooking pots,[28] as unearthed in large quantities at Bryggen and other urban contexts, may thus indicate an economy based on surplus production and interchange.

Compared with the textile implements from Bryggen with its favourable preservation conditions and a far more varied assortment in wood and bone, besides stone, the comparable spindle whorls of stone, varied, as shown, from 6 g to more than 50 g with 62% under 25 g and 11% under 10 g (cf. Fig. 9.2). At Høybøen, the measurable spindle whorls were somewhat heavier. They range from 11 g to 58 g with about a half under 25 g, and lack the lightest group as found in Bergen. The loom weights or possible loom weights from Bryggen have a wider weight spectrum than Høybøen too, from under 100 g to around 1200 g. The majority, *c.* 80% weigh between 300 g and 900 g, and *c.* 10% under 300 g (cf. Fig. 9.3). The medieval loom weights seem, then, to be generally heavier than the Viking Age rural assemblages, indicating a trend of weaving coarser fabrics. The loom weights at Høybøen represent the widest weight range of the six Viking Age rural sites, up to 1 kg, but are generally lighter compared to the general trend in the nearby town of Bergen.

Osa: a farm from the inner zone of Hardanger in Hordafylki

Another medieval farm site with traces of textile production from Hordafylki related to in-fjord Hardanger at the place Osa,[29] with remains of four buildings, of which two, Buildings 2 and 3, date to the centuries between *c.* AD 1000 and 1400, while the other two, 1 and 4, are not dated. Of these, Building 2 comprised a single broken spindle whorl with an estimated weight of 35 g and 28 loom weights, and some bakestones, typical of the Middle Ages. The site included five other spindle whorls, four in the weight spectrum 25–40 g and one over 60 g and generally heavier then than those from Høybøen and the whorls from Bryggen. The 28 loom weights in Building 2 seem to represent a loom, of which 17 are complete and 11 estimated as for weight.

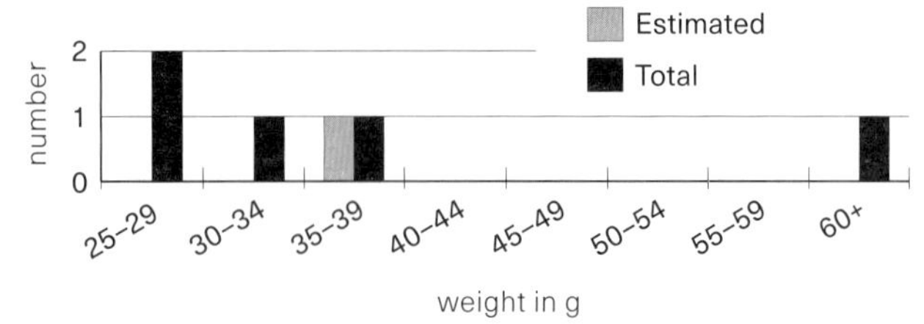

Fig. 9.14: Spindle whorls from the site Osa, by weight in weight groups of 5 g (N:6) (Graphic design: Per Bækken).

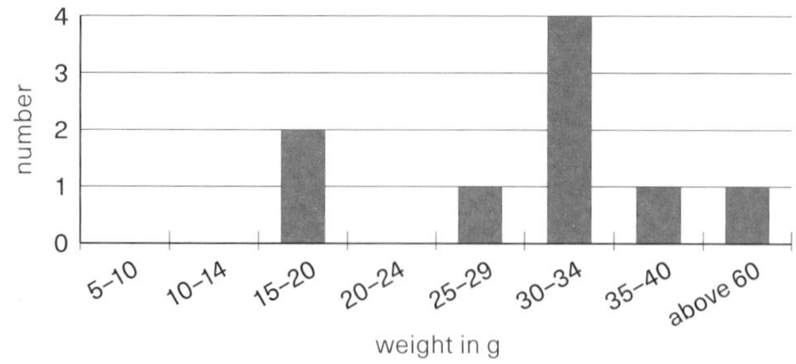

Fig. 9.16: Spindle whorls from Dale, Høyanger (N:9); four with estimated weights (Graphic design: Per Bækken).

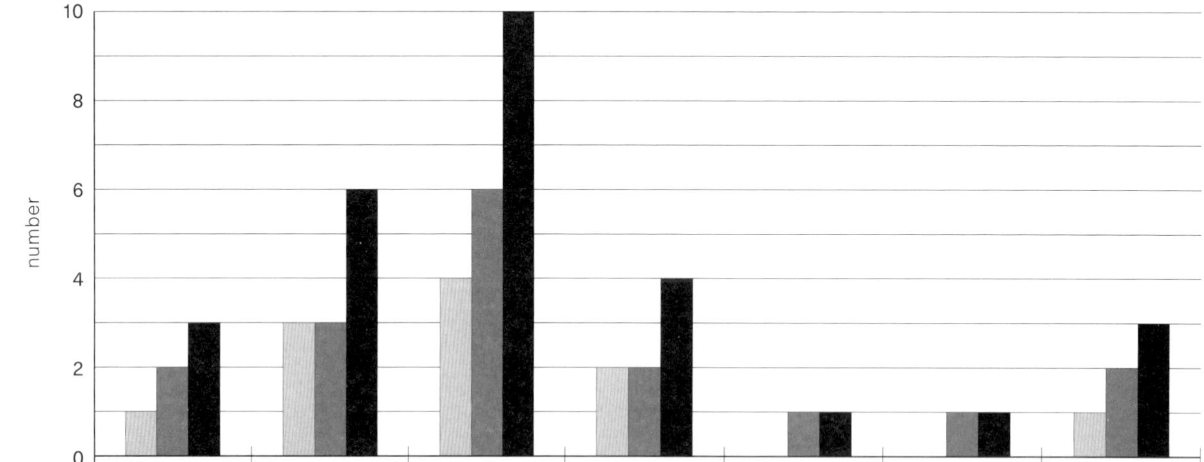

	-200	200–300	300–400	400–500	500–600	600–700	700–800 g
Estimated	1	3	4	2			1
Complete	2	3	6	2	1	1	2
Total	3	6	10	4	1	1	3

Fig. 9.15: Loom weights from Osa, building 2, by weight in weight groups of 100 g (N:28) (Graphic design: Per Bækken).

They weigh from under 200 g to around 800 g, with a peak in weight group 300–400 g. The weights indicate, then, another and finer production than the spindle whorls left or lost at the site (Figs 9.14 and 9.15). Compared with the loom weights from coastal Høybøen, probably represented by two, possibly three looms, the weight spectrum is about the same, from just under 200 g to 800 g, with a peak in weight group 300–400 g. They are also lighter than the loom weights from urban Bryggen.

A longhouse at Øvre Dale in Høyanger, Sygnafylki

Another medieval longhouse site with textile tools stems from Dale, located in the mid-zone of Sygnafylki in a valley stretching from a side arm of the Sognefjord, and with favourable conditions for agriculture and animal husbandry with vast grazing areas. The house site was first discovered when ploughing the fields,[30] and reported and followed up by the museum.[31] The textile equipment at the site comprised at least eight spindle whorls and 10 loom weights of varied weights. The whorls range from 15 g to over 60 g, with a peak in the weight group 30–35 g (Fig. 9.16).

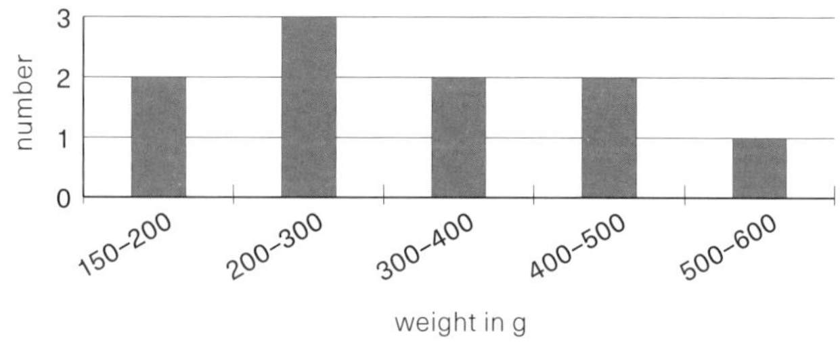

Fig. 9.17: Loom weights from Dale, Høyanger (N:10) (Graphic design: Per Bækken).

The loom weights vary from 170 g to 530 g (Fig. 9.17). Two pieces of a soapstone vessel with drilled holes may have served as loom weights too, possibly for adjusting the weight and balancing the tension. The weights, however, are too few to give a representative impression of a loom and comprise rather commonly used weights for common types of textiles (Fig. 9.17). The other artefacts in this context are also interesting – hones, a key, and shards of soapstone vessels, one with a flat base of medieval type, bakestones, and foreign pottery – and probably acquired in Bergen or through exchange networks.

Osen in Gaular: continuity or change?

At the central place Osen with traces of textile production from the Viking period, the remains of medieval structures and of textile production were less distinct than from the Viking Age longhouse and pit house (cf. Chapter 8). Due to the more blurred find conditions related to the succeeding medieval building, neither size nor structure could be decided precisely. Remains of the floor indicate, however, that it was raised on sill stones and connected to a pit house of about 30 m^2 close by, dated to around AD 1040–1160.[32] Only a single spindle whorl of *c.* 25 g and some loom weights and knives were left when abandoned not long after.

Most of the textile producing equipment at Osen refers then to the Viking period (cf. Chapters 5 and 8). Of the altogether 38 spindle whorls retrieved at the place, all the others related to the Viking Age longhouse II and to a work area close to the hearth. Nearly all the loom weights, 78 of altogether 86, also related to this early building. The rest connected to the two pit houses, and in the later only a fragment – and indeed marking a change in textile production compared with the Viking period. According to written evidence, the farm's high status and central position continued into the Middle Ages, however, the heydays of textile production from the Viking period were over judging by the archaeological evidence. This seems to have been the case for many central places in western Norway marked with earlier traces of tools and textiles and other evidence related to textile production.

Liset in Bremanger: a new place for textile production?

In a more peripheral area by the outer coast in Bremanger, and in the opposite end in terms of power and status but not in terms of opportunities within textile production, the place Liset on the island Frøya in Bremanger now came to the fore in this northern part of Firdafylki. While the textile production faded out at Osen, traces of textile tools and textile production expanded at a location in the outfield of the farm Liset, where a special building for textile production was discovered and excavated.[33] Located close to the sea, only 12 m from the shoreline, the excavations revealed a 15 m long and nearly 8 m wide three-room building, dated broadly to the Middle Age. It was built in stone and wood, paved with flagstones, and all rooms contained textile tools. Room 1 to the north, 6 m long, was demarcated by a stone wall, while the mid-room with the entrance had a corner hearth, while the smaller room 3 to the south as an extension, measured 2.5 m × 2.5 m or 6.25 m^2. The surrounding landscape, with all-year-round grazing land, also showed traces of cultivation.

The textile tools, spindle whorls and loom weights related to all three rooms. However, most of the loom weights connected to the front and mid-rooms – 46 complete or measurable, and 15 fragments. The two spindle whorls in the medium and light weight groups were for spinning fine and medium fine yarn. The 20 loom weights in room 1 weigh from under 200 to around 700 g, of which seven, or 35%, weigh under 400 g – peaking in the weight groups from 400 g to 600 g. The 14 loom weights in room 2 were generally lighter and vary between *c.* 200 g and 600 g and with more than half, or 57%, under 400 g, concentrating between 200 g and 500 g. The smaller assemblage of 12 weights in room 3, vary between *c.* 200 g and 700 g, and with 42% under 400 g (Fig. 9.18). The different weight patterns indicate weaving of different types of fabrics, generally of finer, and medium fine qualities. In total, the weights gather between *c.* 200 g and 600 g. Whether they represent two or three looms is more difficult to decide, as many weights are broken and left as fragments. Since the upright loom was a flexible device, the weights were adapted to the warping and according to the type of textile qualities. The weights related to these deserted coastal environments indicate a multipurpose production of varied qualities, possibly serving as a subordinate workstation in the outskirts of the main farm.

Continued textile production in the Årdal mountains?

At Nyset-Steggje in the Årdal mountains, only two of the areas in Vikadalen and Berdal (cf. Chapter 8), indicate continued use and textile production into the Middle Ages. The building constructions were still of the same type as from the Viking period but generally larger, from 30–35 to 47 m^2 and often with a hearth in the middle. However, fewer remains were found and without the same distinct traces of production, which was only represented by a few spindle whorls and in a few cases, a loom weight or two. One of the five buildings with traces of textile production in the transition to the new millennium (Loc 100, site A), contained four spindle whorls and a few loom weights, which may just as well stem from the Late Viking Age, as later (and listed in Table 8.1). What is more special for the other medieval sites with traces of textile production, is the absence of traces of weaving. Two of the buildings dated to the Early Middle Ages (Loc 26 and 82), each had a single spindle whorl, respectively 12 g and 49 g, for fine to very coarse yarn. The two other buildings represent long-term use too and as for textile production related to different periods – the Early Iron Age to the very end of the Middle Ages with, respectively, two and three spindle whorls, and all in common weight groups of 20–25 g and from 17 g to 32 g (Table 9.3).

Compared with the textile tools from the Viking period in the Årdal mountains, the spindle whorls were both fewer and heavier, weighing from 12 g to 49 g with only three under 25 g, and only with faint traces of the loom indicated by fragments of middle-sized loom weights, possibly

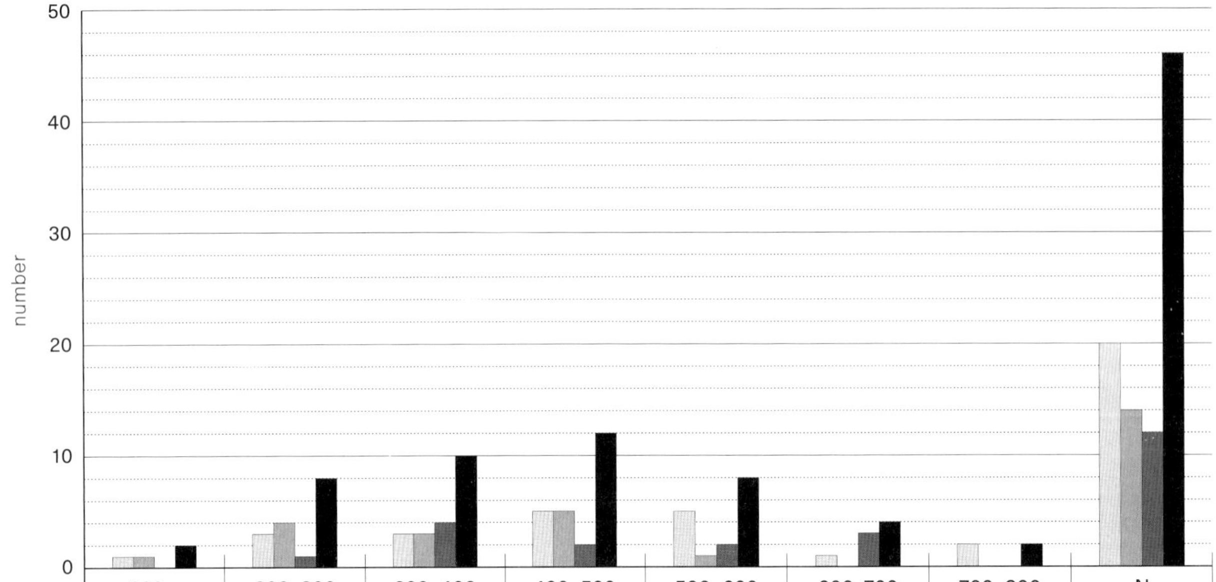

	200 g	200–300	300–400	400–500	500–600	600–700	700–800	N
Room 1	1	3	3	5	5	1	2	20
Room 2	1	4	3	5	1			14
Room 3		1	4	2	2	3		12
Total	2	8	10	12	8	4	2	46

Fig. 9.18: Loom weights from Liset, Bremanger, by weight in weight groups of 100 g (N:46) (Graphic design: Per Bækken).

Table 9.3: Medieval sites at Nyset-Steggje with textile producing implements (= estimated weight).*

Sites	*Contexts*	*Dating*	*Whorls N*	*Weight range in g*	*Loom weights N*	*Other tools*	*Indicators of status*
Loc. 100	House site A	1030–1230	4*	6, 15, 21, 24*	2*		
Loc. 26	House site B	900–1160, youngest phase	1+ ½	12	–	Hones	Key
Loc. 82	House site B	1010–1270	1	49	–		
Loc. 82	House site A/E	1450–1650	2	24, 20*	–	Arrow heads, knives	
Loc. 115	House site A	Phases from AD 200 to AD 1470	3	17, 25, 32	–	Steatite vessel, hones, knives, arrowheads	
Loc. 115	House site C	Iron Age and Middle Ages	1	20*			
6 (5)			8+1/2, 5 (4*)	12–49	2		

from the early 11th century. The finds included tools also for other use areas, among others hunting. Altogether, the activities related to textile production in these areas seem to have declined compared with the Viking period. As seasonal workstations, they were combined with dairying and other activities, such as hunting, involving people of both genders.

Changes in the production

The medieval working places for textile production thus differ from those in the Viking period and in the Early and High Middle Ages, the largest quantitative difference connects to the urban production environments at Bryggen. Although possibly having taken the new horizontal loom into use, the warp-weighted loom still appears to have ruled the ground during the whole period of study both at urban Bryggen and in rural contexts.

Table 9.4 summarises the representation of textile tools from medieval working environments, representing the urban settlement at Bryggen as the largest and areas in the Årdal mountains. It also includes the five rural sites from different zones and counties, two from the coastal and inner zones Hordafylki, two from the mid- and coastal zones in Firdafylki, and one from the mid-zone of Sygnafylki,

Table 9.4: Representation of complete and measurable tools in relation to the assemblages of spinning and weaving implements at the six sites. Weight range and relative representation within the range is marked in brackets.

Sites	*Spindle whorls (measurable weight)*	*Weight range (g)*	*Peak: weight group (% below 25g)*	*Loom weights (measurable weights)*	*Weight range (g)*	*Peak: weight group (% below 400g)*
Bryggen, Bergen	206 (162)	6–49	15–19g (62%)	792 (555)	100–1300	300–600 (27%)
Høybøen, Fjell	24 (18)	11–58	15–19 (61%)	72 (58)	200–600	300–400 (55%)
Osa, Ulvik	6	25–60	35 (0)	28	200–800	300–400 (82%)
Dale, Høyanger	8	15–60+	–	10	150–600	300–400 (80%)
Osen, Gaular	1	25		1	?	?
Liset, Bremanger	5	–	–	61 (46)	200–700	400–600 (70%)
Nyset-Steggje	9	12–49g	20–24			

besides the mountainous areas at Nyset-Steggje in the inner zone.

Qualitatively, however, the table shows that there are no greater differences between the textile producing implements at Bryggen and those at, for instance, the nearest coastal place, Høybøen. The other contemporary medieval sites with textile tools are fewer and less compatible but the tools are generally heavier than those from Viking Age workplaces. The tools for spinning and weaving from Bryggen match well with those from the earliest 'urban' setting at Kaupang, with both able to produce a broad range of qualities from the finest to the coarser, and in fact even finer (cf. Fig. 9.8). However, the tools from Viking Age rural workplaces represent a broader weight spectrum than those from the urban environments of Kaupang and Bryggen. Except the Høybøen assemblage, the smaller medieval rural tool collections are less compatible.

Differences between urban and rural production?

Altogether, these representations indicate that the dichotomy between rural and urban textile production may be exaggerated, at least in the early stages of urbanisation in Norway – and without clear evidence of standardisation in the domestic production. However, a comparison between the textile tools from these periods is inevitably influenced by the major societal changes, and in western Norway, as also in the North Atlantic, by Bergen's expanding role in international trade and within production of crafts. As early as in the late 12th century onwards, scattered sources refer to import of high-quality textiles from England, and in increasing amounts from the early 14th century.[34] The coarser fabrics of 2/2-twill woven textiles of *vaðmál* were imported from Iceland from the late 1100s onwards, and from the 1300s with import in larger scale of coarser textiles and canvas from the Baltic.[35] In time, this expanding import concurs with a decline in the domestic urban production, judging by the tools.

The state formation processes and change of religions, and eventually the abolishment of slavery by the 12th century and thus the use of unfree *Þræll*s as a common work force, must have affected production and its organisation too. While the Viking Age graves with textile producing implements and workplaces largely related to central places or were dislocated in agglomerated separate buildings serving as large-scale production places, such as at Bjørkum and in the high-lying shieling areas as Nyset-Steggje, possibly involved people of different status, including an unfree work force. The finds of socially distinguishing artefacts, such as at Ytre Moa give a nuanced picture in terms of status. The pit houses at central high-status farms and high-status female graves, such as in Vik and Sogndal, however, may reflect the hierarchies in production at each end of the social ladder. The medieval textile working environments had another character, both spatially and socially. By the Middle Ages, literary evidence largely supplements the archaeological traces of production. This is best evidenced in the North Atlantic, related to the production of 2/2 twill, *vaðmál.* These sources may also shed light on earlier practices from their 'mother' country.

Tools, textiles, and textile production environments in the North Atlantic

Wool, yarn, and textiles stand out as trademarks and special properties of the North Atlantic area – a legacy that dates to the Viking period. Throughout this area, *vaðmál* also became a standard economic measure; commodity 'money', used for taxation and exchange and by the Middle Ages in fact became the main export article, representing a different development than in western Norway. This was possible by using the area's comparative advantages of raw materials and product specialisation with its vast areas for sheep breeding and flock management in the

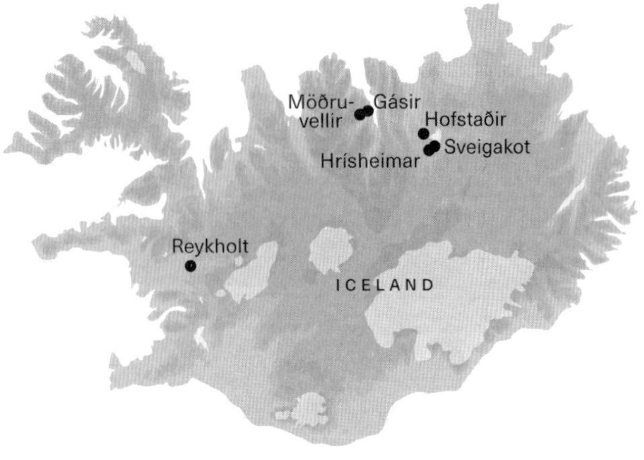

Fig. 9.19: Places in Iceland with traces of textile production, mentioned in the text (Illustration: Per Bækken).

chilly climate, needing sparse feeding and yielding fine wool (cf. Chapter 4).

The production is documented archaeologically and through written sources in terms of qualities and quantities, work processes, gender, and working conditions. The tools in these areas are also better preserved and more numerous than the organic traces of textiles. These tool remains, mainly spindle whorls and to some extent loom weights, generally represent lost and discarded objects at deserted farms, and give direct information about the textile production environments. In Iceland, many of the tools related to pit houses, as in earlier Viking Age contexts in Norway, representing special work zones for weaving and to some degree spinning. Since many of them date back to the *landnám* period – earlier than the written evidence and sometimes also with preserved textiles – they link different sources of information over time.

According to the medieval written sources, and the Icelandic law code *Grágás* from the 13th century, the term *vaðmál* included different qualities both for domestic use in the households and for export; *hafnarvaðmál* and *vöruvaðmál. Hafnarvaðmál,* originally produced for domestic use in the households, later also became an export article, denoted as *gjaldavóð*. *Hafnarvaðmál* could comprise several different qualities, such as *merkurvaðmál, klæðavóð,* and *smávoð,* used for better clothes, carpets, bed covers and the like. *Vöruvaðmál* denoted commercial commodities and was a medium of exchange and measure of value, and as noted, served as commodity 'money', and used for taxation and exchange. The commercial weaves, denoted ON *vóð vaðmáls* or *vörðavaðmál,* as a standard economic measure and means of currency should be '*thri-skefttvi elm breið*', which means woven in 2/2 twill technique and of two ells (47.4 cm or 90 cm as the standard width). The length of such a cloth, is not specified, but according to later customary norms in the earlier referred to *Búalög,* the standard length should be 20 ells or 9.5 m in length and 90 cm wide (=8.55 m^2), evidently not larger than could be woven by one weaver. Judging by the information in *Búalög*, the standard measures and qualities were defined by the number of warp threads per ell, from the coarsest types of *gjaldavóð*, to the finer *klæðavóð* and the finest *smávóð*, and represented 9–11 and 13–14 warp threads per cm, respectively.[36] This information is essential for understanding the archaeological remains of both tools and textiles.

The archaeological finds of textiles in Iceland, mostly dating from the High and Late Middle Ages, give direct insight into the textile qualities, the spinning – twist, spin direction, and thickness of threads – and the weaving techniques. It is therefore interesting to compare the finds with the later recorded norms and descriptions. The textiles show that 2/2 twill, *vaðmál,* clearly dominates, constituting nearly 90% of the textiles found in Iceland. The earliest recorded textiles stem from one of the earlier mentioned pit houses at Hofstaðir, which dates from the late 900s to the early 11th century. Here, only six fragments are identified as for technique or quality: all in 2/2 twill and z/s spun with warp threads estimated to be 8–10 and 10–12 threads per cm.[37] The spindle whorls from the site indicate that they in fact could have produced these mostly medium and coarser yarn for the warp and the weft in the different *vaðmál* qualities – and even finer than the norm for the *smávoð*.

At the high-status farm Reykholt, in south-west Iceland, archaeological remains of textiles are represented over a longer time span, but only *c.* 12% of around the 200 preserved textiles were medieval, covering the whole period but most of them from the layers dated to *c.* AD 1000–1300. The textile remains formed a rather homogenous quality of coarse and medium fabrics, with 4–13 threads per cm, and can be identified as *vaðmál* with its general characteristics – thinner and tighter z-spun warp threads, with a stronger twist than the softer and looser weft threads, but sometimes also with a more balanced thread system.[38] At the Mööruvellir site, which is partly contemporary with Reykholt, *vaðmál* also dominated and showed rather similar qualities of both coarse and finer qualities, with the most frequent thread counts lying in the range of 9–12 × 5–7 threads per cm. Some textiles at Mööruvellir also showed traces of fulling.[39] Similar qualities are also documented at the nearby medieval trading place, Gásir, where all but one of the 50 pieces were of 2/2 twill, *vaðmál*, of varying qualities but with a cluster in the category 9–10 warp threads.[40] In general, the fabrics were coarser than at the high-status farms of Reykholt and Mööruvellir. The finds of spindle whorls and textiles demonstrate that the production of different types of *vaðmál* began at an early stage of the colonisation. The varied range of spindle whorls corresponds with the varied pattern of qualities, all within the wide group of qualities of *vaðmál*, but mostly of medium and coarser types, comparable with *vöruvaðmál* and the coarsest types of *gjaldavóð* (Table 9.5). Spindle whorls from only two medieval sites, Bær and Gröf

Table 9.5: Analysed medieval textile finds from places in Iceland: Hofstaðir, Reykholt, Möðruvellir, and Gásir.

Site	*Fragments, N*	*Date*	*Warp, thread count per cm*	*Weft, thread count per cm*
Hofstaðir	6 of 11	980–1030	8–18	10–12
Reykholt	24 of 200	*c.* 1000–1500	7–13	5–12
Möðruvellir	80	*c.* 1200– *c.* 1300	6–16	5–9
Gásir	50	14th–15th centuries	4–12	3–8

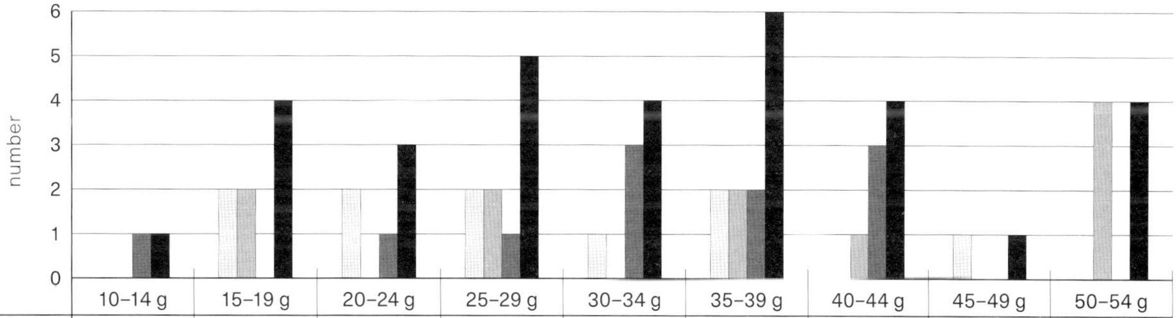

	10–14 g	15–19 g	20–24 g	25–29 g	30–34 g	35–39 g	40–44 g	45–49 g	50–54 g
Sveigakot		2	2	2	1	2		1	
Hrisheimar		2		2		2	1		4
Hofstaðir	1		1	1	3	2	3		
Total	1	4	3	5	4	6	4	1	4

Fig. 9.20: Spindle whorls from three locations in Iceland: Sveigakot, Hrisheimar, and Hofstaðir in weight groups of 5 g (N:32) (Graphic design: Per Bækken).

í Öfarum, in south-west Iceland and abandoned in 1362, could be assessed. The spindle whorls from here are nearly all larger and heavier in size, *c.* 40–50 g or more, but too few have survived to ascertain a trend. Generally, both the spindle whorls and loom weights indicate textile production of medium to coarse textiles, with finer qualities are less common (Fig. 9.20).

As shown in the preceding chapter, the best represented site for textile tools is at Hofstaðir in the Mývatn area, with the spindle whorls measuring from 12 g to more than 40 g, along with other early sites in Iceland with spindle whorls from *c.* 15 g to 50 g, and the sites in the Faroe Islands and Shetland too.

Like whorls from medieval workplaces in western Norway, they indicate production of medium to coarser yarn, and not a standardised production. Only the high-status central place Jarlshof in Shetland, showed more homogenous whorls in shape and size, in the weight group 25–35 g. What is especially interesting is that several sites contained both tools and textiles, mostly dating to the High and Late Middle Ages and give insight into the textile qualities, the twist, spin direction, and thickness of threads, thread count and weaving techniques. The early traces of textile production in Shetland are fewer and less studied, although the excavations at Jarlshof have revealed ample traces of textile production throughout the Norse period, especially from the earliest centuries after the colonisation. Altogether, 31 spindle whorls and 160 loom weights were recovered, of which several were found in clusters, seeming to represent the whole loom.[41] More recent excavations at The Biggings, Papa Stour, have uncovered both textiles and tools from the Early Middle Ages, including 24 spindle whorls of different shapes and material. They weigh from 11 g to *c.* 40 g, capable of spinning yarn of different gauges: fine, medium, to rather coarse threads and largely corresponding with the thread qualities in the fabric remains found at the site.[42] The Jarlshof whorls are more homogeneous in shape and size, commonly within the weight group of 25–35 g. Altogether, they seem to indicate larger-scale and possibly more homogenous textile production than at the other sites.

Although textiles are rare finds in the Faroe Islands, with none yet dated to the Middle Ages, textile tools are common elements. The number of spindle whorls per site is generally higher than from Iceland, while the looms, strangely enough, have left few traces. The finds from Toftanes in the Faroe Islands include more than 50 spindle whorls mostly of steatite and of different shapes, as well as rough-outs,[43] while the other two sites, í Uppistovubeitinum close by, and the shieling area, Argisbrekka, contained 16 and 23 whorls, respectively.[44] Like the Icelandic cases, they show wide weight ranges, from less than 15 g to more than 70 g. More than half of the measurable whorls, however, weigh between 30 g and 50 g, with a peak in the weight group 40–45 g. Although wide weight ranges are represented at

Table 9.6: Analysed medieval textile finds from Iceland and Shetland.

Site	*No. of fragments analysed*	*Date*	*Warp, thread count per cm*	*Weft, thread count per cm*
Reykholt	24 (of 200)	1000–1500s	7–13	5–12
Möðruvellir	80	1200 – *c* 1350	6–16	5–9
Gásir	50	1300–1400s	4–12	3–8
The Biggings	24	1100–1200s	5–12	4–10

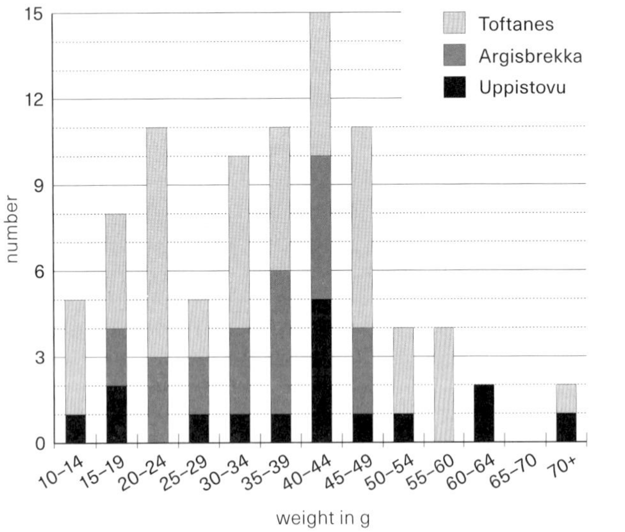

Fig. 9.21: Measurable spindle whorls from Toftanes (N:49), Argisbrekka (N:23), I Uppistovubeitinum (N:16) according to weight groups of 5 g (N:89) (Graphic design: Per Bækken).

all the sites, the whorls from Toftanes in general appear somewhat lighter than those from the other sites, with a higher percentage (35%) lighter than 30 g (Fig. 9.21). Compared with the Icelandic whorls, the Faroese whorls show a wider weight range and more of them occur in the heavier weight groups, but neither shape nor weight reveal standardised types.

Common patterns in production?

Overall, the qualities found at different North Atlantic Viking Age and medieval sites, correspond with the documentary sources from Iceland, showing that 2/2 twill, *vaðmál* was not one standard weave, but was distinguished by different qualities for different uses and of different value. The archaeological finds of textiles and tools also indicate that a variety of qualities were produced from an early stage in the Viking Age, and included coarser, medium, and finer weaves. The majority were woven in the 2/2 twill technique of *vaðmál*, however, of varying qualities and with the most frequent thread counts of 9–12/5–7 threads per cm.[45]

The different cases presented here, indicate that textile production was an important industry in the North Atlantic area from the early colonisation period. The finds of many textile tools from both high status and common households clearly demonstrate the significance of textile production from an early stage. It also involved farming and economic management at a higher level and within wider networks.

Other archaeological remains of medieval textiles also substantiate the written information and show that they were woven in a standardised manner.[46] The workers and working conditions are defined too, where female weavers and 'spinsters' of subordinate status should work on a full day basis for board wages. In Iceland without any towns, a large-scale textile production of *vaðmál* for export was based solely on the use of the upright warp-weighted loom, but still on a professional, specialised basis, but organised within a rural household structure by female weavers of subordinate status, also denoted as a *vefkona*.[47]

The textile remains formed rather homogeneous groups of coarse and medium fabrics, with 4–13 threads per cm, which can be identified as *vaðmál* with its general characteristics – thinner and tighter z-spun warp threads, with a stronger twist than the softer and looser s-spun weft threads, but sometimes also a more balanced thread system (Table 9.6).[48]

Textile fragments showing a similar range of textile qualities, from about the same time as the medieval Reykholt textiles, and found at The Biggings, Papa Stour, in Shetland too. Like the Icelandic textiles, they include coarser, medium, and finer weaves.[49] They seem to concentrate around the qualities defined for the historically documented *vaðmál* types of *hafnarvaðmál*, as well as the coarsest and cheapest quality of *vöruvaðmál*. Overall, then, the qualities found at different North Atlantic Viking Age and medieval sites correspond with the documentary sources from Iceland, showing that *vaðmál* was not one standard weave, but distinguished by different qualities for different uses and value. The archaeological finds of textiles and tools also indicate that all these qualities were produced from a very early stage of the Viking Age.

The different cases presented here, indicate that textile production was an important industry in the North Atlantic area from the early colonisation period. The many textile tools uncovered from both high-status and ordinary households, clearly demonstrate the significance of textile production. The production, based on specialised farm management of sheep breeding, escalated in the 12th and 13th centuries as substantiated by both the zoo-archaeological[50] and written evidence. These cases, revealed through various archaeological contexts and texts from western Norway and the North Atlantic show that larger parts of the textile production in the period *c.* 1000–1300 was still carried out by an overwhelmingly female work force both in rural and urban areas, including all the connected work processes in the countryside. Eventually, the production also directed towards an expanding market, which again added significant income and value. This demonstrates that old technology based on hand spinning and the warp-weighted loom could also contend in the new market economy, as is shown by the specialised production of the multipurpose, functional, and relatively reasonable *vaðmál* in the North Atlantic, and early in the period able to produce the finest qualities of diamond twill in western Norway. By the end of the period around 1300, in contrast, the varied *vaðmál* products were largely transported from the North Atlantic via Bergen for consumption and wider distribution in the expanding international trade.

These cases show, then, that one should be cautious in using dichotomies such as domestic production/specialised crafts and male crafts versus female household activities. The cases from Iceland show that large-scale textile production, on a specialised level even for export, could be carried out and regulated within scattered rural households and mainly carried out by women and based on old technology. This is not unlike the production of finer qualities in the Viking period in western Norway at special places and environments. The cases, from each end of the period, further demonstrate that the production was integrated in larger networks, which also made it possible to specialise in rural settings.

Notes

1 Munro 2003.
2 Helle 2005; Crawford 2013, 16.
3 Rast-Eicher & Windler 2006, 31.
4 Munro 2003, 196; Rast-Eicher & Windler 2006, 32.
5 Hoffmann, *KLNM* XIX, 679.
6 *e.g.* Hoffmann 1964; Kjellberg 1979; Kjellberg & Hoffmann 1991.
7 Øye 2016a.
8 Øye 1988; 2014; 2015c.
9 Rui 1992; Gjøl Hagen [1988] 1994; Øye 2015c.
10 Øye 1988, 74–75.
11 Grenander-Nyberg 1994.
12 Øye 1988, 66.
13 Øye 1988, 66.
14 Øye 1988, 106.
15 Schjølberg 1998.
16 Vedeler 2007.
17 Kjellberg 1979; Gjøl Hagen [1988] 1994.
18 Schjølberg 1998, 209–213.
19 Owen-Crocker et al. 2012, 46 with references.
20 Øye 2015c, with references.
21 Gjøl Hagen [1988] 1994, 83 with references; Pfarr 2005.
22 Lurekalven at Lygra, another medieval coastal site with textile tools, is omitted since the tools were not available for inspection and measuring (cf. Kaland 1979, 71–86).
23 Vedeler Nilsen 1992.
24 Randers 1981, 102.
25 Randers 1981, 99: refers to tax list 1657: 1 horse, 1 ox, 22 cows, 3 heifers, 14 goats, 23 sheep.
26 Andersson 1999, 25 (cf. Chapter 4).
27 Øye 1988, 63.
28 Randers 1981.
29 B15009, Osa, Ulvik Unpublished report by Kjersti Randers 1993.
30 B14968; B14198, Øvre Dale, Høyanger, gnr. 62/4.
31 B14220, Unpublished report by K. Randers. I thank Kjersti Randers for making it available to me.
32 Diinhoff 2002.
33 B9916, excavated by the archeologist Wenche Slomann in 1948, unpublished report University Museum of Bergen.
34 Nedkvitne 2014, 28–31, 60–61, 82.
35 Helle 1982, 317; Blom 1991, 99; Nedkvitne 1991; Þorláksson 1991.
36 Hoffmann 1964, 213.
37 Sverrisdóttir 2009, 318–319.
38 Walton Rodgers 2012, 198–204.
39 Hayeur Smith 2011b.
40 Hayeur Smith 2011a.
41 Hamilton 1956.
42 Walton Rodgers 2012.
43 Stumann Hansen 1991.
44 Mahler 2007.
45 Hayeur Smith 2011a; 2011b.
46 Hayeur Smith 2015.
47 Þorláksson 1991.
48 Walton Rodgers 2012, 198–204.
49 Walton Rodgers 1999.
50 Harrison 2011.

Epilogue

How textile production was carried out and organised in the Viking period and into the High Middle Ages at these fringes of north-western Europe, may perhaps appear as a trivial issue at first sight – expecting it to be a rather static production within the domestic sphere, largely carried out by women in rural settings and basically as a household production at farm level for its own needs, however, diverging as for status. This may also explain why this type of production and its organisation has largely 'gone under the radar', compared with other types of production and technologies of the period that were transforming raw materials into finished products, such as iron and minerals based on new technologies, quarrying, mining, forging, boat building, farming, and others. However, textile production of varied qualities and wide use areas was also of vital importance and a prerequisite for larger societal changes of the period, such as exploring and settling new land in the North Atlantic, furnishing the new elites with prestigious and socially distinguishing outfits, as well as complying with basic needs by providing people with textiles for all practical use areas and keeping the body warm both on land and at sea. These different types of production, based on varied raw material both hard and soft within different production spheres, also seem to have been closely entangled within gendered, largely reciprocal work spheres, as revealed in the Viking Age graves with tools.

As for textile production, the area of study shared many common denominators and opportunities in climates and landscapes with vast coastal and upland grazing land, offering favourable conditions for sheep breeding and wool production, making textile production a central element in the economy. Here, not least, textiles served as prerequisites for many spheres of society – among others, sails, although not physically identified in the present corpus, they clearly existed, triggering maritime communication, trade, and urbanisation.

The production has left many traces, archaeologically and in script, which together constitute an exceptionally rich corpus that has so far been largely studied as separate source categories, and in this part of Europe, either related to the Viking period or the Middle Ages. Transgressing the periods with their varied source categories, acting as independent but also supplementary sources, makes it possible to assess the production from broader perspectives, related to organisation, gender, tools and technology, quality and use areas, structures, and agency. Not least, do the many graves reveal varied status in various environments, close to upland resources or at places of status, power, and communication and often of long time-depth which indicate manors within estates. The grave goods reveal varied status too, where the tools reflect different work processes – as also remains of textiles when preserved under favourable conditions. The newly discovered workplaces in different environments, both in lowlands, often connected to central places, and in upland areas too, indicate production at different levels, and above ordinary household production as more advanced and even specialised.

Socially, the burial sites, often appearing as monuments in various landscapes and at sites of varying status, reveal the funeral customs of the period with graves furnished with textile tools and other grave goods, however, with a social bias as for status. They also reveal patterns and common traits of social organisation and agency, however, not homogeneous, representing various types of environments and social status. As shown in the foregoing chapters, the textile production was complex and tightly integrated in the farming systems of the period. It involved a variety of tools for the various time-consuming work processes from raw material to finished products of varied qualities and use areas. This again required a wide range of competencies, from sorting and combing wool, spinning the threads in both the warp and weft, weaving, and sewing, and, when on a larger scale and of finer textile qualities, a higher competency and larger work force also needed organisation on a higher level.

Remains of the funeral dress reveal many as socially distinguishing outfits in exquisite qualities, woven of thin yarn and in more advanced techniques of diamond twill than the coarser and less lavish textiles. Such weaves have therefore been interpreted as foreign textiles, however, this is still a debated issue. When assessed contextually and related to *in-situ* tools in the same contexts, tools and textiles may substantiate a domestic, however, also advanced production. The 10th-century high-status woman buried in a large mound at Veka in inland Voss, as already presented, is such a case and as one of the best represented graves from western Norway as for grave goods, tools, and textiles. The Veka grave also shows affiliation with other contemporary graves as for status, dress accessories, tools, and textiles, indicating networks too. These graves hardly represent ordinary household production and rather reflect women as organisers, supervising and coordinating the many work operations from preparing fibres to finished products. They may have included high-quality textiles, as documented by remains of the funeral dress, such as in the Veka grave, woven in broken diamond twill, while its band of samitum silk must have been imported. The dress, in contrast, could have been produced on home ground, judging by the tools in the grave. Still, this prominent woman furnished with exquisite grave goods was hardly a manual worker but rather organiser of all these operations, representing production on an advanced level, also substantiated when textiles have been preserved.

Counting some 140 Viking Age graves with textile tools from central western Norway, and with graves in the North Atlantic of the same nature but relatively few with textile tools, they do when all combined offer an exceptionally rich source material, which also from a Scandinavian perspective appears as special. Appearing in different environments, often at central places, the 29 graves with both tools and textiles also make this role of these high-standing women even more plausible.

Although the textiles often signify very high quality, the variation in fineness in the thread systems and thread count do not appear as standardised qualities, and only rarely as fine as the finest from the Viking Age towns of Birka and Hedeby. However, some do, which is the case in the prestigious Veka grave. Hence, Veka has even given name to the diamond twill qualities in the graves from western Norway as recognition of their origins. That some of the graves in Iceland have also left remains of the funeral dress woven in a diamond twill quality and with similar dress accessories, show close relations between the areas but, judging by the local wool, were produced in Iceland.

The grave inventory reflecting production varies in scale and complexity in various settings – seemingly from ordinary household production to a more advanced household production, to an attached specialist production, producing an even broader spectrum of qualities. Still, being symbolic grave goods, they only indirectly reflect production though the tools as concrete manifestations and the social dimension also add clarity.

That many of the high-status women buried with textile tools as grave goods, form a different social pattern that indicates not only higher status but also other roles above ordinary household production, becomes clearer when comparing them with the social representation of all recorded female Viking Age graves from western Norway. These include gendered grave goods and status-indicating dress accessories, however, without textile tools. The latter case, as shown in Chapter 5, exposes a classical hierarchical pyramid, where the highest stratum, status III, represents 16% of the graves, 40% of status II in the middle, and 46% of lower status I, although still provided with grave goods. In contrast, as we have seen, the representation of graves with textile tools forms an upturned pyramid, where as many as 59% of the female graves represent the highest status III, 26% in the middle stratum of status II, and 15% of the lower status I. Indirectly, this uncommon representation implies a large archeologically invisible workforce. This upturned representation clearly demonstrates the importance of textile production, reflecting high-status women commemorated by tools indicating their roles as managers of production rather than as manual workers. An over-representation of such graves at central places substantiates such an interpretation.

The tools for textile production selected as grave goods could in some cases represent the whole chain of work processes but also include other tools related to the initial and final stages of outdoor fibre preparation – wool shearing, weeding, and harvesting of flax and hemp, but also food production, evidently symbolising and representing an overall responsibility as managers and organisers of wider domains. In sum, the female graves of high status with varied tool categories exceeding the individual work capacity of such women, indicate roles as managers and organisers of broader work domains. Archaeological sites of working environments and production places from both the Viking period and the High Middle Ages reveal also other common structures transferred to the North Atlantic. They included also special workplaces in various landscapes and social settings related to ordinary farms and central places, close by and at larger distances related to upland shielings.

The issue of whether men also became involved in textile production, indicating a new role and possibly connected to production of sailcloth, has been raised, since weaving beaters sometimes appear in male weapon graves, although frequently in unclear contexts and often as the only tool. The weaponry and status may again suggest a role as organisers, which in this case could explain the scarcity and lack of other tools, however, the contexts are not clear cut. The graves, of both high-status females and males often appeared in the same contexts and with tools for utilising outland resources, not only wool but also wood, iron, and stone. Overall, weaving beaters of the same kind are more common tools in female graves and generally of high status though the wider role as managers and organisers is a largely under-communicated aspect. The graves thus convey vital information about the organisation of production and women's role and rank as organisers but indirectly also women's technological capacity as the manual workers.

The varied tools represented in the graves sometimes reveal the whole chain of work processes, exposing technological aspects of production, and expose versatile work processes, based on technical skills and expertise, as well as cooperation and larger networks exceeding the household level. The tools discovered in working environments further corroborate the impression gained from burial contexts and funeral clothes, such as at Veka and other high-status graves from western Norway and some from Iceland too, which further underline the scale and extent of production in varied contexts.

Combining information from all the available sources with new approaches and methodologies, not least experimental archaeology, altogether opens for broader understanding of the levels of production – as for work, technology, technological capacity, and qualities, and indirectly its organisation. The tools and textiles reflect a variety of qualities, both within the luxurious segment but also necessities for many use areas. By using replicas of various tools and testing the ability of them, such as the size and weight of spindle whorls to produce certain yarn qualities, has made it possible to assess relations between tools and textiles and indicate range of qualities too. We see then that spindle whorls in rural contexts match the fineness of whorls found in Viking Age urban settings, such as Kaupang, Birka, and Hedeby, among others, and in principle can produce similar fine qualities. However, the levels discerned include both ordinary household production and production in special environments and also an advanced production to meet requirements within the elite, and seemingly also specialised production within the same environments, or being dislocated but probably connected. The production must have included trivial and practical textiles just for keeping the body warm and dry, and products for sailing, rigging, and transport, although to a small extent represented as grave goods in these graves. Some of the tools, such as heavier spindle whorls sometimes occur among the grave goods and may be such indicators.

The finds of tools and textiles stemming from various grave contexts still represent symbolic and prestigious grave goods as well as remains of clothing in Viking Age burials. However, only as parts of the dress, they still demonstrate technologies used and textile qualities and not least technological capacity. In working environments, traces of production are largely traced as tools or easily lost or damaged and discarded artefacts of spindle whorls and loom weights, reflecting workplaces close to the settlements and in outlying areas close to the resources. Finds of textiles in these surroundings, however, are also found but are rather exceptional.

Combining archaeological tools and textiles from the contexts of graves and workplaces with environmental data and written evidence, has made it possible to gain new knowledge about production in various geographical and social landscapes. Both directly and indirectly, they reveal socio-technological systems, modes of production and organisation over time at various spatial levels, from the local to the regional variations and changes. Comparing tools, textiles, and contexts – from graves and rural to urban workplaces and contexts, also extracted from written sources, and from a period of five hundred years – reveals both continuity and changes, reflecting larger overarching societal changes and structures.

Medieval textiles, however, are discovered in urban contexts in Norway, and in western Norway, Bryggen in Bergen, but also in various rural contexts, all basically representing working environments, and in the North Atlantic, from Shetland, the Faroe Islands, and especially Iceland, as an expanding industry. Here, written sources add valuable information about a standardised and specialised production in rural environments and specify the various standardised qualities of homespun wadmal and also the amount of work involved. This information is of transferable value and can be compared with archaeologically preserved tools and textiles. While the production of high-quality textiles of broken diamond twill carried out in high status environments ceased by the end of the Viking period in western Norway, the production of homespun wadmal in the North Atlantic expanded to become a main export article to Bergen and even transhipped to wider areas.

By focusing contextually on the 'small things' – largely underrated and neglected artefacts such as spindle whorls and loom weights – the archaeological remains of tools and textiles are supplied and compared with information from written sources and the overall aim has been to achieve a more holistic understanding of the processes and levels of textile production on these northern fringes of Europe in a long-term perspective too, to tell a 'big story' providing new perspectives on women's role in production, not least related to the Viking period, and their largely under-estimated role as organisers, and also for extent and societal importance. The medieval textile production in western Norway, both rural and urban, seems to have continued basically on a household level in rural areas, but was also organised as special urban crafts, as stated in the Urban Code of Bergen, with weavers using the horizontal treadle loom, but in most tenements still using the old upright loom well into the 1300s, and hand spinning using spindle whorls of various qualities. What is interesting is that the tools in the rural contexts were able to produce just as fine and finer yarn than the urban contexts of Kaupang and the medieval site of Bryggen in Bergen. The dichotomy between rural and urban contexts as well as domestic/foreign may thus be overrated related to quality and perhaps also organisation. In a temporal perspective, we see a change in production, where the production of standardised homespun wadmal became a commodity in demand and imported to Bergen for further distribution, while in western Norway, production was reduced and mainly based on ordinary household production, when being able to reach a broader range of imported qualities via Bergen.

By tracing tools, textiles, texts, and contexts in wide time frames, across periods from the Viking Age to include the High Middle Ages, and at various spatial levels from the local to the regional from western Norway to the North Atlantic, it reveals both similarities and changes, as well as variations within the production of textiles. To a large degree they coincide over larger areas and follow socio-political trends over time, and not least, these varied sources demonstrate the importance of textile production at many levels, as for technologies, use areas, importance, and gender roles. This multidisciplinary approach has also made it possible to reveal actors who otherwise are not in the forefront of storytelling but still played an important but under-communicated role.

Bibliography

Abbreviations

BK	*Bjorgynjar Kalvskinn*, published by P.A. Munch, Christiania 1843.
Bl	Den nyere By-Lov el. Bjarkö-Ret, utgiven af Kong Magnus Haakonsön. *NgL II*, 185–290.
Búalög.	*Verðlag á Íslandi á 12.-19. öld, Útg, Arnór Sigurjónsson*, 1966.
DN	*Diplomatarium Norvegicum*, utg. av C.C.A. Lange, C.R. Unger *et al.* I–XXI, Christiania 1849–Oslo 1976.
Egil's saga	*Egil's saga Skalla-Grímssonar*, published by Sigurður Nordal. Íslenzk fornrit, 2. Reykjavík: Hið íslenzka fornritafélag, 1933.
F	*Den ældre Frostatingslov. NgL* I, 119–258.
Fritzner	Fritzner, J. 1973, *Ordbog over Det gamle norske Sprog I–III. Omarbeidet, forøget og forbedret udgave;* nytt uforandret opptrykk av 2, utg. (1883–96) m. et bd. tillegg og rettelser (eds D.A. Seip and T. Knudsen), Oslo.
G	*Den ældre Gulatingslov. NgL* I, 1–118.
Grágás	*Lagasagn íslenska þjóðveldisins*. 1992. Gunnar Karlsson, Kristján Sveinsson og Mörður Árnason sáu um útgáfu. Reykjavík, Mál og menning.
KLNM	*Kulturhistorisk leksikon for nordisk middelalder. Fra vikingtid til reformasjonstid*, I–XXII, Oslo 1981 (1956–1978).
KVHAA	*Kungliga Vitterhets Historie och Antikvitets Akademien*, Stockholm.
L	*Den nyere Lands-Lov, udgiven af Kong Magnus Haakonsson, NgL II*, 1–178.
NG	*Norske Gaardnavne*, I–XIX, O. Rygh, A. Kjær and H. Falk, Christiania. 1897–1936.
NgL	*Norges gamle Love*, utg. R. Keyser, P.A. Munch, G. Storm, E. Hertzberg. I–V. Christiani (1846–1895).
NTT	*Nordisk Textilteknisk Terminologi. Førindustriell vävnadsproduktion* by E. Strömberg, A. Geijer. M. Hald and M. Hoffmann.
Osebergfundet	*vol. I–V*, A.W. Brøgger and H. Schetelig (eds). Oslo 1917–1928.
Seyðabrevið	Hertug Haakon Magnussöns store Retterbod for Færöerne. Oslo 28de Juni 1298, *NgL III*, 10, Christiania 1846.
UBAS Master/ hovedfag	*University of Bergen Archeological Series*, University of Bergen.
UBAS Nordic/ Nordisk	*University of Bergen Archeological Series*, University of Bergen.

Bibliography

Aasen, I. 1918, *Norsk Ordbog med dansk forklaring*. Alb. Cammermeyers Forlag, Christiania.

Andersen, E. 1995, Woolen material for sails, in *Shipshape. Essays for Ole Crumlin Pedersen on the occasion of his 60th anniversary*, O. Olsen, T. Madsen and F. Reck (eds), 249–270. Vikingeskibshallen, Roskilde.

Andersen, E., J. Milland and E. Myhre 1989, *Uldsejl i 1000 år*. Vikingeskibshallen, Roskilde.

Andersen, R. and A.O. Martinussen (eds) 2016, *The Warp-Weighted Loom. Kljásteinavefstaðurinn, Oppstadveven*. Skald forlag, Leikanger.

Andersson, E. 1998, *Textilproduktion i arkeologisk kontext. En metodstudie av yngre järnåldersboplatser i Skåne*. University of Lund, Institute of Archaeology, Report Series no. 58, Lund.

Andersson, E. 1999, *The Common Thread: Textile Production during the Late Iron Age–Viking Age*. University of Lund, Institute of Archaeology, Report Series no. 67, Lund.

Andersson, E. 2000, Textilproduktion i Löddeköpinge – endast för husbehov?, in *Porten til Skåne: Löddeköpinge under järnålder och medeltid*, F. Svanberg and B. Söderberg (eds), 158–187. *Riksantikvarämbetet, Arkeologiska undersökingar, Skrifter* 32. Riksantikvarieämbetet, Malmö.

Andersson, E. 2003, *Tools for Textile Production, from Birka and Hedeby*, Birka Studies, Vol. 8. Riksantikvarieämbetet & Statens historiska museum, Stockholm.

Andersson, E. 2008, Tools, Textile Production and Society in Viking Age Birka, in *Dressing the Past*, M. Gleba, C. Munkholt and M.-L. Nosch (eds), 68–85. Ancient Textile Series Vol. 3. Oxbow Books, Oxford.

Andersson Strand, E. 2011, Textile Production in Western Europe, in *The Archaeology of Medieval Europe, Vol. 2. Twelfth to*

Sixteenth Centuries, M. Carver and J. Klápste (eds), 315–320. Aarhus University Press, Aarhus.

Andersson Strand, E. 2012, The *Textile Chaîne Opératoire*: Using a Multidisciplinary Approach to Textile Archaeology with a Focus on the Ancient Near East. *Paléorient* 38, 1–2, 21–40.

Andersson Strand, E. 2016, Segel och segelduksproduktion i arkeologisk context, in *Vikingetidens sejl. Festskrift tilegnet Erik Andersen, Arkeologiske Skrifter 14*, M. Ravn and L. Gebauer (eds), 21–26. Saxo-Instituttet, Københavns Universitet, Copenhagen.

Anker, A. 1969, Tekstilfunnene. *Kaupanger stavkirke IV. Foreningen til norske Fortidsminnemerkers Bevaring, Årbok*, 105–110. Fortidsminneforeningen, Oslo.

Anstey H. and T. Westen 2005, *The Anstey Weston Guide to Textile Terms*. Weston Publishing Limited, London.

Arge, S.V. 1991, The Landnám in the Faroes. *Arctic Anthropology* 28(2), 101–125.

Arge, S.V. 1997, Í Uppistovubeitinum. Site and settlement. *Fróðskaparrit 45, bók 1997*, 27–44. Tórshavn.

Arge, S.V., G. Sveinbjarnardóttir, K.J. Edwards and P.C. Buckland 2005, Viking and Medieval Settlement in the Faroes. People, Place, and Environment. *Human Ecology* 33(5), 597–620.

Ashby, S. and S.M. Sindbæk (eds) 2020, *Crafts and Social Networks in Viking Towns*. Oxbow Books, Oxford.

Ashby, S., A.N. Coutu and S.M. Sindbæk 2015, Urban networks and Arctic outlands craft specialists and reindeer antler in Viking towns. *European Journal of Archaeology* 18, 679–704.

Aschehoug, T.H. 1890, *Statistiske Studier over Folkemængde og Jordbrug i Norges landdistrikter i det syttende og attende aarhundrede*. Tillægshefte til Statsøkonomisk Tidsskrift, Christiania.

Åstveit, L.I. and A.S. Melvær 2013, *Arkeologiske undersøkelser av aktivitets- og bosetningsspor fra jernalder/middelalder på Sæbø gnr/bnr 1/24, Vik, Sogn og Fjordane*. Arkeologisk rapport. Seksjon for Ytre Kulturminnevern, University of Bergen.

Bakka, E. 1963, Some English Decorated Metal Objects Found in Norwegian Viking Graves. Contribution to the Art History of the Eighth Century AD. *Årbok for Universitetet i Bergen*, 4–65.

Bakka, E. 1971, Ytre Moa – ein øydegard frå vikingtida, in *Bygdebok for Årdal*, S. Ve (ed.), 149–178. Band I, Bergen.

Ballin Smith, B. 1999, The Artefactual Material. Introduction, in *The Biggings, Papa Stour, Shetland. The history and excavation of a royal Norwegian farm*, B.E. Crawford and B.B. Smith (eds), 126–129. Society of Antiquaries of Scotland, Edinburgh.

Ballin Smith, B. 2007, Stobister, Sinnabist and Starrapund, Three Wilderness Settlements in Shetland, in *West Over Sea: Studies in Scandinavian Sea-Borne Expansion and Settlement before 1300. The Northern World*, B.B. Smith, S. Taylor and G. Williams (eds), 419–430. Brill, Leiden.

Barber, E.J.W. 1991, *Prehistoric Textiles: The Development of Cloth in the Neolithic and Bronze Age with Special Reference to the Aegean*. Princeton University Press, Princeton.

Barber, E. 2007, Weaving the Social Fabric, in *Ancient Textiles. Production, Craft and Society*, C. Gillis and M.-L. Nosch (eds), 173–178, Oxbow Books, Oxford.

Barndon, R. and A.B. Olsen 2018, En grav med smedverktøy fra tidlig vikingtid på Nordheim i Sogndal. En analyse av gravgods, handlingsrekker og symbolikk. *Viking* Vol. LXXXI, 63–88.

Batey, C. 2011, Hofstaðir: Does the Artefact Assemblage Reflect Special Status, in *Viking Settlements and Viking Society. Papers from the 16th International Viking Congress, Reyjkavik*, S. Sigurðsson (ed.), 18–30. The University of Iceland Press, Reykjavík.

Batzer, A. and L Dokkedal 1992, The Warp-Weighted Loom: Some New Experimental Notes, in *NESAT* V. *Tidens Tand*, L. Bender Jørgensen and E. Munksgaard (eds), 231–234. Det Kongelige Danske Kunstakademi, Copenhagen.

Baug, I. 2002, *Kvernsteinsbrota i Hyllestad. Arkeologiske undersøkingar i steinbrotsområdet i Hyllestad i Sogn og Fjordane*. Bergverksmuseet Skrift 22. Norsk Bergverksmuseum, Kongsberg.

Baug, I. 2015, *Quarrying in Western Norway. An Archaeological Study of Production and Distribution in the Viking Period and Middle Ages*. Archaeopress, Oxford.

Baug, I., D. Skre, T. Heldal and Ø.J. Jansen 2019, The Beginning of the Viking Age in the West. *Journal of Maritime Archaeology* 14, 43–80. DOI: https://doi.org/10.1007/s11457-018-9221-3.

Bender Jørgensen, L. 1986, *Forhistoriske tekstiler i Skandinavien,* Nordiske Fortidsminder serie B, Bind 9. Det kongelige Nordiske Oldskriftselskap, København.

Bender Jørgensen, L. 1991, *North European Textiles until AD 1000*. Aarhus University Press, Aarhus.

Bender Jørgensen, L. 2005, Textiles of Seafaring: An Introduction to an Interdisciplinary Research Project, in *Northern Archaeological Textiles, NESAT VII*, F. Prichard and J.P. Wild (eds), 65–74. Oxbow Books, Oxford.

Bender Jørgensen, L. and K. Grömer 2012, The Archaeology of Textiles – Recent Advances and New Methods. Portal, godišnjak Hrvatskog restauratorskog zavoda 3/2012, 91–108.

Birkelid, E. 1938, *Fedrekult i Norge. Et forsøk på en systematisk-deskriptiv fremstilling*. Det Norske Videnskapsakademi i Oslo, Oslo.

Bjørgo, T. 2005, Iron Age House Remains from Mountain Areas in Inner Sogn, Western Norway, in *Fra funn til samfunn. Jernalderstudier tilegnet Bergljot Solberg på 70-årsdagen*, K.A. Bergsvik and A. Engevik (eds), 209–228. *UBAS* Nordisk 1. University of Bergen, Bergen.

Bjørgo, T., S. Kristoffersen and C. Prescott 1992, *Arkeologiske undersøkeser i Nyset-Steggjevassdragene 1981–87*. Arkeologiske rapporter 16. University of Bergen, Bergen.

Bjørkvik, H. 1981, Gård. *Norge. KLNM* 5, 625–631.

Blom, G.A. 1991, *Hellig Olavs by. Middelalderen til 1537*. Trondheims historie bind 1, Trondheim.

Bødal, S. 1998, *Vik i Sogn. Lokalsamfunn med overregionale kontakter*. Unpublished master thesis, University of Bergen.

Bøe, J. 1926, An Ornamented Celtic Bronze Object, Found in a Norwegian Grave, in *Bergens Museum Aarbok 1924–25*, Hist.-antikv. række nr. 4. ‚AS John Griegs boktrykkeri,, 2–36 Bergen.

Braudel, F. 1982, *On History* (translated by S. Matthews). University of Chicago Press, Chicago.

Braudel, F. 1996, *The Mediterranean and the Mediterranean World in the Age of Philip II*. University of California Press, Berkley.

Bray, F. 1997, *Technology and Gender. Ancient Textiles: Fabrics of Power in Late Imperial China*. University of California Press, Berkley.

Bray, F. 2007, Gender and Technology. *The Annual Review of Anthropology* 2007, 36, 37–53.

Brewington, S.D. 2010, *Third Interim Report on Analysis of Archaeofauna from Undir Junkarinfløtti, Sandoy, Faroe Islands*. NORSEC Zooarchaeological Laboratories Report No. 46, 1–18.

Bruen Olsen, A. 2005, Et vikingtids tunanlegg på Hjelle i Stryn – en konservativ institusjon i et konservativt samfunn, in *Fra funn til samfunn. Jernalderstudier tilegnet Bergljot Solberg på 70-årsdagen*, K.A. Bergsvik and A. Engevik (eds), 319–355. *UBAS Nordisk* 1. University of Bergen, Bergen.

Bruen Olsen, A. 2015, Courtyard Sites in Western Norway. Central Assembly Places and Judicial Institutions in the Late Iron Age, in *Viking Worlds. Things, Spaces and Movement*, M. Hem Eriksen, U. Pedersen, B. Rundberget, I. Axelsen and H. Lund Berg (eds), 43–55. Oxbow Books, Oxford.

Bruns, F. 1900, *Die Lübecker Bergenfahrer unf ihre Chronostik. Hansische Gesshichtsquellen*. Pass & Graleb, Berlin.

Byock, J. 1993, *Medieval Iceland. Society, Sagas and Power*. Hisarlik Press/Open Library, Chippingham.

Cartwright, B. 2015, Making the Cloth that Binds Us. The Role of Textile Production in Producing Viking-Age Identities, in *Viking Worlds. Things, Spaces and Movements*, M. Hem Eriksen, U. Pedersen, B. Rundberget, I. Axelsen and H.L. Berg (eds), 160–178. Oxbow Books, Oxford.

Carus-Wilson, E. 1969, Haberget: A Medieval Textile Conundrum. Medieval Archaeology. *Journal of the Society for Medieval Archaeology* XIII, 148–166.

Church, M.J., S.V. Arge, S. Brewington, T.H. McGovern, S. Perdakaris, I.T. Lawson, G.T. Cook, C. Amundsen, R. Harrison, Y. Krivogorskaya and E. Dunbar 2005, Puffins, Pigs, Cod and Barley: Palaeoeconomy at Undir Junkarinsfløtti, Sandoy, Faroe Islands. *Environmental Archaeology* 10, 179–197.

Coatsworth, E. and G.R. Owen-Crocker 2007, *Medieval Textiles of the British Isles AD 450 –1100. An Annotated Bibliography*. BAR British Series 445. Archaeopress, Oxford.

Costin, C. 1991, Craft Specialization Issues in Defining and Explaining the Organization, in *Archaeological Method and Theory*, M.B. Schiffer (eds), 1–56. University of Arizona, Tucson.

Costin, C. 1993, Textiles, Women, and Political Economy in the Late Prehispanic Andes. *Research in Economic Anthropology* 14, 3–28.

Costin, C. 1996, Exploring the Relationship Between Gender and Craft in Complex Societies: Methodological and Theoretical Issues of Gender-Attribution, in *Gender and Archaeology*, R.P. Wright (ed.), 111–140. University Philadelphia Press, Philadelphia.

Crawford, B.E. 2013, *The Northern Earldoms Orkney and Caithness from AD 870 to 1470*. John Donald, Edinburgh.

Crawford, B.E. and B. Ballin Smith 1999, *The Biggings, Papa Stour, Shetland. The History and Excavation of a Royal Norwegian Farm*. Society of Antiquities of Scotland Monograph Series Number 15, Edinburgh.

Crowfoot, E., F. Pritchard and K. Staniland [1992] 2001, *Textiles and Clothing c.1150–1450. Medieval Finds from Excavations in London*. The Boydell Press in association with The Museum of London, London.

Crumlin-Pedersen O. 1997, *Viking-Age Ships and Shipbuilding in Hedeby/Haithabu and Schleswig. Ships and Boats of the North vol. 2*. Viking Ship Museum, Roskilde.

Damsholt, N. 1984, The Role of Icelandic Women in the Sagas and in the Production of Homespun Cloth. Scandinavian Journal of History 9(2–3), 75–90.

Diinhoff, S. 2002, *Yngre jernalder og middelalderbosætning på Osen gard i Dalsfjord. Arkeæologiske ðgivningsundersøgelser i 2001 og 2002 ved Osen gard, gnr. 94, bnr l, Gaular kommune, Sogn og Fjordane*. Bergen Museum, Arkeologisk Rapport, Bergen.

Diinhoff, S. 2016, *Arkeologisk frigivningsundersøgelse af et forhistorisk dyrkning- og bosætningsområde på Henjum bnr.17, bnr. 76, Leikanger kommune, Sogn og Fjordnae, Prosjekt 279*. University of Bergen, Fornminneseksjonen, Arkeologisk Rapport, Bergen.

Diinhoff, S. 2017, *Vikingetids kvindegrav*. Osen gård, Gnr. 94, Bnr. 1. Gaular kommune, Sogn og Fjordane fylke. Rapport Nr. 4. 2017. University of Bergen, Fornminneseksjonen, Arkeologisk rapport, Bergen.

Dobres, M.-A. 2000, *Technology and Social Agency. Outlining a Practical Framework for Archaeology*. Wiley-Blackwell, Oxford.

Dommasnes , L.H. 1982, Late Iron Age in Western Norway. Female Roles and Ranks as Deduced from Analyses of Burial Customs. *Norwegian Archaeological Review* 15(1–2), 70–84.

Downham, C. 2009, Hiberno-Norwegians and Anglo-Danes, anachronistic ethnicities in Viking Age England. *Medieval Scandinavia* 19, 139–169.

Downham, C. 2017, The Earliest Viking Activity in England? *English History Review* 132, 1–12.

Duby, G. 1998, *Rural Economy and Country Life in the West* (1968) (Translated by C. Postan). Penn Press, Pennsylvania.

Edwards, K.J., P.C. Buckland, R. Craigie, E. Panagiotakopulu and S. Stummann Hansen 1998, Landscapes at Landnám: Palynological and Palaeoentomological Evidence from Toftanes, Faroe Islands. *Fróðskaparrit, 46. Bók*, Annales Societatis Scientarium Færoensis, 229–244.

Einarsson, B.F. 1992, Granastaðir – grophuset och andra isländska grophus i ett nordisk sammanhang. *Viking* bd. LV, 95–120.

Ejstrud, B., S. Andresen, A. Appel, S. Gjerlevsen and B. Thomsen 2011, *From Flax to Linen – Experiments with Flax at Ribe Viking Centre*, B. Ejstrud (ed.). Maritime Archaeology Programme. University of Southern Denmark, Esbjerg.

Eldjárn, K. 1956, *Kuml og haugfé úr heiðnum sið á Islandi*. Bókutgafan Nordri, Reykjavík.

Eldjárn, K. and A. Friðriksson 2000, *Kuml ok haugfé úr heiðnum sið á Islandi*. 2nd edition. Màl ok menning, Reykjavík.

Epstein, S.R. 2008, Craft Guilds in the Pre-Modern Economy: A Discussion. *The Economic History Reviews. New Series* 61(1), 155–174. http://www.jstor.org/stable/40057560.

Faith, R. 1999, *The English Peasantry and the Growth of Lordship. Studies in the Early History of Britain*. Leicester University Press, Leicester.

Falk, H. 1919, *Altwestnordische Kleiderkunde mit besonderer Berücksichtigung der Terminologie. Skrifter*. Videnskabsselskapet i Kristiania Historisk-filosofisk klasse 1918:3.

Friðriksson, A. 2009, Social and Symbolic Landscapes in Late Iron Age Iceland. *Archaeologia Islandica* 7, 9–21.

Friðriksson, A. and Vésteinsson, O. 2011, Landscapes of Burial: Contrasting the Pagan and Christian Paradigmas of Burial in Viking Age and medieval Iceland. *Archaeologica Islandica* 9, 50–64.

Gabrielsen, K. 2007, Vestlandets steinkors. Monumentalisme i brytningen mellom hedendom og kristendom, in *UBAS Hovedfag/Master* 2, 189–290. Bergen.

Gardeła, L. 2009, A Biography of the Seiðr-Staffs. Towards an Archaeology of Emotions, in *Between Paganism and Christianity in the North*, L.P. Słupecki and J. Morawiec (eds), 190–219. Rzeszów University, Rzeszów.

Gardeła, L. 2016, *(Magic) Staffs in the Viking Age, Studia Medievalia Septentrionalia*, Band 27. Verlag Fassbaender, Vienna.

Gaut, B. 2011, Vessel Glass and Evidence of Glasswork, in *Things from the Town. Artefacts and Inhabitants in Viking-age Kaupang*, D. Skre (ed.), 169–279, Kaupang Excavation Project Publication Series, Vol. 3, *Norske Oldfunn* XXIV. Aarhus University Press, Aarhus.

Geijer, A. 1938, *Birka III, Die Textilfunde aus den Gräbern*. Almqvist & Wiksells boktrykkeri, Uppsala.

Geijer, A. 1965, Var järnålderens 'frisiska kläde' tilverkat i Syrien? Reflektioner i anslutning till et arbete om tyngdvåvstolen. *Journal of Swedish Antiquarian Research* 60, 112–132.

Geijer, A. 1972, *Ur textilkonstens historia*. 3. opplag. Gidlunds förlag, Stockholm.

Geijer, A. 1979, *A History of Textile Art*. Sotheby Parke Bernet, London.

Geijer, A. and Anderbjörk, J.E. 1939, Two Textile Implements from the early Middle Ages. *Folkliv, Acta Ethnologica and Folkloristica Europae*, 232–241, Stockholm.

Gellein, K. 2007, Kristen innflytelse i hedensk tid? En analyse med utgangspunkt i graver fra yngre jernalder i Hordaland, in *UBAS Hovedfag/Master* 2, 11–113. Bergen.

Gjessing, G. 1934, *Studier i norsk merovingertid, Kronologi og oldsakformer*. J. Dybwad, Oslo

Gjøl Hagen, K. [1988] 1994, *Profesjonalisme og urbanisering. Profesjonalismeproblemet i håndverket belyst ved et tekstil- og vevloddsmateriale fra middelalderens Trondheim, fra 1000-tallet frem til slutten av 1300-tallet.* Universitetets Oldsaksamling Skrifter. Ny rekke. Nr 16, Oslo.

Gjøstein Resi, H. 2011, Gemstones: Cornelian, Rock Crystlal, Amethyst, Fluorspar and Garnet, in *Things from the Town. Artefacts and Inhabitants in Viking-age Kaupang*, D. Skre (ed.), 143–166. Kaupang Excavation Project Publication Series, Vol. 3, Norske Oldfunn XXIV, Aarhus University Press, Aarhus.

Glørstad, Z.T. 2010, *Ringspennen og kappen. Kulturelle møter, politiske symboler og sentraliseringsprosesser i Norge ca. 800–950*. Doctoral Thesis, University of Oslo, Oslo.

Glørstad, Z.T. 2012, Sign of the Times? The Transfer and Transformation of Penannular Brooches in Viking-Age Norway. *Norwegian Archaeological Review* 45(1), 30–51.

Good, I. 2001, Archaeological Textiles: A Review of Current Research. *Annual Review of Anthropology* 30, 209–226.

Goody, J., J. Thirsk and E.P. Thompson 1976, *Family and Inheritance. Rural Societies in Western Europe 1200–1800*. Cambridge University Press, Cambridge.

Goodacre, G., A. Helgason, J. Nicholson, L. Southam, L. Ferguson, E. Hickey, E. Vega, K. Stefánsson, R. Ward and B. Sykes 2005, Genetic Evidence for a Family-Based Scandinavian Settlement of Shetland and Orkney During the Viking Periods. *Heredity* 95, 129–135. DOI: 10.1038/sj.hdy.6800661.

Grant, A. 1992, Animal Resources, in *The Countryside of medieval England*, G. Astill and A. Grant (eds), 150–187. Blackwell, Oxford.

Grenander-Nyberg, G. 1984, Eine Schaftrolle aus Haithabu als Teil eines Trittwebstuhl mit waagerecht gespannter Kette. *Berichte über die Ausgrabungen in Haitahbu* 19, 145–150.

Grenander-Nyberg, G. 1990. Nordens äldste daterade trampvävstol. *Vävläraren* 4, 11–12.

Grenander-Nyberg, G. 1994, Prehistoric and Early Medieval Features in the Construction of the Oldest North European Treadle Looms, in *Archäologische Textilfunde – Archaeological Textiles*, G. Jaacks and K. Tidow (eds), 203–212. Textilsymposium Neumünster *NESAT V* 1993. Textilmuseum Neumünster, Neumünster.

Guckelsberger, M. 2015, Blákledda konan – Ny rannsokna fornu kumli. bundled- up in Blue – the Re-Investigations of a Viking Age Grave. *Publications of the National Museum of Iceland* 38, 25–43.

Guckelsberger, M. and M. Mader 2017, The Woman Dressed in Blue. A textile Find from the 10th c. Icelandic Grave and its Reconstruction. https://www.academia.edu/30706919/The_Woman_Dressed_in_Blue.

Guðjonsson, E. 1985, Nogle Bemærkninger om den islandske vægtvev, 'vefstaður'. *Bygd og by. Festskrift til Marta Hoffmann*. Norsk Folkemuseums Årbok 1983–84: 30, 116–128.

Guðjonsson, E. 1990, Some Aspects on the Icelandic Warp-Weighted Loom, 'vefstaður'. *Textile History* 21(2), 165–179.

Guðjonsson, E. 1994, Warp-Weighted Looms in Iceland and Greenland. Comparison of Medieval Loom Parts in Greenland in 1934 and 1990–92 to Loom Parts from the Eighteenth and Nineteenth Century Warp-Weighted Looms in Iceland. Preliminary Report, in *Archäologische Textilfunde – Archaeological Textiles*, G. Jaacks and K. Tidow (eds), 178–195. Textilsymposium Neumünster *NESAT V*. Textilmuseum Neumünster, Neumünster.

Hägg, I. 1969, Die wikingerzeitliche Frauentracht von Birka. Einige Bemerkungen zur Hemdform. *Tor. Meddelanden från Institutionen för Nordisk fornkunnskap vid Uppsala Universitet* XIII, 13–25.

Hägg, I. 1971, Mantel och kjortel i vikingtidens dräkt. *Fornvännen. Tidsskrift för svensk antikvarisk forskning* 1971/3, 141– 153.

Hägg, I. 1974, *Kvinnodräkten i Birka*, Aun 2, 108. Uppsala University, Department of Archaeology and Ancient History, Uppsala.

Hägg, I. 1985, *Die Textilfunde aus dem Hafen von Haithabu*. Berichte über die Ausgrabungen 20. K. Wachholt, Neumünster.

Hägg, I. 1986, Die Tracht, in *Birka II:2, Systematische Analysen der Graberfunde*, G. Arwidsson (ed.), 51–72. Almquist & Wiksell, Stockholm.

Hägg, I. 1991, *Die Textilfunde aus den Gräbern von Haithabu*. Berichte über die Ausgrabungen 29. K. Wachholt, Neumünster.

Hägg, I. 1994, Friesisches Tuch, in *Archäologische Textilfunde – Archaeological Textiles*, G. Jaacks and K. Tidow (eds), 82–94. Textilsymposium Neumünster *NESAT V*. Textilmuseum Neumünster, Neumünster.

Hafliðasson, H., J. Eriksson and S. van Kreveld 2000, The Tephrochronology of Iceland and the North Atlantic Region During the Middle and Late Quaternary: A Review. *Journal of Quaternary Science* 15(1), 3–22.

Hákonardóttir, G. 2016, Iceland's Settlement and Trade in Woven Goods – 900 Years Working on the Loom, in *The Warp-Weighted Loom. Kljásteinavefstaðurinn, Oppstadveven*, R. Andersen and A.O. Martinussen (eds), 21–58. Skald forlag, Leikanger.

Hamilton, J.R.C. 1956, *Excavations at Jarlshof, Shetland*. Ministry of World Archaeological Reports No. I, Edinburgh.

Hansen, G. and P. Storemyr 2017, *Soapstone in the North. Products and People 7000 BC–AD 1700. UBAS* 9. University of Bergen, Bergen.

Härke, H. 2014, Grave Goods in Early Medieval Burials, Messages and Meanings. *Mortality* 19(1), 41–60. DOI: 10.1080/13576275.2013.870544.

Harrington, S. 2008, *Aspects of Gender Identity and Craft Production in the European Migration Period: Iron Weaving Beaters and Associated Textile Making Tools from England, Norway and Alamannia*. BAR International Series 1797. Archaeopress, Oxford.

Harrison, R. 2011, *Möðruvellir in Hörgurdalur, N. Iceland. General Overview of the Archaeofauna Analyzed from 2006–08 Midden Mound Excavations. HERC/NORSE* Zooarchaelogical Laboratory REPORT No. 59.

Hayeur Smith, M. 2004, Dressing the Dead: Gender, Identity, and Adornment in Viking-Age Iceland, in *Vinland Revisited, the Norse at the Turn of the First Millennium, Viking Millennium International Symposium, Sept.15-24, 2000, Newfoundland and Labrador*, S. Lewis-Simpson (ed.), 227–240. Historic Sites Association of Newfoundland and Labrador, Inc., St Johns.

Hayeur Smith, M. 2011a, Preliminary Textile Report, Gásir, Iceland. Research. *Report of the Circumpolar Laboratory No. 2. Haffenreffer Museum of Anthropology*, Brown University.

Hayeur Smith, M. 2011b, Preliminary Textile Report, Möðruvellir Iceland. *Research Reports of the Circumpolar Laboratory No. 3. Haffenreffer Museum of Anthropology*, Brown University.

Hayeur Smith, M. 2013a, *The Textile Collection from the 1988 Bessastaðir Excavation. In Bessastaðarannsókn II Kirkjugarður og miðaldaminjar, uppgraftarsvæði 12–15 Reykjavík 2013*, G. Ólafsson (ed.). National Museum of Iceland, Reykjavík.

Hayeur Smith, M. 2013b, Thorir's Bargain: Gender, *Vaðmál* and the Law. *Worlds Archaeology* 45(5), 730–746. DOI: 10.1080/00438243.2013.860272.

Hayeur Smith, M. 2015, Weaving Wealth: Cloth and Trade in Viking Age and Medieval Iceland, in *Textiles and the Medieval Economy. Production, Trade and Consumption of Textiles, 8th–16th Centuries*, A.L. Huang and C. Jahnke (eds), 23–40. Oxbow Books, Oxford.

Hayeur Smith, M. 2016, Textiles in Iceland, AD 873–1600, in *The Warp Weighted Loom. Kljásteinavefstaðurinn, Oppstadveven*, R. Andersen and A.O. Martinussen (eds), 209–218. Skald forlag, Leikanger.

Hayeur Smith, M. 2018, Vaðmál and Cloth Currency in Viking and Medieval Iceland, in *Silver, Butter, Cloth. Monetary and Social Economies in the Viking Age*, J. Kershaw and G. Williams (eds). Oxford University Press, Oxford. DOI: 10.1093/oso/9780198827986.001.0001.

Hayeur Smith, M., K.P. Smith and K.M. Frei 2019, 'Tangled in Blue': The Death, Dress and Identity of an Early Viking-Age Female Settler from Ketilstaðir, Iceland. *Medieval Archaeology* 63(1), 95–127. DOI: 10.1080/00766097.2019.1589816.

Heide, E. 2006, Spinning seiðr, in *Old Norse Religion in Long-Term Perspectives. Origins, Changes, and Interactions. An International Conference in Lund, Sweden, June 3-7, 2004*, A. Andrén and K. Jennbert (eds), 164–170. Vägar till Midgård 8. Nordic Academic Press, Lund.

Heide, E. 2014, The Early Viking Ship Types. *Sjøfartshistorisk årbok* 2012, 81–153.

Helgason, A., E. Hickey, S. Goodacre, V. Bosnes, K. Stefánsson, R. Ward and B. Sykes 2001, mtDNA and the Islands of the North Atlantic: Estimating the Proportions of Norse and Gaelic Ancestry. *The American Journal of Human* Genetics 68(3), 723–737. DOI: 10.1086/318785.

Helgason, A., C. Lalueza-Fox, S. Ghosh, S. Sigurðardóttir, M.L. Sampietro, E. Gigli, A. Baker, J. Bertranpetit, L. Arnadóttir, U. Thornorsteinsdottir and K. Stefánsson 2009, Sequences from First Settlers Reveal Rapid Evolution in Icelandic mtDNA Pool. *PLoS Genet* 5(1), e1000343. DOI: 10.1371/journal.pgen.1000343.

Helgen, G. 1982, *Odd og egg: merovingertidsfunn fra Hordaland, Sogn og Fjordane*. Arkeologiske avhandlinger no. 3. University of Bergen, Bergen.

Helle, K. 1982, *Bergen bys historie vol. I. Kongssete og kjøpstad fra opphavet til 1536*. Universitetsforlaget, Bergen, Oslo, Tromsø.

Helle, K. 1995, *Under kirke og kongemakt, 1130–1350. Aschehougs Norgeshistorie Vol. 3*. Aschehoug forlag, Oslo.

Helle, K. 2001, *Gulatinget og Gulatingslova*. Skald forlag, Leikanger.

Helle, K. (ed.) 2003, *The Cambridge History of Scandinavia, Prehistory to 1520* Vol. 1. Cambridge University Press, Cambridge.

Helle, K. 2005, The Position of the Faroes and Other 'Tributary Lands' in the Medieval Norwegian Dominion, in *Viking and the Norse in the North Atlantic*, A. Mortensen and S.V. Arge (eds), 11–21, *Annales Societatis Scientiarium Færoensis Supplementum* XLIV. Førroya Frodskaparfelag, Tórshavn.

Hellesøe, H.P. 2006, *Båtgravene i Nordfjord*. Unpublished Master thesis, University of Bergen, Bergen.

Herlihy, D. 1990, *Opera mulibria, Women and Work in Medieval Europe*. McGraw Hill. New York.

Hjärthner-Holdar, E. 2011, Technological Choices and Changes and the Concept of Innovation. Reflections on the Introduction of Iron in Areas with Different Power Systems, in *The Archaeology of Regional Technologies. Case Studies from the Palaeolithic to the Age of the Vikings*, R. Barndon, A. Engevik and I. Øye (eds), 165–177. The Edwin Mellen Press Lewiston, Queenston, Lampeter.

Høeg, H.I. 1997, *Pollenanlytiske undersøkelser i Østerdalsområdet med hovedvekt på Rødsmoen, Åmot i Hedmark*. Varia, 46, Universitetets Oldsaksamling, Oslo.

Hoffmann, M. 1964, *The Warp-Weighted Loom: Studies in the History and Technology of an Ancient Technology of an Ancient Implement*. Universitetsforlaget, Oslo.

Hoffmann, M. 1966, Erlendur muaðardúkar í islenzkum konukumlun frá vikingaöld. *Árbok hins íslenzka fornleifar félags* 63, árg 1965. Nabu Press, Reykjavík.

Hoffmann, M. 1991, *Fra fiber til tøy. Tekstilredskaper og bruken av dem i norsk tradisjon*. Landbruksforlaget, Oslo.

Holm, P. 1986, The Slave Trade of Dublin, Ninth to the Twelfth Centuries. *Peritia* 5, 317–345.

Holm-Olsen, I.M. 1972, *Tekstilfunn fra Vestlandet fra vikingtid. Bruksstoffene av ull – deres produksjonsmiljø og produksjonssted.* Unpublished Magister thesis, University of Bergen, Bergen.

Holm-Olsen, I.M. 1975, Noen gravfunn fra Vestlandet som kaster lys over vikingtidens kvinnedrakt. *Viking 1975*, 197–206.

Hufthammer, A.K. 1987, Beinmaterialet fra Bryggen og ervervslivet i middelalderen, in *Kystliv. Onsdagskvelder i Bryggens Museum III*, I. Øye (ed.), 59–71. Bryggens Museum, Bergen.

Hughes, T.P. 1987, The Evolution of Large Technological Systems, in *The Social Construction of Technological Systems. New Directions in the Sociology and History of Technology*, W.E.T. Bijker, T.P. Hughes and T.J. Pinch (eds), 51–82. MIT Press, Cambridge, MA.

Indrebø, G. 1935, *Fjordung, Granskingar i eldre norsk organisasjonssoge*. Bergens Museums Årbok 1935.

Ingstad, A.S. 1979, To kvinnegraver med tekstiler fra Kaupang. *Universitets Oldsaksamling* 150 år, 158–164.

Ingstad, A.S. 1999, Tekstilene, in *Gravplassene i Bikjholbergene/Lamøya. Undersøkelsene 1950–1957. Del B Olddsaksformer. Kulturhistorisk tilbakeblikk. Del C. Tekstilene*, C. Blindheim, B. Heyerdahl-Larsen and A.S. Ingstad (eds) , 219–272. Norske Oldfunn, vol. 19. UKM, Oslo.

Ingstad, A.S. 2006, Brukstekstilene, in *Osebergfunnet*, bd. IV *Tekstilene*, A.E. Christensen and M. Nockert (eds), 185–276. Kulturhistorisk Museum, Universitetet i Oslo, Oslo.

Isaksen, E. 2017, *Hvalbeinsplater fra yngre jernalder. En analyse av hvalbeinsplatenes kontekst og funksjon*. Unpublished Master's thesis, University of Tromsø.

Iversen, F. 1999, *Var middelalderens lendmannsgårder kjerner i eldre godssamlinger? En analyse av romlig organisering av graver og eiendomsstruktur i Hordaland og Sogn og Fjordane.* Arkeologiske avhandlinger og rapporter fra Universitetet i Bergen 4, Bergen.

Iversen, F. 2008, *Eiendom, makt og statsdannelse. Kongsgårder og gods i Hordaland i yngre jernalder og middelalder. UBAS.* Nordisk 6. Bergen.

Iversen, T. 1995, Framveksten av det norske leilendingsvesenet i middelalderen – en forklaringsskisse, in *Heimen* bd. 32, Tidsskrift for lokal- og regional histore, 169–180. Landslaget for lokalhistorie, Trondheim.

Iversen, T. 1997, *Trelldommen – norsk slaveri i middelalderen.* Historisk institutt University of Bergen, Bergen.

Jåtten, T. 2006, *Insulære gjenstander fra yngre jernalder funnet på Vestlandet. Betydning og funksjon i ny kontekst.* Unpublished Master's thesis, University of Bergen, Bergen.

Jesch, J. 1991, *Women in the Viking Age*. The Boydell Press, Woodbridge.

Johnston, E. 2016a, The Warp-Weighted loom in Shetland, in *The Warp-Weighted Loom. Kljásteinavefstaðurinn, Oppstadveven*, R. Andersen and A.O. Martinussen (eds), 59–92. Skald forlag, Leikanger.

Johnston, E. 2016b, Spinning, in *The Warp- Weighted L,oom. Kljásteinavefstaðurinn, Oppstadveven*, R. Andersen and A.O. Martinussen (eds), 209–218. Skald forlag, Leikanger.

Johnston, E., M. Juuhl and G. Hákonardóttir 2016, Weaving Instructions for the Warp-Weighted Loom, in *The Warp-Weighted Loom. Kljásteinavefstaðurinn, Oppstadveven*, R. Andersen and A.O. Martinussen (eds), 148–178. Skald forlag, Leikanger.

Juuhl, M.K. 2016, The Warp Weighted Loom in Norway with Emphasis on Western Norway, in *The Warp-Weighted Loom. Kljásteinavefstaðurinn, Oppstadveven*, R. Andersen and A.O. Martinussen (eds), 93–144. Skald forlag, Leikanger.

Júliusson, Á.-D. 1995, *Bønder i pestens tid. Landbrug, godsdrift og social konflikt i semiddelalderens islandske bondesamfund.* Unpublished doctoral thesis, University of Copenhagen.

Júliusson, Á.-D. 1997, *Bønder i pestens tid. Landbrug, godsdrift og social konflikt i semiddelalderens islandske bondesamfund.* University of Copenhagen, Copenhagen.

Júliusson, Á.-D. 2010, Signs of Power. Manorial Demesnes in Medieval Iceland. *Viking and Medieval Scandinavia* 6, 1–29. DOI: 10.1484/J.VMS.1.102135.

Kaland, P.E. 1979, Landskapsutvikling og bosetningshistories I Nordhordlands lynghei-område. *På leiting etter den eldste graden. Nye metoder i studie av tidlig norsk bosettingshistorie*, 71–86. Universitetsforlaget, Oslo.

Kaland, S.H.H. 1973, Westness-utgravningene på Rousay, Orknøyene. *Viking* 37, 77–102.

Kaland, S.H.H. 1979, Lurekalven, en lyngheigård fra vikingtid/middelalder. En økonomisk Studie, in *På leiting etter den eldste garden. Nye metoder i studiet av tidlig norsk bosettingshistorie*, R. Fladby and J. Sandnes (eds), 71– 86. Universitetsforlaget, Oslo.

Kaland, S.H.H. 2006, Kvinnegraven fra Trå i Granvin – en gydjes grav?, in *Samfunn, symboler og identitet – Festskrift til Gro Mandt på 70-årsdagen. UBAS, Nordisk 3*, R. Barndon, K.A. Bergsvik and A. Engevik (eds) 2006, 351–362. University of Bergen, Bergen.

Kjellberg, A. 1979, Tekstilmaterialet fra Oslogate 7. Feltene 'Oslogate' Bebyggelsesrester og funngrupper, in *De arkeologiske utgravninger i Gamlebyen, Oslo* vol. 2, E. Schia and P.B. Molaug (eds), 83–104. Øvre Ervik, Alvheim og Eide Akademiske Forlag.

Kjellberg, A. and M. Hoffmann 1991, Tekstiler, in *Dagliglivets gjenstander*, E. Schia and P. Molaug (eds), 13–80. De arkeologiske utgravninger i Gamlebyen, Oslo vol. 8. Alvheim og Eide Akademiske Forlag, Øvre Ervik.

Lamb, H.H. 1995, *Climate, History and the Modern World.* Routledge, London and New York.

Larsen, K.C. 1995, *Ytre Moa, Ytre Moa . Et gårdanlegg frå vikingtid i Årdal. En studie i gårdsstruktur og byggeskikk.* Unpublished Master's thesis, University of Bergen, Bergen.

Larson, K. 2016, Weaving in the Dark? Hearth-House Weaving on a Warp Weighted Loom, in *The Warp Weighted Loom. Kljásteinavefstaðurinn, Oppstadveven*, R. Andersen and A.O. Martinussen (eds), 243–250. Skald forlag, Leikanger.

Lemonnier, P. 1992, *Elements of an Anthropology of Technology*. Anthropological Papers. Museum of Anthropology, Michigan, University of Michigan No. 88.

Lillehammer, A. 1967, *Innberetning og rapport om utgravningene på Ytre Moa*. Topografisk arkiv, Historisk museum, Bergen.

Lillehammer, A. 1971, *Funn frå forhistorisk tid.* I S. Ve (red.) Bygdebok for Årdal, Band 1. Årdal Sogelag, Årdal.

Lindkvist, T. 2003, Early Political Organisation: Introductory Survey, in *The Cambridge History of Scandinavia, Prehistory to 1520* Vol. 1, K. Helle (ed.), 160–167. Cambridge University Press, Cambridge.

Lønning, S. 1976, *Spinne ull på rokk og håndtein*. Landbruksforlaget, Oslo.

Lorange, A. 1875. *Samlingen av Norske Oldsager i Bergens Museum*. J.D. Beyers bogtrykkeri, Bergen.

Lucas, G. 2009, *Hofstaðir. Excavations of a Viking Age Feasting Hall in North-Eastern Iceland*. Institute of Archaeology, Monograph No. 1. Reykjavík.

Lukešová, H. 2011, Fragmenter av kvinnedrakten fra vikingtiden. Metoder for identifikasjon av gamle tekstilfragmenter. *Viking*, Norsk arkeologisk årbok bd. LXXXIV, 149–164.

Lukešová, H., A.S. Palau and B. Holst 2017, Identifying Plant Fibre Textiles from Norwegian Merovingian Period and Viking Age Graves; The Late Iron Age Collection of the University Museum of Bergen. *Journal of Archaeological Science: Reports* 13, 281–285.

Lukešová, H., H.L. Andersen, M. Kolínová and B. Holst 2019, Is it Hop? Identifying Hop Fibres in a European Historical Context. *Archaeometry* 61(2), 494–505.

Lund, N. 1994, If the Vikings Knew a Leding – What Was it Like?, in *Birka Studies 3, the 12th Viking Congress. Development around the Baltic and the North Sea in the Viking Age*, B. Ambrosiani and H. Clarke (eds), 91–105. The Birka Project for Riksantikvarieämbetet and Statens Historiska Muuseer, Stockholm.

Lund, N. 1996, *Lid, leding og landeværn*. Vikingskibshallen in Roskilde, Roskilde.

Lunde, M. 2018, *Jernspidd og kjønn. Jernspidd i graver fra Norden belyst i relasjon til gender*. Unpublished Master's thesis, University of Oslo, Oslo.

Lunden, K. 1969, Om gardtal og folketal i Noreg ca. 1340–1663. *Historisk Tidsskrift* 1969, 89–113.

Lunden, K. 1978, *Korn og kaup. Studier over prisar og jordbruk på Vestlandet i mellomalderen*. Universitetsforlaget, Oslo.

Lunden, K. 1979, Gardar, bruk og menneske i høgmellomalderen. *Historisk Tidsskrift* 58, 111–158.

Lunden, K. 2004, Recession and New Expansion 1350–1814, in *Norwegian Agricultural History*, R. Almås (ed.) 144–232. Tapir Academic Press, Trondheim.

Madsen, H.J. 1969, Om grubehuse. *Mark og Montre fra sydvestjyske museer*. Kulturhistoriske museer i Ribe amt, Ribe.

Mahler, D. 2007, *Sæteren ved Argisbrekka. Økonomiske forandringer på Færøerne i Vikingetid og tidlig middelalder*. Faroe University Press, Tórshavn.

Magnus, B. 1978, De eldste tider i Gloppen og Breim, in *Soga om Gloppen og Breim. Frå dei eldste tider til om lag år 1800*, P. Sandal (ed.), 103–122. Gloppen Sogelag, Sandane.

Mårtensson, L., E. Andersson, M.-L. Nosch and A. Batzer 2006a, *Technical Report. Experimental Archaeology Part 1. 2005–2006. Tools and Textiles – Texts and Contexts. Research Programme*. The Danish National Research Foundation's Centre for Textile Research (CTR), University of Copenhagen.

Mårtensson, L., E. Andersson, M.-L. Nosch and A. Batzer 2006b, *Technical Report. Experimental Archaeology Part 2:1 Flax 2006. Tools and Textiles – Texts and Contexts. Research PROGRAMME*. The Danish National Research Foundation's Centre for Textile Research (CTR), University of Copenhagen.

Mårtensson, L., E. Andersson, M.-L. Nosch and A. Batzer 2006c, *Technical Report. Experimental Archaeology Part 2:2 Whorl or Bead? Tools and Textiles – Texts and Contexts. Research Programme*. The Danish National Research Foundation's Centre for Textile Research (CTR), University of Copenhagen.

Mårtensson, L., M.-L. Nosch and E. Andersson 2009, The Shape of Things: Understanding a Loomweight. *Oxford Journal of Archaeology* 28(4), 373–398.

Martinsen, L. 1996, Maksimum og minimum. Norsk busetnadshistorie etter DNØ, in *Innsikt og utsyn. Festskrift til Jørn Sandnes*. Skriftserie fra Historisk institutt nr 12, NTNU, K. Haarstad, a. Kirkhusmo, D. Slettan, S. Supphellen (eds), 144–182. University in Trondheim, Trondheim.

Medieval Catalogue 1940, London Museum, London.

Mehler, N. 2015, The Sulphur Trade of Iceland from the Viking Age to the End of the Hanseatic Period, in *Nordic Middle Ages – Artefacts, Landscapes and Society. Essays in Honour of Ingvild Øye on her 70th Birthday*, I. Baug, J. Larsen and S.S. Mygland (eds), 193–212. *UBAS Nordic* 8. Bergen.

Meulengracht Sørensen, P. 1983, *The Unmanly Man. Concepts of Sexual Defamation in Early Northern Society*. Trans. J. Turville-Petre. Odense University Press, Odense.

Militello, P. 2007, Textile Industry and Minoan Palaces, in *Ancient Textiles. Production, Craft and Society*, E. Andersson, B. Burke, C. Gillis and M.-L. Nosch (eds), 36–45. Oxbow Book, Oxford.

Milek, K. 2012, The Roles of Pit Houses and Gendered Spaces in Viking-Age Farmsteads in Iceland. *Medieval Archaeology, 56, Society for Medieval Archaeology* 2012, 85–130.

Moberg, E. 1992, *Vikingtidssverdene. Hjemlig produserte eller importerte. Med utgangspunkt i et lokalt sverdmateriale fra Sogn og Fjordane*. Unpublished Master's thesis, University of Bergen, Bergen.

Mortensen, M. 1993, *B 15005 Bosetningsspor fra jernalder. Stedje gnr. 64, Sogndal kommune, Sogn og Fjordane. Innberetning*. Unpublished Report, University Museum of Bergen, Bergen.

Mortensen, M. 1997, For Women Only? Reflections on a Viking Age Settlement, Stedje in Sogndal in Western Norway. *Studien zur Sachsen Forschung* 10, 196–206.

Müller. M. 2013, Peasant Women, Agency and Status in Late 13th and Early 14th century England, Some Reconsiderations, in *Married Women and the Law, c. 1200–1700*, C. Beattie and M. F. Stevens (eds), 91–114. Boydell and Brewer, Woodbridge.

Mundal, E. 2020, Kva hadde kvinna på tinget å gjere? Kvinneroller på tinget etter norske landskapslover og Landslova, in *Lov og lovgivning i middelalderen. Nye studier av Magnus Lagabøtes Landslov*, A.C. Horn and K.A. Seip (eds), 212–249. Nasjonalbiblioteket, Oslo.

Munro, J. 2000a, Wool and Wool-Based Textiles in the West European Economy, c. 800–1500: Innovations and Traditions in Textile Products, Technology, and Industrial Organisation. Working Paper no.5 UT.ECIPA–MUNRO-00–05.

Munro, J. 2000b, The West European Woollen Industries and their Struggles for International Markets, c. 1000–1500. https://www.economics.utoronto.ca/public/workingPapers/UT-ECIPA-MUNRO-00-04.pdf.

Munro, J. 2003, Medieval Woollen Textiles, Technology and Organisation, in *The Cambridge History of Western Textiles*, D. Jenkins (ed.), 181–227. Cambridge University Press, Cambridge.

Myhre, B. 2002, Landbruk, landskap og samfunn 4000 f. Kr–800 e.Kr, in *Norges landbrukshistorie I, Jorda blir levevei 4000 f.Kr –1350 e.Kr*, B. Myhre and I. Øye (eds), 12–213. Det Norske Samlaget, Oslo.

Myrdal, J. 2012, *Boskapsskötselen under medeltiden: en källpluralistisk studie*. Nordiska museet, Stockholm.

Myrdal, J. and I. Øye 2021, Technological Capacity of Women in the High Middle Ages, in *Routledge Handbook on Medieval Rural Life. The Routledge History of Rural Life*, M. Müller (ed.), 264–281. Taylor & Francis/Routledge, Oxford.

Nedkvitne, A. 1991, *Byen under Eikaberg. Fra byens oppkomst til 1536. Oslo bys historie bd. 1*. J.W. Cappelens forlag AS, Oslo.

Nedkvitne, A. 2014, *The German Hansa and Bergen 1100–1600*. Böhlau Verlag, Köln, Weimar, Wien.

Nielssen, A.R. 2012, *Landnåm fra nord: utvandringa fra det nordlige Norge til Island i vikingtid*. Cappelen Damm, Oslo.

Owen, O. and M. Dalland 1994, Scar, Sanday: A Viking Boat-Burial from Orkney. An Interim Report, in *The Twelfth Viking Congress. Birka Studies. Development around the Baltic and the North Sea in the Viking Age*, B. Ambrosiani and H. Clarke (eds), 159–172. Riksantikvarieambatet, Stockholm.

Owen, O. and M. Dalland 1999, *Scar: A Viking Boat Burial on Sanday, Orkney*. Tuckwell Press, in association with Historic Scotland, East Linton.

Owen-Crocker, G., E. Coatsworth and M. Hayward 2012, *Encyclopedia of Medieval Dress and Textiles of the British Isles c. 450–1450*. Brill, Leiden.

Øye, I. 1986, Sogndal i mellomalderen, in *Bygdebok for Sogndal* bd. 1, P. Sandal (ed.), 239–460. Sogndal Sogelag, Bergen

Øye, I. 1988, *Textile Equipment and its Working Environment, Bryggen in Bergen c. 1150–1500*. The Bryggen Papers. Main Series, vol. 2. Norwegian University Press, Bergen.

Øye, I. 1998a, *Middelalderbyens agrare trekk*. Bryggens Museum, Bergen.

Øye, I. (ed.) 1998b, *Fires in Bergen – Revisited*. The Bryggen Papers, Supplementary Series vol. 6. Fagbokfolaget AS, Bergen.

Øye, I. 2002, *Landbruk Under Press 800–1350. Norges Landbrukshistorie*, vol. 1, 216–414. Det Norske Samlaget, Oslo.

Øye, I. 2003, Outfields as Part of the Medieval Farm – Four Archaeological Case Studies from Western Norway, in *Scandinavian Archaeological Practice – In Theory*, J. Bergstøl (ed.), 400–411. Oslo Archaeological Series vol. 1. University of Oslo, Oslo.

Øye, I. 2004, Agricultural Conditions and Rural Societies c. 800–1350, in *Norwegian Agricultural History*, R. Almås (ed.), 79–140. Tapir Academic Press, Trondheim.

Øye, I. 2005, Farming and Farming Systems in Norse Societies of the North Atlantic, in *Viking and Norse in the North Atlantic. Select Papers from Proceedings of the Fourteenth Viking Congress, Tórshavn, 19–30 July 2001, Annales Societas Scientiarium Færoensis Supplementum XLIV*, A. Mortensen and S. Arge (eds), 359–370. Foroya Fródskaparfelag, Tórshavn.

Øye, I. 2006, Kvinner som tradisjonsformidlere – rom og redskaper, in *UBAS Nordisk* 3. *Samfunn, symboler og identitet. Festskrift til Gro Mandt på 70-årsdagen*, R. Barndon, S. Innselset, K.K. Kristoffersen and T.K. Lødøen (eds), 439–453. *UBAS Nordisk 3. Samfunn, symboler og identitet*. Bergen.

Øye, I. 2009, Settlement Patterns and Field Systems in Medieval Norway. *Landscape History* 30(2), 37–54.

Øye, I. 2010, Women in Early Towns, in *The Viking Age. Ireland and the West. Proceedings of the XVth Viking Congress*, J. Sheehan and Donnchadh Ó Corráin (eds), 298–308. Four Courts Press, Dublin.

Øye, I. 2011, Textile-Production Equipment, in *Things from the Town. Artefacts and Inhabitants in Viking-age Kaupang*, D. Skre (ed.), 339–372. Kaupang Excavation Project Publication Series, Vol. 3, Norske Oldfunn XXIV. Aarhus University Press, Aarhus.

Øye, I. 2013, Utmarka i vestnordisk perspektiv, in *From Nature to Script, Reykholt, Environment, Centre and Manuscript Making*, H. Þorlaksson and Þ. Sigurðsdotttir (eds), 7–77. Reykholt Publikasjon 4. Museum Tusculanum Press, Copenhagen.

Øye. I. 2014, Bergen and the German Hansa in an Archaeological Perspective, in *Lübeck und der Hanseraum. Beiträge zu Archäologie und Kulturgeschichte. Festschrift für Manfred Gläser*, A. Falk, U. Müller and M. Schneider (eds), 481–490. Schmidt-Römhild Verlag, Lübeck.

Øye, I. 2015a, Production, Quality, and Social Status in Viking Age Dress. Three Cases from Western Norway" in *Medieval Clothing and Textiles 12*, F. Pritchard, G.R. Owen-Crocker, R. Netherton, C.L. Dahl and G. Pac (eds), 1–28. Boydell Press, Woodbridge.

Øye, I. 2015b, Technology and Textile Production from the Viking Age and the Middle Ages: Norwegian Cases, in *Textiles and the Medieval Economy: Production, Trade, and Consumption of Textiles, 8th–16th Centuries*, A. Ling Huang and C. Jahnke (eds), 41–63. Oxbow Books, Oxford.

Øye, I. 2015c, Tekstilproduksjon i middelalderbyene – mer enn bare husflid?, in *En aktivist for Middelalderbyen Oslo: Festskrift til Petter B. Molaug*, L.-M.B. Johansen, J. Brendalsmo, K. Paasche and E. Marstein Bauer (eds), 25–51. , Novus Forlag, Oslo.

Øye, I. 2016a, When did Weaving Become a Male Profession? *Danish Journal of Archaeology* 5(1–2), 37–54.

Øye, I. 2016b, Tools and Textile Production in the North Atlantic, in *Proceedings from the 17th Viking Congress in Shetland*, V. Turner, O. Owen and D. Waugh (eds), 245–250. Shetland Heritage Publications, Lerwick.

Øye, I. 2016c, The Loom in the Grave – A Source to Textile Production in the Viking Period, in *The Warp-Weighted Loom. Kljásteinavefstaðurinn, Oppstadveven*, R. Andersen and A.O. Martinussen (eds), 219–234, Skald forlag, Leikanger.

Øye, I., R. Barndon and A. Engevik. 2011, Technology and Regions in a Long Time Perspective – An Introduction, in *The Archaeology of Regional Technologies. Case Studies from the Palaeolithic to the Age of the Vikings*, R. Barndon, A. Engevik and I. Øye (eds), 1–14. The Edwin Mellen Press Lewiston, Queenston, Lampeter.

Petersen, J. 1928, *Vikingtidens smykker*. Stavanger Museum, Dreyers grafiske anstalt, Stavanger.

Petersen, J. 1940, *Viking Antiquities V. British antiquities of the Viking Period found in Norway*, H. Shetelig (ed.) . Aschehoug forlag, Oslo.

Petersen, J. 1951, *Vikingetidens redskaper*. Skrifter utgitt av Det norske videnskaps-akademi i Oslo. II. Hist.-filos. Klasse 1951:4. Oslo.

Pétursdóttir, Þ. 2007, *Deyr fé, deyja frændr. Re-Animating Mortuary Remains from Viking Age Iceland*. Unpublished Master's thesis, University of Tromsø, Tromsø.

Pétursdóttir, Þ. 2009, Viking Age Graves. Lack of Material – Lack of Interpretation? *Archaeologica Islandica* 7, 22–40.

Pfarr, E.-M. 2005, Handwerk oder Industrie? Erfahrungen bei Herstellung eines hochmittelalterlichen Wollgewebes auf dem Gewichtswebstuhl, in *Northern Archaeological Textiles NESAT VII*, F. Pritchard and J. P. Wild (eds), 58–64. Oxbow Books, Oxford.

Pilø, L. and D. Skre 2011, Introduction to the Site, in *Things from the Town. Artefacts and Inhabitants in Viking-Age Kaupang*, D. Skre (ed.), 17–26. Kaupang Excavation Project Publication Series, Vol. 3, Norske Oldfunn XXIV. Aarhus University Press, Aarhus.

Plather, U. 2011, Analyses of Jet-like Objects, in *Things from the Town. Artefacts and Inhabitants in Viking-age Kaupang*, D. Skre (ed.), 129–142. Kaupang Excavation Project Publication Series, Vol. 3, Norske Oldfunn XXIV. Aarhus University Press, Aarhus.

Postan, M.M. 1972, *The Medieval Economy and Society: An Economic History of Britain in the Middle Ages*. Orion Publishing Group, London.

Power, E. 1941, *The Wool Trade in English Medieval History*. Oxford University Press, Oxford.

Price, N. 2002, *The Viking Way: Religion and War in Late Iron Age Scandinavia*. Department of Archaeology and Ancient History, Uppsala University.

Price, N., T. Douglas and H. Gestsdóttir 2006, The First Settlers in Iceland. An Isotope Approach to Colonization. *Antiquity* 80, 130–144.

Price, N., C. Hedenstierna, T. Zachrisson, A. Kjellström, J. Stora, K Kezwinska, V. Sobrado, M. Jakobsson and A. Groterstrom 2019, Viking Warrior Women? Reassessing Birka chamber grave Bj. 581. *Antiquity* 93(367), 181–198.

Rabben, A. 2002, *Med vevsverd og stekepanne. Tekstilredskaper og kjøkkenredskaper i vestnorske mannsgraver fra yngre jernalder.* Unpublished Master's thesis, University of Bergen. Bergen.

Ramstad, M. 2010, Vikingtidplassen Bjørkum – utgraving og formidling. *Årbok for Universitetsmuseet i Bergen* 15, 56–67.

Ramstad, M. 2018, Vikingstaden Bjørkum – en gård, en samlingsplass, litt av begge deler eller noe helt annet? *Sogeskrift frå Lærdal, Art tie* 2, 1–29.

Ramstad, M., L.S. Halvorsen and A.B. Olsen 2011, Bjørkum: Feasting, Craft Production and Specialisation on a Viking Age Rural Site in Norway. *Antiquity* Project Gallery 85(328): https://www.antiquity.ac.uk/projgall/ramstad328/.

Randers, K. 1981, *Høybøen – en ødegård på Sotra. En undersøkelse av bruksperioder og erverv basert på bosetningsspor fra eldre jernalder og middelalder.* Unpublished Magister thesis, University of Bergen, Bergen.

Rasmussen, A.K. 1969, Grubehuse i Vilslev. *Mark og Montre. Fra sydvestjyske museer*, 22–25.

Rast-Eicher, A. and R. Windler 2006, Mit den Füssen weben. *Tradition und Innovation in der mittelalterlichen Webstuhlgeschichte. Zeitschrift für Kunst und Kultur in swr Schweiz*, 29–35.

Rognø, L. 2019, *Hesteutstyr i graver i yngre jernalder i Sogn og Fjordane. Bruks- eller statusgjenstander?* Unpublished Master's thesis, University of Bergen, Bergen.

Rosenquist, A.M. 1966, Undersøkelse av fibre fra Snæhvammur 3931. *Àrbók hins Íslenszka fornleifafélags*, 96–108.

Rosenquist, A.M. 2006, Analyser av Osebergtekstiler, in *Osebergfunnet, bd. IV Tekstilene*, A.E. Christensen and M. Nockert (eds), 171–184. Kulturhistorisk Museum, Universitetet i Oslo, Oslo.

Rosser, G. 1997, Crafts, Guilds and Negotiation of Work in the Medieval Town. *Past and Present* 154, 3–31.

Rui, L.M. 1992, Kljåsteiner – vevlodd, in *Dagliglivets gjenstander – del II*, E. Schia and P.B. Molaug (eds), 113–131. De arkeologiske utgravninger i Gamlebyen, Oslo vol. 8. Alvheim og Eide Akademiske Forlag, Øvre Ervik.

Ryder, M.L. 1984, Medieval Sheep and Wool Types. *Agricultural History Review* 32(1), 14–28. http://www.jstor.org/stable/40274301.

Ryder, M.L. 2005, The Human Development of Different Fleece-Types in Sheep and its Association with the Development of Textile Crafts, in *NESAT VII*, F. Prichard and J.P. Wild (eds), 122–128. Oxbow Books, Oxford.

Rygh, O. 1885, *Norske Oldsager ordnede og forklarede af O. Rygh*. Cammermeyer, Christiania.

Sandnes, J. 1979, Ødegårdsprosjektet og tallet på gårdsbruk i Norge i høgmellomalderen. *Historisk Tidsskrift* 58, 397–409.

Sandnes, J. 1983, 'Tolv kyr, to hester og tre træler'. Litt om omfanget av træleholdet i Norge i vikingtid og tidlig kristen tid. *Historisk Tidsskrift* 62, 79–83.

Schjølberg, E. 1984, The Hair Products, in *The Bryggen Papers. Supplementary Series No. 1*, A. E. Herteig (ed.), 73–92. The Norwegian University Press, Bergen.

Schjølberg, E. 1988, Cordage and Similar Products from Bryggen in Bergen, in *Bryggen Papers. Supplementary Series No. 3*, A.E. Herteig (ed.), 69–139. Norwegian University Press, Bergen.

Schjølberg, E. 1992, The Wools in Some Medieval Textiles from Bryggen in Bergen, Norway, in *Tidens Tand Nr.5 Archaeological Textiles in Northern Europe. Report from the 4th NESAT Symposium 1–3. May 1990 in Copenhagen*, L.B. Jørgensen and E. Munksgaard (eds), 151–157. Konservatorskolem/Det kgl. Danske Kunstakademi, Copenhagen.

Schjølberg, E. 1998, 12th Century Twills from Bergen, Norway, in *Textiles in European Archaeology. Report from the 6th NESAT Symposium, 7–11th May 1996 in Borås*, L. Bender Jørgensen and C. Rinaldo (eds), 209–215. GOTARC Series A, Vol. 1. University of Gothenburg, Gothenburg.

Shetelig, H. 1904, *Bergens Museums Årbok 1904*. Bergens Museum, Bergen.

Shetelig, H. 1905, Gravene på Myklebostad paa Nordfjordeid, in *Bergens Museums Aarbok 1905*, no. 7, 3–53. Bergens Museum, Bergen.

Shetelig, H. 1912, *Vestlandske graver i jernalderen*, Bergens Museums Skrifter. Ny rekke. Bd. II. No. 1. Bergens Museum, Bergen.

Shetelig, H. (ed.) 1940, *Viking Antiquities in Great Britain and Ireland. Part V. British Antiquities of the Viking Period Found in Norway*. Aschehoug, Oslo.

Shetelig, H. 1947, Traces of the Custom of the 'Suttee' in Norway during the Viking Age, in *Arkeologi - Historie - Kunst - Kultur, utgitt til syttiårsdagen 25. juni 1947*, J. Bøe (ed.), 29–39. Sagabook of the Viking Club VI, Bergen.

Sigurðsson, J.V. 2011, *Norsk historie 800-1536. Frå krigerske bønder til lydige undersåttar*. Oslo.

Sindbæk, S.M. 2005, *Ruter og rutinifisering: Vikingetidens fjernahandel i Vesteruropa*. Multivers Academic, Copenhagen.

Sindbæk, S.M. 2007, Networks and Nodal Points: The Emergence of Towns in Early Viking Ages Scandinavia. *Antiquity* 81, 119–132.

Sindbæk, S.M. 2011, Re-Assembling Regions. The Social Occasions of Technological Exchange in Viking Age Scandinavia. In *The Archaeology of Regional Technologies. Case Studies from the Palaeolithic to the Age of the Vikings*, R. Barndon, A. Engevik and I. Øye (eds), 263–287. The Edwin Mellen Press Lewiston, Queenston, Lampeter.

Sindbæk, S.M. 2012, Urban Crafts and Oval Brooches. Style, Innovation and Social Networks in Viking Age Towns, in *Papers from the Proceedings of the Viking Congresses, 2011*, S. Sigmundsson (ed.), 407–421. University of Iceland Press, Reykjavik.

Sindbæk, S.M. 2017, Urbanism and Exchange in the North Atlantic/Baltic, 600–1900 CE, in *The Routledge Handbook of Archaeology and Globalization*, T. Hodos (ed.), 553–565. Routledge, London.

Sjøvold, T. 1974, *The Iron Age Settlements of Arctic Norway*. Tromsø Museums Skrifter Vol X:2, Oslo.

Skipstad, E.M. 2009, *Kvinner og kristendom på Vestlandet. En undersøkelse med utgangspunkt i graver frå yngre jernalder i Sogn*. Unpublished Master's thesis, University of Bergen, Bergen.

Skoglund, G., M. Nockert and B. Holst 2013, Viking and Early Middle Ages Northern Scandinavian Textiles proven to be of Hemp. *Scientific Reports* 3, 2686 (2013). https://doi.org/10.1038/srep02686.

Skre, D. 1998, *Herredømmet. Bosetning og besittelse på Romerike 200–1350 e. Kr.* Acta Humaniora, Oslo.

Skre, D. 1999, Eiendom og hierarki i det før- og tidlig-statlige norske samfunnet. *Heimen* 2, 123–137.

Skre, D. (ed.) 2011, *Things from the Town. Artefacts and Inhabitants in Viking-Age Kaupang*. Kaupang Excavation Project Publication Series, Vol. 3, Norske Oldfunn XXIV. Aarhus University Press, Aarhus.

Slomann, W. 1986, Tre vestnorske spenner fra jernalderen. In *Universitets Oldsaksamlings Skrifter Nr. 6 Ny rekke. Festskrift til Wenche Slomann. Hjemlig tradisjon og fremmede innslag i norsk jernalder*, I. Martens, B. Myhre and E. Straume (eds), 17–40. Universitets Oldsaksamling, Oslo.

Solberg, B. 1985, Social Status in the Merovingian and Viking Periods in Norway from Archaeological and Historical Sources. *Norwegian Archaeological Review* 18(1), 61 –76.

Solberg, B. 1986, Sogndal i førhistorisk tid, in *Bygdebok for Sogndal* bd. 1, P. Sandal (ed.), 120–235. Sogndal sogelag, Bergen.

Solberg, B. 2010, *Jernalderen i Norge. 500 før Kristus til 1030 etter Kristus*. Cappelen Akademisk, Oslo.

Solberg Andreassen, I.S. 2001, *Fortellinger om makt og ære – identitet gjennom dyreornamentikk i islandske vikingtidsgraver*. Unpublished Master's thesis, University of Bergen, Bergen.

Soly, H. 2008, The Political Economy of European Craft Guilds: Power Relations and Economic Strategies of Merchants and Early Master Artisans in the Medieval and Early Modern Textile Industries. *International Review of Social History* 53, 45–71.

Sømme, A. 1954, *Jordbrukets geografi i Norge* A. Tekstbind, Skrifter fra Norges handelshøyskole, i rekken *Geografiske avhandlinger* 3. J.W. Eides forlag, Bergen.

Sørheim, H. 2011, Three Prominent Norwegian Ladies with British British Connections. *Acta Archaeologica* 18(82), 17–54.

Sørheim, H. 2018, *Sentralbygd, tettsted, knutepunkt, by. Bosetningshierarkier og sentraldannelser fra jernalder til middelalder. UBAS* 10. University of Bergen, Bergen.

Stefánsson, M. 1986, Bergen – Islands første hovedstad, in *Kjøpstad og rikssentrum. Onsdagskvelser i Bryggens Museum II*, I. Øye (ed.), 70–87. Bryggens Museum, Bergen.

Stefánsson, M. 2003, The Norse Island Communities of the Western Ocean, in *The Cambridge History of Scandinavia, Prehistory to 1520, Vol. 1*, K. Helle (ed.), 202–220. Cambridge University Press, Cambridge.

Steinnes, A. [1936] 1982, *Mål, vekt og verderekning i Noreg i millomalderen og ei tid etter*, *Nordisk Kultur* XXX. Aschehoug & co. forlag, Oslo.

Steinsland, G. 2005, *Norrøn religion: Myter, riter, samfunn*. Pax forlag, Oslo.

Stephenson, M.G. 1988, Wool Yields in the Medieval Economy. *Economic History Reviewer* 2nd ser. XLI.3, 368–391.

Štolcová, T. and K. Grömer 2010, Loom-Weights, Spindles and Textiles – Textile Production in Central Europe from the Bronze Age to the Iron Age, in *North European Symposium for Archaeological Textiles, NESAT* X, E.B. Andersson Strand, M. Gleba, U. Mannering, C. Munkholt and M. Ringgard, 9–20. Oxbow Books, Oxford.

Stumann Hansen, S. 1991, 'Toftanes'. A Faroese Viking-Age Farmstead From the 9th and 10th Centuries AD. *Acta Archaeologica* 61, 44–53.

Stumann Hansen, S. 2013, 'Toftanes'. A Viking-Age Farmstead in the Faroe Islands. Archaeology, Environment and Economy. *Acta Archaeologica* 84, 1–239.

Stylegar, F.-A. 2007, The Kaupang Cemetries Revisited, in *Kaupang in Skiringsal*, D. Skre (ed.), 65–126. Kaupang Excavation Project. Publication Series, Vol. 1, Norske Oldfunn XXII. Aarhus University Press, Aarhus.

Stylegar, F.-A., 2010, Hvorfor er det færre kvinne- enn mannsgraver fra vikingtiden I Norge? *Primitive tider* 12. Årg, 71–80.

Summanen, H. 2002, Grophus, hantverk ock Sigtuna stad - om romsliga føraendingar i feodalismens tecken. MA-thesis in archaeology, Arkeologiska institutusjonen, Stockholms universitet.

Sveinbjarnardóttir, G. 1991, Shielings in Iceland: An Archaeological and Historical Survey. *Acta Archaeologica* 62, 93–96.

Sveinbjarnardóttir, G. 2012, *Reykholt. Archaeological Investigations at a High-Status Farm in Western Iceland*. Snorrestofa & The National Museum of Iceland, Reykjavik.

Sverrisdóttir, Á. 2009, Textiles, in *Hofstaðir. Excavations of a Viking Age Feasting Hall in North-Eastern Iceland*, G. Lucas (ed.), 318–320. Institute of Archaeology, Monograph No. 1. University of Iceland, Reykjavík.

Troldtoft Andresen, S. and S. Karg 2011, Retting Pits for Textile Fibre Plants at Danish Prehistoric Sites Dated Between 800 B.V and AD 1050. *Veget Hist Archaeobot* DOI 10.1007/s000334-011-0324-0.

Tveiten, O. 2012, *Mellom aust og vest. Jarnvinna kring Langfjella i yngre jarnalder og mellomalder*. Unpublished PhD thesis, University of Bergen, Bergen.

Tveiten, O. 2015, Tale of Two Technologies / Late Medieval Irin Productionin Oppland, Norway, in *Nordisc Middle Ages / Artefacts, Landscapes and Society. Essays in Honour of Ingvild Øye on her 70th Birthday, UBAS University of Bergen Archaeological Series 8*, I. Baug, J..Larsen and S.S Mygland (eds), 331–338. University of Bergen, Bergen.

Þorláksson, H. 1991, *Vaðmál og verðlag. Vaðmál í utanlandsviðskiptun og í búskap Íslendinga á 13. og 14. öld*. Fjölföldun Sigurjóns, Reykjavík.

Undheim, P. 1985, *Osteologisk materiale fra Dreggen. En økologisk studie fra middelalderens Bergen*. Unpublished Master's thesis, University of Bergen, Bergen.

Vedeler, M. 2004, Er dette rester etter klær? Problemer knyttet til funksjonsbestemmelse av arkeologiske tekstiler. *Collegium Medievale. Interdisciplinary Journal of Medieval Research* 17, 75–86.

Vedeler, M. 2007, *Klær og formspråk i norsk middelalder.* Acta humaniora 280. Unipub forlag, Oslo.

Vedeler Nilsen, M. 1992, *Draktmaterialet fra Guddal, funksjon og visuell kommunikasjon.* Unpublished Master's thesis, University of Bergen, Bergen.

Vickers, K., J. Bending, P.C. Buckland, K.J. Edwards, S.S. Hansen and G. Cook 2005, Toftanes. The Paleoecology of a Faroese Landnám Farm. *Human Ecology* 33(5), 685–710.

Waetzoldt, H. 2007, The Use of Wool for the Production of Strings, Ropes, Braied Mats, and Similar Fabrics, in *Ancient Textiles. Production, Craft and Society*, C. Gillis and M.-L. Nosch (eds), 112–121. Oxbow Books, Oxford.

Walton, P. 1988, Dyes and Wools in Iron Age Textiles from Norway and Denmark. *Journal of Danish Archaeology* 7, 144–158.

Walton Rogers, P. 1997, *Textile Production at 16-22 Coppergate.* The Archaeology of York 17(11). York Archaeological Trust, York.

Walton Rogers, P. 1999, Textile, Yarn and Fibre from the Biggings', in *The Biggings, Papa Stour, Shetland. The History and Excavation of a Royal Norwegian Farm*, B.E Crawford and B.B. Smith (eds), 194–199. Society of Antiquities of Scotland Monograph Series Number 15, Edinburgh.

Walton Rogers, P. 2012, Textiles, Wool and Hair, in *Reykholt. Archaeological Investigations at a High-Status Farm in Western Iceland*, G. Sveinbjarnardóttir (ed.), 197–217. The National Museum of Iceland, Reykjavik.

Wamers, E. 1984, Eine Zungenfibel aus dem Hafen von Haithabu, in *Berichte über die Ausgrabungen in Haithabu, Bericht 19. Das archaeologische Fundmaterial IV*, K Schietzel (ed.), 63 –127. Wachholts Verlag, Neumünster.

Wamers, E. 1985, *Insularer Metallschmück in wikingerzeitlichen Gräbern Nordeuropas. Untersuchungen zur skandinavischen Westexpansion*, Neumünster.

Wamers, E. 1991, Til omarbeidelsen av insulære metallsmykker; bemerkninger till en recension. *Fornvännen. Journal of Swedish Antiquarian Research* 86, 110–120.

Wamers, E. 1998, Insular Finds in Viking Age Scandinavia and the State Formation of Norway, in *Ireland and Scandinavia in the Early Viking Age*, H.B. Clarke, M.N. Mhaonaigh and R. Ó Floinn (eds), 37–72. Four Courts Press, Dublin.

Warburg, L. 1976, *Spindebog.* Borgen, København.

Wild, P.J. 1970, *Textile Manufacture in the Northern Roman Provinces.* Cambridge University Press, Cambridge.

Williams, H. 2007, The Emotive Force of Early Medieval Mortuary Practices. *Archaeological Review from Cambridge* 22(1), 107–123.

Zagal-Mach Wolfe, U.I. 2013, Grasping Technology, Assessing Craft, Developing a Research Method for the Study of Craft-Tradition. *Acta Archaeologica Lundensia*, Series in 8°, Vol. 63.

Appendices

Abbreviations: F = Firdafylki; H = Hordafylki; S = Sygnafylki
R = Rygh 1885; JP = J. Petersen 1928 and 1951
F = Female M = Male
VJG = Vestlandske graver fra jernalderen, Shetelig 1912

Appendix 5.0: Some selected graves with textile tools from the study area, referred to in the text, predating the period of study (N:7)

Location	*No.*	*Textile tools*	*Dress accessories*	*Other gender indicators*	*Context*	*Status*
Sandal, Jølster, F	B6656 I	Shears (R172), 1 weaving beater (R150), 10 loom weights	Silver brooch, 2 silver rings.	1 bucket-shaped vessel of clay, 1 strike-a-light	Large mound, 15–18 m across, double grave covered by slabs	FIII/ MIII
Årheim, Stryn, F	B4408	1 spindle whorl	–	1 sword, 1 axe R153	Mound, double grave?	MII/F?
Kvåle, Sogndal, S	B6516	1 short weaving beater (R150), 34 cm, 1 blade, wood handle, 16 cm, 1 spindle whorl of iron, 31 spindles with hooks	1 relief brooch (R262), 2 smaller gilt silver brooches (R256), 2 equal- armed brooches of bronze, beads of glass	2 keys of bronze in a ring, 1 awl, chest w/lock	Remains of a cairn, in a burial chest with two rooms, one for the tools	FIII
Ølmheim, Sogndal, S	B12215	Shears, 2 spindle whorls (*c. 10g, *15–20g), 6 loom weights,1 weaving beater (R440), 62 cm long, 1 'vevreell' (R413)	Button-on-bow brooch, 7 beads of glass and 1 of bronze, silver band	1 key, 1 casket with lock, 1 sickle, 1 hoe, 1 horse bit, 1 knife	Cremation burial in a flat grave	FII–III
Nornes, Sogndal, S	B9688	1 weaving beater (R150), 1 spindle whorl, 4 cm across	1 relief brooch of gilt copper alloy, 1 small cruciform brooch of bronze, 2 clasps, 1 belt ring (R321)	1 hooked key,1 knife, 1 bucket-shaped vessel of clay.	Cairn, chest	FIII
Hovland I, Ullensvang, H	B4719 I	Shears, 1 weaving beater		2 swords, 2 arrowheads, 1 axe R553, 1 sickle	Mound, cremation *c.* 4 m distance from grave no. II	MIII/FII?
Hovland II, Ullensvang, H	B4719	2 weaving beaters in either grave I or II: a spindle whorl (*20–25g), 1 loom weight	In either I or II: 1 bronze brooch	2 swords (R499), 2 arrowheads, 2 shield bosses,1 frying pan, iron and soapstone vessel; in either I or II	Inhumation grave in a stone chamber	MIII/FII?

Appendix 5.1: Female graves with textile tools and dress accessories (N:87) (alphabetically)

Location	*No.*	*Textile tools*	*Dress accessories*	*Possible female indicators*	*Context*	*Status*
Arnestad, Gloppen, F	B7653	Shears (R443), wool combs, 2 spindle whorls (1 retrieved, 11g + *20–25 g), 1 weaving beater (R440), *c.* 80 cm long, 1 'vevreell' (R413), *c.* 11 cm long	2 oval brooches, JP fig. 47, Borre style (Montelius fig.563), 3. brooch (R669), 3 beads, 2 round pendants of bronze (Jelling style)	1 sickle, 1 long iron stick/staff with a handle, more than 83 cm (VJG. fig. 442), frying pan (R730), iron cauldron (R731), fragments of a bucket of wood, 1 vessel of iron, mountings to a casket, melted glass, 1 two-part horse bit	Cremation, long mound, but unburnt artefacts	FIII
Beinnes, Olden, Stryn, F	B7184	Tines from wool combs, 1 weaving beater (R440), 68 cm	2 oval brooches (R647), 1 ring-headed pin	–	Mound	FIII
Belle, Aurland, S	B5773	7 loom weights (200–400 g), 1 weaving beater (R440), *c.* 72 cm long	1 Irish brooch, reused ornamental mounting	–	Cremation burial in boat (*c.* 60 rivets and spikes) in a mound	FI–II
Belle, Aurland, S	B5815	Wool combs, 1 weaving beater (R440), 75 cm long, 'vevreell' (R413)	2 S-shaped brooches, 1 ring-headed pin	-	Cremation grave in a long-mound, boat burial (125 rivets and 22 spikes)	FIII
Berdal, Feios, S	B594, 596-603	1 spindle whorl (g?), 5 loom weights	2 oval brooches (R644) (JP fig. 15), 1 Hiberno-Saxon gilt bronze cruciform brooch, an arm ring, 5 beads	1 bronze mounting (to a casket?), parts of an urn of clay	Boat burial in a large cairn	FIII
Bjørnar-heim, Breim, Gloppen, F	B9502	1 spindle whorl (26 g), 1 weaving beater (R440)	1 oval brooch (JP16), 1 bead, mosaic	Mounting to a casket	Cremation, disturbed grave (40 rivets and spikes, probably a boat grave	FII
Bolstad, Evanger, Voss, H	B13951	2 spindle whorls (*10–15 g; *36–40 g)	Small bronze ring, gilt bronze ring, 2 beads	–	Cremation, probably in a boat (37 rivets and 22 spikes), in a large cairn with several graves	FII
Brekke, Oppstryn, F	B6433	Wool combs (34 tines in two rows), 3 loom weights	1 oval brooch (R654), a 3. brooch, 1 bead	–	Flat gave	FIII
Brekke, Lavik, S	B6637, B6649	1 spindle whorl (21 g), 1 weaving beater	Circular Irish brooch	1 iron ladle, 1 hoe	1 of 2 cremation burials in a mound	FI–II
Brekke, Lavik, S	B6062–6063	1 spindle whorl (*25–30 g)	1 oval brooch (R649)	–	Cairn	FII

(Continued)

Appendix 5.1: Female graves with textile tools and dress accessories (N:87) (alphabetically) (Continued)

Location	No.	Textile tools	Dress accessories	Possible female indicators	Context	Status
Brekke, Lavik, S	B8117–B8118	1 loom weight?	1 rectangular silver brooch, 1 oval brooch (R652), 3 beads	1 hone	Cremation, secondary burial in a mound, boat burial (B8118)	FII–III
Bu, Ullensvang, H	B4599	19 *c.* 12.5 cm long tines of wool combs, 2 loom weights, 1 small spindle whorl (lost)	2 oval brooches (R663, 664. Jelling type), 2 glass beads, *c.* 85 cm long chain of iron	1 horse bit (R567), 1 key? Parts of a sickle	Cremation in mound	FIII
Dale, Fjaler, F	B5910	Small shears (R443), pair of wool combs, with *c.* 13 cm long bent tines, 2 spindle whorls (33 g + 1 of lead), 1 loom weight, a smoothing pellet of glass, 8 cm across	2 oval brooches (R652), 3. brooch of silver (JP 138), armlet of silver, 2 silver finger rings, 8 beads of glass in different colours	1 key? sickle, 3 cooking equipment: 2 cauldrons (of iron and soapstone), a small vessel of iron, 1 simple horse bit	Inhumation, flat grave	FIII
Eide, Gloppen, F	B14583	1 weaving beater, 1 flax hackle (R433)	2 oval brooches (R647)	–	Grave	FIII
Fana prg, Fana, H	B322-326	2 spindle whorls (14 g, 16 g)	2 beads, 1 silver chain	Bronze cauldron of Insular origin, 33 cm across, bronze bowl, 18 cm across; melted glass	Cremation in a large mound	FII–FIII
Fitjar, Fitjar, H	B4231	Shears (R442), 2 loom weights, 1 spearheaded weaving beater (R150)	2 oval brooches (R647; 648), 1 bronze armlet, 1 glass bead	–	Flat grave	FIII
Fladen/Jamno, Vaksdal, H	B4432	Shears, 1 spindle whorl (30 g), 2 loom weights	1 oval brooch (R652), 2 ring-headed pins of bronze, tin and silver-coated (Glørstad type III b), 2 beads	1 steatite vessel, hones, horse-bits	Cremation Mixed find: weaponry from the 8th century, (Helgen 1972, 39)	FIII
Folkedal, Granvin, H	B7114+B7149, 7867	Shears (R442) 25.6 cm long, 9 *c.*12 cm long tines from wool combs, 4 spindle whorls (2: *10–15 g; 2: *26–30 g), parts of 1 weaving beater (R440), 1 smoothing plaque of whalebone (R449)	8 beads	3 keys (2 hooked, 1 large), 1 frying pan (R453), 1 iron vessel, strike-a-light (R426), 1 sickle, 2 horse-bits (R567), special knife (R408), burnt bones	Cremation, boat grave in a long mound (10 × 4 m), *c.* 100 rivets, *c.* 20 spikes	FIII
Folva, Oppstryn, F	B4752	Shears (R442), 1 spindle whorl (15 g)	2 oval brooches (R647), 1 Irish 3. brooch, 3 beads	1 key (R459), mounting to a casket, 1 hone, 1 axe	Inhumation, grave chamber in a mound	FIII
Fosnes, Oppstryn, F	B12695	1 weaving beater (R440)	1 oval brooch (R648), 1 armlet of copper alloy (R721)	Traces of imprints of textiles at the brooch	Flat grave?	FII

(Continued)

Appendix 5.1: Female graves with textile tools and dress accessories (N:87) (alphabetically) (Continued)

Location	No.	Textile tools	Dress accessories	Possible female indicators	Context	Status
Fosse, Meland, H	B12012	Fragments of wool combs, 3 spindle whorls (2: *11–15 g; 1: *26–30 g), 4 loom weights, 1 weaving beater (R440), 1 smoothing plaque of whalebone	2 oval brooches (R643), 58 beads, 2 bronze armlets (R709; JP191), 1 circular gilt reworked bronze mounting of Insular origins	4 keys, 3 of bronze, 1 of iron, 1 sickle, 2 hones, 1 knife, 1 strike-a-light, traces of textiles	Inhumation, flat grave	FIII
Gimmeland, Fana, H	B5884	1 shears R442, 1 spindle whorl (*c.30g), 1 'vevreell' R413	2 oval brooches (R644), 1 penannular brooch	1 hoe	Cremation in a mound	FIII
Gjelsvik, Førde, F	B11246	6 loom weights, 1 weaving beater (straight)	2 oval brooches (R655), remains of string of copper alloy	Fragments of an iron cauldron, one of copper alloy, a third of soapstone, and a small vessel of iron, 1 frying spade, a horse bit, possibly part of a harness	Cremation, flat grave	FIII
Gjerland, Førde, F	B10878	2 spindle whorls (23 g, 30 g)	1 trefoil brooch (JP fig. 109)	2 hones	Cremation, flat grave	FI
Gjerland, Førde, F	B12043	1 shears (R443), 1 spindle whorl (weight group *25–30 g)?	2 round brooches; 1 as fragments of copper alloy (JP fig. 116), another of silver (R665?), 7 glass beads, 1 silver foliated	Cauldron of iron, 1 sickle, part of a knife, handle to, a casket, 1 hone, a two-part horse bit	Cremation, flat grave	FIII
Grodås, Hornindal, F	B8000	1 needle-case of copper alloy, with preserved threads	2 oval brooches (R657), textiles	A hooked key, parts of a casket, and a knife	Inhumation	FIII
Gutdal, Oppstryn, F	B5030	2 shears (1 R442; 1 R443), 20 and 24 cm long, tines of wool combs, 1 spindle whorl (20 g), 1 weaving beater, *c.* 43 cm long (with a spit, and handle of bone)	3 oval brooches (R647; one variant D), 2 equal-armed brooches (JP fig. 83), 13 beads (glass, amber, jet), 2 gold earrings	1 key (R459), parts of a casket, frying pans, 1 fire-steel, 1 sickle, 1 horse bit	Chamber grave in mound	FIII
Gutdal, Oppstryn, F	B5144	Shears (R442), 24 cm, 30 *c.* 10 cm long tines of wool combs, 1 spindle whorl (27 g)	2 brooches of iron, 6 beads	1 key (R459), handle and mountings of a casket	Cremation, secondary grave in a mound	FII–III
Gutdal, Oppstryn, F	B5909	1 loom weight	1 oval bronze brooch (R643) without ornamentation	Parts of a casket, 1 small knife	Chamber grave in a small mound	FII

(Continued)

Appendix 5.1: Female graves with textile tools and dress accessories (N:87) (alphabetically) (Continued)

Location	No.	Textile tools	Dress accessories	Possible female indicators	Context	Status
Gutdal, Oppstryn, F	B6846	Shears (R442), 28 cm long, 14 tines 9.7 cm long of wool combs, 1 spindle whorl (24 g), 1 loom weight, plus parts of 3 others, 1 weaving beater (R440) *c.* 57 cm	1 brooch of iron (similar as in B5144), 1 chain of iron	Handles to a casket	Cremation, flat grave	FII
Gutdal, Oppstryn, F	B5525	2 small spindle whorls of fired clay	1 large oblong brooch of copper alloy (R664), remains of a second, 1 pendant of gold? (JP167), 2 similar pendants of copper alloy (R690), beads of glass, carnelian, crystal, and a long comb	1 long iron stick/staff with a handle and ring, 1 hooked key (R459), a grip for a vessel of steatite, various unidentified remains of iron, remnants of a hen	Cremation burial, little info	FIII
Hilde, Innvik, Stryn, F	B5717, B5755	Shears (R443), 23.5 cm, 2 spindle whorls of fired clay (*10–15 g) and steatite (*26–30 g), 1 loom weight, 1 beater of whalebone (R439)	2 oval brooches (R652), 5 beads of glass: 2 red, 1 chrystal, 1 mosaic and 1 silver bead (R695), 9 small Thor's hammers	1 key (R459), 1 casket, 1 lamp, 73 cm long iron stick/staff of iron with a ring and handle, 1 sickle, 1 cooking equipment, bronze cauldron of Insular origins, 1 hone of schist	Cremation in a boat, boat-shaped mound, rivets, and spikes	FIII
Hodneland, Lindås, H	B4704	2 spindle whorls (31 g, 42 g), 8 loom weights	2 oval brooches (R648), 1 equal armed small brooch (JP fig. 59; R658), 7 cm long of copper alloy		Secondary grave in an older grave mound	FIII
Hopperstad, Vik, S	B4511	2 spindle whorls (g?), not retrieved), 9 loom weights, 1 weaving beater (R440)	Various beads, silver, 1 filigree, agate, glass.	Remains of a bronze cauldron (R726), remains of another bowl, 1 cup-shaped bowl, a ladle of copper alloy and with long shaft, a bucket of wood, two glass vessels, bowl- and flask-shaped, white and yellow, 1 long iron spit/staff with handle of copper alloy; 1 frying pan with shaft lacking, casket of Insular origins, Kufic dirham (AD 738/39), a scale, 5 tin-coated weights of lead, 1 push key (R456), casket, horse bit, parts of a strike-a-light, 2 quern stones, 1 hoe	Boat burial, rivets and nails connected to wood, traces of a mound Inhumation	FIII
Hopperstad, Vik, S	B4637	Wool combs, 1 spindle whorl (*26–30 g). 1 weaving-beater, 1 smoothing plaque of whalebone	1 oval brooch (R652)	1 frying pan of iron (R430), 1 iron ladle with long shaft (R431)	Grave close to B4511, inhumation	FII

(Continued)

Appendix 5.1: Female graves with textile tools and dress accessories (N:87) (alphabetically) (Continued)

Location	*No.*	*Textile tools*	*Dress accessories*	*Possible female indicators*	*Context*	*Status*
Hopperstad, Vik, S	B7761	2 shears (R443), 16 cm, 2 cm long, 1 spindle whorl (*26–30 g), 1 loom weight, 1 weaving beater blade 57 cm, 83 cm long, wooden handle 1 'vevreell' (R413), *c.* 13 cm long, wooden handle	2 oval brooches (R652), 1 trefoil brooch (cf. R673), 107 beads, textiles	1 sickle, 1 hoe, 1 casket, horse gear?	Inhumation grave, covered by stones	FIII
Hopperstad, Vik, S	B9060	Wool combs, parts of 2 spindles, 12 loom weights, 1 needle case of bronze (VJG 492), part of a smoothing plaque of whalebone, 1 'vevreell' (R413), *c.* 15 cm long and with a wooden handle	2 oval brooches (R654), 1 Irish as a 3. brooch, 21 beads, textiles	1 sickle, 1 frying fork (R428)	Inhumation burial above grave B9061	FIII
Hopperstad, Vik, S	B9061	1 spindle whorl (18 g), 1 weaving beater (58 cm), part of a spindle?	1 oval bronze brooch (R640), 2 rectangular brooches, 60 mostly multi-coloured beads, + 8 of bronze, textiles	3 keys (R459), 1 sickle	Inhumation, boat burial below no. B9060, boat *c.* 6 m, flat grave	FIII
Hove, Vik, S	B4841	Wool combs (35 tines), 1 'vevreell' (R413)	1 bronze chain	2 special type of horse bits (R476), and remains of a casket, 1 iron cauldron	Mound	FII
Hønsi, Vik, S	B708–17; B2734+2739	Wool combs, 1 weaving beater, 1 smoothing pellet of glass	Round brooch of silver, 1 gilt silver pendant, both Jelling style (fig. JP157), textiles with gold threads	1 iron cauldron, parts of 2 horse bits, 1 of iron and 1 of bronze of Insular origins, 1 frying pan with long shaft	Long-mound Nails (boat?) B4390 in the same mound and a weapon grave	FIII
(Kirk)eide, Stryn, F	B8060	Shears, 1 spindle whorl (*c.* 30 g)	2 bronze pendants (Borre style, JP155), 30 beads	–	Flat grave	FII–III?
(Kirk)eide, Stryn, F	B8953	Shears (R443), 1 spindle whorl of fired clay (*11–15 g)	2 oval brooches (Jelling style, JP57) Textiles	1 casket, 2 food related equipment: a large cauldron of soapstone and a frying pan, parts of a sickle, a small hone, a lamp of iron, 1 two-part horse bit and parts of the harness	Flat grave, inhumation burial in a wooden coffin, surrounded by flat stones	FIII
Koppen, Haus, H	B6089	1 spindle whorl (*c.* 30 g), 14 loom weights	1 penannular bronze brooch (Type I, Eastern variant), 3 beads	–	Cremation burial, flat grave	FI
Korsvoll, Gaular, F	B9769	Wool combs (21 tines, plus 20 as fragments), 4 loom weights	1 or 2 brooches (R652 or 654), 3 beads	1 sickle, 1 frying pan, 1 ladle, a two-part horse bit	Cremation burial, flat grave	FII–III

(Continued)

Appendix 5.1: Female graves with textile tools and dress accessories (N:87) (alphabetically) (Continued)

Location	*No.*	*Textile tools*	*Dress accessories*	*Possible female indicators*	*Context*	*Status*
Kvåle, Sogndal, S	B3456	Shears (R443), wool combs, parts of wool combs or hackle, 3 spindle whorls (5 g, 5 g of fired clay, 1 stone, *30–35 g), 1 weaving beater (67 cm long)	2 oval brooches (R652), 1 trefoil brooch (R671) as a third brooch, 2 bronze fibulae	3 keys, 1 casket with lock, 2 sickles, 4 food related equipment: 1 bronze cauldron of Insular origins, 1 ladle and vessel of iron and part of a frying pan, 1 spit/staff, 67 cm long, and with an iron ring, part of a horse bit	Cremation burial, flat grave	FIII
Langenes, Selje, F	B5806	1 spindle whorl (17 g), 3 loom weights	1 oval brooch (R647), 20 beads	1 vessel of steatite	Inhumation burial in a cairn	FII
Langeset, Innvik, Stryn, F	B6035	Wool combs, 1 loom weight	1 oval brooch (R643), 32 glass beads, of which 2 gold foliated, and some mosaic	1 iron ladle (R430), 1 frying pan, 1 horse bit (R567)	Inhumation burial in a small mound	FII
Loen, Stryn, F	B8818	Shears (R443), 1 weaving beater (R440)	3 oval brooches, (R652–4), parts of a chain between two of them and remains of leather, textiles	Remains of a frying pan, 1 sickle, part of a casket?	Flat grave	FIII
Lomelde, Sogndal, S	B9864	3 loom weights	1 equal armed brooch, 9.2 cm long	–	Disturbed grave in a cairn	FI
Lygra,Alver, H	B6185	1 spindle whorl (12 g), 8 loom weights, fragment of weaving beater (R440)	1 armlet of bronze (R721)		Flat grave, spread out	FI
Lyslo, Innvik/ Stryn, F	B283/496/504	1 spindle whorl (28 g)	1 rhombic-shaped bronze brooch (JP141), *c.* 50 beads to a necklace		Burial in a mound	FII
Mardal, Gloppen, F	B4605–06?	1 shears (R443), *c.* 30 12–13 cm long tines to wool combs, 2 spindle whorls (g?) 2 loom weights	2 oval brooches (R654), 1 penannular brooch	1 frying pan (R430), 1 sickle, strike-a-light (R426), parts of a mounting	Inhumation burial in a mound	FIII
Maurset, Hornindal, F	B7524, B7611	1 needle-case of bronze with iron needles and textiles	2 oval brooches, (R648), 1 equal armed brooch (R660), 57 beads, plus 3 others: mosaic, jet and amber.		Inhumation burial in a cairn?	FIII
Mindresunde, Stryn, F	B4505	1 spindle whorl (37 g), 2 loom weights	2 oval brooches (Berdal type A, earlier than R648), 1 Irish penannular brooch (Glørstad Type 1A, Insular), *c.* 50 beads, some gold foliated, and multi-coloured glass, plus a stone pendant (JP170)	1 bronze key	Inhumation burial, flat grave, stone chamber	FIII

(Continued)

Appendix 5.1: Female graves with textile tools and dress accessories (N:87) (alphabetically) (Continued)

Location	*No.*	*Textile tools*	*Dress accessories*	*Possible female indicators*	*Context*	*Status*
Myklebust III, Eid, F	B5807	Shears (R443), 12 cm long tines from wool combs or hackles	2 oval brooches: (R647 and R644) part of a third horse-shaped brooch. Textiles	1 bronze cauldron (R726), 1 iron cauldron (R731), part of a frying pan	Secondary burial in a mound	FIII
Myklebust IV, Eid, F	B5764	Shears (R443), *c.* 20 tines of a hackle (R433), 6 loom weights, 1 weaving beater of whalebone (R439), 60 cm long, 1 bone needle, 9 cm	2 oval brooches R652, 30 beads, of crystal, carnelian and multicoloured glass, 1 of silver filigree and with spiral décor, 1 silver chain	Mountings of a chest, 1 iron cauldron with a long handle (R431), 1 frying pan (R430), 1 long iron stick/staff with a handle, 88 cm long, 1 sickle	Cremation burial in a boat	FIII
Myklebust, Davik, F	B7005	3 loom weights, 1 broken and incomplete, 1 broken weaving beater (R440)	1 equal armed brooch (R661, Oseberg style)	1 strike-a-light of flint	Inhumation burial, flat grave	FI–FII
Mølmesdal, Høyanger, S	B7366	1 weaving beater (R440), 72 cm long	2 oval brooches (R652)	–	Grave	FII–III
Målsnes, Balestrand, S	B12131	Shears (R443), 22 *c.* 12 cm long tines from wool combs, 2 spindle whorls (27 g, 47 g), 1 needle-case of iron	2 oval brooches (R652), 1 penannular bronze brooch), 1 bronze chain, 3 beads, textiles	1 hooked key (R254), remains of a casket, 1 small hone for sharpening needles, 1 sickle, strike-a-light (R426)	Long mound (10 × 3 m)	FIII
Naustdal, Hauganes, F	B9626	1 spindle whorl (26 g)	2 oval brooches (Berdal type, type B (fig. JP12), 1 equal armed brooch (R661), 1 bronze armlet (R721; JP fig. 184), textiles	Remains of bridles?, 1 small hone	Flat grave, inhumation grave	FIII
Nes (Helgheim), Sogndal, S	B9064	Shears, 3 spindle whorls (22, 29 g, *15–20 g), 1 weaving beater (81 cm+)	1 oval brooch (R657), but probably originally two, 5 glass beads, textiles	1 casket, 1 cooking vessel of iron, 1 horse bit	Burial in a cairn	FIII
Nes (Helgheim), Sogndal, S	B9065	Shears (25 cm long+), tines for two combs, 11 cm long and 18 in each, 15 loom weights, 1 weaving beater (more than 55 cm long), 1 'vevreell' (R413)	2 Berdal brooches (JP13, type C), 2 bronze armlets (JP190; single find), 134 glass beads, Irish penannular brooch in gilt silver (Glørstad Type 1C, Insular), textiles	1 iron ladle, a two-part horse bit of bronze, 16.5 cm long	Flat grave	FIII
Osen, Gaular, F	B15813	1 spindle whorl (14 g), 1 loom weight	13 beads monochrome glass, blue and red, textiles	1 frying spade, 1 strike-a-light (flint)	Flat grave?, inhumation, Plundered	FI?
Rykkje, Voss, H	B7836	1spindle whorl (19 g), 8 loom weights	1 bronze pendant, 4 glass beads, 1 silver foliated	1 sickle (R384), part of a mounting	Cremation burial, grave chamber in a mound, more than one grave in the mound, unclear	FI

(Continued)

Appendix 5.1: Female graves with textile tools and dress accessories (N:87) (alphabetically) (Continued)

Location	*No.*	*Textile tools*	*Dress accessories*	*Possible female indicators*	*Context*	*Status*
Sandal, Breim, Gloppen, F	B9014	Shears (R442), tines from two different wool combs, 1 spindle whorl (11 g), 6 loom weights	2 oval brooches (R643), 80 beads, of which 1 silver and 10 gold foliated, amethyst etc., textiles	1 key (R459), two-part horse bit (R570), 1 sickle, 1 small knife, 1 awl, 1 strike-a-light, a small glass vessel, enamel decorated bronze mounting bronze and another bronze mounting on wood of Insular origins	secondary grave in a mound, inhumation	FIII
Sandal, Jølster, F	B11413	Tines from 2 wool combs with handles of wood, 4 spindle whorls (18 g, 21 g, 38 g, 40 g), 10 loom weights, 1 weaving beater R440 (82 cm long), 1 smoothing plaque of whalebone	2 oval brooches (R657), 1 silver chain, 39.9 cm long, with rings, textiles	1 casket with loom weights therein, 1 two-part horse bit, 1 strike-a-light, an axe, 1 sickle, 1 awl	Inhumation burial in a boat grave in a long cairn	FIII
Saue, Voss, H	B1544-1559	1 spindle whorl (12 g)	Mosaic beads to a chain		Probable demolished grave	FI
Seim, Granvin, H	B8635, B9091	Shears, 25 *c.* 13 cm long tines from wool combs, 1 spindle whorl (14 g), 1 whalebone plaque	2 oval brooches (R647)	2 keys, 1 frying pan, handle to a cauldron, 2 two-part horse bits (R570), 1 with part of the harness, strike-a-light (R426), 1 mounting, 1 sickle, 2 small knives	Cremation burial, as a secondary grave in a mound (B3460)	FIII
Skjervum, Vik, S	B6500 a-g	1 weaving beater (R440), 67 cm long	2 oval brooches: (1 R643), and 1 R640, plus an Irish reworked bronze brooch, 4 arm rings (R709), 5 beads, textiles		1 of 3 graves in the area, these objects as one context, a flat grave a–g; a male in a boat grave with weapons and a horse	FIII
Stalheim, Vossestrand, H	B4736	1 weaving beater, 72+ cm, 1 'vevreell' (R413)	1 oval brooch (R647), plus 1 of Berdal type, 2 bronze armlets (R721)	1 frying pan, 1 hoe	Flat grave, sided by slabs	FII–III
Svingeset, Stryn, F	B6483	1 spindle whorl of lead (2.3 cm, g?), 2 small sewing needles of bronze	2 oval brooches (R649, variant F; R647), 57 beads of multicoloured glass, of which 44 small blue, 4 beads, textiles	1 large key (R459), 1 chest, 1 small knife, 1 miniature bronze axe (Irish parallel)	Small long mound, covered by slabs	FIII
Sørheim, Etne, H	B12058	1 spindle whorl (*20–25 g)	1 needle of bronze	1 cooking vessel of steatite	Boat grave	FI?
Sårheim, Breim, F	B10975	1 weaving beater	2 oval brooches (R647), 3. oval brooch, 18 beads	3 keys in a ring, 1 sickle	Inhumation	FIII

(Continued)

Appendix 5.1: Female graves with textile tools and dress accessories (N:87) (alphabetically) (Continued)

Location	*No.*	*Textile tools*	*Dress accessories*	*Possible female indicators*	*Context*	*Status*
Terum, Aurland, S	B9063	Shears (R442), 3 spindle whorls (22g, 28g and 1 *30–35 g), 1 'vevreell' (R413)	1 oval brooch (R647), 1 armlet of bronze (fig. JP186–88), 6 beads, textiles	1 casket, lock, 1 sickle, 1 hoe, 1 file	Partly disturbed cairn, burnt bones	FII–III
Terum, Aurland, S	B8991	Tines of wool combs, 10.3 cm long, 'vevreell' (R413)	1 mosaic bead, 1 iron pin to a bronze brooch	3 keys (R459), 1 handle + mounting to a casket, 1 sickle, 1 hoe (R401), 1 frying pan, 1 tweezers, 1 earpick	Mound	FI
Tokvam I, Aurland, S	B8669	1 spindle whorl (**c.* 25 g)	2 oval brooches (R650), 1 cruciform Irish brooch (R624),	Fragments of a sickle or lea and a knife (R405)	Central grave in a cairn	FIII
Tokvam II, Aurland, S	B8669	Shears (R442), tines from wool combs, 2 spindle whorls (*20–25 g, *30–35 g), 28 loom weights, 1 weaving beater (R440) (80 cm long), 1 'vevreell' (R413)	38 beads of different shape and size, 1 of them of amber, 34 of glass, and the rest glass-mosaic	1 iron ladle with a long shaft (R431), 1 handle to a casket (R451)	Secondary grave in the same cairn, NE part	FI
Tonning, Stryn, F	B584-93	1 spindle whorl (33 g), 5 loom weights	1 penannular brooch, 2 armlets of copper alloy (R709?), 28 beads, of which 10 glass/mosaic	1 frying pan, 1 horse bit	Grave (old finding; 1842)	FII–III
Toranger, Austevoll, H	B5874	Shears (R442), 24 cm long, 1 spindle whorl (10 g), 4 loom weights, 1 weaving beater (R440), 61 cm long, 1 whalebone plaque (R449)	1 oval brooch, R652	1 key (R459), 1 sickle	Cremation grave in a cairn	FII–III
Trå, Granvin, H	B6657	Remains of shears and wool combs, 1 spindle whorl (10 g), 12 loom weights, 1 weaving beater (R440) (79 cm), 1 black glass pellet (R446), 7.6 cm in diameter	2 oval brooches, R652, 1 trefoil gilt bronze brooch in Borre style (R671), a chain with pendants of 11 silver rings combined with 15 beads of coloured glass, carnelian, rock crystal and mosaic-glass and amber, textiles	1 casket with lock, 2 horse bits, 1 frying pan with a long shaft (R430), 1 ladle with a long shaft (R431), 1 iron stick/staff, 88 cm long, finds of Insular origins: a ladle, remains of a bronze bowl (R726), a brooch and glass	Secondary grave inhumation in an older mound, buried in a wooden coffin	FIII
Tveiti, Vangsnes/Vik, S	B5702	1 spindle whorl (*26–30 g)	2 oval brooches (R644)	–	Flat grave	FII
Vangdal, Kvam, H	B5627	Shears (R442), tines of wool combs, 1 weaving beater (R440), 74.5+ cm	3 oval brooches (R652)	1 handle to casket (R451), 1 cooking vessel of steatite with shaft (R728)	Grave	FIII

(Continued)

Appendix 5.1: Female graves with textile tools and dress accessories (N:87) (alphabetically) (Continued)

Location	*No.*	*Textile tools*	*Dress accessories*	*Possible female indicators*	*Context*	*Status*
Vatne/Votno, Alversund, H	B3308	Shears (R443?), 2 weaving beaters, 1 or iron (R440), the other of whalebone (both as fragments)	1 oval brooch (Berdal type D), 1 rectangular gilt bronze brooch of Insular origins	1 ladle with a long shaft, 1 sickle, 1 horse bit	Cremation burial in a long mound	FIII
Veka, Voss, H	B6228	A pair of wool combs, 2 spindle whorls (9 g, 17 g) 11 loom weights preserved, needle-case of copper alloy + iron sewing needles, 1 'vevreell', R413	2 oval brooches (R652), 1 oblong brooch of bronze (JP140), 1 silver and 1 armlet of silver, (JP fig. 192) and 1 of copper alloy, 118 beads, mostly glass, Anglo Saxon silver coin (792/6), textiles	1 small chest, 1 74 cm long iron stick/staff with a ring and handle, 1 sickle, 2 hoes, 1 brush, textiles	Inhumation burial in a wooden chest, large cairn	FIII
Vinjum, Aurland, S	B7731	2 large shears (R442 and R443), wool combs, 2 spindle whorls (19 g, 28 g), 48 loom weights, 1 'vevreell', R413	2 oval brooches (R657), four-sided silver brooch Irish (Type I?), 37 beads, of which 15 glass/mosaic, 21 monochrome glass, 4 of them silver foliated, 2 armlets of bronze, mounting to a drinking horn (Irish), 1 cup and lid of tinplated bronze, textiles	2 keys (R459 & R454), 1 frying pan with a long shaft (R430), parts of a long iron spit/staff, 96 cm long, 1 sickle, 1 hoe, 1 knife	Inhumation burial in a boat, possibly originally in a mound	FIII
Vinnes, Fusa, H	B4977	1 spindle whorl (11 g), 1 weaving beater (fragment)	2 oval brooches (Oseberg style)	Vessel of soapstone	Cremation, stone setting in burial in a large cairn	FIII
(Ytre) Eide, Stryn, F	B5804	7 loom weights, 1 weaving beater	1 bronze armlet (R719), 11 beads	Parts of a horse bit	Flat grave, damaged mound? covered by 3 slabs	FII

Appendix 5.2: Probable/possible female graves: Graves with textile tools, without dress accessories and weapons (N:16)

Location	*No.*	*Textile tools*	*Work operations*	*Other possible female indicators*	*Context*
Amla, Sogndal, S	B10390	Shears (R442), 1 weaving beater (R440) 64 cm+	2	1 sickle, 1 hoe	Uncertain grave, little information
Brekke, Lavik, S	B10426	1 weaving beater (R440)	1	1 hoe, 1 ard share	Small cairn
Bygd, Vossestrand, H	B8598	1 spindle whorl (*20–25 g)	1	1 small knife	Burial in a small boat (90–100 rivets, *c.* 25 spikes) in a mound, burnt and unburnt bones
Eivindvik, Gulen, S	B143–145	3 spindle whorls	1	–	Grave mound
Fretheim, Aurland, S	B8390	1 weaving beater (R440), 50 cm long	1	–	Uncertain grave
Holmedal, Fjaler, F	B7893	7 loom weights, textiles	1	–	Boat grave, inhumation
Førde, Sveio, H	B6212	Tines from wool combs	1	1 frying pan (R430), 1 hoe (R401)	Inhumation burial, flat grave on a terrace
Gjerdskvål, Voss, H	B5914	Shears (R442), tines from wool combs, 2 spindle whorls (15 g, 26 g)	3	–	Burial in a mound
Hopperstad, Vik, S	B4381	Part of a weaving beater (R440)	1	1 hoe	Found in an earlier excavated mound
Langenes, Selje, F	B6327	1 spindle whorl? (6 cm across), 11 loom weights	2	Grindstone (R622)	flat grave
Osland, Høyanger, S	B14183	1 spindle whorl (*25–30 g), 2 loom weights	2	–	Found in masses from a cairn
Revikja, Vågsøy, F	B5741	1 weaving beater of whalebone (R439)	1	1 undefined tool of whalebone, R450, 14 cm long, 11 cm wide	Inhumation burial in a mound
Setre II, Breim, Gloppen, F	B7810 II	Part of shears (R443), a few remains of tines from wool combs, 2 spindle whorls (20 g, 31 g), 2 loom weights, remains of a weaving beater (R440)	4	1 frying pan, 1 horse bit	Inhumation burial, 1 of 3 burials at the farm, close to each other in a row
Setre III, Breim, Gloppen, F	B7810 III	7 loom weights	1	–	Graves, no info about structure
Tryti, Vik, S	B4160	Remains of wool combs, shears, and a weaving beater (now lost)	3	Parts of mountings to a casket	Mound, FI?
Øvreset, Breim, Gloppen, F	B6662	Remains of shears (R442), 1 tine from wool combs, 1 spindle whorl (*20–25 g), 16 loom weights, 2 weaving beaters (2 types and sizes, 1 type (R440), the other short (R150) with a spit, blade 24 cm long, 1 'vevreell' (R413)	4	1 key (R459), 1 glass bead, 1 knife, 11.5 cm long, a horse bit, parts of harness (R577)	Cremation burial in a cairn, FII?

Appendix 5.3: Double graves with dress accessories, weapons, and textile tools (N:12)

Location	*No.*	*Textile tools*	*Weapons*	*Dress accessories*	*Possible female indicators*	*Context: earlier assessment*	*Status F/M*
Bø, Breim, Gloppen, F	B9501	Shears (R442), 27 cm long, 44 tines from wool combs or a hackle, some complete, 12 cm long, 2 spindle whorls (19g, 46 g), 9 loom weights, 1 weaving beater (R440) in three pieces, *c.* 75 cm in total length	1 double-edged sword, 1 axe, 5 arrowheads	2 oval brooches (R647, Oseberg style), 4 beads, 1 of amber, textiles	1 frying pan, remains of (R430), mounting to a casket, 1 sickle, remains of horse harness (R577), balance scale of Insular origins	Inhumation burial in a mound, boat grave (276 rivets and iron spikes)	FIII/MIII
Flo, Oppstryn, F	B5865/ B10176	Tines from wool combs	1 sword (R491), 1 axe and tools for wood-working	Bronze piece part of a brooch (R647)?	1 key, a sickle, a two part horse bit.	Cremation burial in a boat, 140 rivets and spikes	FI/MII
Fyling, Gaular, F	B11470	Shears (R442), 20 *c.* 13 cm long tines from wool combs, 2 spindle whorls (*20–25 g), 1 weaving beater (R440), *c.* 58 cm long	1 double-edged sword, 1 spear-head, an axe, 14 arrow-heads, 2 shield bosses	2 oval brooches (R652), 2 trefoil brooches, 2 beads, Hiberno-Saxon bronze buckle and buckle plate, ornamented of Insular origins	1 key, remains of a chest and 5 horse-bits, all two-part bits (R570), 2 sickles, knives, hones	Cremation burial in a small cairn, boat burial	FIII/MIII
Hauge krk, Lærdal, S	B2151	21 loom weights	1 axe, 2 spearheads.	1 oval brooch (Berdal group D; R648–9), beads	1 key, bronze piece part of jewellery?, Insular penannular brooch, Type 1A	Cremation grave in a mound, assessed as double grave	FIII/MII
Hyrt, Vossestrand, H	B4864	A pair of wool combs, with 11 cm long tines, 1 spindle whorl, 22 g, originally 20 loom weights, 8 preserved weight range 250–450 g, total weight 7050 g	1 double-edged sword, spearhead, 1 axe, 1 shield boss, 2 knives	2 oval brooches (R652), 20+75 beads, textiles	1 key, casket, mounting of a copper alloy to a drinking horn of Insular origin	Inhumation in a mound	FIII/MIII
Lunde, Voss, H	B2233	Shears, 1 spindle whorl (33 g)	2 double-edged swords, 4 axes, 3 arrowheads, 1 shield boss	2 oval brooches (R649), 1 third round Irish brooch), 3 beads, pieces of a bronze chain	3 horse bits (R570 and R569)	Inhumation burial in a mound	FIII/MIII
Myklebust I, Eid, F	B5807	1 hackle, 1 weaving beater 49 cm long blade, with a spit, 20 cm long handle of wood	1 single-edged sword, 4 arrowheads, 1 shield boss	1 oval brooch (early type), 1 bead, textiles	Handle and mounting to a chest, small knives	Inhumation burials in a mound, the male in a chest, female outside	FII/MIII
Nes, Sogndal, S	B3971-72/86	Wool combs, 2 spindle whorls, 46 g, plus 1 of fired clay, 1 weaving beater, 52 cm long	1 double-edged sword, 1 spearhead, 2 axes	2 oval brooches	Frying pan, strike-a-light, 1 horse bit	Not a closed context, but probably several graves 2 graves?	FIII+MIII

(Continued)

Appendix 5.3: Double graves with dress accessories, weapons, and textile tools (N:12) (Continued)

Location	*No.*	*Textile tools*	*Weapons*	*Dress accessories*	*Possible female indicators*	*Context: earlier assessment*	*Status F/M*
Store Kongs-vik, Tysnes, H	B7639	Shears, 6 loom weights	1 single-edged sword, 1 spear, 1 axe	2 oval brooches (R640), 1 rectangular brooch? needles? 1 large bead of jet, textiles	1 frying pan (R430), sickle or lea	Inhumation burial in a cairn	FIII/MIII
Vik prg/ Grov, Vik, S	B8143	1 weaving beater (R440)	1 single-edged sword, 1 axe	1 oval brooch (Type JP 7)	–	Mound Inhumation?	FII/MII
Vik, Nedstryn, F	B4756	2 shears (R442, R443), 'vevreell' (R413)	1 single-edged sword, arrowheads, spearheads	2 oval brooches (R647), ring brooch, beads	3 horse bits, 1 complete, 2 as parts, 1 key, mounting to a chest, a hoe, parts of an iron cauldron, tools for wood-working and forging	Cremation, ship-shaped flat grave, boat grave	FIII/MIII
Ytre Arna, Haus, H	B5800	Shears (R443), 16 cm long, 1 needle-case of copper alloy, 1 'vevreell' (R413), 13 cm long	2 double-edged swords, 2 axes	2 brooches (R652), 2 round brooches of copper alloy (R666; R683, Jelling style), 1 pin, 7 beads of carnelian and glass	2 keys, 1 chest by its mountings, 3 sickles, 1 two-part horse bit and remains of harness	Cremation burial in a large cairn	FIII/MIII

Appendix 5.4: Textile tools in weapon graves without female dress accessories but with other female indicators – possible asymmetrical double graves? (N:12)

Location	*No.*	*Textile tools*	*Possible female indicators*	*Weapons*	*Earlier assessment*	*Status*
Bryn, Voss, H	B3987	Tines from wool combs, 1 weaving beater, 67 cm long	2 keys, 1 mosaic bead, mountings to a casket, frying pan, bronze cauldron of Insular origins, 1 horse-bit, 1 bead, strike-a-light	2 swords, 21 arrowheads, 2 axes, 2 shield bosses. Tools for wood-working and forging	Cremation, double grave one of several possible graves and grave chambers in a mound.	MIII/FI?
Gryten, Gulen, S	B8272	1 spindle whorl (*25–30 g), 1 smoothing pellet (glass)	–	1 arrowhead (R539), 1 possible spearhead	Flat grave? Burnt bones	MII/FI
(Ytre) Hauge, Gloppen, F	B5150	1 spindle whorl (g?)	Bronze fragment, 3 beads, 1 casket and lock, 1 hoe	1 sword, 3 spearheads (2 R518), 1 axe (R553)	Mound, boat grave, weapons found together	MIII/F?
Hilde, Innvik Stryn, F	B5588	Tines from wool combs, 1 spindle whorl (*20–25 g), 2 loom weights, 1 weaving beater (R440)	1 sickle, frying pan, two rings from a horse-bit	2 swords (R489; R491), 1 spearhead, 1 axe, 1 knife	Cremation burial in a mound –all artefacts found together, probably a double grave	MIII/F1?
Hillestad, Lærdal, S	B11242	1 spindle whorl (*25–30 g)	1 two-part horse-bit (R657), 1 casket?	1 sword, 1 spearhead R520, 7 arrowheads + 3 others, 1 shield boss, 1 axe	Long cairn found during house construction	MIII/F1?
Hus, Jølster, F	B6797	Tines from wool combs	1 horse-bit, 1 iron vessel, 1 fire-steel	Double-edged sword, 1 axe, parts of 1 knife, 1 large hone	Flat grave, double grave?, inhumation	MII/F?
Lekve, Ulvik, H	B5490	1 weaving beater (R440)	1 ladle, 1 frying pan, 1 horse bit	1 axe (R559)	Found together without visible superstructure	MI/FI
Naterstad, Kvinnherad, H	B5927/5928	Shears, remains of wool combs, 1 spindle whorl (g?), 17 loom weights (+5 lost)	1 casket, 1 horse bit	1 spearhead, 1 axe	Boat grave ?	MII+F?
Skarde, Odda, H	B5028	1 loom weight, 1 weaving beater (R440)	1 iron vessel	1 axe (R561), 10 arrowheads	All found in a stone chest, possible double grave (Shetelig 1912, 202)	MII/F1
Totland, Samnanger, H	B6609	Shears (R443), 9 tines of wool combs, 12.5–13.5 cm long, 1 weaving beater (R440)	1 handle to a casket, 1 sickle, 1 horse gear	1 sword JP494, 5 arrowheads, 1 axe (R555)	Cremation, double grave	MIII/FI?
Vardberg, Eidfjord, H	B4924	1 spindle whorl (g?), 1 needle case of bone	1 hooked key, 1 bone comb	6 arrowheads R539+ 2, 1 spear head, 1 shield boss	Cremation, mound, boat grave, 35 nails, 40 rivets	MII–III/F?
Øvre Strand,Gaular, F	B13628	Tines from wool combs	2 glass beads, blue green and a melon-shaped	Remains of a sword or spear, 2 arrowheads, 1 shield boss	Double grave, inhumation, a wooden chest(?) in a mound	MIII/F?

Appendix 5.5: Unclear contexts as separate graves (N:15)

Location	*No.*	*Textile tools*	*Weapons*	*Other indicators*	*Earlier assessment*	*Status*
Farnes, Evanger, Voss, H	B7835	17 tines from wool combs	1 spearhead (R532), 1 arrowhead (R539), a lea	strike a light	Found at separate times at nearly same location. The remains of the wool combs differ from the weapons, being fire exposed	MII & F?
Fitje, Voss, H	B3708	1 spit-shaped weaving beater (R150), 57 cm long	1 double-edged sword	.	Found in a mound, together with various iron objects and appear as an incomplete find; a horse bit was taken into use, and other unrecorded items shared between the finders, possibly a male grave	MII & F?
Gjelle, Voss, H	C1313-17	Shears, 1 spindle whorl (g?), 5 loom weights	2 swords, 2 axes	1 frying pan, 1 horse bit, several unidentified objects	Found at different places at the farm, without clear contexts	M I & F?
Hollekve, Sogndal, S	B10438	1 defect weaving beater (R440), blade more than 42 cm	1 spearhead		Found together, but no grave according to the finder	MI & F?
Horvei, Evanger, Voss, H	B6924	1 weaving beater (R440)	1 double-edged sword (R494)		Found during arable farming. Stemming from one or two demolished graves	MI & F?
Jernes, Voss, H	B5485/ B5516	1 needle case of copper alloy	1 shield boss (R562)		Found at the same farm, but not the same context according to the catalogue	MI & F?
Moa, Årdal, S	B8248	Shears, possible remains of wool combs, 1 loom weight	1 arrowhead (R537)		Found during farming, no structure is mentioned	MII & F?
Netteland, Kvinnherad, H	B5775/ B13229	1 spindle whorl	1 sword (R492), 1 spearhead, 1 axe (R561), 1 shield boss (R562)	The spindle whorl is found a as stray find during farming	Cairn, the spindle whorl from the same area but no clear context	MII & F?
Revikja,Vågsøy, F	B11239	1 spindle whorl (*20–25 g)	1 sword, 1 axe		Bones, found in relation to a cairn, seemingly without any order	MIII & F?
Saue, Voss, H	B6380	1 spindle whorl, (*30–35 g)	1 sword, 1 spearhead (R529), 5 arrow heads (R539), 2 shield bosses (R564), 1 axe (R561)	1 simple horse bit, 1 casket	Cremation burial in a cairn	MIII & F?
Sekse, Ullensvang, H	B4424	Shears (R443) Remains of a weaving beater	2 swords, 2 spearheads (R530), 1 arrowhead, 1 axe (R559)	Remains of an iron vessel, 1 sickle, and a horse bit	Cremation, burnt bones and ashes, found over the years in an earlier destroyed cairn	MII & F?
Skiple, Voss, H	B8545 II	1 loom weight	1 arrowhead (R539)		Small mound, artefacts found spread around	MI & F?
Vedå, Osterøy, H	B3326	1 spindle whorl (g?)	1 sword, 1 axe	Remains of a vessel	Cremation in a grave	MII+F?
Øvsthus, Voss, H	B10855	1 loom weight (lost)	1 sword (lost), spearhead		Mound, sparse information. Uncertain connections to a nearby mound	MII+F?
Årøyni, Sogndal, S	B4878, 35a-f	1 weaving beater 54+ cm long, possible remains of wool combs	1 spearhead	Trefoil brooch (30a)	Found in a hollow space in a grave cairn with several graves, both cremation, and inhumation	MI+F?